A Data Engineering Approach to Wave Scattering Analysis

IEEE Press
445 Hoes Lane
Piscataway, NJ 08854

A Data Engineering Approach to Wave Scattering Analysis

with Applications in Radar, Sonar, Medical Diagnostics, Structural Flaw Detection and Intelligent Robotics

Mark K. Hinders
William & Mary, Williamsburg
USA

IEEE Press
WILEY

Published by John Wiley & Sons, Inc., Hoboken, New Jersey.
Published simultaneously in Canada.

Library of Congress Cataloging-in-Publication Data applied for:
Hardback ISBN: 9781394271221

Cover Design: Wiley
Cover Image: © Andriy Onufriyenko/Getty Images

Set in 9.5/12.5pt STIXTwoText by Straive, Chennai, India

To my former and future students.

Contents

About the Author

Mark K. Hinders holds BS, MS, and PhD in Aerospace and Mechanical Engineering from Boston University and is Professor of Applied Science at the College of William & Mary in Virginia. Before coming to Williamsburg in 1993, Professor Hinders was senior scientist at Massachusetts Technological Laboratory, Inc., and research assistant professor at Boston University. Before that Dr. Hinders was an electromagnetics research engineer at the USAF Rome Laboratory located at Hanscom AFB, MA. Professor Hinders conducts research in wave propagation and scattering phenomena, applied to medical imaging, intelligent robotics, security screening, remote sensing, and nondestructive evaluation. He and his students study the interaction of acoustic, ultrasonic, elastic, thermal, electromagnetic, and optical waves with materials, tissues, and structures.

About the Author

[illegible]

Preface

I like to work on real-world problems that matter. I also like that there are new problems all the time, with the caveat that the basic math and physics doesn't change so we get to continue to exploit our hard-won mastery of difficult subjects. In my lab I have a Monty Python poster. It's John Cleese in a suit and tie sitting at a desk on a beach. It's not captioned, but he would be saying, "And now for something completely different."

20 September 2024

Mark K. Hinders
Williamsburg, Virginia

Acknowledgments

The author would especially like to thank his research mentors, the late Profs. Asim Yildiz and Guido Sandri, and *their* research mentors, Prof. Julian Schwinger and Dr. J. Robert Oppenheimer. Asim Yildiz (DEng, Yale) was already a Professor of Engineering at the University of New Hampshire (UNH) when Prof. Schwinger at Harvard told him that he was "doing good physics" already so he should "get a union card." Schwinger meant that Yildiz should get a PhD in theoretical physics with him at Harvard, which Yildiz did while still keeping his faculty position at UNH and mentoring his own engineering graduate students, running his own research program, and so on. He also taught Schwinger to play tennis, having been a member of the Turkish national team and all the best tennis clubs in the Boston area.

When Prof. Yildiz died at age 58, his genial and irrepressibly jolly Boston University (BU) colleague took on his orphaned doctoral students, including yours truly, even though the students' research areas were all quite distant from his own. Prof. Sandri had done postdoctoral research with Oppenheimer at the Princeton Institute for Advanced Study and then was a senior scientist at Aeronautical Research Associates of Princeton for many years before spending a year in Milan at Instituto Di Mathematica del Politecnico and then joining BU. In retirement, he helped found Wavelet Technologies, Inc. to exploit mathematical applications of novel wavelets in digital signal processing, image processing, data compression, and problems in convolution algebra.

Acknowledgement

Introduction

In this book, scattering analysis is applied to many seemingly different things, including but not in any way limited to:

- **Radar**: The atomic bomb may have ended WWII, but radar won it. Radar kept Britain in the war long enough for the arsenal of democracy to get fully into the game. After the Cold War ended, radar-scattering folks ended up making video games realistic and cell phones reliable.
- **Sonar**: For decades, humans have been hard at work trying to duplicate the innate abilities of echolocating mammals like dolphins, whales, and bats. Acoustic scattering is more than just a scalar version of electromagnetics.
- **Medical diagnostics**: Ultrasound images, mammograms, etc. are 2D "cuts" of three-dimensional anatomy. Doctors are expert at interpreting them, but the diagnosis is still quite subjective. Machine learning can assist doctors by highlighting suspicious features in signals.
- **Structural flaw detection**: Technicians are not as highly trained at diagnosis as are doctors, plus there is no standard "anatomy" and the structure can't tell where it hurts. For many applications, machine learning can take the lead by automatically identifying flaws that could lead to structural failure.
- **On-line inspection**: Process engineers don't want to interpret images. They want the instrumentation to give a green light if the process is OK, and a red light if it's out of spec. Automatic, real-time interpretation of complex process-monitoring signals is now doable.
- **Intelligent robotics**: The key to useful robots is a combination of imaging sensors and the on-board intelligence to interpret them in real time. You want to simply tell the robot to turn left at the big tree, not feed it GPS coordinates.

The modeling techniques, and applications of them, we'll be discussing allow one to implement new and better measurements with both novel instrumentation and artificial intelligence that automates the interpretation of the various (and multiple) imaging data streams. There's rather a lot of high-level math, of course, but that's good for us because we're not the sort of laymen who Einstein said "have a secret grudge against arithmetic."

The underlying mathematical and computational methods we'll discuss transcend any particular application(s). If you can do radar, you can do sonar, seismology, nondestructive evaluation, and so on. That turns out to be pretty important because lifequakes come along about every six to eight years, and any particular thing that you happen to be expert at might go from hot to not in an instant. That has happened to me repeatedly since 1986 and so all these years later, I've accumulated a seemingly diverse set of subject matter expertise with a mindset of always being on the lookout for new applications of what I know and can do.

It has recently come to my attention that the current cohort of graduate students are Gen Z, who did at least some college via Zoom and are fundamentally different from the generations that came before them. They've never not had all the world's knowledge in a portable, semidisposable device that they carry with them at all times. One goal of this book is to help connect this generation with the vast scientific literature that has long existed in musty libraries but now is available for download at places like Internet Archive, but only if you know what sources to seek out and use as the basis for your personal reference library. It's also important to pay homage to those scientists and mathematicians who spent their professional lives developing tools that you can now use to solve problems. Tea will be spilled; the index is designed to allow a quick Hollywood read.

This book is based on a multisemester sequence of graduate classes that I've been giving for three decades now. I started by typing up my own hand-written notes, and as I was going along, many of my students made plots for this book as exercises in class. Thanks for that, BTW. I've also drawn examples whenever possible from the dissertation research of my graduate students, with discussion of the real-world problems that motivated their research. I've deliberately included anecdotes of the sorts of issues that can arise in collaborative research that includes multiple investigators and multiple institutions. Much of our work is done in close collaboration with small companies, several of whom I've shepherded from *de novo* startups on through VC investment or M&A, but I've always avoided taking any equity stake so I'd be at arm's length and could focus on doing what's best for my students' career progression.

This isn't intended to be a comprehensive reference work on scattering for radar, sonar, etc. Indeed, I've deliberately downplayed techniques that aren't all that important now that ubiquitous computing of sufficient power means we can simulate realistic 3D scattering scenarios. I do try to point the reader to classic texts where those mature areas of research are discussed in great detail by those notables who developed the methods, doing the best they could with the computational power they had available. Once upon a time, of course, computer was a job title.

1

Background

1.1 Some History

As a child you probably played the swimming pool game of tag where you close your eyes and then repeatedly call out "Marco" with the response "Polo" each time from the other players by which you're supposed to locate them and tag someone who is then it. Biurnal hearing allowed you to tell which direction to lunge, and you could guesstimate who was nearby and who was farther away. The game is a little trickier to play at indoor pools because the acoustic scattering from walls and such can be confounding.

Marco Polo may or may not have traveled from Venice to China, although he did spend a couple of decades travelling and trading along the Silk Road, where he would certainly have picked up a lot of information, such as how to make power smoothies, and trade goods like yak hair [1]. He mistook rhinos for chubby unicorns, which is now a meme. In his book, he also claimed to have been besties with the emperor Kublai Khan. Christopher Columbus brought a copy of that book along with him on his 1492 trip to the Orient, but it turned out to be unhelpful. Navigating the world based on maps pieced together from stories of other travelers is always going to be a bit iffy. Fortunately, over the last century or so there has been a rapid development of navigation technologies.

The Submarine Signal Company, established in 1901 in Boston, was the first commercial enterprise organized to conduct underwater sound research and to develop equipment to be used for increasing the safety of navigation [2]. "Our invention relates to a method of ringing or sounding a bell and also to a system and apparatus for transmitting intelligence between ships at sea and between the shore and any ship by means of sound-signals made in the water at the transmitting-station by electrical means. These sounds are picked up from the water at the receiving-station by means of electrical or mechanical devices." The initial product line included underwater bells for shore-based stations, buoys, and lightships as well as encased microphones for sound detection on the ships [3, 4].

1.1.1 The Titanic Disaster

In 1912, the unsinkable Titanic struck an iceberg and sank [5]. Not long after, Sir Hiram Maxim self-published a short book and submitted a letter to Scientific American [6] in which he asked, "Has Science reached the end of its tether? Is there no possible means of avoiding such a deplorable loss of life and property? Thousands of ships have been lost by running ashore in a fog, hundreds by collisions with other ships or with icebergs, nearly all resulting in great loss of life and property." Maxim noted that collisions often take place in a fog at night when a searchlight is worse than

A Data Engineering Approach to Wave Scattering Analysis: with Applications in Radar, Sonar, Medical Diagnostics, Structural Flaw Detection and Intelligent Robotics, First Edition. Mark K. Hinders.

useless because it just illuminates the haze. It was (becoming) known that bats used some form of sound that was outside the range of human hearing in order to echolocate and feed, but he thought it was infrasound rather than ultrasound [7, 8]. Maxim described a concept for a very low-frequency directional steam whistle or siren that could be used to (echo)locate icebergs during foggy nights when collisions were most likely to occur. Whether Maxim's patented apparatus would have been effective at preventing collisions at sea is a question that's a little like whether Da Vinci's contraptions would have flown. He got the general idea right, and can be credited with stimulating the imaginations of those who subsequently worked out all the engineering details.

His sketch, reproduced as Figure 1.1, is quite remarkable. The key idea is that the time delay of the echoes determines distance because the speed of sound is known, but more importantly, the shape of the echoes gives information about the object that is returning those echoes. Analysis of those echo waveforms can, in principle, tell the difference between a ship and an iceberg, and even differentiate large and small icebergs. He even illustrates how clutter affects the echoes differently from backscattering targets. Science has not, in fact, reached the end of its tether, even after a century of further development. This is exactly how radar, sonar, and ultrasound work [9, 10].

Maxim's suggested apparatus embodies a modified form of "siren," through which high-pressure steam can be made to flow in order to produce sound-waves with about 14–15 vibrations per second, and consequently not coming within the range of the human ear. These waves, it is asserted, would be capable of traveling great distances, and if they struck against a body ahead of the ship, they would be reflected toward their source, "echo waves" being formed [11]. This self-published pamphlet was discussed in [12].

1.1.2 Das Unterseeboot

The first submarine to successfully dive, cruise below the water surface, and emerge to the surface again on its own was the Sub Marine Explorer of the German American engineer Julius H. Kroehl, which already comprised many technologies that are still essential to modern submarines [13]. The first submarine built in Germany, the three-man Brandtaucher, sank to the bottom of Kiel harbor on 1 February 1851 during a test dive [14]. The Confederate States of America fielded several human-powered submarines, including CSS H. L. Hunley. The first Confederate submarine was the 30-foot-long Pioneer, which sank a target schooner using a towed mine during tests on Lake Pontchartrain, but it was not used in combat. It was scuttled after New Orleans was captured and in 1868 was sold for scrap, but the similar Bayou St. John submarine is preserved in the Louisiana State Museum. CSS Hunley was intended for attacking Union ships that were blockading Confederate seaports. The submarine had a long pole with an explosive charge in the bow called a spar torpedo. The sub had to approach an enemy vessel, attach the explosive, move away, and then detonate it. It was extremely hazardous to operate, and had no air supply other than what was contained inside the main compartment. On two occasions, the sub sank; on the first occasion, half the crew died, and on the second, the entire eight-man crew (including Hunley himself) drowned. On 17 February 1864, CSS Hunley sank USS Housatonic off the Charleston Harbor, the first time a submarine successfully sank another ship, although it sank in the same engagement shortly after signaling its success. Submarines did not have a major impact on the outcome of the American War Between the States,[1] but did portend their coming importance to naval warfare and increased interest in their use in naval warfare.[2]

1 A more accurate name is War of the Rebellion. As we reevaluate our history, monuments, and heroes, naming things properly is important [15].

2 See, for example, https://en.wikipedia.org/wiki/History_of_submarines.

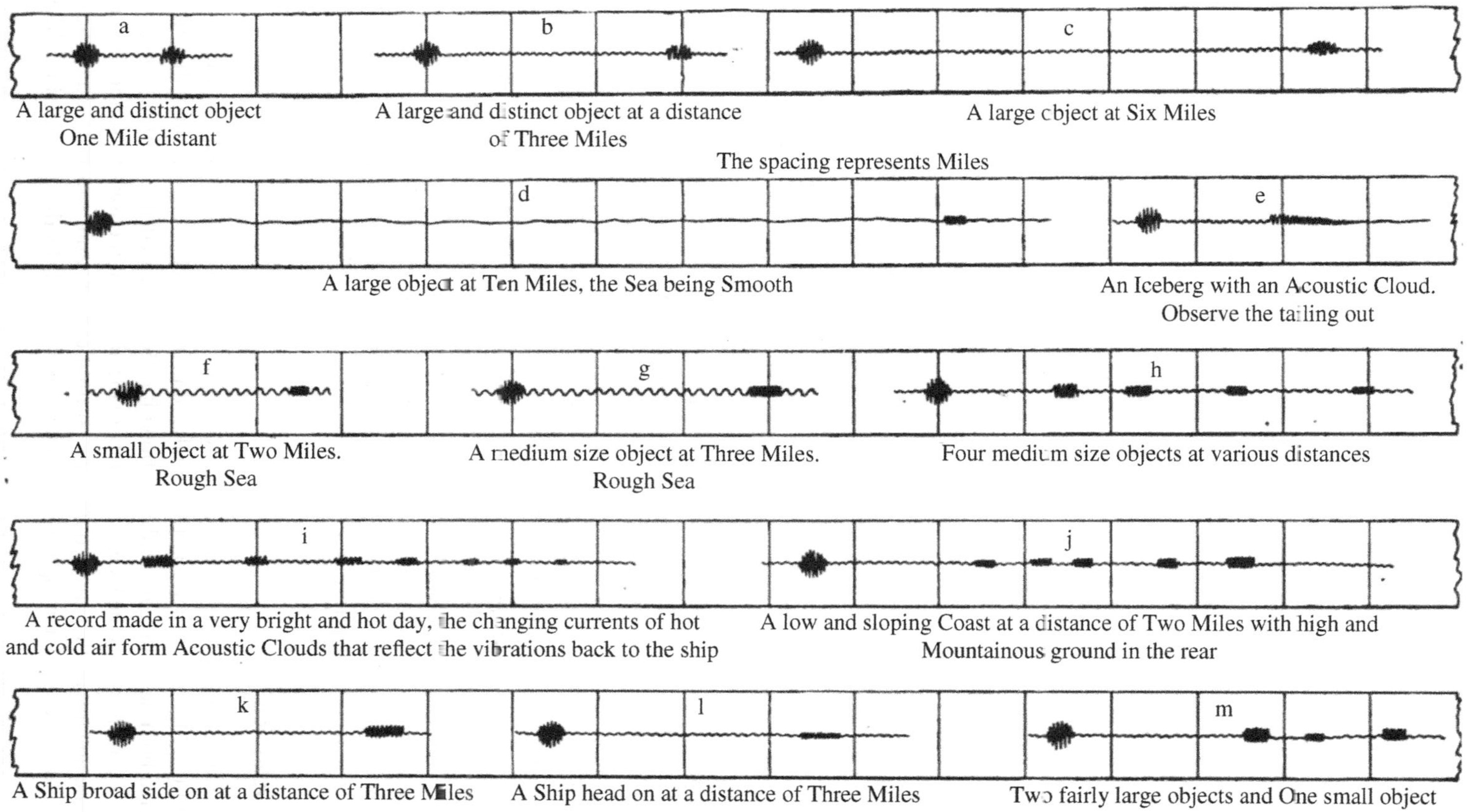

Figure 1.1 The infrasonic echo waves would be recorded by a stretched membrane that the infrasound waves would vibrate, and those membrane vibrations could jingle attached bells or wiggle pens tracing lines on paper as Maxim illustrated. Maxim's concept was discussed in Nature, a leading scientific journal yet today.

You almost certainly know Captain Nemo's submarine Nautilus from Jules Verne's *Twenty Thousand Leagues Under the Sea* (1870), but you may not have read *The Mysterious Island* (1874) or know that Nemo's fictional craft was named after Robert Fulton's real-life submarine Nautilus (1800). Verne was inspired by the French Navy submarine Plongeur, a model of which he saw at the 1867 Exposition Universelle. The fictional Nautilus was battery powered, not nuclear like today's boomers that can travel way more than 20,000 leagues (not quite three laps around the earth) under the sea, sneaking around for six months at a time just in case they need to destroy the world because reasons.

The Battle of Hampton Roads was the most important naval battle of the American Civil War. It was fought over two days in March 1862, where the Elizabeth and Nansemond rivers meet the James River just before it enters the Chesapeake Bay adjacent to the city of Norfolk, Virginia. The battle was a part of the effort of the Confederacy to break the Union blockade, which had cut off Virginia's largest cities and major industrial centers, Norfolk and Richmond, from international trade. The battle was the first meeting in combat of ironclad warships: USS Monitor and CSS Virginia. USS Monitor was a semisubmersed, iron-hulled steamship and was the first ironclad warship commissioned by the Union Navy. Her remains were found upside down 16 miles off Cape Hatteras in 1973 at a depth of about 240 ft. In 1987, the site was declared a National Marine Sanctuary, the first shipwreck to receive this distinction. Because of Monitor's advanced state of deterioration, recovery of any remaining significant artifacts and ship components was quite urgent. Numerous fragile artifacts, including the innovative turret and its two Dahlgren guns, an anchor, steam engine, and propeller, were recovered. They were transported to the Mariners' Museum in Newport News, where a full-scale copy of USS Monitor, the original recovered turret, and a variety of artifacts and related items are on display.[3] Also in Newport News is the largest military shipbuilder in the United States and sole designer, builder, and refueler of nuclear-powered aircraft carriers. HII is responsible for building more current aircraft carriers than the rest of the world's navies put together. Virginia would be a superpower if they seceded today.

Most consider French physicist Pierre Curie's discovery of piezoelectricity in 1877 to be the moment that sonar was conceived. Thirty-five years later, also inspired by the sinking of the Titanic, Physicist Paul Langevin was commissioned to invent a device that detected objects at the bottom of the sea. Langevin invented a hydrophone –what the World Congress of Ultrasound in Medical Education refers to as the "first transducer" in 1915. Langevin built an echo-ranging system using quartz crystals placed between two steel plates to generate sound. In 1918, for the first time, echoes were received from a submarine at distances as great as 1500 m. WWI came to an end, however, before underwater echo ranging could meet the German U-boat threat (See http://rsnr.royalsocietypublishing.org/content/66/2/141, https://phys.org/news/2008-02-inventor-sonar-history.html and http://journals.sagepub.com/doi/pdf/10.1177/0968344516651308).

1.1.3 Aircraft Detection

Acoustic location was used until the early years of WW2 for detection of aircraft by picking up the noise of their engines (Marco!), which was then analyzed to determine the direction of the aircraft (Polo!). Horns give both acoustic gain and directionality; the increased interhorn spacing compared with human ears increases the listener's ability to localize the direction of a sound. Acoustic techniques had the advantage that they could "see" around corners and over hills, due to sound refraction. The technology, shown in Figure 1.2, was rendered obsolete by the introduction of radar,

3 See, for example, https://en.wikipedia.org/wiki/Battle_of_Hampton_Roads.

Figure 1.2 Aircraft engines produced unprecedented sound, so in order to hear them at a distance, the war efforts developed listening devices. A two-horn system at Bolling Field, USA, 1921. Sound location equipment sometimes had four acoustic horns, a horizontal pair and a vertical pair, connected by rubber tubes to stethoscope type earphones worn by two technicians that enabled one to determine the direction and the other the elevation of the aircraft. Source: Unknown author/Wikimedia Commons/Public Domain.

which travels at the speed of light instead of the speed of sound. Japanese acoustic locators were colloquially known as "war tubas," which I hope sounds as funny in Japanese as it does in English.

Practical radar technology was ready just in time to turn the tide during the Battle of Britain, thanks to the cavity magnetron [16] and amateur scientist and Wall Street tycoon Alfred Lee Loomis, who personally funded an enormous amount of scientific research at his private estate before leading radar research efforts during WW2 [17, 18]. The atomic bomb may have ended the war, but radar and sonar won it. Then, during the entirety of the Cold War, uncounted billions were spent continuing to refine radar and sonar technology, countermeasures, counter-countermeasures, etc. with much of that high-level and mathematically esoteric scientific work quite highly classified. The result is virtually undetectable submarines that patrol the world's oceans and stand ready to assure mutual destruction as a deterrent to sneak attack. More visibly, but highly stealthy, radar-evading fighters and bombers carrying satellite-guided precision weapons can destroy any fixed target anywhere on the planet with impunity while minimizing collateral damage. It's both comforting and horrifying at the same time.

What humans have been trying to figure out since Maxim's pamphlet, is how to interpret the various radar blips, sonar pings, and ultrasound images [9]. The fundamental issue is that the shape, size, orientation, and composition of the object determines the character of the scattered signal, so an enormous amount of mental effort has gone into mathematical modeling and computer simulation to try to understand enough about that exceedingly complex physics in order to detect navigation hazards, enemy aircraft and submarines, tumors, structural flaws, etc. Much of that work has been what is called by mathematical physics *forward scattering*, wherein I know what I transmit, and I know the size and shape and materials and location and orientation of the scattering object. I then want to predict the scattered field so I'll know what to look for in my data. The true problem is mathematically much more difficult, called *inverse scattering*, wherein I know what I transmit, and I measure some of the scattered field. I then want to estimate the size and shape and materials and location and orientation of the scatterer. Many scientific generations have been spent trying to solve inverse scattering problems in radar, sonar, and ultrasound. There has been a fair amount of success [19–23].

1.1.4 Medical Ultrasonography and NDE

X-rays were discovered entirely by accident. Professor Roentgen was messing around in his laboratory with some new-fangled cathode ray tubes and he noticed that photographic plates were being fogged even though they were still wrapped tightly in paper and had never been exposed to light. He shut himself into his laboratory and worked tirelessly to explore the behavior of these unknown rays (which he gave the symbol "X") before announcing his discovery to the world and winning the first Nobel Prize. His first X-ray image is still used in textbooks today. It's his wife's hand with her large ring. Presumably she was bringing him a sandwich or something and he said, "Put your hand here, bitte."

Getting the occasional X-ray isn't a problem, but if you X-ray your hand each time you tune up the machine, you're going to lose that hand. Edison recognized the adverse health effects of X-rays pretty early on, and personally stepped back from the development efforts in his laboratory. He ended up losing his head X-ray technician bit by bit, starting with fingers and then hands.

It's a little hard to overstate how quickly X-rays became important; it took a matter of several weeks. In those days, doctors made house calls, of course, but they couldn't really do all that much. They would often do their diagnosis without even having their female patients disrobe, so it's not all that surprising that a common mansplainy diagnosis was "women's troubles." X-rays were a revelation. Some considered them shockingly improper in that they could see through clothing and make photographs of bones and other private things.

Personally, I find X-rays kind of boring. All they do is go in a straight line getting absorbed along the way according to the tissue density. CT scans are cool, of course, but the radiation dose is a concern. They're also expensive.

I like diagnostic ultrasound quite a lot. It's so safe we use it on pregnant women. It's cheap. It's portable. It's real time. The physics is really complicated. It works for medical imaging and structural health monitoring, with surprisingly few differences. The same equations also describe underwater sound and bat echolocation. Medical ultrasonography dates from the early 1950s[4] although ultrasound for structural inspection is just a bit older.

In my research group, we do both medical imaging and structural health monitoring using ultrasound. We also do quite a lot of other things because every so often the world changes dramatically and unpredictably. For example, when the Space Shuttle Challenger exploded in January 1986 (Figure 1.3). I was a second-semester senior, aerospace engineering major, about to be commissioned as an officer in the US Air Force with orders to go to Space Command. I was in my apartment in Boston doing propulsion homework when the anomaly happened and the entire space program imploded.

---

4 John Julian Wild was born in Kent, England and received BA and MA degrees in Natural Sciences (with honors) from Cambridge University. In 1944, he joined the Royal Army Medical Corps and attained the rank of major and then after World War II, he was a fellow in the Department of Surgery at the University of Minnesota. He had become interested in treating bowel distention, which was often fatal, following bomb blasts from buzz-bombs. He needed to measure the changes in thickness of the bowel wall and pulse-echo ultrasound was considered a possibility, but commercial nondestructive testing (NDT) equipment developed by Donald Sproule in England and Floyd Firestone in the United States for detecting cracks in tank armor plate, operated at too low a frequency to achieve the resolution required for bowel wall measurement. A much more sophisticated piece of ultrasonic equipment developed during wartime to train flyers to read radar maps of enemy territory operated at 15 MHz. Using this equipment Wild quickly confirmed the possibility of measuring living bowel wall thickness at that frequency. Experiments with a surgical specimen of cancer of the stomach wall proved the concept of using pulse-echo ultrasound for tumor diagnosis and detection, although the use of ultrasound in medicine was skeptically received. http://www.washingtonpost.com/wp-dyn/content/article/2009/09/23/AR2009092304474.html.

Figure 1.3 We've grown used to wonders in this century. It's hard to dazzle us. But for 25 years, the United States space program has been doing just that. We've grown used to the idea of space, and perhaps we forget that we've only just begun. We're still pioneers. They, the members of the Challenger crew, were pioneers. We'll continue our quest in space. There will be more shuttle flights and more shuttle crews and, yes, more volunteers, more civilians, and more teachers in space. Nothing ends here; our hopes and our journeys continue [24]. The loss of the Space Shuttle Challenger may have effectively ended a generation's hopes of going to space, but it did cause an explosion of R&D. Source: NASA/Public Domain.

Since the Air Force suddenly didn't have anything for me to do, they let me stay at the university for a year and get a master's degree. I wrote a thesis on elastic wave attenuation due to scattering from inhomogeneities, with applications in seismology. That went well enough that I requested to delay my active duty service commitment for three more years and get a PhD. The Air Force didn't need second lieutenants with doctorates, so they issued me orders to report to San Bernadino, CA where, for four years, I would do something or other entirely unrelated to either of my freshly minted engineering degrees.

From his hospital bed where he was recovering from colon cancer surgery, my advisor phoned a three-star general and got my orders changed so that I would be assigned to the Air Force base outside of Boston and could then continue on with graduate school in my spare time. Even back then, I knew how unusual that was. So, I cut my hair, put on my uniform, and reported for duty on 21 July 1987. They had more lieutenants than they knew what to do with because this was just about at the peak of the Reagan military buildup. In modern terms, it was a lieutenant bubble.

The commanding general had a policy of making the various program offices on base compete for the new talent, so I was told to take two weeks and interview around base and then come back and tell them where I wanted to be assigned. It was an easy choice because I had just finished a master's thesis with applications in seismology, and the one place in the Air Force where that kind of research was done was the Air Force Geophysical Laboratory up on the Hill with an excellent library and grass and trees and such. Recall that the whole point of getting my orders changed was so that I could both serve on active duty and finish my PhD.

It turned out that AFGL was a tenant organization on the base, and so I could choose to be assigned anywhere except there. If you're familiar with the way bureaucracies function, that outcome should give you comfort because my story was going far too well to be believable. So, there was a little cinder-block building up on the Hill near the library, where a small group of people were doing electromagnetic scattering for counter low-observables, that is, stealth. They didn't need me or have anything for me to do, but I had myself assigned there and they gave me a big gray steel desk in a corner where I could try to come up with something for myself to do related to electromagnetic scattering.

It worked out reasonably well. I finished my PhD on schedule, despite my advisor dying from colon cancer. Along the way, I taught myself quite a lot about electromagnetic scattering for stealth at a time when the big aerospace companies were paying enormous salaries to anybody who could recognize Maxwell's equations in neon lights. If you're doing the math in your head, you may have calculated that the Cold War ended just before my four years in the Air Force were up. I learned early on in my career to focus on the fundamental mathematics and physics that can be applied to a variety of applications so that you can reinvent yourself as needed. You'll need to.

Meanwhile, the Challenger disaster and a sequence of aircraft accidents caused by structural flaws (Aloha Airlines Flight 243 and United Flight 232) caused the field of nondestructive evaluation[5] to explode, in a good way.

Flight 243 was caused by cracks around rivets in fuselage lap joints, which developed due to metal fatigue after many years of pressurization for innumerable short hops between islands. Flight 232 was caused by a crack in the titanium engine rotor hub that developed due to a hard alpha inclusion in the titanium billet the rotor hub was forged from. Both tragedies should have been prevented by proper inspections. In 232, only 100 people died; in 243, only one person died. Only is the key word, except that *nobody* should have died dammit!

At NASA Langley Research Center in Hampton, VA, the Nondestructive Evaluation Sciences Branch grew much faster than their allotment of civil service slots would allow. The answer was to bring on board most of the scientific researchers as contractors rather than government employees, and The College of William & Mary in Virginia was one of those contractors. I was excited to come to Williamsburg to help build the Applied Science Department and oversee the NDE graduate program, with very close ties to NASA LaRC because we had a number of graduate students and research scientists who worked on the Center. We also often work with the people building aircraft carriers, just across the peninsula from NASA. There seems to be a never-ending sequence of scientifically interesting and technically challenging NDE problems both places. I've been at W&M for 62 semesters and counting. I plan to stay here for an even 100 semesters, so I guess you could count down: T-minus 38 semesters. You may note that many of the notables who developed the scattering we'll be discussing had quite long and productive scientific careers.

5 Donald O. Thompson recognized that there wasn't an adequate science base for NDT to be more quantitative, so he convinced the Defense Advanced Research Projects Agency (DARPA) to fund university researchers in a coordinated manner. A key feature of the program was the Annual Review of Progress in QNDE, the proceedings of which now represent 40+ years of the progress literature for QNDE. I had a chance to talk with Don briefly a week before he died at the age of 86. It was at the 2013 QNDE meeting in Baltimore and after the dinner and speeches, where he was recognized as a leader in the development of the global NDE community, I knelt down in front of his wheelchair and we chatted a bit. Later I commented to my graduate students who weren't there that they missed a chance to see me genuflect. Don deserved that level of respect from the NDE community. He was responsible, perhaps more than anyone, for turning NDT from a collection of technician-level engineering methods into a stand alone scientific field. He had a particular talent for recognizing scientific talent, combined with the political savvy to convince funders to fund the research of the scientists he had convinced to cooperate instead of compete. I have a complete set of the QNDE proceedings in my laboratory, and have been attending the conference with some regularity since 1993. The origin of the meeting is that Don didn't want to have to write long technical reports every year, so instead he arranged for everybody to come to a conference and submit progress reports on their work for that year, and the collection of these papers were lightly edited and served as both conference proceedings and technical report. It seems obvious in retrospect, but it is really genius. One of Don's long-time colleagues said of him, "Don ruled with an iron hand, but was compassionate and the best boss I have ever known."

1.2 Ultrasound Immersion Tank Scans

So I ended up doing radar scattering for a living by accident, and then ended up doing ultrasound scattering for NDE by accident. I am also accidentally a leading expert in dental ultrasonography and developed a sequence of prototype machine-learning-based prostate cancer diagnosis systems that used both trans-rectal and trans-urethral ultrasound scanners. It's new things all the time, which is part of the fun, especially since the math and physics we've worked hard to master doesn't change. We'll get to all that, but first I think it would be helpful to describe some of the ways that the data we'll be modeling with those equations is acquired. We'll start with ultrasound, which is a subset of acoustics.

Acoustics describes the phenomenon of mechanical vibrations and their propagation in solid, liquid, or gaseous materials. Sound waves above 20 kHz or so are inaudible and are referred to as ultrasonic or ultrasound. Ultrasound is useful for medical imaging and structural flaw detection because it propagates well in many solids and liquids, and because it is sensitive to local changes in material properties. Ultrasonic wavelengths are of the same order as flaw sizes that are often of interest.

Here's a simple question for you: What are ultrasonic wavelengths at 100 kHz, 1 MHz, 10 MHz, 100 MHz, and 1 GHz for several common structural materials? This question is going to come up again throughout the book, so I'd suggest taking a few minutes and tabulating the numbers. More importantly, keep in mind about what the answers turn out to be.

Here's another question: What can I do to keep the sound from my neighbor's stereo from coming through the wall of my apartment? *The Straight Dope* explains nicely a few key concepts that are in play in response to apartment-dweller-Derek's question, so I think I'll just include three paragraphs here from Cecil Adams' overnight staff reporters[6]:

Everyone knows intuitively – it barely counts as an observation – that sounds get weaker (i.e. become attenuated) as they travel farther from their source. But why? Much of it is scattering and absorption. A sound wave passing through any medium – air, say, or drywall – does so by causing the medium's molecules to vibrate. Scattering is the extent to which the wave gets fragmented and redirected upon striking an obstacle in its path. (Picture ripples on the surface of a pond breaking into subripples when they encounter a rock or stick poking out of the water.) Absorption is the drop in volume caused by energy loss in the form of heat – the result of making all those molecules move around. And the effects of both scattering and absorption increase with the frequency of the wave – the higher the frequency, the greater its tendency to die out. Thus, the treble and midrange sounds coming from your neighbor's apartment get scattered and absorbed more thoroughly as they pass through the various matter surrounding it, leaving the big, dumb low-end waves to lumber along till they find you.

If the frequency and volume are right, sound waves can cause entire objects to vibrate sympathetically – surely you've heard of those souped-up car stereos that turn their host vehicles into gigantic joy buzzers. Because of their relatively large mass, things like walls and floors resonate more at low frequencies than at high ones, and thus can help to pass the bass notes along, particularly if the speaker is touching the potentially resonant surface. Long-term resonance can be pretty destructive: Thanks in part to vibrations caused by the wind that regularly swept over it, in 1940, the old Tacoma Narrows Bridge[7] shook itself to pieces and collapsed into Puget Sound.

---

6 https://www.straightdope.com/21343871/why-is-bass-so-boomy-and-what-can-i-do-about-it.

7 I always have my structures class look up "who died" when Tacoma Narrows collapsed. It was a cocker spaniel named Tubby who was left in the car you can see in the video.

Low-frequency noise is weird stuff. Years back, I noted that infrasound – sound pitched below the hearing range of most humans, which stops at around 20 Hz – can cause dizziness. Some recent research suggests it may do more than that. After taking spectrum analysis readings at a couple of UK sites repeatedly described by visitors as "haunted," Vic Tandy and Tony Lawrence of Coventry University have argued that the presence of 18.9 Hz infrasound is responsible for the creepy feelings described. (In one case, they concluded that a terrifying, seemingly paranormal experience of Tandy's had likely resulted from the whirring of a laboratory extractor fan causing his eyeballs to resonate.) And in 2003, the use of 17 Hz infrasound at London concerts of experimental electronic music correlated with audience reports of "unusual experiences" including nausea, momentary anxiety, tingling, and a sense of coldness. Ideally, Derek, by the time your neighbor has traded his bass for an ultra-low-end tone generator, one of you will have found someplace else to live.

Here's another question: Is there some sort of semiaudible acoustic weapon causing concussion-like symptoms among US diplomats around the world? Havana Syndrome caused quite a lot of noise a few years ago, although it turned out to be a fairly typical case of mass psychogenic illness. The sounds that diplomats were hearing were crickets and cicadas, but it's not true that the ill effects were all in their heads. Psychogenic illnesses can, in fact, result in physiological symptoms [25]. To be clear, infrasound generated by wind turbines does not manifest ghosts in your basement or bats in your belfry [26]. Bats do use ultrasound to echolocate, though. Typical frequencies are around 50 kHz and it's a good thing that those frequencies are inaudible to us because bats are really loud and they screech COVID into the miasma while I'm trying to sleep.[8]

The last time I went to the eye doctor to get my prescription updated, I was very excited to have him scan my eyeball via Optical Coherence Tomography. He was equally excited to have a patient who knew the technology and how to interpret the images. We may have chatted for a while and backed up his other appointments. My bad.

OCT uses low-coherence interferometry to produce a two-dimensional image of optical scattering from internal tissue microstructures in a way that is analogous to ultrasonic pulse-echo imaging. Indeed, the display looks rather a lot like ultrasound B-mode, but OCT has longitudinal and lateral spatial resolutions of a few micrometers and can detect reflected signals as small as 10^{-10} of the incident optical power. I've been following this technology development from the earliest day. The first NSF review panel I ever served on, back in the 1990s, was for OCT. The most unwieldy review panel I ever served on was when NIH was pushing hard to get this technology out of the laboratory and into clinical usage. That panel had 50-ish reviewers on it, and we were trying to do the premeeting streamlining of the noncompetitive proposals via an old-fashioned

---

8 As a student, Donald Griffin and a fellow student, Robert Galambos, found that bats could use reflected ultrasound to detect objects. In 1944, Dr. Griffin coined the term echolocation to describe the phenomenon. To many, the idea was outrageous. Dr. Griffin once wrote, "One distinguished physiologist was so shocked by our presentation at a scientific meeting that he seized Bob by the shoulders and shook him while expostulating, 'You can't really mean that!' " While echolocation is well accepted today, Dr. Griffin's pleas that animal thinking and consciousness become standard fare for research have met with more mixed success. The numerous and vocal critics of the growing field of cognitive ethology include both scientists and philosophers. Scientists complain the field is too dependent on anecdote, highly subjective and anthropomorphic, more akin to the way a dog owner envisions his pet's day than the way a scientist typically approaches the study of animal behavior.

Whitlow W.L. Au was born in Honolulu in 1940 and earned his bachelor of science in electrical engineering from U. Hawai'i at Manoa and his PhD from Western Washington University, and then joined the U.S. Navy's Naval Undersea Center. Following his seminal paper in 1974 on the echolocation signals of Atlantic bottlenose dolphins, Au went on to methodically quantify the performance and signal characteristics of dolphins and other small whales that echolocate under water. His book, "The Sonar of Dolphins" published in 1993, remains as the primary source for describing dolphin echolocation. He died in 2020 at age 79.

conference call. The problem was that the panel chair was on his cell phone driving in his car, and when his call would drop, we'd all have to wait on the phone until he dialed back in.

Because OCT uses near-IR wavelengths, there's sufficient penetration of opaque soft tissues to use it for a variety of superficial applications. Many are being developed in dentistry these days. Since the light can be delivered via fiber optics, there are all manner of endoscopic applications of OCT. You'd be surprised to know what urologists will jam where and still call it noninvasive.

Ultrasonic frequencies used for medical imaging and NDT of structural materials tend to be in the low MHz range. At these frequencies, the wavelengths are quite small and attenuation is usually a concern. The trade-off is that you'd like to go to higher frequencies in order to get better resolution, but attenuation often scales like the square of frequency, so doubling the frequency quarters the depth of penetration. As a personal aside, I think all this is pretty interesting because it requires a fair amount of understanding of the propagation properties of the ultrasound in order to design sensible measurement schemes. The first ultrasound measurement scheme we should talk about is immersion tank scanning.

Figure 1.4 is a cartoon of an ultrasound immersion tank. At MHz frequencies ultrasound doesn't propagate through air much at all, so if you can immerse your test specimen in water, then you can use that water bath to couple the ultrasound from your transducer into and back out of the specimen. The transducer is typically held by a search tube which is connected to a scanning bridge that moves it to and fro and back and forth under computer control.

Immersion tanks come in a variety of sizes (Figure 1.5), but the electronics is all pretty much the same. The computer that manages the motion control also communicates with the ultrasonic pulser-receiver, which sends and receives the voltage signals to and from the transducer as well as with the analog/digital conversion unit, which turns analog voltage signals into digital computer values and vice versa.

In Figure 1.6, I've isolated just the part that we'll need to model, at least initially. The transducer emits an ultrasound pulse which is incident upon the test specimen, which for simplicity I've drawn as a slab. The acoustic properties of the slab are, of course, different from the water it's immersed in.

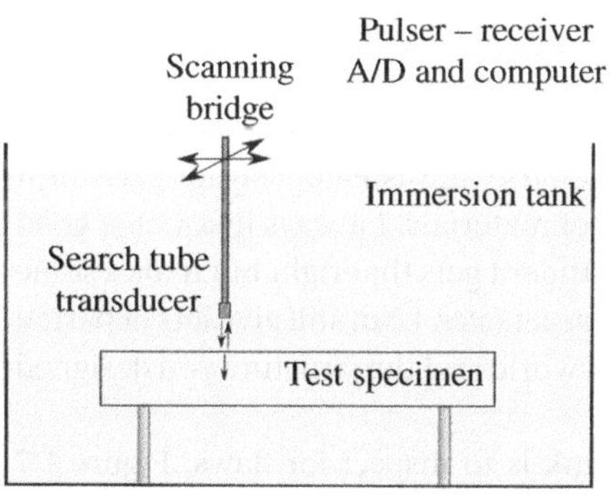

Figure 1.4 Ultrasound immersion tank provides a convenient way to scan a specimen for flaws, if the specimen can be immersed in water. The ultrasound transducer is attached to the end of a search tube, which is rastered across an area via stepper motor in a scanning bridge. At each point in the scan, the pulser–receiver sends a voltage spike to the transducer, which excites an ultrasonic wave pulse at the resonant frequency of the PZT crystal in the transducer. That ultrasound wave both reflects from the test specimen and is transmitted into it. The resulting echoes are received by the same PZT crystal, which then sends an analog voltage trace back to the receiver, where it is amplified and filtered before being converted to a digital signal and stored on the computer, which also controls the scanning and triggers the pulser. Each digitized voltage trace, called an A-scan, carries information about the various material interfaces and discontinuities because time delay of the echoes corresponds to depth, scaled by the speed of sound.

(a)

(b)

Figure 1.5 Immersion tanks can be quite large, as in this one (a), which holds 6 × 6 × 6-feet of water. Note a smaller tank that holds 2 × 2 × 3-feet of water at the lower left of the image behind the window. In acoustic microscopy where very high frequencies are used to find tiny flaws in small samples, an immersion tank may be a small water container on a high-precision scanning bridge atop a vibration-isolation table. Ultrasound can also be done in contact mode (b), either with hand-held transducers or with scanners configured to have the transducer(s) follow any curvature of the surface.

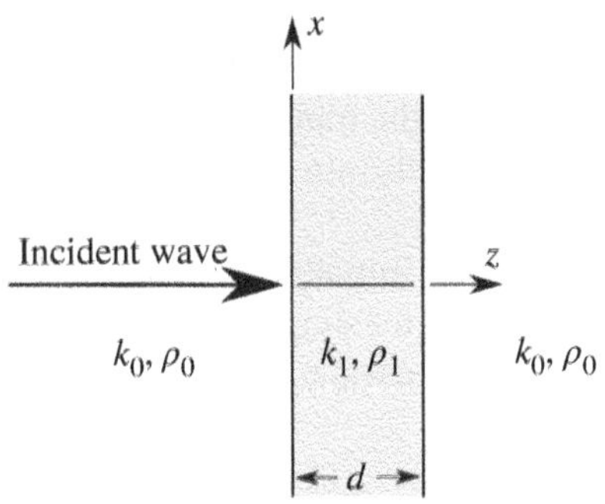

Figure 1.6 Consider an incident plane wave interacting with a slab of thickness d. For an acoustic model, we need to know the speed of sound and density in order to characterize the wave behavior, but since $k = \omega/c$ it's also convenient to use wave number and density. Here we assume that the material to the left and to the right of the slab are the same fluid, that is, water in an immersion tank, but the slab is some other material, so the subscripts are different.

I hope it doesn't throw you off that I've drawn it sideways because one of the key skills we're going to need is the ability to analyze a physical measurement scenario and isolate just the parts of it that are important for our models. When I teach mechanics of materials, I always insist on a good free-body diagram as the first step in solving a problem. If a student gets that right but makes some mathematical errors during the solution that leads to the wrong answer, I can still give lots of partial credit with the caveat that if you screw up the math in the real world and the structure you designed falls down, that's all on you.

Since the point of scanning a sample in an immersion tank is to inspect for flaws, Figure 1.7 shows the ultrasound inside of the sample reflecting from a flaw. The whole name of the game is going to be analyzing such reflections from flaws in order to tell as much as you can about the flaw. It's not good enough to just find a flaw, we want to know how deep it is, how large it is, what kind of flaw it is, how a flaw of that sort could impact the viability of the structural element, whether the flaw is likely to get worse (they never just magically heal themselves), or can I still use this structural element for a while longer.

If putting the sample in an immersion tank isn't practical, ultrasound measurements can be done with the transducers placed directly onto the surface of the structure, typically with some sort of couplant. Medical ultrasound uses hypoallergenic gels that can even be warmed for patient comfort.

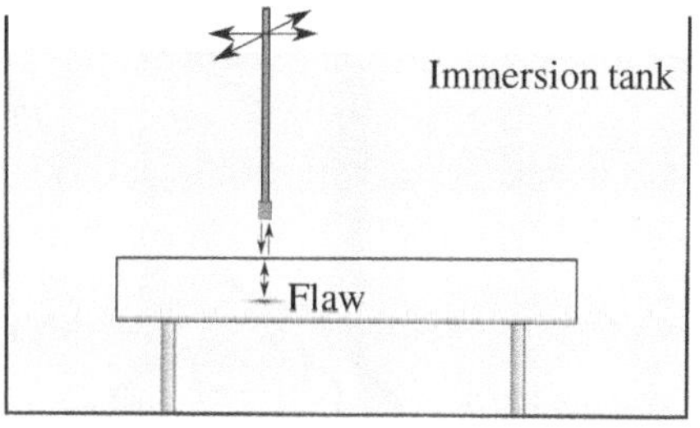

Figure 1.7 Reflections from a flaw in the test specimen will give echoes in the A-scan that are between the front-face and back-face echoes. Indeed, the time delay of the flaw echo tells its depth accurately, and as the transducer is raster scanned across the test specimen, the peak corresponding to that flaw echo can be tracked to determine the lateral extent of the flaw.

NDT uses a variety of specialized couplants, which don't need to be warmed, but sometimes just a little film of water can work fine. Honey also works, but it gets so sticky so just don't. The function of the couplant is to make sure that there's no air gap between the transducer face and the surface. Oh, and for structural health monitoring applications, we sometimes just glue the transducers to the surface and leave them there. Figure 1.5 shows a few contact scanning apparatus in the laboratory, both for a cylindrical storage tank in the foreground and two plate scanners in the background. Figure 1.8 shows two examples of portable ultrasound systems, where the ultrasound pulser-receiver and A/D functions are in the small black box connected to either a tablet or a laptop via USB. It has been quite interesting over the years watching ultrasound electronics shrink down from two racks of equipment to a pocket-sized unit connected to a tablet. Visitors to the lab are often a little disappointed that there aren't more flashing lights and whirring equipment to show them. Field trips to newspaper offices and telephone exchanges also used to be more interesting.

The pulser-receiver exchanges analog voltage pulses with the transducer. Figure 1.9 shows a cut-away view of a typical ultrasound transducer. Sorry that it's not more exciting than a piezoelectric crystal disc metallized on the two flat faces so that a voltage across it causes it to vibrate and emit ultrasound, and then when it's vibrated by returning ultrasound echoes, it gives a voltage response. I have some cut-away transducers for visitors to handle.

Figure 1.8 In recent years,both the computers and the ultrasonic instrumentation have shrunk dramatically. A tablet or laptop controls a USB-powered *nanopulser*, which is a combined pulser-receiver and A/D unit connected to a contact transducer or handpiece.

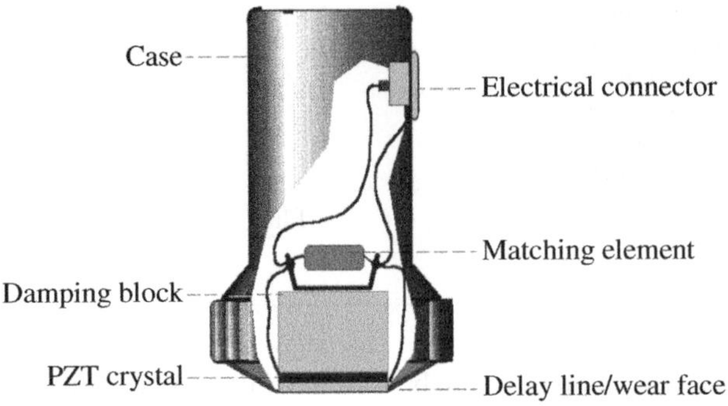

Figure 1.9 Cutaway view of an ultrasound transducer. The PZT crystal has metallized faces, so it vibrates in thickness mode when excited via a voltage spike. The damping block is typically tungsten-loaded epoxy, and keeps the transducer from ringing so that reflections can be recorded. The delay line is used to couple the ultrasound more optimally to water for immersion scanning or metals for contact scanning.

1.3 A-, B-, C-Scans, M-Mode

We're going to have to introduce some jargon. I've included some cartoons that I hope will be helpful, but I made these figures while learning a drawing package so they're not fancy. Most people get these concepts best when they collect some data themselves, or at least see some animations/demonstrations so don't be shy about looking up some of those.[9]

Figure 1.10 shows an A-scan or A-line. A stands for amplitude, and this representation is pretty much just the voltage of the transducer as a function of time. Back in the day, this is what you would see on the little green (or orange) screen of your oscilloscope and then you'd take a polaroid picture of it to paste into your laboratory notebook with rubber cement.

If you send a spike voltage to the transducer (Figure 1.10), it will give a shape kind of like I've drawn because the transducers are damped (so as to not reverberate like a bell) and reflections from material interfaces will give reflections that are about that same shape. The nearest interfaces will

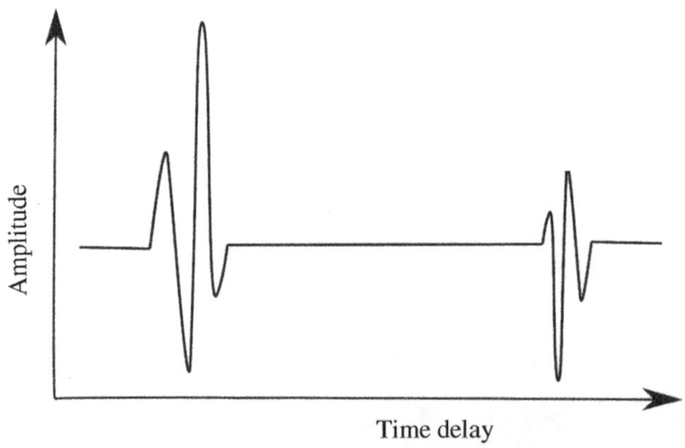

Figure 1.10 A pulse-echo waveform is called an A-line, which shows amplitude vs. time delay. In this cartoon, the first pulse is the front-face reflection from a sample and the later pulse is the back-face reflection. Knowing the speed of sound in the material allows the thickness of a sample to be measured.

9 Here are some books you'd find in my lab: [27–48].

give the first reflections, and deeper interfaces will give later reflections. Assuming a fairly constant speed of sound, the time delay of the reflections gives the depths of the interfaces. The amplitude of the reflection is related to the difference in acoustic properties between the materials at the interface because some of the incident wave is reflected and some of it is transmitted. Subsequent interfaces will only see that transmitted signal, so their reflections will be smaller. Attenuation will also tend to reduce the amplitude of deeper reflections. Again, playing with this in the lab for a few minutes makes this all intuitive. Assuming you know what material you're testing and you look up the speed of sound, as long as you include a factor of 2 because the reflection goes to the interface and then comes back, it's a pretty simple matter to verify that the two pulses in Figure 1.10 are the front face and back face reflections from the slab. If you're doing these tests in an immersion tank, you can also reach in and lift up the slab and watch both peaks in the A-line move toward you. Try to remember to roll up your sleeve first, though.

Once you've verified that the peaks you're looking at are the front and back face reflections, and of course dried off your sleeve, you can look for any echoes between them, which might be due to flaws, as Figure 1.11 shows. As before, the time delay of the echo from the flaw tells how deep it is, and the amplitude of that echo carries information about how reflective the flaw is. The ultrasound beam always has some lateral dimension to it, more like a flashlight than a laser pointer, but the A-line doesn't give you much spatial, that is, lateral, information. Focused transducers will give a larger reflection if the flaw is at the focal depth, of course, so again it really helps to get a feel for this if you can play around in the lab for a while. Roll up your sleeve this time.

Since you can just save the data to your computer and don't have to take a ruler to that Polaroid in your lab notebook, the first thing you're going to do with the A-lines is to window out the front and back face reflections. We usually call that process gating (Figure 1.12) and there used to be knobs on your ultrasound instrument that you would use to adjust the gates while you were recording data. It was always disappointing when you set the gate(s) wrong and recorded useless data. You and your soggy sleeves often had to take a few tries at things to get the gate and amplification and such just right before you got the data you needed. The point is that you don't care about the front and back-face reflections; you care about the often much smaller echoes from the flaws and you want to zoom in on those to analyze the details.

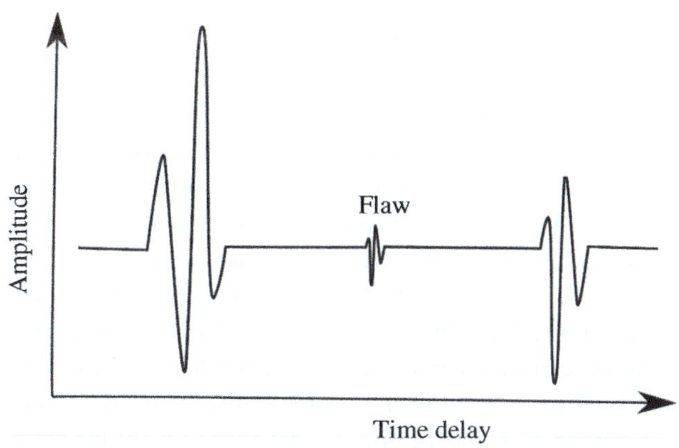

Figure 1.11 A flaw in a sample will occur between the front-face and back-face reflections. Time delay of the flaw echo tells the depth of the flaw, and analysis of the detailed character of the flaw echo is the essence of quantitative nondestructive evaluation.

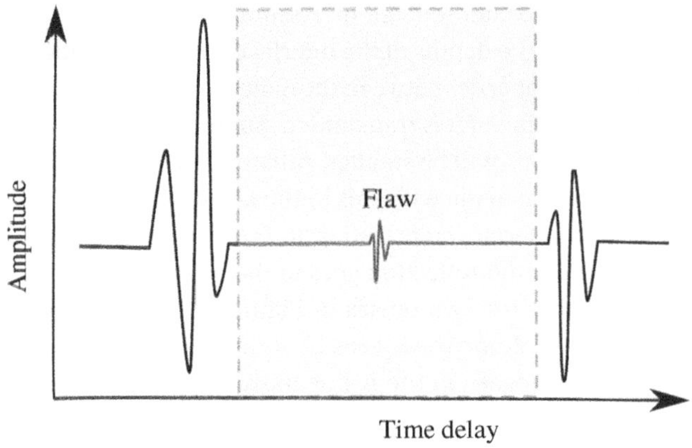

Figure 1.12 Since it's the flaw that's of interest, typically one gates the A-lines to exclude the front- and back-face echoes. The width of the gate, shown in the shading, can be adjusted to isolate particular depth slices of interest by ignoring any echoes that are not inside that time-delay window.

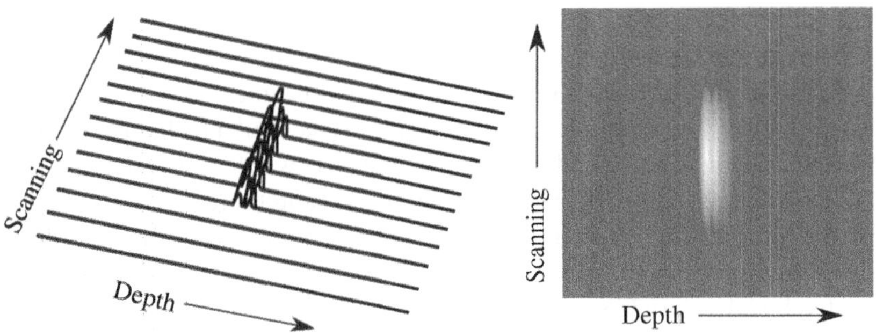

Figure 1.13 As the transducer is scanned laterally, a number of A-scans can be registered next to each other in order to determine the lateral extent of the flaw. If the amplitudes of the registered A-lines are shown as brightness, the two-dimensional representation is called a B-scan. In medical imaging, the A-lines are connected in a fan shape, so the familiar ultrasonographic B-Scan will be fan shaped. For a small intra-cavity probe, the transducer array may be annular and the B-scan image is a hollow circle.

If the search tube that holds your transducer is attached to a stepper motor system, then while your soggy sleeve is drying, a sequence of A-lines could be recorded as the transducer is moved laterally. Figure 1.13 shows several A-lines spread out next to each other, and since I've gated out the front- and back-face reflections, it becomes obvious what the lateral extent of the flaw probably is. Figure 1.13 also shows this same 2D data set but with amplitude of the echoes converted to brightness of the corresponding pixels. Since you're doing this with a scanner and you want to make pretty B-mode images, you record lots of A-lines to get lots of pixels. I hope you noticed that B stands for *Brightness*. In addition, if you're reading the printed version of this book or the ebook on a tablet, please rotate it so Figure 1.13 shows scanning across the top and depth pointing down.

So B-scans are just a bunch of A-scans lined up next to each other because your computer-controlled scanner does that without you having to pedal it, and we get these lovely intuitive images that tell us the depth and size of flaws. Surely you've noticed that most scanners can scan in two directions, so you can make a bunch of B-scans and stack them up adjacent to each other as in Figure 1.14. Old timers who remember gluing Polaroids into their lab notebooks with rubber

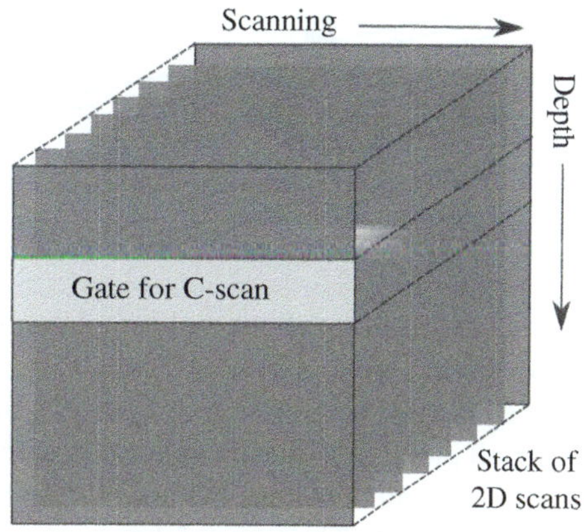

Figure 1.14 The 3D volume of data can be sliced vertically to give a B-scan or horizontally to give a C-scan. Typically, the A-scans are all gated to analyze the echoes from a particular depth range and then the C-scan image is rendered by using the highest amplitude echo within that gate at each location as the image pixel value there. The depth gate can be as large or as small as you like, and you can form as many C scans as you like. If you prefer you can just acknowledge that you've got a 3D volume of data that can be sliced in a variety of ways to make 2D images. Be careful using words like slice when you're talking to expectant parents about their fetal ultrasounds, though.

cement are all jealous that you can acquire this much data.[10] It seems too easy to people who had to get up off the sofa and walk across the room to flip between the three stations on the television or turn the record over.

Back to gating. Once you've got that 3D volume of data, you can slice it in a variety of ways. A C-scan is a horizontal slice as shown in Figure 1.14. Recall that the point is to locate the flaws and then to tell something about them, so a C-scan allows you to isolate just the echoes from the depth range that includes the flaws of interest. This used to be tricky. High-resolution scans can take a long time, and before computers could handle 3D volumes of data you had to set the gates for the C-scan *before* you started the scan. It wasn't unusual to have to try a few different scans to get the data that you needed because if the gate was at the wrong depth or the gate was too narrow you would end up with a garbage image. Sometimes you stood there shaking the Polaroid waiting for the image to come into view only to find you had your thumb over the lens.

The C-scan is going to end up being a 2D image, with pixel brightness corresponding to echo amplitude, but it's the orthogonal view to a B-scan. Typically, the pixel values are the largest amplitude echoes of each A-line inside the gate, but you can also set the gate to be quite narrow and make a bunch of C-scans for a bunch of different depths. Since you've got a 3D volume of data, you can also go back and form B-scans by slicing the data in the other vertical plane. No rubber cement required. You can scroll through a whole bunch of different B- and/or C-scans without even getting up off the sofa.

B-modes are common in medical imaging, but they are often sector scans rather than rectangles. This is because medical ultrasound uses phased arrays to sweep the beam electronically, and to update the images at video frame rates so we can see motion. The images are still formed from a bunch of A-scans next to each other, with echo amplitudes converted to brightness. Speckle is just the manner that inherent tissue inhomogeneities present in brightness mode.

---

10 Professor John M. Reid was born in Minneapolis and received the BS (1950) and MS (1957) degrees in Electrical Engineering from the University of Minnesota, as well as the PhD in Electrical Engineering from the University of Pennsylvania (1965). Reid developed the first clinical ultrasonic scanner with John J. Wild via a grant from the National Cancer Institute. Reid was the sole engineer to build and operate Wild's ultrasonic apparatus. They built the first linear B-mode instrument, a formidable technical task, in order to visualize tumors by sweeping from side to side across breast lumps. In May 1953, this instrument produced a real-time image at 15 MHz of a 7 mm cancer of the nipple in situ along with A-mode differential reflections. Based on technology from WWII radar, Reid devised important circuitry to compensate for the attenuation of ultrasound in tissues by setting the receiver gain as a function of the tissue depth. Similar mechanisms were deployed in most medical and NDT ultrasound systems that followed. I'm just old enough to have used these analog time-gain compensation units, and to have met Prof. Reid once briefly.

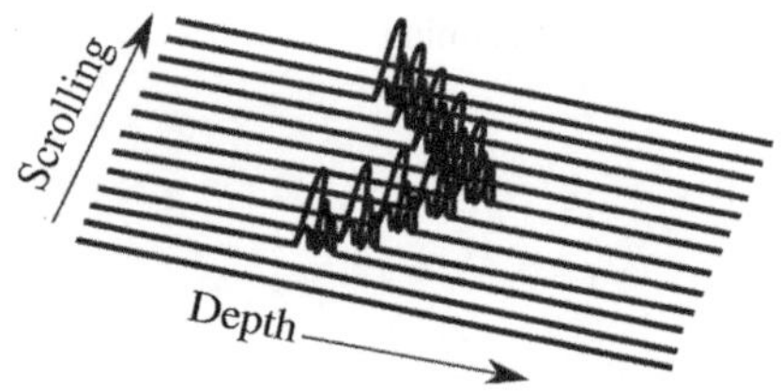

Figure 1.15 In addition to A-, B-, and C-mode images, there's also M-mode, which stands for motion. It's basically just a scrolling B-scan, but can be quite useful assessing motion in cardiology and other areas. As the sequence of A-lines scroll, the interface causing the echo moves distally and then back toward the transducer. It's much more exciting when it's animated, sorry.

The coolest thing about real-time imaging is that you can watch things moving inside a living being, for example, a beating heart. M-mode images are also formed from a sequence of A-lines with amplitude converted to brightness, but here the A-lines scroll across the screen in a sort of waterfall plot. I'm a little disappointed with my cartoon in Figure 1.15 though. Of course, M stands for *Motion*, and of course, the amplitudes are converted to pixel brightness values so that M-mode images are pretty. The name of the game in medical imaging has always been to present the clinician a set of high-resolution images to interpret in making a diagnosis. These days machine learning can often assist the licensed medical professional by highlighting suspicious features in images, but artificial intelligence algorithms are typically prohibited by regulators from actually making any diagnosis.

The above measurement schemes are all what we call pulse-echo because the same transducer both transmits the ultrasound pulse and then records the returning echoes. In many situations, particularly for structural health monitoring, it is useful to have one transducer transmit the ultrasound pulse and another transducer record the ultrasound transmissions at some other location. We call this a pitch-catch measurement, but don't have some sort of alphabetical nomenclature to categorize them so I'll describe a few.

A common scheme is to transmit the ultrasound beam into a specimen and then record what comes out the back side with a transducer along the line of sight. Flaws or other material imperfections that reflect some energy will reduce the energy in the transmitted wave that is recorded, and if the two transducers are attached to a yoke system, they can be scanned across an area to look for things like disbonds and delaminations that reflect strongly and hence give a pronounced shadow effect in the transmission. Pitch-catch schemes can also be quite useful for guided waves because large areas of extended structures can be screened for imperfections quickly. Guided waves also often have the useful feature that their speed of propagation depends on a structural property, like plate thickness, so a pitch-catch measurement can look for subtle changes in arrival time of a wave packet that indicates a thickness change due to corrosion along the path of travel. There are also a variety of tomographic geometries where a large number of criss-cross pitch-catch measurements can be used to reconstruct an image of the interior tissue or material variations [49–56].

Of course, the transmitting and receiving transducers can both be on the same side of the sample. Perhaps they are offset just a bit and at angles to the surface so that the echoes are recorded at specific angles. As we'll see, this simple generalization makes things much more complicated because ultrasound waves in solid materials can be either longitudinal or transverse modes, and interaction with interfaces, imperfections, etc. causes mode conversion between them. Angles of reflection and refraction are also different for longitudinal and transverse modes, so you can set up a measurement to transmit longitudinal modes at angles that refract and/or reflect specific wave modes that allow you to inspect parts for specific flaw types. Any good ultrasound lab will have a whole cabinet of transducers of different frequencies, sizes, shapes, etc. Some will be for immersion scanning and some will be for use in direct contact with the specimen surface using a gel or oil (but not honey) couplant. They all look pretty similar and the distinguishing numbers on them are hard to read. Let's pause here for a few minutes while you go gather up the transducers you've been using and put them away properly.

1.4 Monostatic, Bistatic, Doppler

The radar equivalent to pulse-echo ultrasound is called monostatic. An antenna transmits the radar pulses and also records the backscattering.[11] Because the speed of light is so much larger than the speed of sound, the electronics necessary to simply record the RF waveforms and process them afterward is pretty recent. Figure 1.16 shows a typical setup with a software-defined radio connected to a tablet for transmitting and/or receiving electromagnetic waves. With one of these setups, you're doing monostatic measurements, but if you have a couple of them and transmit with one and receive with another that is called bistatic. Three would be tristatic, I suppose. The point is that these are quite portable and it's now a simple matter to go out and about and collect RF waveforms in monostatic or bistatic configurations, etc. without getting your sleeves wet. You should wear a hat, though.

The textbook example of the Doppler effect is the change in pitch of a train as it passes you. I'm pretty old and I don't recall ever waiting at a railroad crossing while this happened. I've ridden Amtrak a couple of times, but that was a miserable experience where we seemed to mostly wait while freight trains used the tracks that Amtrak borrows. It didn't smell great.

The good news is that doppler imaging typically doesn't work according to the Doppler effect. I'll explain in terms of doppler ultrasound that is used to map blood flow, especially in cardiology. These color images are typically overlaid onto a higher-resolution B-mode gray-scale image. It's much more exciting when you can see the motion, so feel free to look some up on the internet and then I'll describe how it's done.

The basic idea is that you send in a rapid sequence of pulses and then compare the various echoes in the several A-lines to calculate how much those peaks in the A-lines have shifted from one

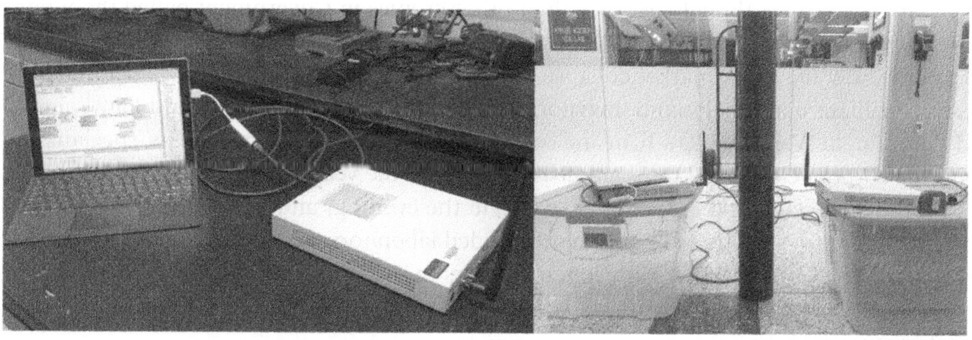

Figure 1.16 The portable, inexpensive, high-performance USRP is a scalable software defined radio platform that features a customizable field programmable gate array for high-performance digital signal processing and full duplex capabilities.

---

11 Merrill Skolnik served as superintendent of the radar division of the U.S. Naval Research Laboratory in Washington, D.C., for more than 30 years. While there, he made significant contributions including helping to develop high-frequency, over-the-horizon radar; a system that can identify friend or foe during combat; and high-resolution radar techniques. He is best known for his introductory text "Introduction to Radar Systems" and for editing the "Radar Handbook." He died in 2022 at age 94.

pulse to the next. That's straightforward to do because the shifts from one A-line to the next are small if the pulse sequence is rapid enough. Color-code the peaks approaching the transducer one color and code the peaks retreating from the transducer another color. Usually, the colors are red and blue.

In addition, you may have noticed that the color doppler part of the image is just for a certain region. That's because the sequence of doppler A-lines are often done in between the video frame rate B-mode image pulses, so there's a limit on the number of doppler pulse sequences that can get done and still update the whole image fast enough that it looks all natural to the human eye.

Doppler radar that weather guessers use to mislead you about whether you should grab an umbrella on your way out the door could work according to the simple scheme I just described for doppler ultrasound, but there's this issue with the speed of light compared to the speed of sound and the need to digitize RF waveforms and process them.... In practice, pulsed doppler radar employs a number of signal processing tricks to map out the velocity of whatever backscatters the radar beam, but again it's not really the doppler effect like an enormously long coal train that blows its whistle to taunt you as you're waiting at the RR crossing with your windows up because you've heard breathing coal dust might be bad for you.

"The Big Short" is a book by Michael Lewis and film starring Christian Bale, Steve Carell, Ryan Gosling, and Brad Pitt which tells the story of hedge fund bros who saw the collapse of the housing bubble coming and made quite a lot of money betting on that. You may or may not remember that the Great Recession was caused by the logical fallacy that because people who own homes are better off, everybody should own a home. Home mortgages were given to people who had no business buying homes because the risk of default on loans was disconnected from the decisions to write the loans by converting them to mortgage-backed securities that could be traded on Wall Street. The PhD Quants understood that there was an assumption of statistical independence in their models, but their MBA Banker bosses didn't and the bankers' CPA cronies at the accounting firms formally assured everybody that the risk was minimal because the only way things could ever go wrong was if housing prices all fell at once and that would never happen. It was a happy ending for the Wall Street Bankers, though, because their other cronies in Government bailed them out. Act two in this play was student loans, BTW. Act three is subprime car loans.

Alfred Loomis saw the stock market collapse coming and liquidated his holdings beforehand. Loomis was a brilliant math major and inventor at Yale, but after attending Harvard Law School, he joined a prominent Wall Street law firm and became one of its brightest young stars, specializing in complicated corporate financial transactions. Throughout the years, while he traded vast sums of money on the financial markets during the week, in the evenings and on weekends he worked with the world's greatest scientists at his own self-funded laboratory. Loomis had bought a second mansion, a giant mock Tudor home with concrete floors in Tuxedo Park, where he invited many of the world's most distinguished scientists to come and spend time [18]. Albert Einstein called it "a palace of science."

In the late 1930s, Alfred Loomis became increasingly convinced war was coming in Europe and that the United States would be drawn into it. Then, in early 1940 Winston Churchill organized the Tizard Mission to introduce U.S. researchers to several new technologies the British had been developing. Loomis welcomed them to his hotel suite in New York, where he learned about the cavity magnetron, which was a key piece of technology that made radar small enough to put on airplanes. Loomis was instrumental in establishing the Radiation Laboratory at MIT, arranging for funding and clearing bureaucratic roadblocks. The Rad Lab developed the radar systems that gave the Allies control of the skies, enabled the sinking of U-boats, spotted incoming German bombers, and provided cover for the D-Day landing. The Rad Lab became the Air Force Cambridge Research

Laboratory after the war, with many activities eventually moving to Hanscom, AFB where I spent four years and a day. I even met R.W.P. King once.[12]

1.5 Didey Wagon vs. War Wagon

I learned to drive at age 13. It was in the orange 1968 VW bug with a pumpkin stem on top shown in Figure 1.17. I've never gotten a straight answer from my mother about whether she was in on the joke when it came back from the paint shop garish orange instead of light blue. My brother occasionally got in trouble when that car was reported having been seen at some place he shouldn't have been. Tracking teenagers is much easier these days.

Figure 1.17 You'd be pretty unlikely to get a speeding ticket in a pumpkin car even going downhill. The highly rounded front hood and fenders won't backscatter radar all that much, and it's rear-engine and air-cooled, so there's no radiator to reflect the radar. There's no heat in the winter, but the engine is directly above the narrow rear wheels so it's really good in the snow even without studded snow tires. If back in the day, I'm tooling down the road in my wine-colored Volvo 240 Wagon blowing past a candy-apple red Corvette, there's no way In hell I'm getting a speeding ticket, even though any radar reflections are coming from me and my dad-mobile and not from that hopelessly insecure, compensating-for-something, former frat-star in the slow lane.

---

12 Professor Ronold W. P. King's academic tenure at Harvard University spanned all or parts of eight decades, and his legacy is perhaps best defined by the fundamental nature of his research, his exceptional insight into the physics of problems in applied electromagnetics, and his mentorship of over 100 PhD students. He became a Gordon McKay Chaired Professor in 1946, and was Professor Emeritus from 1972 to 2004. He remained active in his research efforts until late in his life, completing his last book at age 87, producing his last PhD student at age 94, and publishing his last paper at age 99. He was honored by his past students on his 100th birthday, and died in 2006. For the 67-year period from 1938 to 2004, he directed an active research group at Harvard investigating the applications of electromagnetic theory to problems in antennas, radiowave propagation, and subsurface communications. The 1960s were characterized by work on scattering and diffraction of electromagnetic waves from spheres, cylinders, strips, and disks, some with dielectric coatings. In his autobiography, he describes when he reached the proscribed retirement age of 67 in September 1972. "President Pusey, as required by Harvard rules, came to my office in McKay Laboratory to thank me for my services and announce my becoming a Professor Emeritus. I requested permission to continue for five more years, pointing out that I was well able to do so." The rules permitted an additional five years, but the Dean "had requested that this privilege not be granted." King had five or six graduate students at the time and they appealed to the President, who denied the request. Mandatory retirement ages for faculty were outlawed nationwide in 1994, BTW.

Figure 1.18 The Horten [65] flying wing would have been quite stealthy even if its wings weren't made from plywood. After the war, Reimar Horten said he mixed charcoal dust with wood glue to absorb radar and shield the aircraft from detection by British radar that operated at 20–30 MHz. Source: HawkeyeUK/Wikimedia Commons/CC BY-SA 2.0.

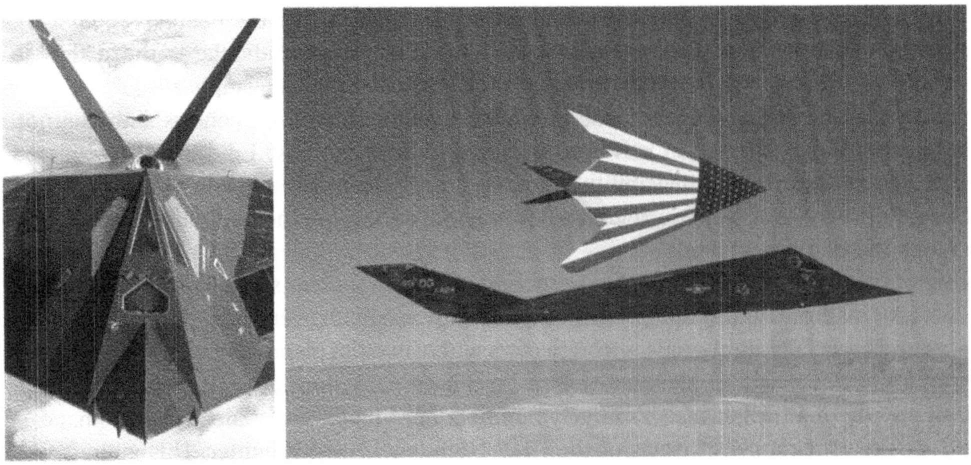

Figure 1.19 Most people first became aware of stealth aircraft when the F117A Nighthawk was used to great effect in Iraq in 1991. The faceted shape is designed to redirect radar, not absorb it. Source: (a) Master Sgt. Lance Cheung/U.S. Air Force/Wikimedia Commons/Public domain. (b) Senior Master Sgt. Kim Frey/U.S. Air Force/Wikimedia Commons/Public domain.

top secret operational stealth fighter. I now understand that the faceted nature of the design was due to the limitations of computers and methods for calculating radar cross section back in the day.

The main thing to notice from the pictures in Figure 1.19 is that there are no flat vertical surfaces or right angles. The facets do actually line up, so that the incoming radar beam is reflected off into a few carefully chosen directions, with only diffraction from corners and edges sending radar energy back toward the transmitter. It's also a relatively small aircraft, and it turned out to be quite effective at sneaking unnoticed all the way downtown and using precision munitions to turn off the radars and antiaircraft installations of adversaries. The infrared camera that it used to sneak about at night also made for compelling press conferences. You can see an F117 up close in the Cold War Gallery at the National Museum of the United States Air Force in Dayton, OH. The Testors F19 model kit might still be available on eBay.

Teenagers these days are all walking around with semidisposable supercomputers in their pockets, which they mostly use to post their brunch to *Instagram* and make *TikTok* videos. I'm not quite old enough to remember when computer was a job description rather than an electronic machine, and I missed punch cards by about a year, but I do remember assiduously backing up my computer files onto reel-to-reel tape. I still kind of miss programming in FORTRAN77. I don't miss waiting in line at the computer center for a terminal until 4 a.m., so I could get my programming assignment done.

Back in the day, you had to first identify the components of the scattering that were contributing to your total radar cross section and then analyze those individual scattering centers. The elegance of the design of the F117 was not its beauty or aerodynamic performance (it had neither); it was that from a radar perspective, it was flat plates, edges where flat plates joined at a minimum number of angles, and corners at the intersection of three plates. Those were doable with the kinds of computers I remember fondly from my twenties.

Figure 1.20 shows a standard airframe geometry with typical scattering centers and mechanisms identified. Decades of mental effort documented in innumerable doctoral dissertations was expended to develop the mathematics and computational methods to understand these sorts of contributions. Any locally flat surfaces will give rise to specular reflections where the angle of reflection is equal to the angle of incidence, measured relative to the surface normal at that point. The facets on a 117A are carefully angled so as to control reflections. Any surfaces at right angles to each other will give retro-reflections that send the energy right back to the transmitter for a large range of incident angles. So if you're going to have horizontal stabilizers, you don't want anything vertical. Once those two contributions are minimized, you might think we're good. Nope. The tip of the nose of the fuselage will give tip diffraction in pretty much all directions. Both the leading and trailing edges of the wings and tails will give edge diffractions, which will depend on their sweep angles. Any corners will give corner diffraction, and any discontinuity in curvature will also diffract. Any gaps or seams around a door or hatch or control surface will scatter. Engine inlets and outlets will give cavity returns and maybe even a characteristic doppler signal if the rotating fans are in view. Some of those contributions are small, of course, but to be truly stealthy they need to be knocked down systematically. Here's a couple more that are rather surprising. At certain angles of incidence, surface waves will be generated. We call them traveling waves if they're going down the long axis of a body and creeping waves if they're going around the short axis. They'll scatter if they hit a discontinuity or edge or the back end or whatever. Oh, and if you change materials, that will also scatter.

Figure 1.21 is a cartoon of my wife's Volvo, with scattering centers identified. Figure 1.22 shows what you could do to minimize its radar cross section. The main contribution to radar backscattering is the radiator and engine block, so you'd want to put an angled reflector in front of that because tilting the radiator reduces the cooling efficiency. Once that's taken care of then the next level of stealthiness comes from preventing scattering inside the passenger compartment; aircraft canopies solve that problem with a thin metallization layer that is optically transparent but reflective in the radar threat band. I also noted a few additional steps one could take.

The car I wanted to buy back in the late 1980s was a blue Saab 900S four-door sedan. I thought the Saab and Volvo would complement each other nicely in my garage. I do understand that there was a healthy Yankees–Sox sort of rivalry between Volvo and Saab back in the day before they were sold to China and killed, respectively, by American auto companies, but unlike baseball, you don't have to choose sides. Volvos have always been very safe because the founder's wife was an ER doctor and presumably she told him over dinner about the mangled after-effects of car crashes she had tried to humpty-dumpty back together again at work that day. Saabs have always been quirky because they held onto their form-follows-function aerospace design heritage, which as an aerospace engineer, I liked quite a lot. If you do the same things for the Saab that I suggested for the Volvo in Figure 1.22,

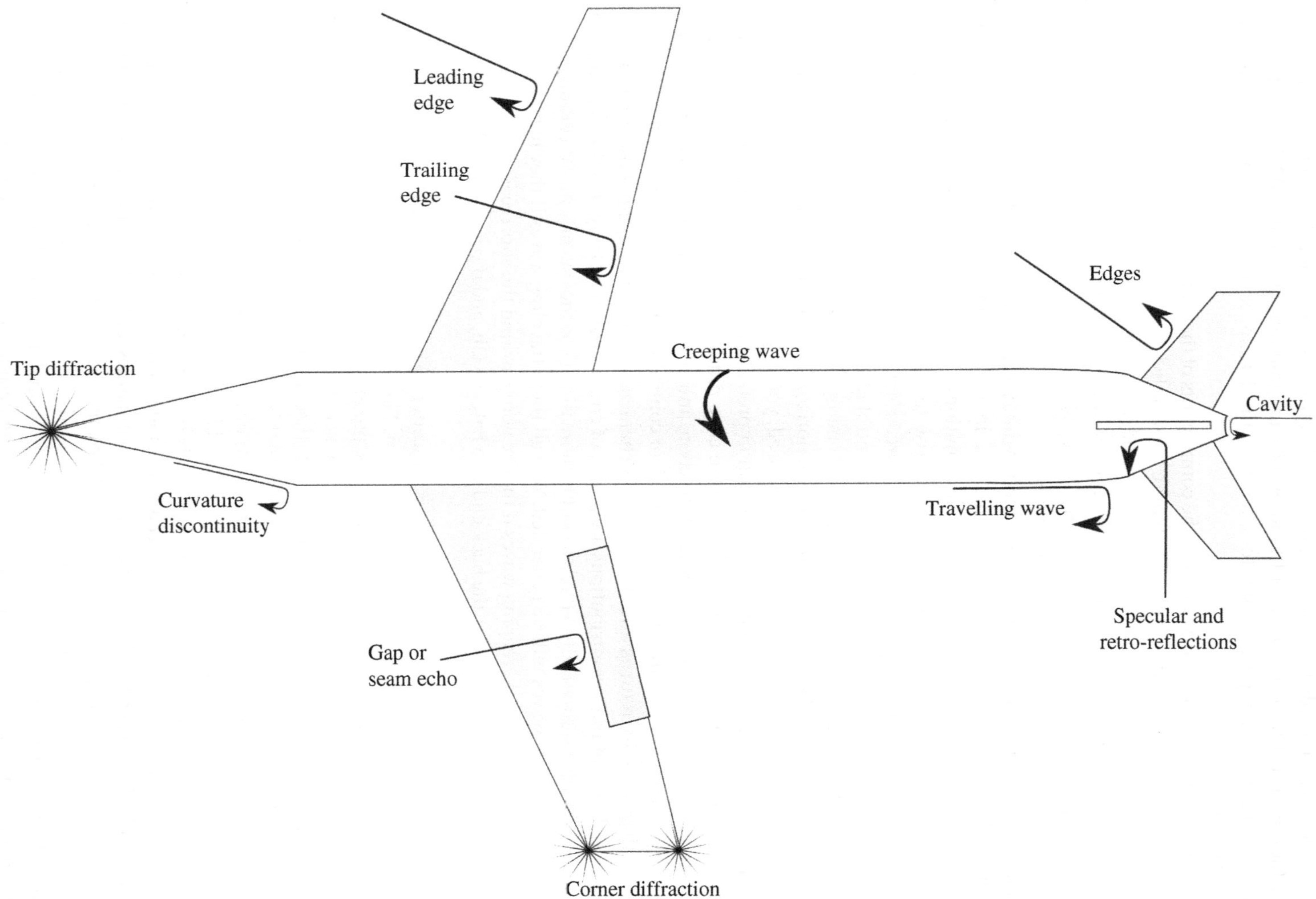

Figure 1.20 A simplified cartoon of an air vehicle allows us to consider what geometric features cause scattering.

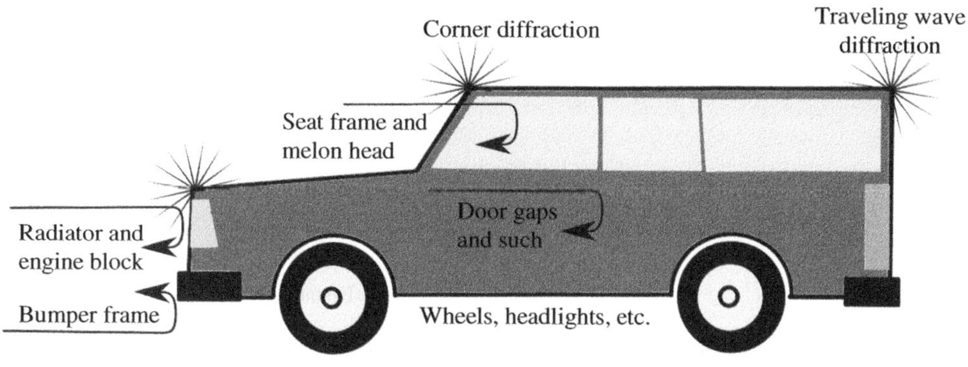

Figure 1.21 The primary contribution to radar backscattering for a Volvo 240 wagon is going to be the radiator and engine block or firewall. The windshield is transparent, of course, so next are probably metal seat frames and maybe your melon head, especially if it's covered in tin foil. Metalized coatings on headlight reflectors might be pretty significant, and of course, the steel bumpers under the rubber coverings will reflect even though they're just a bit curved. Then what? I see some corners, which will diffract. Traveling waves on the long flat roof will diffract from the gap at the tailgate, and the same goes for the hood. Gaps around doors and body panels will scatter. Wheels would give a Doppler signal if police radar was that sophisticated, but it's not.

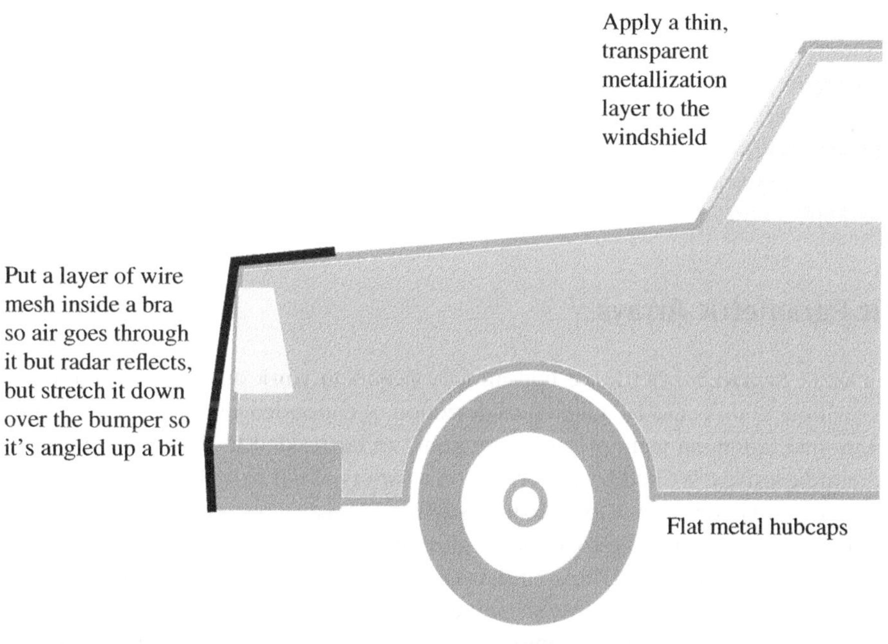

Figure 1.22 Step number one, get yourself a windshield with the Quickclear (https://en.wikipedia.org/wiki/Quickclear) option. The metallic content of the glass has been shown to degrade the performance of certain windshield-mounted accessories, such as GPS navigators, telephone antennas, and radar detectors, so we know it will work nicely to keep your melon head from backscattering police radar. Step number two, get your car a bra which is typically used to protect the paint from pebbles. Carefully peel apart the vinyl and felt layers and add a layer of wire mesh in between them with a grid spacing less than about a cm, and put that on the car. Step number three, get yourself some boring flat metal hubcaps and if you really want to go for broke, replace the side mirrors with cameras and tape over panel gaps with that foil tape HVAC pros use, hidden under PTA-approved bumper stickers of course. Step number four, drive like a maniac while rocking that dad bod. In addition, don't forget to hit a curb to round over your front license plate just a bit.

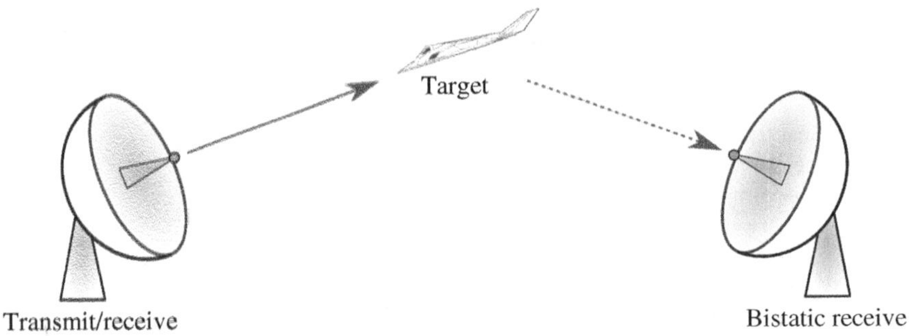

Figure 1.23 The highly swept planform, carefully aligned facets, screened inlets, metalized glass, etc. ensure that the head-on RCS of the F117A Nighthawk is minimal because very little of the radar beam is backscattered. Bistatic radar is a countermeasure to this by trying to pick up those redirected radar echoes with a second receiver.

you would have a very stealthy daily driver that is as good in the snow as an orange VW Bug but also has heat. If you're not familiar with the Saab 900S, look it up. The windshield is quite curved, so if it's metallized to reflect radar, it will backscatter almost nothing. Also note how much the bumper protrudes ahead of the curved nose, so if a mesh-enhanced bra is stretched from below the ground-effects feature, up over the bumper and partway up the bonnet….

All of the above discussion is based on the assumption that the radar illuminating you is monostatic. The radar isn't so much absorbed as it is redirected. An obvious countermeasure to this is sketched in Figure 1.23 where you transmit from one antenna and record the scattered radar signals with another antenna someplace else. Stealth is a game of countermeasures and counter-countermeasures and so on. What you really need is to be able to understand both the operational mission(s) and the scattering behavior of your vehicle, so you can drive to college and not get speeding tickets.

1.6 Acoustic Parametric Arrays

By mid-afternoon of 11 September 2001, we had already begun to work out how the kinds of things that we knew how to do could be adapted to detection of concealed weapons. It was clear that metal detectors had failed, so technologies were going to be needed to detect nonmetallic weapons such as ceramic knives. We had been working on a variety of tomographic reconstruction projects using acoustics, and had begun to explore how sophisticated air-coupled ultrasound in the 50-kHz range might be useful as a mobile-robot navigation sensor. I wrote up a one-page description of my thoughts, a bit of which said, "Ultrasound has an inherent advantage over magnetic and radar-based imaging techniques which are only sensitive to hidden metallic items. Ultrasound can just as easily detect hard plastic items hidden about the body. Indeed, ultrasound is just as sensitive to unfilled gaps as it is to discrete hard objects. Every material – solid, liquid or gas – affects ultrasound in a distinct way. Although that physics is extremely complex, our expertise lies in precisely the esoteric science and technology necessary to harness the complexity and sort it out in computers that are used to automatically process the scan results and identify the suspicious items in real time."

We also modified one of our 3D ultrasound scanners to accept transducers in the 20–200 kHz frequency range and covered a dress-maker's dummy with tissue-mimicking gel, so that we could perform experiments in the lab by putting different types of clothing on "no head" with various simulated weapons underneath. We came to two very important conclusions. First, it's hard to focus the ultrasound beam at any appreciable stand-off distance with even a fairly large transducer because the wavelength is rather big. Second, the acoustic impedance mismatch between air and cloth is large enough that most of the ultrasound beam simply reflected from the surface of the garment. A third conclusion was that the backscattering was really complicated. Part of the answer turned out to be employing a new technology.

Nonlinear acoustics allows one to make a compact, highly directional loudspeaker called a parametric array.[16] If you plug your iPod into it it plays music, but you only hear the music if the speaker is pointed at you. Standard ultrasound transducers are used to transmit an acoustic beam at, for example, 50 kHz along with another signal at 49 kHz. Directionality will be determined by the size of the transducer array compared to the wavelength of the ultrasonic beams. Nonlinear effects in air will then give both the sum and the difference frequencies, 99 and 1 kHz. Since attenuation of sound goes like frequency squared, the 99 kHz beam will attenuate fairly rapidly while the 1 kHz audio beam will propagate much, much farther than the original ultrasound beam, while still being as directional as that ultrasound beam. In order to play music, one of the two ultrasonic beams is simply modulated so that the difference frequency follows the intended melody. The nonlinear acoustics is complex, but the resulting devices are quite simple except that the nonlinear effect itself is a function of frequency, so you might need some pretty sophisticated electronics to make the music sound good. You could also hide a subwoofer under the table when making demonstrations, Woody.

Our approach to stand-off concealed weapons detection was to use parametric arrays to direct a narrow, low-frequency beam of sound toward a person's body. Although the beam of sound is in the audible frequency range because it's not pointed at the person's head, they won't hear it. The low frequencies will penetrate the clothing like your neighbor's bass notes do your apartment's walls, and then if there's a concealed weapon that will backscatter a signal that can be picked up with a microphone. Figure 1.24 shows two versions.

Figure 1.24 Waveform synthesized via Matlab on laptop and sent via headphone jack to parametric array which transmits it via nonlinear ultrasonics. Dish collects back-scattered sound from target and sends it to laptop via microphone jack where Matlab records the digitized waveform for analysis.

---

16 The parametric array was invented by Peter Westervelt [66] although Woody Norris likes to take credit for it since he seems to be the first one to make and sell devices commercially, which we bought and used in the aughts for concealed weapons detection. Joe Pompei started Holosonics after his PhD at MIT on this subject. He built us a system or two after LRAD took off big and Norris' company wasn't interested in R&D anymore. Sennheiser tried to hire Dr. Pompei and he claims they stole his ideas and then didn't hire him. More recently, there was a kickstarter-funded project that called their small unit the Soundlazer.

As this work progressed and the problem of most interest was suicide bombers approaching checkpoints, we did systematic experiments and simulations to detect concealed explosives. We built a custom system with a curved parametric array to focus the beam at 50 ft because you need to detect the explosive from beyond what is called “the lethal radius,” which I took to mean the distance at which if a suicide bomber blows himself up, he doesn’t get entrails all over me. Once we got that system working and were ready to begin collecting systematic data of backscattering from a variety of simulated weapons and explosives, as well as benign scattering objects under clothing, our sponsor said, “we meant 50 m, not 50 ft” but then didn’t have the budget for us to build that prototype. One member of the development team did volunteer to strap on a test article that simulated an explosive and ball bearing-filled jockstrap, and then be scanned with the parametric array system. I made some excuse to not be available to be there for those tests. We’ll do scattering from spheres in Chapter 6, BTW.

We later mounted the parametric array and speaker dish on a mobile robot and took *rMary* out and about to collect backscattering data from approaching vehicles, as shown in Figure 1.24. The parametric array projects a narrow acoustic beam down the street and records with the dish microphone the echoes that backscatter from oncoming vehicles up to 50 m away. Since scattering is strongly frequency dependent, we transmit a linear chirp signal and then use the differences in backscattering as a function of frequency to distinguish different classes of vehicles. It’s important when you’re sending a robot out to run an errand for you that the robot watches for traffic and not get squished.

1.7 Forward to Scattering

What we’ll see in the coming chapters is that radar and sonar and ultrasound and seismology and optics and a wide variety of other physical phenomena, which are usually considered as distinct fields of inquiry, can instead be considered as different aspects of the same thing. The field equations that describe these various wave phenomena are different, of course, but in all cases we ultimately end up solving wave equations to model wave propagation. That seems obvious, but there are some subtle differences in wave modes and boundary conditions that will add richness to wave scattering behavior necessary for our mathematical models to predict actual physical behavior.

The equations and methods of solution that we’re going to be discussing are not new. I didn’t invent them. I’m not a mathematician. There will be no lemmas, proofs, etc. My goal in this book is to pass along a way of thinking about all these seemingly disparate subjects, along with some of the mathematical toolkit that’s necessary to exploit models to understand scattering behavior so as to be able to exploit them to solve problems in the real world.

We’re mostly going to be doing *forward scattering* problems and will focus initially on those canonical geometries where exact solutions are in hand. What we’ll see is that the scattering behavior for even simple situations has a surprising richness to it, and if we try to make the models just a bit more realistic, we will often end up with equations that are not solvable and thus not all that useful. We’ll start at about the time Americans were having a Civil War and trying not to drown in hand-cranked submarines.

In the middle of the 19th century, Alfred Clebsch was trying to mathematically model the refraction of light by lenses. The electromagnetic theory of light hadn’t been developed yet, so he was using an elastodynamic theory of the aether to model light propagation and scattering. Much of the mathematics of what’s now called Mie Scattering hadn’t been worked out quite yet, for example, the necessary special functions and their properties, so Clebsch had to invent the math to solve the problem. In addition, the elastodynamic model of light is unnecessarily cumbersome, particularly the boundary conditions which couple the longitudinal and transverse wave modes at an interface.

In the late 1980s when I was doing scattering research for the Air Force, the USAF Geophysics Library had a robust enough collection that when I needed to read Clebsch's 1861 paper, I simply fetched that journal volume from the stacks and gingerly photocopied it. I then manually transliterated it (and Mie's famous 1908 paper) from German so I could follow the mathematical development. The scattering literature over the last half of the 19th century was surprisingly interesting to read, as people realized that the electromagnetic theory of light worked so much better than elastodynamics. There's even a 1900 paper by Lamb [67] with a disclaimer to the effect: I worked on this elastodynamic model for a very long time and I realize that the electromagnetic model is better in every way, but I'm publishing this anyway in case someday somebody might find it useful. I guess publish or perish has always been a thing. I do hope you find some of the rest of this book helpful.

References

1 Mental Floss. 15 Surprising Facts About Marco Polo. https://www.mentalfloss.com/article/84526/15-surprising-facts-about-marco-polo (accessed 14 September 2024).

2 Howarth, T.R. (2015). The submarine signal company. *The Journal of the Acoustical Society of America* 137: 2273. https://doi.org/10.1121/1.4920303.

3 (a) Gray, E. and Mundy, A.J. (1899). Transmission of sound. US Patent 636,519, filed 14 April 1899, granted 7 November 1899; (b) See also: Mundy, A.J. (1905). Submarine signal. US Patent 636,519, filed 14 February 1903, granted 4 July 1905. Assigned to Submarine Signal Co. Boston, MA and US Patent 1,207,388, Reginald Aubrey Fessenden. Method and apparatus for submarine signaling. filed 29 January 1913 also assigned to Submarine Signal Co.

4 Sawyer, F.L. (1914). Submarine signaling and a proposed method of safe navigation in fog. *Transactions of the Society of Naval Architects and Marine Engineers* 22: 115–128.

5 1,340 Perish as Titanic Sinks; Only 886, Mostly Women and Children, Rescued. New York Tribune (New York, NY), April 16, 1912, page 1.

6 Maxim, H.S. (1912). Preventing collisions at sea: a mechanical application of the bat's sixth sense. *Scientific American* 107 (4): 80–82.

7 Dijkgraaf, S. (1960). Spallanzani's unpublished experiments on the sensory basis of object perception in bats. *Isis* 51 (1): 9–20. JSTOR www.jstor.org/stable/227600.

8 Griffin, D.R. (1958). *Listening in the Dark: The Acoustic Orientation of Bats and Men'*. Yale University Press. (Paperback April 1, 1986 ISBN-13: 978-0801493676).

9 Denny, M. (2007). *Blip, Ping, and Buzz: Making Sense of Radar and Sonar*. Johns Hopkins University Press. ISBN-13: 978-0801886652.

10 Au, W.W.L. (1993). *The Sonar of Dolphins*. New York: Springer-Verlag.

11 Maxim, H.S. (1912). *A New System of Preventing Collisions at Sea*. London: Cassel and Co.

12 WH, W. (1912). A new system of preventing collisions at sea. *Nature* 89 (2230): 542–543.

13 Delgado, J.P. (2006). Archaeological Reconnaissance of the 1865 American-Built Sub Marine Explorer at Isla San Telmo, Archipielago de las Perlas, Panama. *International Journal of Nautical Archaeology Journal* 35 (2): 230–252.

14 Showell, J.M. (2006). *The U-boat Century: German Submarine Warfare, 1906–2006*. Chatham Publishing. ISBN 1-59114-892-8.

15 Nowitall, I.B. (2024). *Post-Pandemic Adulting*, W&N Edutainment Press.

16 Brittain, J.E. (1985). The magnetron and the beginnings of the microwave age. *Physics Today* 38 (7): 60. https://doi.org/10.1063/1.880982.

17 Buderi, R. (1998). *The Invention That Changed the World: How a Small Group of Radar Pioneers Won the Second World War and Launched a Technical Revolution*. Touchstone; Reprint edition.

18 Conant, J. and Park, T. (2002). *A Wall Street Tycoon and the Secret Palace of Science That Changed the Course of World War II*. New York: Simon and Schuster.

19 Bowman, J.J., Senior, T.B.A., Uslenghi, P.L.E., and Asvestas, J.S. (1970). *Electromagnetic and Acoustic Scattering by Simple Shapes*. North-Holland Pub. Co. Paperback edition: CRC Press (September 1, 1988) ISBN-13: 978-0891168850.

20 van de Hulst, H.C. (1981). *Light Scattering by Small Particles, Dover Books on Physics*. Corrected Edition. ISBN-13: 978-0486642284.

21 Kerker, M. (1969). *The Scattering of Light and Other Electromagnetic Radiation*. New York: Academic Press.

22 Bohren, C. and Huffman, D. (2007). *Absorption and Scattering of Light by Small Particles*. New York: Wiley. ISBN: 9780471293408.

23 Knott, E.F., Tuley, M.T., and Shaeffer, J.F. (2004). *Radar Cross Section, Scitech Radar and Defense*, 2e. SciTech Publishing.

24 Reagan, R.W. Explosion of the Space Shuttle Challenger. Address to the Nation, January 28, 1986 by President Ronald W. Reagan. https://history.nasa.gov/reagan12886.html (accessed 14 September 2024).

25 Baloh, R.W. and Bartholomew, R.E. (2020). *Havana Syndrome: Mass Psychogenic Illness and the Real Story Behind the Embassy Mystery and Hysteria*. Springer. https://link.springer.com/book/10.1007/978-3-030-40746-9.

26 Jaekl, P. (2017). Why People Believe Low-Frequency Sound is Dangerous. The Atlantic, June 19, 2017. https://www.theatlantic.com/science/archive/2017/06/wind-turbine-syndrome/530694/ (accessed 14 September 2024).

27 Krautkramer, J. and Krautkramer, H. (1990). *Ultrasonic Testing of Materials*, 4e. Springer Science+Business Media.

28 Bray, D.E. and Stanley, R.K. (1997). *Nondestructive Evaluation: A Tool in Design, Manufacturing and Service*. CRC Press. https://doi.org/10.1201/9781315272993.

29 Giurgiutiu, V. (2007). *Structural Health Monitoring: With Piezoelectric Wafer Active Sensors*. Elsevier.

30 Kundu, T. (2003). *Ultrasonic Nondestructive Evaluation: Engineering and Biological Material Characterization*. Boca Raton, FL: CRC Press.

31 Langenberg, K.-J., Marklein, R., and Mayer, K. (2012). *Ultrasonic Nondestructive Testing of Materials: Theoretical Foundations*. CRC Press.

32 Adams, D. (2007). *Health Monitoring of Structural Materials and Components: Methods with Applications*. Wiley.

33 Mix, P.E. (2005). *Introduction to Nondestructive Testing: A Training Guide*. Wiley.

34 Schmerr, L.W. (2016). *Fundamentals of Ultrasonic Nondestructive Evaluation*, vol. 122. New York: Springer.

35 Reese, R.T. and Kawahara, W.A. (1993). *Handbook on Structural Testing, Society for Experimental Mechanics*. Fairmont Press.

36 ASM Handbook (1989). *Nondestructive Evaluation and Quality Control*. Metals Park, OH: American Society of Metals.

37 Liu, G.-R. and Han, X. (2003). *Computational Inverse Techniques in Nondestructive Evaluation*. CRC Press.

38 Kremkau, F.W. (2006). *Diagnostic Ultrasound: Principles and Instruments*, 7e. Saunders.

39 McGahan, J.P. and Goldberg, B.B. (1998). *Diagnostic Ultrasound: A Logical Approach*. Lippincott-Raven.

40 Webb, S. (1996). *The Physics of Medical Imaging-Medical Science Series, IOP Medical Science Series*. Taylor & Francis.

41 Shung, K., Smith, M., and Tsui, B.M.W. (2012). *Principles of Medical Imaging*. Academic Press.

42 Bushberg, J.T. and Boone, J.M. (2011). *The Essential Physics of Medical Imaging*. Lippincott Williams & Wilkins.

43 Cho, Z.-H., Jones, J.P., and Singh, M. (1993). *Foundations of Medical Imaging*. New York: Wiley.

44 Suetens, P. (2017). *Fundamentals of Medical Imaging*. Cambridge University Press.

45 Harisinghani, M.G., Chen, J.W., and Weissleder, R. (2018). *Primer of Diagnostic Imaging (E-Book)*. Elsevier Health Sciences.

46 Postema, M. (2011). *Fundamentals of Medical Ultrasonics*. CRC Press.

47 Meire, H.B. and Farrant, P. (1995). *Basic Ultrasound*. Wiley.

48 McDicken, W.N. (1976). *Diagnostic Ultrasonics*. Crosby Lockwood Staples.

49 Kak, A.C. and Slaney, M. (2001). *Principles of Computerized Tomographic Imaging*. Society for Industrial and Applied Mathematics. https://www.slaney.org/pct/pct-toc.html.

50 McKeon, J.C.P. and Hinders, M.K. (1999). Parallel projection and crosshole Lamb wave contact scanning tomography. *The Journal of the Acoustical Society of America* 106: 2568–2577. https://doi.org/10.1121/1.428088.

51 Malyarenko, E.V. and Hinders, M.K. (2000). Fan beam and double crosshole Lamb wave tomography for mapping flaws in aging aircraft structures. *The Journal of the Acoustical Society of America* 108 (4): 1631–1639.

52 Malyarenko, E.V. and Hinders, M.K. (2001). Ultrasonic Lamb wave diffraction tomography. *Ultrasonics* 39 (4): 269–281.

53 Leonard, K.R., Malyarenko, E.V., and Hinders, M.K. (2002). Ultrasonic Lamb wave tomography. *Inverse Problems* 18 (6): 1795–1808.

54 Leonard, K.R. and Hinders, M.K. (2003). Guided wave helical ultrasonic tomography of pipes. *The Journal of the Acoustical Society of America* 114 (2): 767–774.

55 Hou, J., Leonard, K.R., and Hinders, M.K. (2004). Automatic multi-mode Lamb wave arrival time extraction for improved tomographic reconstruction. *Inverse Problems* 20 (6): 1873–1888.

56 Leonard, K.R. and Hinders, M.K. (2005). Multi-mode Lamb wave tomography with arrival time sorting. *The Journal of the Acoustical Society of America* 117 (4): 2028–2038.

57 Ferrel, O.P. (1956). *Radar on the Highway*. Popular Electronics. http://www.rfcafe.com/references/popular-electronics/radar-highway-popular-electronics-may-1956.htm (accessed 14 September 2024).

58 Richards, M.A. (2014). *Fundamentals of Radar Signal Processing*. McGraw-Hill Education.

59 Skolnik, M. (2008). An introduction and overview of radar. In: *Radar Handbook*, 3e. McGraw Hill.

60 Bedard, P. (1979). Smoking Out Old Smokey. A battlefield comparison of twelve radar detectors. *Car and Driver* 24 (8): 69–80.

61 Richardson, D. (1989). *Stealth: Deception, Evasion, and Concealment in the Air*. Orion Books. ISBN-13: 978-0517573433.

62 Sweetman, B. (1986). *Stealth Aircraft: Secrets of Future Airpower*. Motorbooks International. ISBN-13: 978-0879382087.

63 Zale, K. (2016). *Stealth Aircraft Technology*. CreateSpace Independent Publishing Platform. ISBN-13: 978-1523749263.

64 Hoffman, B.G. (2012). *American Icon: Alan Mulally and the fight to save Ford Motor Company*. New York: Random House.

65 Giangreco, L. (2020). *Why the Experimental Nazi Aircraft Known as the Horten Never Took Off*. Smithsonian Magazine.

66 Muir, T.G. and Korman, M.S. (2006). Professor Peter Westervelt and the parametric array. *The Journal of the Acoustical Society of America* 119: 3231. https://doi.org/10.1121/1.4785967.

67 Lamb, H. (1900). Problems relating to the impact of waves on a spherical obstacle in an elastic medium. *Proceedings of the London Mathematical Society* s1–32 (1): 120–151.

2

Field Equations

2.1 Index Notation

We will need to introduce tensor notation in order to conveniently write the equations that describe elastic waves in solids. Tensors aren't anything all that mysterious, but are just a convenient notation to keep track of things with multiple components and then write complicated equations without having to write everything out in their components. If you find yourself starting to develop a mental block, think back to when you first learned about vectors in elementary physics. Once you grasped the concept, there was no turning back because vector notation is so incredibly useful. If you don't believe me, go try reading a 19th century book on elasticity or something and see if you can resist the temptation to scribble in the margins the vector notation versions of all the equations that are written out fully in components. Of course, not everybody gets the concept of vectors; some of your classmates in elementary physics did quite poorly because of that. That there are two different ways to multiply vectors is a pretty strange concept, after all, but don't get too smug because the animal that mathematicians were using prior to Gibbs' vectors were called quaternions.[1] They really suck.

Since you long since came to love them, consider the vectors $\vec{a}$ and $\vec{x}$ where

$$\vec{a} = \begin{Bmatrix} a_1 \\ a_2 \\ a_3 \end{Bmatrix} \qquad \vec{x} = \begin{Bmatrix} x_1 \\ x_2 \\ x_3 \end{Bmatrix} \tag{2.1}$$

Now form the dot product $\vec{a} \cdot \vec{x} = p = a_1x_1 + a_2x_2 + a_3x_3$, where p is a scalar. Instead of the arrow to signify a vector, we can use an index "i" for the vectors

$$a_i = (a_1, a_2, a_3) \qquad x_i = (x_1, x_2, x_3) \tag{2.2}$$

and the dot product is $\sum_{i=1}^{3} a_ix_i = p$. Since in 3D vectors will always have three components, we don't really have to write the summation

$$a_ix_i \equiv a_1x_1 + a_2x_2 + a_3x_3 = p \tag{2.3}$$

Einstein gets credit for both recognizing that it's not necessary to write the summation and for being so highly regarded that people went along with it. The "Einstein Summation Convention"

1 Not everyone loved vectors. Peter Guthrie Tait said that "Professor Gibbs must be ranked as one of the retarders of Quaternion progress, in virtue of his pamphlet on Vector Analysis, a sort of hermaphrodite monster, compounded of the notations of Hamilton and of Grassmann."

A Data Engineering Approach to Wave Scattering Analysis: with Applications in Radar, Sonar, Medical Diagnostics, Structural Flaw Detection and Intelligent Robotics, First Edition. Mark K. Hinders.

says that pairs of repeated indices are summed over the range 1–3. It works for all sorts of vector operations, such as the magnitude of a vector $|\vec{u}|^2 = u_1^2 + u_2^2 + u_3^2 = u_i u_i$.

Now flashback to learning about matrices and linear algebra, which probably involved lots of painful proofs, but matrices are useful enough that you made it through and don't have to do proofs anymore. Index notation makes linear algebra actually pretty fun. Consider the matrix $\underline{\underline{\mathbf{A}}}$:

$$\underline{\underline{\mathbf{A}}} = \begin{bmatrix} A_{11} & A_{12} & A_{13} \\ A_{21} & A_{22} & A_{23} \\ A_{31} & A_{32} & A_{33} \end{bmatrix} = A_{ij} \qquad (i, j = 1, 2, 3) \tag{2.4}$$

and write

$$A_{ij}u_j = \begin{bmatrix} A_{11} & A_{12} & A_{13} \\ A_{21} & A_{22} & A_{23} \\ A_{31} & A_{32} & A_{33} \end{bmatrix} \begin{Bmatrix} x_1 \\ x_2 \\ x_3 \end{Bmatrix} = \begin{Bmatrix} A_{11}u_1 + A_{12}u_2 + A_{13}u_3 \\ A_{21}u_1 + A_{22}u_2 + A_{23}u_3 \\ A_{31}u_1 + A_{32}u_2 + A_{33}u_3 \end{Bmatrix} \tag{2.5}$$

Note that we have summed the repeated index and that the resulting quantity is a vector. It has three components.

An index that is repeated (and hence summed) is called a dummy index. An index that is not repeated is called a free index. The number of free indices determines what the quantity is:

$$\begin{aligned} &\textit{scalar}: \ 0 \text{ free indices} \\ &\textit{vector}: \ 1 \text{ free index} \\ &\textit{matrix}: \ 2 \text{ free indices } (3 \times 3) \\ &\textit{array}: \ 3 \text{ free indices } (3 \times 3 \times 3) \end{aligned}$$

The cool part is you can have as many indices as you want as long as the repeated ones are summed. Here's a correct equation

$$A_{ijk} = B_{ijklmn}C_{lmn} = x_i y_j z_k \tag{2.6}$$

In any equation, each term must be the same tensor animal, that is, have the same number of free indices. It's also very important to only repeat indices twice! I mean once. Repeat once, which means using them twice. Don't over think it, like I just did.

We need two special symbols to be able to do vector and tensor analysis: δ_{ij} and ε_{ijk}. The Kronecker delta is defined such that

$$\delta_{ij} = 1 \ \text{for}\, i = j \qquad\qquad \delta_{ij} = 0 \ \text{for}\, i \neq j$$

Watch the following $u_i\delta_{ij} = u_1\delta_{1j} + u_2\delta_{2j} + u_3\delta_{3j}$ but j takes on the values 1, 2, 3

$$j = 1: \ u_i\delta_{i1} = u_1\delta_{11} + u_2\delta_{21} + u_3\delta_{31} = u_1$$

$$j = 2: \ u_i\delta_{i1} = u_1\delta_{12} + u_2\delta_{22} + u_3\delta_{32} = u_2$$

$$j = 3: \ u_i\delta_{i3} = u_1\delta_{13} + u_2\delta_{23} + u_3\delta_{33} = u_3$$

We could have merely written $u_i\delta_{ij} = u_j$. Hence the Kronecker delta is just the identity matrix

$$\underline{\underline{\mathbf{I}}} = \begin{bmatrix} \delta_{11} & \delta_{12} & \delta_{13} \\ \delta_{21} & \delta_{22} & \delta_{23} \\ \delta_{31} & \delta_{32} & \delta_{33} \end{bmatrix} = \begin{bmatrix} I_{11} & I_{12} & I_{13} \\ I_{21} & I_{22} & I_{23} \\ I_{31} & I_{32} & I_{33} \end{bmatrix} \equiv \delta_{ij} \tag{2.7}$$

The permutation symbol ε_{ijk} is defined as:

$$\varepsilon_{123} = \varepsilon_{231} = \varepsilon_{312} = +1 \qquad \varepsilon_{321} = \varepsilon_{132} = \varepsilon_{213} = -1$$

$$\varepsilon_{ijk} = 0 \text{ for any indices the same}$$

The cross product is written in index notation as $w_k = \varepsilon_{ijk} u_i v_j \equiv \varepsilon_{kij} u_i v_j$.

Exercise 2.1 Expand out the components to show that this is identical to $\vec{w} = \vec{u} \times \vec{v}$.

Note that i and j are repeated so we can pedantically write

$$w_k = \sum_{i=1}^{3} \sum_{j=1}^{3} \varepsilon_{ijk} u_i v_j \equiv \varepsilon_{kij} u_i v_j \tag{2.8}$$

which is a pain to do, but then most of the terms will be hard zero.

We can now write divergence and curl by thinking of the gradient as a "vector operator" $\vec{\nabla} \rightarrow \frac{\partial}{\partial x_i}$ and thus get

$$\vec{\nabla} \cdot \vec{v} \Rightarrow \frac{\partial v_i}{\partial x_i} = \frac{\partial v_1}{\partial x_1} + \frac{\partial v_2}{\partial x_2} + \frac{\partial v_3}{\partial x_3} \tag{2.9}$$

$$\vec{\nabla} \times \vec{v} \Rightarrow \varepsilon_{ijk} \frac{\partial}{\partial x_j} v_k \tag{2.10}$$

$$\nabla \phi \Rightarrow \frac{\partial}{\partial x_i} \phi \tag{2.11}$$

We also have one identity that often helps to simplify expressions

$$\varepsilon_{ijk} \varepsilon_{ist} = \delta_{js} \delta_{kt} - \delta_{jt} \delta_{ks} \tag{2.12}$$

You should remember this identity, or at least remember about it and where to find it.[2] I put it in bold font in case you're trying to find it here later.

Some authors use a comma to indicate the gradient operator: $\frac{\partial u_k}{\partial x_j} \equiv u_{k,j}$ which I never do because my handwriting is messy and my commas look just like a j or sometimes even an i if I get really sloppy. The shorthand that I especially like is to write ∂_i instead of $\frac{\partial}{\partial x_i}$ which is what I'll usually do.

2.2 Stress Is Force per Unit Area

2.2.1 Two-Question Pop Quiz, Pass–Fail

Q#1: I'm going to poke you with a pencil. Which end do you choose? If you chose the eraser, you pass. If you chose the pointy end, you're an idiot. Assuming I poke you with the same force either way, the eraser is better for two reasons. The main one isn't that the eraser is soft, although that does matter a bit. It's that the eraser has a large cross-sectional area compared to the pointy end, and so

2 "Div, Grad, Curl, and All That" by H. M. Schey is a lovely little book of vector calculus, helping science and engineering students master gradient, curl, and Laplacian operators without the required knowledge of advanced mathematics. It's been around for decades, and if you're a little fuzzy on vector calculus you should find yourself a copy. I know students like this book a lot because they keep borrowing my copy and so I buy another one. In addition, don't be shy about looking up video lectures online when some bit of math is a little fuzzy. What we're going to do is quite mathy, and you don't want that to hold you back.

the force of the poke is spread out over a larger area. The sharpened end will concentrate the force of the poke over a much smaller area, and if it's really sharp, or I poke you just a bit too hard, the pointy end will puncture your skin and the lead might even break off inside of you. It would hurt, but don't worry about lead poisoning because pencil lead is just graphite and not lead. I used to be concerned about that because I have had a piece of pencil lead under the skin on the heal of my right hand since junior high school. In the 1970s, gasoline had lead in it, BTW.

Q#2: When constructing a bed of nails, what's the primary safety consideration? I'll give you a hint, it's not making sure that the nails are disinfected or that the nails are all blunted. Ready? The answer is to make sure that all the nails are at exactly the same height so that your weight is spread out over them evenly. One nail that sticks up higher than the rest would take too much of your weight and would injure you. When lying on a bed of nails it's pretty important to spread your weight out as much as possible. It also helps to not have the Jimmy legs.[3] You should try not to sneeze, and think carefully about how you're going to get back up again.

If you passed the pop quiz, then you have some understanding that stress might be a more important quantity than force when what you care about is the deformation of materials, and I hope you now have some mental images to remind you that: *stress is force per unit area.* Consider a block of some material in static equilibrium under the action of equal and opposite forces (Figure 2.1). If the magnitude of the forces is F and the area of the faces on which they're acting is A, then we can define the stress on each of those two faces as $\sigma = F/A$.

We're going to call this stress the "normal" stress when the forces F are normal to the surfaces with area A. At this point, I'm hoping that you didn't just skim over the mathematical preliminaries because we're about to need tensors. The reason is that in the definition of stress, we have to keep track of the orientations of both the force and the area we're talking about. To wit, consider the same block of material which is under the action of equal and opposite forces that are still of magnitude F but now are tangent to the surfaces A instead of perpendicular to it (Figure 2.1, right). We can still talk about force per unit area, but the stress is fundamentally different. Go find a piece of something soft like some foam or a marshmallow and play with it a bit if this isn't yet obvious. The equal and opposite tangential forces will shear the block instead of squishing it, so we're going to call this force per unit area the "shear" stress, $\tau = F/A$. The units are still going to be pounds per square inch (psi), which are familiar from air pressure in tires and such, or Newtons per square meter (N/m^2), which are Pascals (Pa) in the metric system.

"Without the slightest warning he leaped at me, snatching an object from the table. Before I could take a backward step I felt a needle plunge deep into my arm, and cried out with the pain of

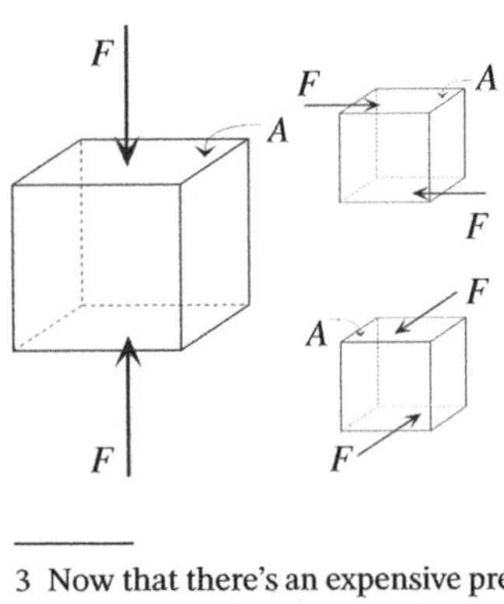

Figure 2.1 Stress is force per unit area. Normal stress on the left. Shear stress is also force per unit area, but there are two tangential directions for F.

3 Now that there's an expensive prescription drug to treat Jimmy legs, we're supposed to call it by the medical name of Restless Leg Syndrome or RLS.

it. Things became hazy, distorted. A wave of vertigo swept over me. Then it passed, and my vision cleared. The Professor stood leering before me."

That's from "He Who Shrank" by Henry Hasse, a science fiction story that first appeared in the August 1936 issue of Amazing Stories magazine. The story's narrator tells how his employer, the Professor, the world's greatest scientist, has devised a formula that will cause a man to continuously shrink in size, allowing him to explore the subcosmic universe. The idea is that each of the atoms that make up the stuff in your world are in fact the solar systems of some galaxy at the next level down. That marshmallow you were just squishing is somebody else's Milky Way galaxy, etc. Of course, the Bohr model of the atom isn't used anymore now that we have proper quantum mechanical descriptions of energy levels, shells, etc. Curiously, most of the solid stuff around you seems to be emptiness, so that inner space actually is kind of like outer space. You'd be right if you thought the Professor's research assistant was freaked out about this when he started to shrink, but the Professor had thought of this: "You will be quite safe in airless space," he went on. "In the thirty years I have worked on the problem, I would not be likely to overlook that point – though I will admit it gave me much trouble. But as I said, 'Shrinx' is all the more marvelous in the fact that its qualities are many. After many difficulties and failures, I managed to instill in it a certain potency by which it supplies sufficient oxygen for your need, distributed throughout the bloodstream. It also irradiates a certain amount of heat, and inasmuch as I consider the supposed sub-zero temperature of space as being somewhat exaggerated, I don't think you need worry about any discomfort in open space."

I often tease my materials science colleagues by telling them that I don't believe in atoms. I say it often enough and mention it casually with apparent seriousness; I'm pretty sure some of them aren't really sure about me. What I'd like you to do at this point is to pretend that there's no such thing as atoms and that the solid stuff around you isn't in fact mostly empty (inner) space. Then I want you to imagine the blocks of stuff in equilibrium under the action of equal and opposite forces (Figure 2.1) starting to shrink. Of course, the areas A will shrink in proportion, and if you like imagine that the forces F are reduced during this process so as to make the stresses σ and τ stay at the same numerical value. Keep going, so the blocks are now more appropriately described as *infinitesimal rectangular parallelepipeds* (which I'm pretty sure you can't say twelve times fast) and then keep on going down past the length scale were the lumpiness of atoms would start to show up if one believed in that sort of thing. If you wish, you can imagine everything shrinking down to where an individual atom becomes a solar system and then down into some marshmallow on that world where an atom there becomes a solar system and on that earth there's a marshmallow, etc. until you run out of marshmallows. The little blocks of stuff are now mathematical points, and the cool part is that ignoring the atomic nature of our world allows us to define stress at a mathematical point. Since stress is force per unit area, you have to formally do some limiting process because you obviously can't divide by a zero area, but that limit turns out to work just fine.

In general, all six faces of our infinitesimal rectangular parallelepiped will be subjected to one normal and two shear stresses, with each of the faces designated by their outer normal vectors. Since there are nine stress components to keep track of, it makes sense to do it in a 3×3 matrix:

$$\tau_{ij} = \begin{pmatrix} \tau_{11} & \tau_{12} & \tau_{13} \\ \tau_{21} & \tau_{22} & \tau_{23} \\ \tau_{31} & \tau_{32} & \tau_{33} \end{pmatrix} \qquad (i,j = 1,2,3) \tag{2.13}$$

You may have noticed that I've used τ for both shear and normal stresses. Some books use σ for both. Some are careful to use σ for normal stresses and τ for shear stresses. I reserve the right to use

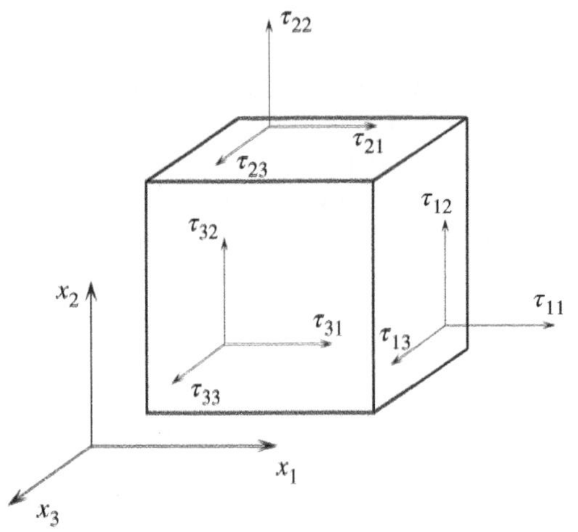

Figure 2.2 The indices on the stress components indicate the direction of the surface normal and the direction of the stress component, respectively.

whichever happens to suit my fancy at a particular moment, trusting that you'll pay attention to the indices. If they're the same, then the direction of the force and the surface normal are the same, so we're talking about a normal stress. If the two indices are different, then the force is tangential to the surface and we're talking about shear stresses. In the matrix, the normal stresses are obviously on the diagonal, while the shear stresses are the off-diagonal terms.

Now let's use this to derive something useful. Consider again an infinitesimal rectangular parallelepiped (Figure 2.2), but now subjected to three stress components on each face as well as some body force like gravity or magnetic pull or something. Let's generalize it just a bit by defining the lengths of each of the three sides as dx_1, dx_2, and dx_3 and assume for completeness that the body force (per unit volume) also has three components X_1, X_2, and X_3. At this point you probably want to think of this as an internal bit of a larger chunk of stuff that is being deformed somehow. I'm assuming you already ate all the marshmallows, so look around for something else squishy that's being squished in a complicated manner. If you're sitting in a comfy chair, it's the cushion. If you're sitting on a wooden bench, it's your buttocks. If you're lying on a bed of nails, you're taking this all too literally. Each little piece of that larger object is being pressed upon in a complicated way by the adjacent parts, and each independently feels gravity and other body forces. If you've drawn a blank, I want you to imagine a great big fat guy[4] riding a roller coaster next to you. I mean really fat. It took two attendants to push with all their might to get the restraint to snap shut which will hopefully hold him securely throughout the ride because he's really wedged in there. As the coaster goes through loops and corkscrews that part of your seatmate that is spilling over into your personal space is going to be pressing on you in a complex manner. Gravity and inertia are also going to be affecting both of you, which is pretty much the point of roller coasters after all.

The point of that mental exercise was for you to get a mental picture of a deformable body with a complex three-dimensional stress distribution. That little block of material (e.g. undigested marshmallow) you imagine deep inside somewhere will have stresses on each face, but opposite faces will be just a bit different from each other. So, if $\tau_{11}(x_1, x_2, x_3)$ is the normal stress on the "left" face as

4 I was a big fat guy while in my early 40s, but then stopped eating crap like marshmallows and dropped 30% of my body weight in four months. My wife and her friends were all mad at me because I apparently made losing weight look too easy. After more than 15 years, my wife is still butthurt that I don't eat pie. Since alcohol is also empty calories, I gave up drinking more than a decade ago, so I'm pretty much no fun. My wife can attest to that.

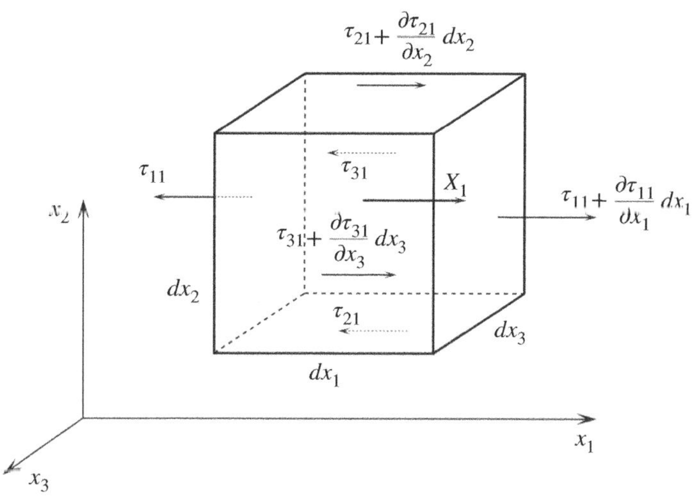

Figure 2.3 Force is stress times area. The surface stresses in the x_1-direction are shown. There is also a body force per unit volume X_1 indicated as well.

shown in Figure 2.3, then the normal stress on the opposite face is only a little bit different because the rectangular parallelepiped is infinitesimal

$$\tau_{11}(x_1 + dx_1, x_2, x_3) = \tau_{11}(x_1, x_2, x_3) + dx_1\frac{\partial\tau_{11}}{\partial x_1}(x_1, x_2, x_3) + H.O.T. \tag{2.14}$$

with the higher-order terms small enough that we can neglect them. Since stress is force divided by area, force is stress times area. The diagram in Figure 2.3 shows the stress components that act in the x_1 direction, along with that component of the body force which we have to multiply times volume, of course. For equilibrium, forces sum to zero in each direction, and for the x_1 direction this gives

$$\left(\tau_{11} + \frac{\partial\tau_{11}}{\partial x_1}dx_1\right)dx_2dx_3 - \tau_{11}dx_2dx_3 + \left(\tau_{21} + \frac{\partial\tau_{21}}{\partial x_2}dx_2\right)dx_1dx_3$$
$$-\tau_{21}dx_2dx_3 + \left(\tau_{31} + \frac{\partial\tau_{31}}{\partial x_3}dx_3\right)dx_1dx_2 - \tau_{31}dx_2dx_3 + X_1dx_1dx_2dx_3 = 0$$

Dividing by $dx_1dx_2dx_3$ is no problem because infinitesimally small is not zero, so doing that gives

$$\frac{\partial\tau_{11}}{\partial x_1} + \frac{\partial\tau_{21}}{\partial x_2} + \frac{\partial\tau_{31}}{\partial x_3} + X_1 = 0 \tag{2.15}$$

and repeating the procedure for the x_2 and x_3 directions gives

$$\frac{\partial\tau_{21}}{\partial x_1} + \frac{\partial\tau_{22}}{\partial x_2} + \frac{\partial\tau_{32}}{\partial x_3} + X_2 = 0 \tag{2.16}$$

$$\frac{\partial\tau_{31}}{\partial x_1} + \frac{\partial\tau_{23}}{\partial x_2} + \frac{\partial\tau_{33}}{\partial x_3} + X_3 = 0 \tag{2.17}$$

or, in our beloved index notation

$$\frac{\partial\tau_{ij}}{\partial x_j} + X_i = 0 \tag{2.18}$$

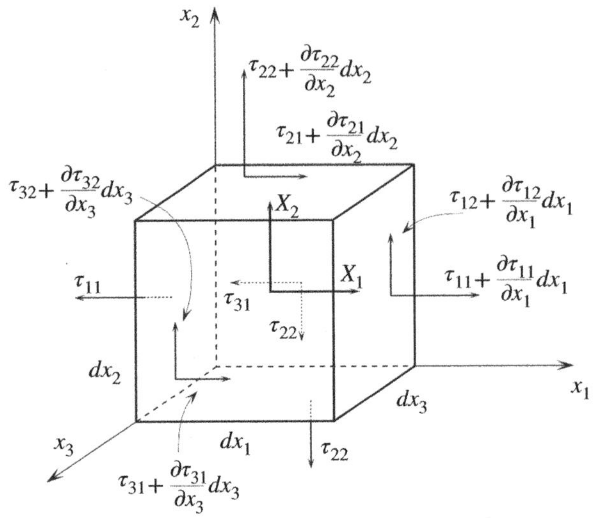

Figure 2.4 Moment is force times distance, so it's then distance times stress times area. The surface stresses in the x_1- and x_2-directions are shown, which will give moments about the x_3-axis. There are also body forces per unit volume X_1 and X_2 indicated as well that will give moments about the x_3-axis.

which is important enough that it gets a name: *The Equations of Equilibrium*. Of course I prefer to write them as:

$$\partial_j \tau_{ij} + X_i = 0 \tag{2.19}$$

D'Alembert's principle says that any little piece of a fat guy on a rollercoaster in motion may be considered to be in equilibrium if we write $X_i^{inertial} = -\rho\ddot{x}_i$ so we can include this as one part of the body force to get our equation of motion for elastodynamics

$$\rho\ddot{x}_i = \partial_j \tau_{ij} + X_i^{body} \tag{2.20}$$

which is the version of $\vec{F} = m\vec{a}$ that we will use, although $X_i^{body} \equiv 0$ for almost all of the applications we'll consider.

Exercise 2.2 Write out the equation of motion in Cartesian coordinates. Be amazed at how much easier our life is with index notation.

Exercise 2.3 By summing moments ($\sum M = 0$) about Ox_1, Ox_2, and Ox_3 show that the stress tensor is symmetric: $\tau_{ij} = \tau_{ji}$. Hint: The components of stresses that contribute moments about the x_3-axis are shown in Figure 2.4.

2.3 Strain Is Dimensionless

In everyday language, the words stress and strain mean the same thing. Not for us. For most people, being strained means doing something that is hard or tires them out or otherwise pushes their limits, causing them stress and needing a neck rub or perhaps a vacation at a spa. Here, a body is said to be strained (or deformed) when the *relative* positions of points in the body are changed. The displacement of a point is defined as the vector distance from the initial to the final location of the point. If $\vec{u} = (u, v, w)$ is the displacement (vector) field, then (x, y, z) is displaced to $(x + u, y + v, z + w)$. The one-dimensional case is shown in Figure 2.5, but you can experience this for yourself by tying three rubber bands together and naming the two knots A and B.

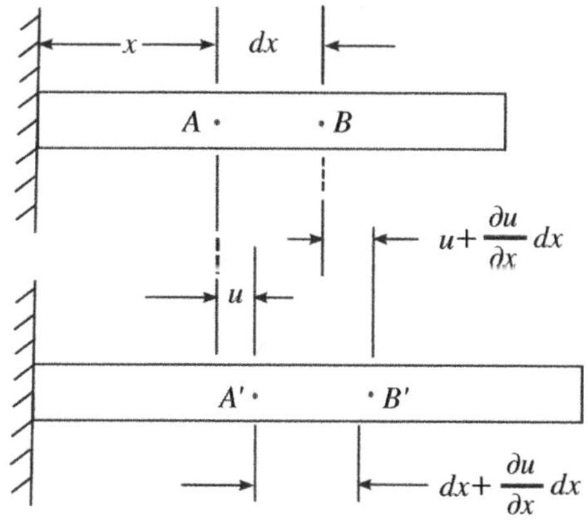

Figure 2.5 Normal strain is change in length per unit length.

After a stress is applied (e.g. by stretching the tied rubber bands from the far ends), the points A and B move to new positions A' and B' and the length AB changes from dx to $dx + \frac{\partial u}{\partial x}dx$. We *define* strain as the unit change in length

$$\epsilon_x = \frac{\left(dx + \frac{\partial u}{\partial x}dx\right) - dx}{dx} \equiv \frac{\partial u}{\partial x} \tag{2.21}$$

Now consider the 2D case of plane strain, where $u = u(x, y)$, $v = v(x, y)$, and $w = 0$ and where an infinitesimal rectangle $ABCD$ is deformed to a rotated parallelogram $A'B'C'D'$ as shown in Figure 2.6. Note that the sides change length; the sides rotate with respect to each other.

Normal strain is a change in length per unit length as before, but now we have to define a shear strain to describe the sort of distortion caused by shear stresses. At this point, please shoot the rubber band across the room and forget all about it. Also, forget about the marshmallows and fat guys on roller coasters because those sorts of things deform in ways that are too complicated for us

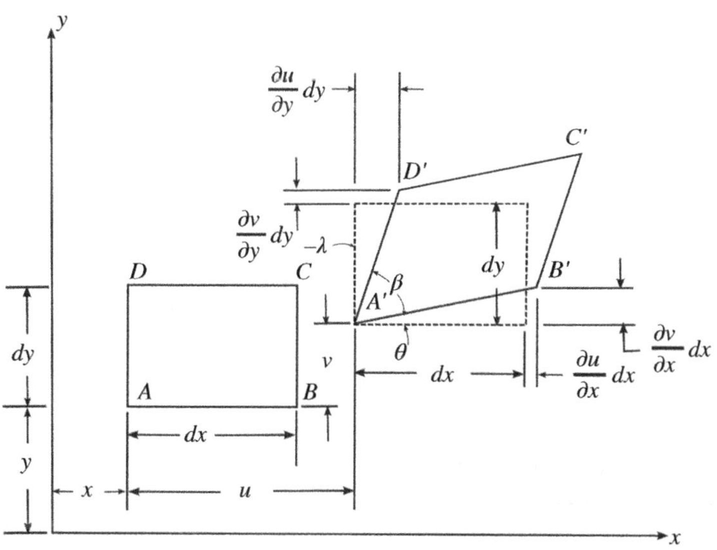

Figure 2.6 Shear strain is change from a rectangle to a parallelogram.

at this point. In a word, they exhibit *nonlinear* strain and we're about to linearize our equations, so we'll end up with something useful. Later on, I promise we'll add back in some nonlinearity when particular applications need it, but you have to trust me on this one because it would be really frustrating if a couple of pages from now we derived equations that turned out to be too complicated to be usable. Here are our normal strains in the x and y directions

$$\epsilon_x = \frac{A'B' - AB}{AB} = \frac{A'B' - dx}{dx} \tag{2.22}$$

$$\epsilon_y = \frac{A'D' - AD}{AD} = \frac{A'D' - dy}{dy} \tag{2.23}$$

which allow us to write the lengths squared

$$(A'B')^2 = [dx(1 + \epsilon_x)]^2 = \left(dx + \frac{\partial u}{\partial x}dx\right)^2 + \left(\frac{\partial v}{\partial x}dx\right)^2 \tag{2.24}$$

$$(A'D')^2 = [dy(1 + \epsilon_y)]^2 = \left(dy + \frac{\partial v}{\partial y}dy\right)^2 + \left(\frac{\partial u}{\partial y}dy\right)^2 \tag{2.25}$$

or

$$\epsilon_x^2 + 2\epsilon_x + 1 = 1 + 2\frac{\partial u}{\partial x} + \left(\frac{\partial u}{\partial x}\right)^2 + \left(\frac{\partial v}{\partial x}\right)^2 \tag{2.26}$$

$$\epsilon_y^2 + 2\epsilon_y + 1 = 1 + 2\frac{\partial v}{\partial y} + \left(\frac{\partial v}{\partial y}\right)^2 + \left(\frac{\partial u}{\partial y}\right)^2 \tag{2.27}$$

For small strains of the type you might get with piano wire instead of rubber bands and rocks instead of marshmallows, we can happily drop the second order terms to get

$$\epsilon_x = \frac{\partial u}{\partial x} \qquad \epsilon_y = \frac{\partial v}{\partial y} \tag{2.28}$$

Shear strain is here defined as the change in the right angle between any two sides of $ABCD$ measured in radians. Referring to the diagram, we can write it

$$\gamma_{xy} = \frac{1}{2}\left(\frac{\pi}{2} - \beta\right) = \frac{1}{2}(\theta + \lambda) \tag{2.29}$$

where we again invoke the smallness of strains to write

$$\tan\theta \approx \theta = \frac{\frac{\partial v}{\partial x}dx}{dx + \frac{\partial u}{\partial x}dx} = \frac{\frac{\partial v}{\partial x}}{1 + \frac{\partial u}{\partial x}} \tag{2.30}$$

Neglecting $\frac{\partial u}{\partial x}$ compared to unity, we have $\theta = \frac{\partial v}{\partial x}$ and similarly $\lambda = \frac{\partial u}{\partial y}$ gives

$$\gamma_{xy} = \frac{1}{2}\left(\frac{\partial u}{\partial y} + \frac{\partial v}{\partial x}\right) \tag{2.31}$$

In the 3D case, the strains are

$$\epsilon_x = \frac{E'F' - EF}{EF} = \frac{A'B' - AB}{AB} = \frac{H'G' - HG}{HG} = \frac{D'C' - DC}{DC} \tag{2.32}$$

$$\gamma_{xy} = \left(\frac{\pi}{2} - \angle F'E'H'\right) = \left(\frac{\pi}{2} - \angle F'G'H'\right) = \cdots \tag{2.33}$$

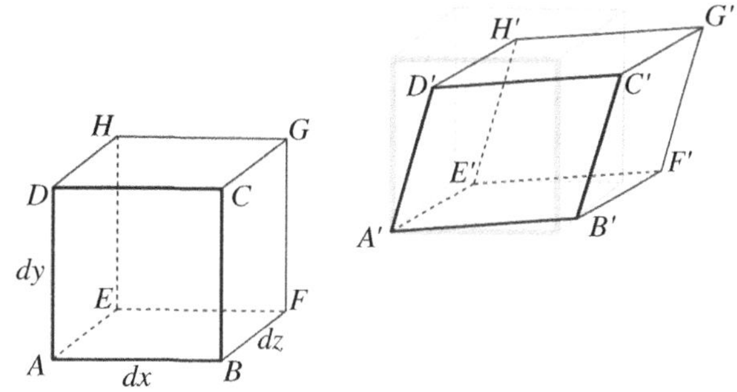

Figure 2.7 Shear strain in 3D is changed from a cube to a rhomboid, which may or may not be a word. Note that rigid body motion – translation and/or rotation – is not included in strain because strain is just normalized deformation.

Exercise 2.4 Draw the many triangles and such to go along with Figure 2.7 and then go through the analysis step by step to show that the nine components of strain are

$$\epsilon_x = \frac{\partial u}{\partial x} \qquad \gamma_{xy} = \frac{1}{2}\left(\frac{\partial u}{\partial y} + \frac{\partial v}{\partial x}\right) = \gamma_{yx}$$

$$\epsilon_y = \frac{\partial v}{\partial xy} \qquad \gamma_{yz} = \frac{1}{2}\left(\frac{\partial v}{\partial z} + \frac{\partial w}{\partial y}\right) \equiv \gamma_{yx}$$

$$\epsilon_z = \frac{\partial w}{\partial z} \qquad \gamma_{xy} = \frac{1}{2}\left(\frac{\partial w}{\partial x} + \frac{\partial u}{\partial z}\right) \equiv \gamma_{yx}$$

We can use index notation to write a strain tensor ϵ_{ij}, which is related to the displacement vector u_i by

$$\epsilon_{ij} = \frac{1}{2}\left(\frac{\partial u_i}{\partial x_j} + \frac{\partial u_j}{\partial x_i}\right) \tag{2.34}$$

Note that stress and strain are both second-rank tensors.

2.4 Stress Is Proportional to Strain

Remember that rubber band you shot across the room not that long ago? If not, go find another one and play with it a little bit. If you happen to have a spring at hand that would be even better, but don't get distracted rooting around for one. Recall from elementary physics that we classify springs according to their "spring constants" and that stiffer springs are harder to stretch than wimpier ones because their spring constants are larger. Recall also that if you pull twice as hard on a spring, it stretches twice as far and pulling four times as hard stretches it four times as much. Sorry if you just broke your rubber band, but it's OK because we're done with it for real this time. All you need to remember is that for *linear* springs, the force and displacement are proportional, with a constant of proportionality given by the spring constant. In Latin that's **Ut tensio, sic vis** meaning "As the extension, so the force," which British physicist Robert Hooke first published in 1660 as a Latin anagram, but we call it Hooke's law anyway.

There's a lovely discussion of the discovery of Hooke's law and related concepts in the book by J.E. Gordon, "Structures: Or Why Things Don't Fall Down," which I highly recommend. It's for anyone who has ever wondered why suspension bridges don't collapse under eight lanes of traffic, how dams hold back – or give way under – thousands of gallons of water, or what principles guide the design of a skyscraper, a nightgown, or a kangaroo. It's a quite readable explanation of the basic forces that hold together the ordinary and essential things of this world – from buildings and bodies to flying aircraft and eggshells. For architects and engineers there are cogent explanations of the concepts of stress, shear, torsion, fracture, and compression, and chapters on safety design and the relationship of efficiency to aesthetics. If you are building a house, a sailboat, or a catapult, here is a handy tool for understanding the mechanics of joinery, floors, ceilings, hulls, and masts – or flying buttresses. In my structures class, I have the students buy this book and then ask them on the final exam what their favorite part was, which is easy to answer if you've read the book.

Back to stress and strain, which turn out to be proportional as well, although because they are tensors, there will be a tensor of proportionality constants. The generalized Hooke's law is

$$\sigma_{ij} = C_{ijkl}\epsilon_{kl} \tag{2.35}$$

where C_{ijkl} is obviously a fourth-rank tensor because it has four indices. It turns out not to be quite so terrible because most of the complexity is there to allow for materials with different stiffnesses in different directions. Wood has different properties along the grain than it does across the grain, for example, but even that anisotropy isn't especially extreme. Modern aircraft and expensive sports equipment are often made from graphite-epoxy composites instead of metals, and these engineered materials often have their stiffness properties tailored to optimize performance. In extreme cases, there might be as many as nine different stiffnesses to measure and then keep track of in C_{ijkl}.

Much of the time, however, we don't have to worry about stiffnesses being different in different directions and can happily assume that the materials we're dealing with are isotropic. In this case, we only have to worry about two stiffnesses and C_{ijkl} takes on the form

$$\boldsymbol{C}_{ijkl} = \delta_{ij}\delta_{kl}\lambda + \left(\delta_{ik}\delta_{jl} + \delta_{il}\delta_{jk}\right)\mu \tag{2.36}$$

where λ, μ are the Lamé parameters. They are related to the commonly used Poisson's ratio, ν, Young's modulus, E, and bulk modulus, K according to

$$\lambda = \frac{2\mu\nu}{1-2\nu} = \frac{\mu(E-2\mu)}{3\mu - E} = K - \frac{2}{3}\mu = \frac{E\nu}{(1+\nu)(1-2\nu)}$$

$$= \frac{3K\nu}{1+\nu} = \frac{3K(3K-E)}{9K-E}$$

$$\mu = \frac{\lambda(1-2\nu)}{2\nu} = \frac{3}{2}(K-\lambda) = \frac{E}{2(1+\nu)} = \frac{3K(1-2\nu)}{2(1+\nu)} = \frac{3KE}{9K-E}$$

$$\nu = \frac{\lambda}{2(\lambda+\mu)} = \frac{\lambda}{3K-\lambda} = \frac{E}{2\mu} - 1 = \frac{3K-2\mu}{2(3K+\mu)} = \frac{3K-E}{6K}$$

$$E = \frac{\mu(3\lambda+2\mu)}{\lambda+\mu} = \frac{\lambda(1+\nu)(1-2\nu)}{\nu} = \frac{9K(K-\lambda)}{3K-\lambda} = 2\mu(1+\nu)$$

$$= \frac{9K\mu}{3K+\mu} = 3K(1-2\nu)$$

$$K = \lambda + \frac{2}{3}\mu = \frac{\lambda(1+\nu)}{3\nu} = \frac{2\mu(1+\nu)}{3(1-2\nu)} = \frac{\mu E}{3(3\mu - E)} = \frac{E}{3(1-2\nu)}$$

$$\frac{\mu}{\lambda+\mu} = 1 - 2\nu \qquad \frac{\lambda}{\lambda+2\mu} = \frac{\nu}{1-\nu}$$

The first of the two Lamé[5] parameters, λ, doesn't have a physical interpretation, sorry. It's also kind of hard to find values for it in books, presumably for the reason that engineers tend to measure things that make sense physically rather than those parameters that show up due to fundamental mathematical properties of tensors. Hence the reason for the conversion table is that these are the parameters that you'll find most easily and then you'll have to convert to the ones that go into the formulas. Note that the second of the two Lamé parameters, μ, is often called the modulus of rigidity and many books will use the symbol G for it. If you subject a rod or shaft to equal and opposite torques at its ends, the modulus of rigidity is the constant of proportionality between the magnitude of the torques and the amount of twist. Poisson's ratio, ν, comes into play when you stretch a rubber band or squish down on a marshmallow. The rubber band contracts laterally and the marshmallow bulges out. Lateral strain is the contraction/bulging compared to the original width. Axial strain is the squished/stretched length compared to the undeformed state.

Poisson's ratio is the lateral strain divided by the axial strain, and it's always a number between zero and one-half unless you construct some weird meta-material that bulges when you stretch it and contracts when you squish it in which case the Poisson's ratio would be negative. Go ahead and get yourself another marshmallow. You don't need to feel guilty about empty calories in the name of science.

The bulk modulus, K, of a substance measures its resistance to uniform compression. It is defined as the pressure increase needed to decrease the volume by a factor of $1/e$. For example, if you took your marshmallow down deep in the ocean where the pressure is immense, it will retain the same shape but will get tiny and then presumably soggy and salty. Conversely, if you took your marshmallow up (or down) into empty space where there is no atmospheric pressure, it will enlarge. Unless you injected it with Shrinx, of course.

2.5 Elastic Waves

We can take our three main equations

$$\rho \partial_t^2 u_i - \partial_j \sigma_{ij} = 0 \qquad \sigma_{ij} = C_{ijkl} \epsilon_{kl}$$

$$\epsilon_{kl} = \frac{1}{2} \left(\partial_k u_l + \partial_l u_k \right)$$

where

$$C_{ijkl} = \delta_{ij} \delta_{kl} \lambda + \left(\delta_{ik} \delta_{jl} + \delta_{il} \delta_{jk} \right) \mu$$

and plug and chug to derive Navier's equation

$$\rho \partial_t^2 u_i - \mu \partial^2 u_i - (\lambda + \mu) \partial_i (\partial_j u_j) = 0 \tag{2.37}$$

In vector notation, this becomes

$$\rho \partial_t^2 \vec{u} - \mu \nabla^2 - (\lambda + \mu) \nabla (\nabla \cdot \vec{u}) = 0 \tag{2.38}$$

5 Gabriel Lamé was a French mathematician whose most significant contribution to engineering was to accurately define the stresses and capabilities of a press fit joint, such as that seen in a dowel pin in a housing. His is one of the 72 names inscribed on the Eiffel Tower, which is held together by dowel pins in housings. https://en.wikipedia.org/wiki/Gabriel_Lam%%;C3%%A9.

Navier's equation[6] is the basic wave equation for elastodynamics in isotropic media. Since we care most about wave behavior, it's usually convenient to suppress the $e^{-i\omega t}$ harmonic time variation, and do our analysis in frequency domain. This allows us to write a vector wave equation of the form

$$(\nabla^2 + K^2)\,\vec{u} - \left(1 - \frac{k^2}{K^2}\right)\nabla(\nabla \cdot \vec{u}) = 0 \tag{2.39}$$

where $K = \omega/c_T$ and $k = \omega/c_L$ are transverse and longitudinal wave numbers, respectively. The wave speeds are defined in terms of the Lamé parameters as:

$$c_L^2 = (\lambda + 2\mu)/\rho \qquad\qquad c_T^2 = \mu/\rho \tag{2.40}$$

This wave equation is kind of a pain, so we do a Helmholtz decomposition. Clebsch's theorem[7] tells us that any vector field can be decomposed into two parts, one which has no curl and one which has no divergence. We therefore write

$$\vec{u} = \vec{u}_L + \vec{u}_T = \nabla\phi + \nabla \times \vec{A} \qquad\qquad (\nabla \cdot \vec{A} = 0) \tag{2.41}$$

Note that $\nabla \times \vec{u}_L \equiv 0$ and $\nabla \cdot \vec{u}_T \equiv 0$ by definition. Plugging into the vector wave equation thus gives us scalar wave equations, one for the scalar potential function ϕ and two for the components of the vector potential function $\vec{A}$

$$(\nabla^2 + k^2)\,\phi = 0 \qquad\qquad (\nabla^2 + K^2)\,A_{I,II} = 0 \tag{2.42}$$

Note that these are both the same equation, named the Helmholtz equation, but with different wave numbers. Often called the scalar wave equation, we can conclude at this point that elastic waves come in two types. We'll often call them *Longitudinal* and *Transverse waves* because in longitudinal waves the vibration of media is in the direction of propagation, whereas for transverse waves, the vibration of media is perpendicular to the direction of propagation (Figure 2.8). Seismologists call them the *Primary* and *Secondary* waves, since the faster *P*-wave gets to you first and is followed by the slower *S*-wave. We could also call the *L*-wave a compressional wave since the physical deformation is an alternating compression and rarefaction of the solid medium. The *S*-wave is often called a shear wave since the physical deformation is a back and forth shearing of the solid medium, which explains why the shear wave velocity is directly dependent on the shear rigidity of the solid medium. Note that gasses, liquids and soft tissues don't support the propagation of shear waves, something that we can take as a fundamental distinction between elastic solids and fluids.

Exercise 2.5 Go back and fill in the algebra that I skipped in this section.

You may be surprised to know that before James Clerk Maxwell[8] unified the fields of electricity and magnetism, it was Navier's equation that was used to model the propagation of light waves.

---

6 Claude-Louis Navier was a French engineer who formulated the general theory of elasticity in a mathematically usable form (1821). His major contribution, however, remains the Navier-Stokes equations (1822), central to fluid mechanics. His is another of the 72 names inscribed on the Eiffel Tower. https://en.wikipedia.org/wiki/Claude-Louis_Navier.

7 Rudolf Friedrich Alfred Clebsch was a German mathematician who made important contributions to scattering theory, including the Clebsch-Gordan coefficients for spherical harmonics, which are now widely used in quantum mechanics. https://en.wikipedia.org/wiki/Alfred_Clebsch.

8 James Clerk Maxwell was a Scottish scientist whose most notable achievement was to formulate the classical theory of electromagnetic radiation, bringing together for the first time electricity, magnetism, and light as different manifestations of the same phenomenon, leading to the prediction of radio waves. https://en.wikipedia.org/wiki/James_Clerk_Maxwell.

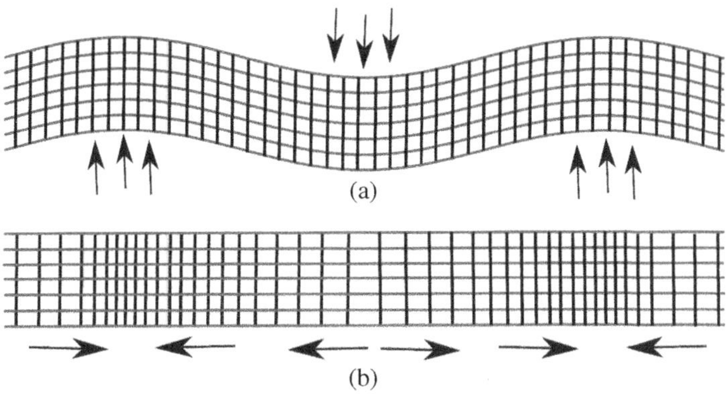

Figure 2.8 Shear waves and longitudinal waves are propagating to the right. Note in the shear waves (a) the displacement is up and down and that the rectangular elements are sheared into parallelograms. Note in the longitudinal waves (b), the rectangular elements remain rectangular, but they are alternately compressed and elongated. Shear waves distort the solid without a change in density whereas longitudinal waves represent a change in density.

The elastic medium that supported the propagation of light waves was called the aether, and it had some rather unusual properties. The obvious one is that we can pass through the aether freely without noticing it. It got worse when measurements demonstrated conclusively that light is a transverse wave with a known speed of propagation, since then, the ratio of the rigidity and density are specified by $c_T^2 = \mu/\rho$ but of course the aether can't be too dense or rigid because then we'd be able to notice it. That's bad enough, but somehow we have to get rid of the longitudinal waves. The simplest way is to make the aether incompressible by setting $\lambda \to \infty$, which gives us the Greenian aether.[9] That works, but now it's an incompressible medium that we are able to move through freely. The alternative is to assume that the reason there are no longitudinal waves is because their velocity is zero, which is what we'd have if $\lambda = -2\mu$. The liable aether never seemed to quite catch on, somehow.

Let's consider Navier's equation again

$$\rho\partial_t^2\vec{u} - \mu\nabla^2 - (\lambda + \mu)\nabla(\nabla \cdot \vec{u}) = 0 \tag{2.43}$$

We said before that the thing that distinguishes a fluid from a solid is that a fluid cannot support shear deformation, that is, it has vanishing shear rigidity. Letting $\mu \to 0$ in Navier's equation gives

$$\rho\partial_t^2\vec{u} - \lambda\nabla(\nabla \cdot \vec{u}) = 0 \tag{2.44}$$

but my favorite vector identity is

$$\nabla \times \nabla \times (\cdot) = \nabla(\nabla \cdot (\cdot)) - \nabla^2(\cdot)$$

For fields with no curl, we can then write $\rho\partial_t^2\vec{u} - \lambda\nabla^2\vec{u} = 0$ or, integrating to get $\rho\partial_t^2\phi - \lambda\nabla^2\phi = 0$ and then if we suppress the harmonic time convention this can be rearranged to give

$$\left(\nabla^2 + \frac{\omega^2\rho}{\lambda}\right)\phi = 0 \tag{2.45}$$

Clearly we can think of acoustic waves in fluids as a subset of elastic waves (longitudinal waves) in solids.

9 See, Whittaker, E.T. (1910). *A History of the Theories of Aether and Electricity from the Age of Descartes to the Close of the Nineteenth Century*. Longmans, Green, and Company, which you should be able to find available for download.

Exercise 2.6 What is λ for several liquids and gasses, and how do those values compare to various solids?

2.6 Electromagnetic Waves

Let's next consider electromagnetics using Maxwell's equations so that we can compare and contrast what we get for acoustics and elastodynamics. It may or may not be surprising to you that we'll keep coming back to the same basic wave equations over and over. We seem to be modeling quite different physics, but then waves are waves and scattering is scattering. I won't go so far as some radar-scattering types who think of acoustics as merely a "scalar" version of electromagnetics, but both physical intuition and mathematical techniques developed in various aspects of wave propagation and scattering are pretty broadly applicable.

Maxwell's equations can be written as:

$$\nabla \cdot \vec{D} = \rho \qquad \nabla \cdot \vec{B} = 0 \tag{2.46}$$

$$\nabla \times \vec{E} = -\partial_t \vec{B} \tag{2.47}$$

$$\nabla \times \vec{H} = \partial_t \vec{D} + \vec{J} \tag{2.48}$$

which are the laws of Gauss, Faraday, and Ampere. In these four equations, $\vec{E}$ is the electric field, $\vec{B}$ is the magnetic flux density, $\vec{D}$ is the electric field displacement, and $\vec{H}$ is the magnetic field. In addition, ρ and $\vec{J}$ are the (electric) charge density and current density.

Constitutive relations are

$$\vec{B} = \mu \vec{H} \qquad \vec{D} = \epsilon \vec{E} \qquad \vec{J} = \sigma \vec{E} \tag{2.49}$$

where ϵ, μ are the permittivity and permeability and σ is the conductivity. With these constitutive equations, Maxwell gives

$$\nabla \cdot (\epsilon \vec{E}) = \rho \qquad \nabla \cdot (\mu \vec{H}) = 0 \tag{2.50}$$

$$\nabla \times \vec{E} = -\partial_t (\mu \vec{H}) \tag{2.51}$$

$$\nabla \times \vec{H} = \partial_t (\epsilon \vec{E}) + \sigma \vec{E} \tag{2.52}$$

Taking the curl of the last two and again using my favorite vector identity

$$\nabla \times (\nabla \times (\cdot)) = \nabla(\nabla \cdot (\cdot)) - \nabla^2(\cdot)$$

we can write, with $\rho \equiv 0$, wave equations

$$\nabla^2 \vec{E} - \mu\epsilon \partial_t^2 \vec{E} - \mu \partial_t (\sigma \vec{E}) = 0 \tag{2.53}$$

$$\nabla^2 \vec{H} - \mu\epsilon \partial_t^2 \vec{H} - \mu \partial_t (\sigma \vec{H}) = 0 \tag{2.54}$$

Assuming harmonic time dependence $e^{-i\omega t}$ gives

$$(\nabla^2 + k^2)\vec{E} = 0 \qquad (\nabla^2 + k^2)\vec{H} = 0 \tag{2.55}$$

where $k^2 = \omega^2 \mu\epsilon[1 + i\sigma/\omega\epsilon]$. In the absence of conductivity, the wave number is real $k = \omega\sqrt{\epsilon\mu}$ and the wave velocity is $1/\sqrt{\mu\epsilon}$.

For scattering, it is often convenient to define relative permittivity and permeability as $\epsilon_r = \epsilon/\epsilon_0$ and $\mu_r = \mu/\mu_0$, where ϵ_0 and μ_0 are the free space values and the speed of light in a vacuum is

$$c = \frac{1}{\sqrt{\mu_0\epsilon_0}} = 3 \times 10^8 \text{ m/s}$$

I happen to like using the relative permittivity and permeability rather a lot because I find the absolute units to be nonintuitive. It also makes it easy to guess the material properties for common materials since many dielectrics have a relative permittivity of about 3 and nonmagnetic materials have a permeability of 1.

Exercise 2.7 How good are the guesses $\epsilon_r \approx 3$ and $\mu_r \approx 1$ for various materials? How much do permeability and permittivity depend on frequency and temperature and can you think of some situations where it would be important to consider such variations in our modeling?

Note that the inclusion of conductivity made the wave number complex, which means that there will be attenuation. Rather a lot actually, although that isn't really obvious from the way we've defined the wave number. If we write it as:

$$k = \omega\sqrt{\epsilon\mu}\left[1 + i\frac{\sigma}{\omega\epsilon}\right]^{\frac{1}{2}} \tag{2.56}$$

it's perhaps a bit more obvious that something odd will happen unless $\sigma/\omega\epsilon$ is small compared to unity. Take a minute and look up some values of conductivity for a few metals and check whether $\sigma/\omega\epsilon$ is small compared to unity. Don't forget that $\omega = 2\pi f$ and consider frequencies appropriate to a few different applications like radar, WiFi, cell phones, radio, etc.

Another way to introduce attenuation is to simply make the permeability and/or the permittivity complex, $\epsilon = \epsilon' + i\epsilon''$ and $\mu = \mu' + i\mu''$, which obviously makes the wave number complex even without explicitly including conductivity. Each material is then simply characterized by its complex μ and ϵ, which means there are four parameters that need to be measured for each frequency band of interest. For the sorts of meta-materials favored by stealth aircraft designers, for example, there will often be strong frequency dependence and so if the other side migrates to radars that are outside of the nominal "threat band," there is always the fear that invisible airplanes will suddenly light up like a Christmas tree. It also means that the optimal placement of cell phone towers (so painstakingly modeled and then argued for at city council) might not give good coverage for the next generation of smart(er) phones. And please don't show up at city council meetings to argue that 5G caused COVID-19.

The Milstar satellite communications system was under development at Hanscom, AFB when I was there in the late 1980s.[10] When it came time to design the radome(s) that cover the antenna(s) that aircraft use to communicate with the satellite, the higher (EHF) frequency band used by Milstar meant that nobody was quite sure what the complex μ and ϵ values were for typical radome materials. Initially, they asked us to go check the library to see if anybody knew the answer, but they had never been measured, so a contract was issued to hurry up and go measure them. A recurring theme when we talk about scattering phenomena is that changing the frequency just a bit can dramatically alter the scattering behavior.

10 https://www.spaceforce.mil/About-Us/Fact-Sheets/Article/2197755/milstar-satellite-communications-system/.

2.7 Acoustic Waves

Now let's derive the acoustic wave equation as it's normally done for liquids and gasses. The basic equations for fluid mechanics are the Navier–Stokes equations, not to be confused with Navier's equation despite the obvious confusion often caused by the too-similar names. It's probably best if we just use descriptive names for them. Conservation of mass and momentum are written

$$\frac{\partial \rho}{\partial t} + \nabla \cdot (\rho \vec{v}) = 0 \tag{2.57}$$

$$\rho \frac{D\vec{v}}{Dt} = -\nabla p \tag{2.58}$$

where the convective derivative is defined as:

$$\frac{D}{Dt} \equiv \frac{\partial}{\partial t} + \vec{v} \cdot \nabla$$

In these ρ is the density, $\vec{v}$ is velocity, and p is pressure. Pressure and density are related by an equation of state $\nabla p = c^2 \nabla \rho$, where c is the speed of sound. Note that we could have included viscous and thermal effects, but that would have made things very complicated.

Sound waves are small fluctuations in a fluid otherwise at rest, so we linearize all of the field variables

$$\rho = \rho_0 + \rho' \qquad p = p_0 + p' \qquad \vec{v} = \vec{v}_0 + \vec{v}'$$

with ρ_0 and p_0 constants and $\vec{v}_0 \equiv 0$ because there's no mean flow. Now plug in and rearrange terms

$$\text{(mass)} \qquad \underbrace{\frac{\partial \rho_0}{\partial t}}_{\text{zero}} + \frac{\partial \rho'}{\partial t} + \nabla \cdot \left[(\rho_0 + \rho')\vec{v}' \right] = 0 \tag{2.59}$$

$$\text{(momentum)} \qquad (\rho_0 + \rho') \left(\frac{\partial}{\partial t} + \vec{v} \cdot \nabla \right) \vec{v}' = - \underbrace{\nabla (p_0}_{\text{zero}} + p') \tag{2.60}$$

$$\text{(state)} \qquad \nabla p' = c^2 \nabla \rho' \tag{2.61}$$

Rearrange these one more time and neglect terms which are second order in small quantities to get the linearized equations

$$\frac{\partial \rho'}{\partial t} + \rho_0 \nabla \cdot \vec{v}' = 0 \tag{2.61}$$

$$\rho_0 \frac{\partial \vec{v}'}{\partial t} = -\nabla p' \tag{2.62}$$

$$p' = c^2 \rho' \Rightarrow \frac{\partial \rho'}{\partial t} = \frac{1}{c^2} \frac{\partial p'}{\partial t} \tag{2.63}$$

Do the following algebra: plug the linearized equation of state into the linearized conservation of mass equation and then take the time derivative of the result. Then take the gradient of the linearized momentum equation and solve for an equation in p'. We then write, dropping the primes, a linear wave equation for pressure

$$\nabla^2 p - \frac{1}{c^2} \partial_t^2 p = 0 \tag{2.64}$$

or, in frequency domain

$$\left(\nabla^2 + \frac{\omega^2}{c^2}\right) p = 0 \tag{2.65}$$

We could also take the curl of the linearized momentum equation to write

$$\rho_0 \frac{\partial}{\partial t} \nabla \times \vec{v} = 0 \Rightarrow \vec{v} = \nabla \phi \tag{2.66}$$

Then the linearized mass and momentum equations give

$$\frac{1}{c^2}\frac{\partial p}{\partial t} + \nabla^2 \phi = 0 \tag{2.67}$$

$$\rho_0 \frac{\partial}{\partial t} \nabla \phi = -\nabla p \Rightarrow \frac{\partial p}{\partial t} = -\rho_0 \frac{\partial^2 \phi}{\partial t^2} \tag{2.68}$$

Combining these two gives

$$\nabla^2 \phi - \frac{1}{c^2} \partial_t^2 \phi = 0 \tag{2.69}$$

or, in frequency domain

$$\left(\nabla^2 + \frac{\omega^2}{c^2}\right) \phi = 0 \tag{2.70}$$

If we write $k = \omega/c$ we then have

$$\left(\nabla^2 + k^2\right) \phi = 0 \qquad \text{or} \qquad \left(\nabla^2 + k^2\right) p = 0 \tag{2.71}$$

which are exactly the same scalar wave equations that we had for elastic waves.

Consider the one-dimensional case where the scalar wave equation is

$$\left(\frac{\partial^2}{\partial x^2} + k^2\right) \phi(x) = 0 \tag{2.72}$$

and has the general solution $\phi(x) = Ae^{ikx} + Be^{-ikx}$, which we interpret as plane waves traveling to the right and left, with amplitudes A and B, respectively. The scalar wave equation is so important in so many different areas, that its solutions have been studied extensively in various coordinate systems.

Exercise 2.8 Derive via separation of variables the general solutions of the scalar wave equation in Cartesian, spherical and cylindrical coordinates. What are the special functions that would show up if we did this in spheroidal and elliptic cylindrical coordinates?

2.8 Anisotropic Elastic Solids

Recall our field equations

$$\rho \partial_t^2 u_i - \partial_j \sigma_{ij} = 0 \qquad \epsilon_{ij} = \frac{1}{2}\left(\partial_j u_i + \partial_i u_j\right)$$

with Hooke's Law $\sigma_{ij} = C_{ijkl}\epsilon_{kl}$ from which we can derive Navier's equation if the medium is isotropic and homogeneous. This is an excellent approximation in many cases, but then there are many materials where the stiffness, rigidity, etc. are directional and that must be accounted for. Wood is the obvious example, since the properties along vs. across the grain can be quite different. Bone is also pretty strongly anisotropic, although we're not used to thinking of bones as having "grain." Of course, most layered materials are quite likely to have very different properties in the plane of the laminations as compared to through the thickness of the laminate. Seismologists have long had to contend with "ray bending" due to the stratified nature of the earth's crust, but for decades now, engineered metals like steel and aluminum have been so dominant that elastic wave propagation in solids hasn't usually had to deal with waves wanting to travel at different speeds in different directions. Similarly, most electromagnetic wave theory, underwater sound, acoustics, and medical imaging only rarely encounter anisotropy. That's a good thing, since as we're about to see, things are going to get complicated very quickly and even after mastering a new, rather clumsy notation and trying to solve rather simple scattering problems, we'll end up frustrated.[11]

My first set of golf clubs had steel shafts, and then either metal or wood clubheads depending on whether they were irons or woods. They were modern in the sense that they didn't have wooden shafts, but not anywhere near as sophisticated as a set I got as a teenager, which had clubheads that were metal for the driver and the fairway woods. I still have steel shafts for my current irons, but my titanium-head woods all have composite shafts made from graphite fibers consolidated in an epoxy matrix. Composites like graphite-epoxy have many advantages over metals. They can be stiffer and lighter and their stiffness can be tailored by arranging the graphite fibers in order to give stiffness primarily in the directions needed to carry the expected loads without wasting it in other directions. This is presumably the whole point of the grain in a tree, but in engineering composites, the differences in properties along vs. against the fiber direction can be much higher. The fibers are sometimes woven into fabrics or pre-preg tape to ease in manufacturing, and very often many layers are used to build up a part's full thickness with fiber direction of each laminæ alternating in a particular fashion to build up a laminate with chosen effective properties.

We account for material anisotropy in the constitutive relations. Because the stress and strain tensors are symmetric, and thus have only six independent components each, there are *only* 21 independent constants in C_{ijkl} instead of 3^4. Of course, C_{ijkl} is still a fourth-rank tensor, so if you think about it, it's a rather unhelpful way to keep track of the material constants. This is especially true if you want to try to understand how the different kinds of anisotropy we're about to encounter relate to each other. I'm not a particularly big fan of the notation we're about to use since I happen to like Cartesian tensor notation quite a lot. You'll probably hate it too.

First, let's forget about the fact that stress and strain are tensors, but let's use the fact that they are symmetric tensors with six components each. We'll then just write stress and strain as column vectors, and while we're at it will be extremely careless with a factor of two in shear strain. Then we'll use an "index" notation that isn't an index notation and not really do the whole Einstein summation convention thing. So that everything will be crystal clear and we'll never get mixed up between the elegant tensor notation and the new contracted notation, we'll simply use little letters

---

11 There's a lot to be said about wave propagation in anisotropic media. B.A. Auld has a two-volume book "Acoustic Fields and Waves in Solids" Krieger Publishing (1990), which is an excellent reference. The punchline isn't very funny, though. Slog your way through both volumes and you'll find that including anisotropy makes things so complicated that you can't really do much with the equations. My rule is to never make things more complicated than you have to, modeling wise, because the problems we're trying to solve are hard enough.

for indices in the tensor notation and capital letters for them in the contracted notation. I told you you'd hate it. Here's a table that makes it about as clear as it's going to get.

Stress	Contracted	Strain	Contracted
σ_{11}	σ_1	ϵ_{11}	ϵ_1
σ_{22}	σ_2	ϵ_{22}	ϵ_2
σ_{33}	σ_3	ϵ_{33}	ϵ_3
$\tau_{23} = \sigma_{23}$	σ_4	$\gamma_{23} = 2\epsilon_{23}$	ϵ_4
$\tau_{31} = \sigma_{31}$	σ_5	$\gamma_{31} = 2\epsilon_{31}$	ϵ_5
$\tau_{12} = \sigma_{12}$	σ_6	$\gamma_{12} = 2\epsilon_{12}$	ϵ_6

with γ used for the engineering shear strain but ϵ used for the tensor shear strain. The factor of two floating around is because we had originally defined shear strain to be a proper tensor but the commonly used engineering version didn't. Then we can write

$$\sigma_I = C_{IJ}\epsilon_J \qquad (I, J = 1, 2, 3, \ldots, 6) \tag{2.73}$$

and C_{IJ} will be a 6×6 matrix. Writing it all out gives

$$\begin{Bmatrix} \sigma_1 \\ \sigma_2 \\ \sigma_3 \\ \sigma_4 \\ \sigma_5 \\ \sigma_6 \end{Bmatrix} = \begin{bmatrix} C_{11} & C_{12} & C_{13} & C_{14} & C_{15} & C_{16} \\ C_{21} & C_{22} & C_{23} & C_{24} & C_{25} & C_{26} \\ C_{31} & C_{32} & C_{33} & C_{34} & C_{35} & C_{36} \\ C_{14} & C_{24} & C_{34} & C_{44} & C_{45} & C_{46} \\ C_{15} & C_{25} & C_{35} & C_{45} & C_{55} & C_{56} \\ C_{16} & C_{26} & C_{36} & C_{46} & C_{56} & C_{66} \end{bmatrix} \begin{Bmatrix} \epsilon_1 \\ \epsilon_2 \\ \epsilon_3 \\ \epsilon_4 \\ \epsilon_5 \\ \epsilon_6 \end{Bmatrix} \tag{2.74}$$

This is sometimes called triclinic by crystals people. There are 21 constants in C_{IJ} because it is a symmetric matrix. So far, we haven't really gained anything, but most situations that we're likely to deal with are going to have some symmetries, which will mean that we won't need quite so many of the stiffnesses. If there is one plane of symmetry, only 13 of the C_{IJ}'s are independent. If there are two orthogonal planes of material symmetry, then there will always exist a third and there are then only nine C_{IJ}'s.

Generally speaking composites people won't ever use any of the more complicated stiffness matrices for the simple reason that the C_{IJ}'s each have to be measured. Remember that they are just numbers for any given material that relate stress to strain. That means to measure the nine C_{IJ}'s, one has to do nine independent measurements, which are carefully aligned with the natural directions of the material. Those measurements are typically done on representative coupon samples in specially designed fixtures, but of course, if you want to understand the material and/or manufacturing variations you'll have to do enough measurements to understand the relevant statistics. Then, anytime you change something about your materials and/or your manufacturing processes you'll have to do another set of measurements. I suppose it wouldn't be so bad if the measurements were quick and/or interesting, but they're time consuming and not particularly exciting. The fun part about putting things in a hydraulic load frame is when they fail catastrophically, for example, a full-size composite airliner wing in a million-pound load frame, but what we're talking about here are just the simple stiffnesses, so nothing usually fails violently and embeds shards of itself in the wall clear across the hangar. Hence, composites folks pretty much always assume that their materials are orthotropic, unless there is some relevant simplification. Often it's an excellent approximation. Wood, for example, can often be considered to be orthotropic, with the three natural directions being defined as along the grain, and then radially and circumferentially across the grain. An orthotropic material that has the same stiffnesses in the three perpendicular directions is called cubic, but isn't isotropic because going diagonally will be different and the amount of the difference will depend on the angle of the diagonal.

If at a point in the material, there is one plane in which the mechanical properties are the same in all directions, then the material is called transversely isotropic. Wood can often be described this way, as can unidirectional composites where the fiber reinforcements are all aligned in the same direction. The stress–strain relationship then has only five independent constants. Finally, for isotropic materials, we have only two independent C_{IJ}'s and thus:

$$\begin{Bmatrix} \sigma_1 \\ \sigma_2 \\ \sigma_3 \\ \sigma_4 \\ \sigma_5 \\ \sigma_6 \end{Bmatrix} = \begin{bmatrix} C_{11} & C_{12} & C_{12} & 0 & 0 & 0 \\ C_{12} & C_{11} & C_{12} & 0 & 0 & 0 \\ C_{12} & C_{12} & C_{11} & 0 & 0 & 0 \\ 0 & 0 & 0 & \frac{(C_{11}-C_{12})}{2} & 0 & 0 \\ 0 & 0 & 0 & 0 & \frac{(C_{11}-C_{12})}{2} & 0 \\ 0 & 0 & 0 & 0 & 0 & \frac{(C_{11}-C_{12})}{2} \end{bmatrix} \begin{Bmatrix} \epsilon_1 \\ \epsilon_2 \\ \epsilon_3 \\ \epsilon_4 \\ \epsilon_5 \\ \epsilon_6 \end{Bmatrix}$$

Exercise 2.9 How are the other material properties we introduced previously related to C_{11} and C_{12} for isotropic materials? What are c_L and c_T in terms of C_{IJ}'s?

Let's go back and reconsider the field equations. We can write

$$\partial_i C_{ijkl} \partial_k u_l = \rho \partial_t^2 u_j \tag{2.75}$$

which reduces to Navier's equation if the isotropic Hooke's tensor is used. Instead, Fourier transform in time and space ($\partial_t \to -i\omega$, $\partial_i \to ikl_i$) to get

$$(ikl_i) C_{ijkl} (ikl_k) u_l = -\omega^2 \rho u_j$$

or $k^2 \left(l_i C_{ijkl} l_k\right) u_l = \rho\omega^2 u_j$. We now define the Christoffel matrix as $\Gamma_{jl} = l_i C_{ijkl} l_k$ and we can write the dispersion relation in its most general form as:

$$k^2 \Gamma_{ijl} u_l = \rho\omega^2 u_j \tag{2.76}$$

In the reduced notation, we write the Christoffel matrix as $\Gamma_{ij} = l_{iK} C_{KL} l_{Lj}$ where

$$l_{iK} = \begin{bmatrix} l_x & 0 & 0 & 0 & l_z & l_y \\ 0 & l_y & 0 & l_z & 0 & l_x \\ 0 & 0 & l_z & l_y & l_x & 0 \end{bmatrix} \qquad l_{Lj} = \begin{bmatrix} l_x & 0 & 0 \\ 0 & l_y & 0 \\ 0 & 0 & l_z \\ 0 & l_z & l_y \\ l_z & 0 & l_x \\ l_y & l_x & 0 \end{bmatrix}$$

Note that I, J, K, L go over the range 1–6 while i, j, k, l just go over 1–3 and $l_x = k_x/k$, $l_y = k_y/k$, $l_z = k_z/k$ are the direction cosines of the propagation direction. Told you you'd hate this notation.

In the most general anisotropic material with 21 independent C_{IJ}'s, the Christoffel equation is

$$k^2 \begin{bmatrix} \alpha & \delta & \epsilon \\ \delta & \beta & \zeta \\ \epsilon & \zeta & \gamma \end{bmatrix} \begin{Bmatrix} u_x \\ u_y \\ u_z \end{Bmatrix} = \rho\omega^2 \begin{Bmatrix} u_x \\ u_y \\ u_z \end{Bmatrix} \tag{2.77}$$

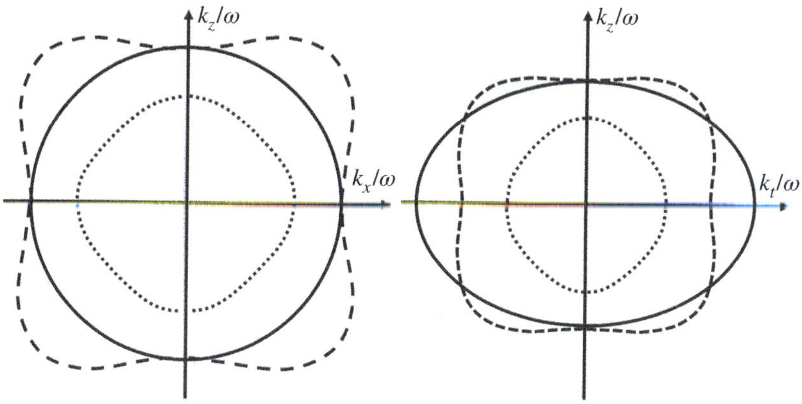

Figure 2.9 Slowness curves for cubic anisotropy. Faster modes have smaller slowness, so the dotted curves in each are the quasi-longitudinal waves. The solid circle on (a) is the pure shear mode whose slowness does not depend on direction for this particular orientation. (b) is propagation in a cubic diagonal plane.

$$\alpha = C_{11}l_x^2 + C_{66}l_y^2 + C_{55}l_z^2 + 2C_{56}l_yl_z + 2C_{15}l_zl_x + 2C_{16}l_xl_y$$
$$\beta = C_{66}l_x^2 + C_{22}l_y^2 + C_{44}l_z^2 + 2C_{24}l_yl_z + 2C_{46}l_zl_x + 2C_{26}l_xl_y$$
$$\gamma = C_{55}l_x^2 + C_{44}l_y^2 + C_{33}l_z^2 + 2C_{34}l_yl_z + 2C_{35}l_zl_x + 2C_{45}l_xl_y$$
$$\delta = C_{16}l_x^2 + C_{26}l_y^2 + C_{45}l_z^2 + (C_{46} + C_{25})l_yl_z + (C_{14} + C_{56})l_zl_x + (C_{12} + C_{66})l_xl_y$$
$$\epsilon = C_{15}l_x^2 + C_{46}l_y^2 + C_{35}l_z^2 + (C_{45} + C_{36})l_yl_z + (C_{13} + C_{55})l_zl_x + (C_{14} + C_{56})l_xl_y$$
$$\zeta = C_{56}l_x^2 + C_{24}l_y^2 + C_{34}l_z^2 + (C_{44} + C_{23})l_yl_z + (C_{36} + C_{45})l_zl_x + (C_{25} + C_{46})l_xl_y$$

Now rewrite the Christoffel equation as $\left(k^2\Gamma_{ij} - \rho\omega^2\delta_{ij}\right) u_j = 0$. A dispersion relation is then found by setting the characteristic determinant equal to zero. At fixed ω, this defines a surface that gives k as a function of its direction, $\hat{l}$ called a wave vector surface. We write

$$\left| \frac{k^2}{\omega^2}\Gamma_{ij}(l_x, l_y, l_z) - \rho\delta_{ij} \right| = 0 \tag{2.78}$$

and note that $k/\omega = 1/v_{phase}$ is the inverse of the phase velocity, which is typically called the slowness. A slowness surface shows us graphically how the propagation depends on direction. For an isotropic material, the waves propagate the same in all directions and we get circles. Examples for cubic anisotropy are shown in Figure 2.9.

Exercise 2.10 Use the parametric plotting capability of your favorite symbolic manipulation software to reproduce the slowness curves for GaAs.

2.9 Summary

I do hope you noticed that for all the different kinds of waves, we sooner or later ended up with the Helmholtz equation. That shouldn't have been surprising to you. The most-happy-excellent news is thus that we really only have one equation to deal with. One very important way that the different kinds of waves are different, though, is the boundary conditions that must be satisfied when waves interact with a discontinuity. Chapter 3 is going to discuss boundary conditions because the whole point of scattering analysis is to model the interaction of waves with discontinuities. I promise we'll get to some actual scattering after that, though.

Part of what I had in mind with this chapter was giving you a sense of the level of mathematical acrobatics that's going to follow. I thought that would be better than springing things on you as the book went along. There will be some special functions and whatnot in the chapters that follow, but if you're reasonably comfortable with the maths in this chapter, you're good to go. You may have expected there to be an extensive references for this chapter. Nope. There are innumerable sources for this mathematics. Assuming you're a digital native, you're probably used to looking things up on line, which is why I was trolling you by including so many biographical snippets of mathematicians from Wikipedia. I'll make a point of pointing you toward classic reference books for the highly esoteric math we'll be using, but for the basic (but still pretty advanced) stuff in this chapter, I want you to gather for yourself the reference materials you find useful, whether those are hyperlinks to on-line resources or pdfs that you save locally or old-fashioned actual books that you keep handy on a bookshelf. You do you.

Regarding that extensive references you may have been expecting me to include, I thought about doing that here, but my worry is that some readers would feel the need to go read/master all those books, papers, etc. before continuing on. That's the opposite of what I want you to do. There are innumerable books on elastodynamics, acoustics, and electromagnetics, of course, so my advice is to begin accumulating your own personal library. Many of the classic texts and monographs continue to be reprinted or are available online. For example, I have two copies of Stratton's "Electromagnetic Theory" that I like quite a lot. One I bought when I was a graduate student and one I got from my advisor when he died. I also just downloaded a PDF version from Internet Archive, but I see that it's available in a few different forms for quite modest prices at Amazon and such. Another bit of advice I'll offer up at this point is to take advantage of used bookstores, library sales, etc. where you can pick up for next to nothing classic books on the sorts of esoteric technical subjects we're talking about here. You may get some pushback from your significant other about it, but there will be times in your professional life, where having the right reference materials at hand will be the difference between getting past a roadblock and continuing to be stuck. My wife and I had an agreement back in the day; I wouldn't ask her how much money she was spending on clothes and she wouldn't ask me how much money I was spending on books. I still have quite a lot of books in my home office, campus office, and lab, as well as a small sliver of a shared walk-in closet for my clothes. Classic books never seem to go out of style. Classic clothes that you can wear for more than one season are an investment.

There's no excuse these days for not doing a proper literature review, except for perhaps the excuse that not including references for this chapter is intended to nudge you to getting started on accumulating your own reference library. In some ways, it used to be simpler. First get access to a good library and then go there to see what reference materials the professionals had curated on your behalf. I have fond memories of going to various libraries back in the day as I tracked down the relevant literature for some problem I was working on. I used to be quite good at photocopying. I still have my Boston Library Consortium ID card, even though it expired on 19 September 1993. These days even NASA is worried: "The current state of knowledge discovery includes use of available internet-based search tools and interlibrary loan processes to aid literature review and discovery of potential partners/collaborators. However, budget reductions have led to a decrease in access to knowledge portals (e.g. journal portals and database platforms). Hence, continued access to knowledge sources is an area of concern." That quote is from a report[12] coauthored by one of my former PhD students, Dr. Cara Leckey, who just spent 18 months as the NASA LaRC Center Transformation Portfolio Manager but is now back to her day job as Nondestructive Evaluation Sciences Branch Head.

In the olden days when you used to have to physically go to the library, there would be a whole wall of books called Science Citation Index where you could look up the listing for a paper and it would list what other papers had cited it. It was updated each year, of course. Aside from the vanity associated with who cited whom, the main utility was seeing who else was citing classic papers in your field these days so that you could look up their papers and see what they were working on and whether that was relevant to what you were working on. When you're writing a proposal to get funding for your clever new idea, you have to make sure that your clever idea is new. That requires scouring the literature to see who has already done what and who else is now doing similar things. These days it's a simple matter of clicking on a few buttons in Google Scholar, but don't neglect the important task of systematically exploring the literature so you can stand on the shoulders of the giants who have come before you. Don't forget to keep publishing, even if you're a tenured full professor. The same goes for NASA Branch Heads, which reminds me that you might like Dr. Leckey's recent book.[13]

I do want to make the point that it's not just the number of publications that matters, it's whether the things you publish matter. A simple measure of that is h-index, where h is the number of your publications that have been cited h or more times. The free software I use to calculate h-index is named *Publish or Perish*. That reminds me of a cartoon I've had on my bulletin board for decades. The caption reads, "Surely you were aware when you accepted the position, Professor, that it was publish or perish." Google that caption and you'll find the cartoon, which is all over the internet, or don't because we can all agree that school shootings are bad. In the cartoon, the Professor is getting up from his desk as the hitman standing next to the Dean is screwing the silencer to his pistol.

Journals can also be assigned an impact factor so that you can argue that the places you publish are way better than where your frenemies publish. Your Provost can have one of her many Associate Provosts keep track of the number of publications in high-impact journals that affect US News rankings, which reminds me of "Adam Ruins Everything - Why College Rankings Are A Crock" which you can find on YouTube.[14]

---

12 https://ntrs.nasa.gov/citations/20230007724.

13 Banerjee, S. and Leckey, C.A.C. (2020) *Computational Nondestructive Evaluation Handbook: Ultrasound Modeling Techniques*. CRC Press. https://books.google.com/books?id=ElzsDwAAQBAJ.

14 https://youtu.be/2unU4vCume0.

3

Boundary Conditions: Continuous and Discretized

The good news is that electromagnetic and acoustic and elastic waves all obey about the same wave equation, or at least we can manipulate the various field equations so that when push comes to shove the equation that we're left with each time is Dr. Helmholtz's. That means we don't have to invent the solutions, we just have to learn to properly use the solutions developed, over the last couple dozen decades or so, to the Helmholtz equation. I was surprised to learn that Helmholtz actually was a real doctor, that is, a professor of anatomy and physiology because that's what his parents made him choose for his major. After two decades of doing physics on the side, he finally switched departments.[1]

The bad news is that the boundary conditions are different for electromagnetic [1–6] acoustic [7–14] and elastic [15–26] waves. It's not so bad, though. We simply have to treat these three families of waves using proper boundary conditions, which I promise will make intuitive sense to you, eventually.

3.1 Boundary Conditions for E&M

Consider a boundary between two media, which we'll cleverly name *Medium 1* and *Medium 2*. They could be any two materials and the boundary could have any shape, but we'll assume that the boundary is locally sort of flat in order to be able to define a unit normal there, $\hat{n}$. The boundary conditions are then

$$\hat{n} \cdot (\vec{B}_2 - \vec{B}_1) = 0 \tag{3.1}$$

$$\hat{n} \times (\vec{E}_2 - \vec{E}_1) = 0 \tag{3.2}$$

$$\hat{n} \times (\vec{H}_2 - \vec{H}_1) = \vec{K} \tag{3.3}$$

$$\hat{n} \cdot (\vec{D}_2 - \vec{D}_1) = \omega \tag{3.4}$$

1 Hermann Ludwig Ferdinand von Helmholtz was a German physician and physicist who made significant contributions in several scientific fields. In physiology and psychology, he is known for his mathematics of the eye, theories of vision, ideas on the visual perception of space, color vision research, and on the sensation of tone, perception of sound, and empiricism in the physiology of perception. In physics, he is known for his theories on the conservation of energy, work in electrodynamics, chemical thermodynamics, and on a mechanical foundation of thermodynamics. As a philosopher, he is known for his philosophy of science, ideas on the relation between the laws of perception and the laws of nature, the science of aesthetics, and ideas on the civilizing power of science. https://en.wikipedia.org/wiki/Hermann_von_Helmholtz.

A Data Engineering Approach to Wave Scattering Analysis: with Applications in Radar, Sonar, Medical Diagnostics, Structural Flaw Detection and Intelligent Robotics, First Edition. Mark K. Hinders.

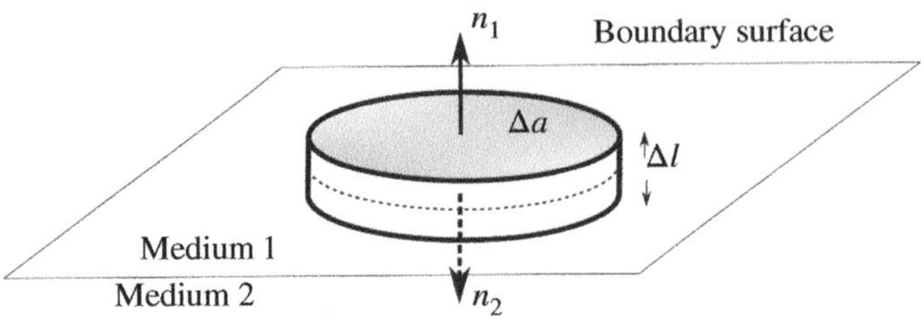

Figure 3.1 Pillbox that straddles the boundary between *Mediums 1* and *2* with surface normals $\hat{n}_1$ and $\hat{n}_2$. The thickness of the pillbox is Δl and the area of the top and bottom surfaces is Δa. Please note that these two media could be any combinations of solids, liquids, and/or gasses.

where $\vec{K}, \omega$ are surface current and charge densities. Note that normal magnetic flux and tangential electric fields are continuous, while the normal electric displacement and tangential magnetic fields are discontinuous.

To derive the first boundary condition relations, consider the pillbox shown in Figure 3.1. The infinitesimal pillbox spans the boundary between the two media, so that the top surface is just a bit on one side of the boundary and the bottom surface is just a bit on the other side. Now recall that $\nabla \cdot \vec{B} = 0$ holds for any point(s) and thus integrating $\nabla \cdot \vec{B}$ over the entire volume of the pillbox still gives hard zero because a volume integration is really nothing more than a summing up of the contributions of all the points in the volume, each of which is zero. Of course, a volume integral isn't all that helpful in writing boundary conditions, so we use the divergence theorem to convert the volume integral to a surface integral, which gives

$$\oint \vec{B} \cdot \hat{n} da = 0 \tag{3.5}$$

and this is approximately equal to

$$\left(\vec{B} \cdot \hat{n}_1 - \vec{B} \cdot \hat{n}_2\right) \Delta a + \text{Contribution from the sides} = 0 \tag{3.6}$$

Now shrink the already tiny pillbox and note that as it gets smaller and smaller the contribution from the sides will tend to zero both because the pillbox is getting really thin and because they will cancel themselves out as the pillbox shrinks down to the limit of a mathematical point. Please don't get hung up on atoms vs. no atoms. Take some Shrinx and just breathe. We then get

$$\left(\vec{B}_1 - \vec{B}_2\right) \cdot \hat{n} = 0 \tag{3.7}$$

Exercise 3.1 Go through the derivations for the other three boundary conditions. If you don't feel like doing that, look them up in some of the E&M books you've begun to collect for your personal library. I like Stratton.

That was pretty simple, assuming you're comfortable with the divergence theorem and the limiting process. The electromagnetics boundary conditions are also pretty simple to remember because the usual situation is simply that the tangential components of $\vec{E}$ and $\vec{H}$ are continuous across any boundary.

3.2 Boundary Conditions for Acoustics

Next we consider the boundary conditions for acoustics, where we have two different fluids called fluids 1 and 2, but we're still going to refer to Figure 3.1. This could be a boundary between air and water or between two liquids or whatever. It could even be a "boundary" inside of a large body of fluid where the temperature or salinity or some other property changes abruptly enough to look like a sharp discontinuity to an acoustic wave.

Be forewarned that some aspects of the boundary conditions for acoustics[2] won't make sense to you unless you have a background in fluid mechanics and especially aerodynamics. The issue has to do with viscosity and, as I may have mentioned, introducing viscosity makes things really complicated in a hurry. One of our most important rules is going to be to never make the model any more complicated than needed in order to capture the physics of interest because we'll often run the risk of formulating problems that are intractable. In much of aerodynamics, it isn't necessary to include things like compressibility and viscosity of the air. That's not to say that air is incompressible (because then sound waves wouldn't exist) and air is certainly viscous. For subsonic aerodynamics, it is very often the case that compressibility doesn't need to be included until the flow gets near Mach 1, which we call the transonic regime. Viscosity usually only has to be included in the so-called boundary layer right next to the surface of the airplane, and the rest of the flow field around an airplane behaves as if it isn't affected by viscosity at all. Ludwig Prandtl, the inventor of boundary layer theory, may have been a stereotypical absent-minded scientist, but he came up with viscous flow theory in order to help explain how a vacuum cleaner picks up a straw so that a better vacuum cleaner could be designed. The "drag" on a straw being sucked into a vacuum cleaner is mathematically the same problem as the "drag" on an airplane flying through the air. Also, we're not talking about the kind of plastic drinking straw that got stuck up a sea turtle's nose and caused plastic straws to be banned in Chicago so now you have to use paper straws, which sucks.

OK, so here's the part that makes sense about the boundary conditions for acoustics. Even though sound waves are vibrations, fluctuations, or whatever, we're going to insist that the two fluids not separate. That should make good sense to you because if they did separate that would create an empty volume between them (of zero pressure) and the pressure in the surrounding fluids would cause the fluid to flow back in to fill that void. Keep in mind that we're talking sound waves here, which are pretty mild in the scheme of things. Maybe if we were concerned about what was happening right at the lightning bolt that generates the clap of thunder, we'd have to deal with such situations, but probably not. Even sonic booms are simply large changes in density and the air doesn't actually come apart and clap back together. Now back to a pillbox (Figure 3.1) as we did for electromagnetics. Start by shrinking it down to a mathematical point so that we can talk about the fluid motion on either side of the boundary and since the two fluids don't separate, the motion on one side of the boundary has to match the motion on the other side. I hope you're with me so far because here's the part that won't make sense. We're going to only enforce the boundary condition on the normal component of the fluid velocity:

$$\hat{n} \cdot \vec{v}_1 = \hat{n} \cdot \vec{v}_2 \tag{3.8}$$

---

2 Allan D. Pierce is known for a large variety of fundamental research on the mechanics of waves, acoustics, and structural vibrations. His work encompasses atmospheric acoustics, nuclear test detection, sonic booms, interaction of sound with structures, waves on fluid immersed shells, diffraction and scattering of sound, the mechanics of marine sediments, and noise control. You should get yourself a copy of his landmark book, "Acoustics: An Introduction to its Physical Principles and Applications," which is widely considered the definitive graduate-level acoustics textbook. Before coming to Boston University in 1993 to take the position as Department Chair, where I had been since I was a frosh in 1982, Dr. Pierce held the Leonhard Chair in Engineering at Penn State, and prior to that he was Regents Professor at Georgia Tech. After graduate work in physics at MIT in quantum mechanics, Dr. Pierce went to work for the RAND Corporation where he began working in acoustics.

We aren't going to say anything about the tangential components of the velocities. As far as we're concerned, they can do whatever they like. They can even be discontinuous if they really want to. It's because we haven't included viscosity that the two fluids can slide tangentially across each other. With viscosity, they would stick and all three components of the velocity would be continuous. Adding viscosity is something we'll do from time to time, but in many acoustics problems, it gives excellent results to forget about it, so as long as we don't brainlock on the whole stick-slip tangential boundary condition, we'll be fine. Just keep in mind that the normal fluid velocity is continuous at the interface and forget the other components for now.

The other boundary condition turns out to make good sense, so consider the small pillbox in Figure 3.1 again. The pillbox surface isn't a real surface, but consider the pressure in the fluids outside of the pillbox. The pressure in fluid 1 is designated as p_1 and the pressure in fluid 2 is cleverly called p_2. Pressure is a scalar, so there's no issue with tangential components and such, but pressure varies spatially, so it's a scalar field. Unit of pressure is force per unit area so if we consider that the pressure acts all over the surface of the pillbox we can calculate the net force on the pillbox as pressure times surface area. Of course, force is a vector, but the pressure acts normal to the surface so we still don't have to worry about tangential components.

Now shrink the pillbox as before and since it's in equilibrium the net force is zero. The forces on the sides will cancel each other out as the pillbox shrinks and we'll be left with the equal and opposite forces on the top and bottom surfaces, which approach a mathematical point on the interface as the limiting process proceeds. The end result is that pressure is continuous across the boundary: $p_1 = p_2$.

Kind of a long story with a simple result that the normal fluid velocity and the pressure are continuous. Two special cases are worth mentioning, though, because they will come up quite a lot. First, if we consider a fluid in contact with a solid material then it's often the case that the sound wave will just reflect from the wall. Of course, it's an approximation which may not always hold, as you know well if you've lived in a dorm or apartment with neighbors who crank up the base on their stereo. Nevertheless, in many cases, the sound doesn't really penetrate into the solid and so the boundary condition is just that the normal fluid velocity is zero at the surface. In this case, there is no condition on the fluid pressure at the boundary and as usual we try to not think about what the tangential fluid velocity might be doing at the surface. Second, no matter how loud you scream under water, your friends standing by the pool can't hear you. The air is so much less dense than the water that it works just fine to approximate the boundary condition at the interface as $p = 0$. Like the wall, this free surface reflects all of the acoustic waves, but it does it in a different way because the pressure is zero instead of the normal fluid velocity. Indeed, for the free surface there is no boundary condition on any of the components of the fluid velocity.

It's often convenient to formulate our acoustic equations in terms of the velocity potential, in which case the boundary conditions become

$$\rho_1 \Phi_1 = \rho_2 \Phi_2 \qquad \hat{n} \cdot \nabla \Phi_1 = \hat{n} \cdot \nabla \Phi_2 \tag{3.9}$$

This makes it easy to see the difference between the "hard" and "soft" boundary conditions. For the rigid and/or infinitely dense wall where the acoustic wave doesn't disturb the solid surface at all the boundary condition is

$$\hat{n} \cdot \vec{v} = 0 \quad \text{or} \quad \hat{n} \cdot \nabla \Phi = 0 \tag{3.10}$$

which is a Neumann boundary condition. If the surface has zero density, that is, a vacuum, no acoustic wave can propagate in it because the pressure is zero and we have $p = 0$ or $\Phi = 0$, which is a Dirichlet boundary condition.

Exercise 3.2 Electromagnetic scattering folks like to say that acoustics is just a scalar version of E&M. Why do they think that? Hint: for most of the Cold War, most of the things that radar was scattering from were metals like aluminum, and were well approximated as perfect conductors.

3.3 Boundary Conditions for Elastodynamics

Boundary conditions for elastodynamics are more complicated than either acoustics or electrodynamics. We have both longitudinal and transverse wave fields and the two types are coupled at the boundary. Moreover, there are a variety of different approximations that we might want to make. Oh, and stress has more than three components which is why we have to write it as a tensor.

We'll start with the simplest case where the two elastic media, still cleverly named *Mediums 1* and *2* as in Figure 3.1, are in welded contact. That means that they remain in perfect contact and so all three components of the displacement vector must be continuous across the interface. Think of them as glued to each other if that makes more sense than welded contact, but no matter what words you use, it's pretty straightforward and we don't have to go through the odd mental process of ignoring some components of displacements. Also, don't be alarmed that in acoustics, we track the velocity field, while in elastodynamics we track the displacements. In frequency domain, they're just different by a factor of $i\omega$, which will show up on both sides of the boundary condition equations and so will cancel. In time domain, the velocity field is just the time derivative of displacement field so it's not really a problem.

OK, so continuous displacement gives us three boundary conditions, but it turns out we need three more. No big deal, as always we draw a little pillbox (Figure 3.1) straddling the boundary and require that it be in equilibrium such that the net force on it is zero. Of course in elastodynamics, we have stress instead of pressure, but similarly, force is stress times area. Not so similar is that stress is a tensor instead of a simple scalar field like pressure. When the pillbox shrinks down to a point the contributions from the sides are going to cancel out as before, so let's forget about those. We're then left with three components of stress acting on the top surface, with corresponding equal and opposite stress components acting on the bottom surface. One pair will be the normal stress that happily corresponds to the role of pressure in the acoustic case, so let's note for the record that the normal stress gets to be continuous across the interface. The other two pairs are the tangentially oriented shear stress components on the top and bottom surfaces. When the pillbox shrinks to zero size, those shear stresses are also continuous. That gives three boundary conditions for stress components, and the name for those three components is *normal surface tractions*. Now, I know what you're thinking. The stress tensor is symmetric, which means that there are really six independent components instead of a full nine, but we've only used three of them in the boundary conditions and so the other three stress components are free to do whatever they want at the interface. You're also a little worried that there's something lurking behind the scenes akin to fluid viscosity that I'm about to spring on you. Fear not, there isn't. It's a good thing, too, because the tensorial nature of the stress boundary conditions is going to make things pretty complicated. So complicated, in fact, that reflection and refraction is the particular place where the elastodynamic theory of the aether collapsed under its own weight in favor of the much better electromagnetic theory of light.

Exercise 3.3 Write the boundary condition equations for two half-spaces when (i) the boundary is the xy-plane, (ii) the boundary is the xz-plane, and (iii) the boundary is the yz-plane. Also write the boundary condition equations for a cylindrical interface $r = a$, where the coordinates are (r, θ, z) as well as a spherical interface $r = a$, where the coordinates are (r, θ, ϕ).

Now for the special cases that simplify things. For an elastic solid in welded contact with a material that is way more rigid and/or dense, the elastic waves will not penetrate into the second medium because it can't be deformed at all. Hence all three displacement components in the elastic solid will be zero at the interface. As you now expect from the similar acoustic reflection from a rigid wall, there is no constraint on the stress components. They can do whatever they like, I guess. The other special case is an elastic solid with a free surface, that is, a seismic wave getting to the surface of the earth. Air is so undense compared to rock that it might as well not even be there. To put it another way, a moonquake would reflect from the airless moon surface essentially the same way that an earthquake does here.[3] The boundary condition then is that the normal surface tractions are zero, which you'll recall is the normal stress and the two components of the shear stress that have the surface's unit normal in them. Please refer back to the previous exercise so you're clear on this point. We'll get in the habit of calling such an interface a free surface or just stress-free, but it's important to remember that doesn't mean all six components of the stress tensor are zero at the free surface. The in-plane normal and/or shear stress components might even be large. They just don't show up in the boundary conditions.

Of course, there are a variety of variations that one could introduce to model specific cases. For example, there might be situations where the two materials are held tightly together such that the normal stress component is continuous, but the shear stress components are not, or are even zero, because the two materials can slide tangentially across one another just a bit. Sounds like that would be unlikely, but "kissing bonds" are actually quite difficult to detect. They happen when a glue joint has disbonded over part of the interface, but the two sides are still held together. There are also a wide variety of situations where a glue joint has weakened somewhat due to manufacturing problems or unexpected aging and so some of the stresses won't be transmitted across the joint properly. That's obviously an issue for the overall strength of the structure, but might be something that can be detected because elastic waves will reflect and/or refract at the interface differently. There are ongoing academic arguments about the right way to write down the boundary conditions for such situations, but the problem is very real. Having a practical and reliable way to non-destructively assess bond strength would allow adhesive joints to be used much more commonly. Spot-welds, rivets, etc. are a much less efficient way of joining components, but adhesive bond strength is usually something that we have to test destructively by pulling on the joint until it fails.

At an interface between an elastic solid and a fluid, we'll find that it's often the case that ultrasonic waves both reflect and refract. In this case, the normal stress in the solid is balanced by the pressure in the fluid. The shear stress components of the normal surface traction in the solid vanish at the surface, unless it's a viscous fluid, of course, but let's not go there. Finally, the normal displacement in the elastic solid is balanced by the normal displacement in the fluid, or if you prefer, the normal velocity in the elastic solid is balanced by the normal velocity in the fluid. Do it either way, but be consistent. Sonar people tend to do it one way, ultrasonic NDT folks the other. In frequency domain, you'll just have a factor of $i\omega$ hanging around that will cancel. And, as you now have come to expect, there won't be any conditions on the tangential components of the displacements or velocities at the interface.

3 The same goes for Mars where there's so little atmosphere that nobody can hear you scream. https://www.science.org/content/article/mars-no-one-can-hear-you-scream.

3.4 Finite Difference Time Domain

In acoustics, electromagnetics, and elastodynamics, we have differential equations (plus boundary conditions and initial conditions) that we want to solve for the "scattered" field. What we'll find is that there are a fairly restricted set of problems that we can actually solve analytically, and even when we are able to grind out the solutions, we often find the "answers" to be such complex mathematical expressions that we have to code them up and evaluate them systematically to get much understanding about the physical behavior. That begs the obvious question of why we don't just do scattering problems via numerical simulations. I have to admit that's a fair point, but offer the following arguments for consideration.

- No matter how big of a computer you have access to for simulating wave propagation and scattering sorts of problems, you're going to want a bigger and faster one, and if it's a shared supercomputer, you'll complain about how long your jobs wait in the queue before they run.
- The physical behavior we're trying to understand is inherently complex enough that we often won't be able to tell whether simulation results make sense or are due to some sort of typo or bug in our code.

I've watched with special interest since the late 1980s (Figure 3.2) as the sorts of computers used for scattering calculations have increased their capabilities dramatically. Back in the day, a VAX computer with a footprint the size of a conference room was just as fast as a 286 PC, except that VAX typically had several (or many) users while you could sometimes have a PC all to yourself without even having to share it. In a moderate-sized research lab, it was common practice for people to politely ask everybody else to "please not run any jobs this weekend" so you could have all of the computing power to try to get your job(s) to finish. Long holiday weekends were often especially nice for this, although if you were trying to run several distinct cases, you often had to physically go to the lab throughout the weekend to check the status of the jobs you hoped would finish and then start new ones. Being able to queue up lots of jobs in a scheduler and then track their progress from home seems too easy somehow, especially since simulations always seem to demand attention at time intervals that are incompatible with a traditional human sleep-wake cycle. At least we don't

Figure 3.2 Here I am in my Air Force office about 1990 or so. It took me a couple of years of steady effort to scrounge up a complete set of matching gray steel office furniture.

have to pedal the computers, but we do always seem to be spending a lot of wall-clock time waiting for results. The reason these sorts of simulations require so much attention – even after we think we have our codes debugged and ready for systematic use – is that we have to "look at" the results to see if the code is running properly. Even a code that we've benchmarked and tested thoroughly can behave strangely when we use it in a new scattering regime. Not only that, often the actual scattering behavior can closely mimic the results you would get from a variety of numerical bugs. Getting too greedy with the spatial and/or temporal simulation grid(s) can also cause rather serious problems, and more worrisome is that they can change the results in subtle but significant ways. That means we must have something to compare our simulation results to, and analytic scattering solutions for canonical shapes turn out to be the answer.

As of 22 July 2024 my sense is that computers that researchers can afford to use routinely have gotten to the point that we can simulate real 3D problems. That's good news, of course, because wave scattering is inherently three dimensional. It is a bit of a game changer, though, because there are lots and lots of numerical techniques predicated on 2D somehow being the true limit of such things. Those techniques were boring and unintuitive anyway, so I'm not really going to talk about them much. Instead, let's start with a conceptually simple, straightforward way of simulating wave propagation and scattering. You may recall first learning about the concept of differentiation as a limit of a finite difference. I happen to remember that particular lecture. I was sitting in about the back row of a lecture hall that sat several hundred students, with the professor down in front writing on an overhead projector with the clear plastic rolls that he could scroll backward to clarify some prior point. The lecture hall was a domed theater built in 1906 as Temple Israel and intended by the architect as a replica of Solomon's Temple. The building is clad in white marble and today much of it is covered in ivy, although BU is on the other side of the Charles River from the Ivy League. We're simply going to do that differences-to-derivatives lecture backward, that is, we'll replace the *derivatives* in our field equations and boundary conditions with the corresponding *differences*. We'll grid up the space in which the waves are propagating, including the scatterer, and then march away in time. Alan Taflove cleverly called this sort of method finite-difference time-domain (FDTD).[4]

Here is the definition of derivatives you learned way back in calculus.

$$
\begin{aligned}
\frac{df(x)}{dx} &= \lim_{h\to 0}\frac{f(x+h)-f(x)}{h} \\
&= \lim_{h\to 0}\frac{f(x)-f(x-h)}{h} \\
&= \lim_{h\to 0}\frac{f(x+h)-f(x-h)}{2h}
\end{aligned}
\tag{3.11}
$$

where $h = x_{i+1} - x_i$ and in the limit these three different ways of doing the difference are equivalent. They're called the forward-difference, backward-difference, and central-difference approximations, for the obvious reasons (Figure 3.3).

4 Allen Taflove received the BS, MS, and PhD degrees in electrical engineering from Northwestern University in 1971, 1972, and 1975, respectively. Since 1988, he was a full professor in the Department of Electrical Engineering and Computer Science of Northwestern's McCormick School of Engineering. He died at age 71 on 25 April 2021. A dedicated teacher and adviser of undergraduate students, Allen was the first McCormick School faculty member to be named both Teacher of the Year and Adviser of the Year in the same academic year (2005–2006). Allen was inducted into the Amateur Radio Hall of Fame by CQ Magazine in recognition of his achievements in computational electrodynamics. He was an FCC-licensed amateur radio operator since 1963, and credited amateur radio with spurring his interest in electrical engineering in general, and electromagnetic fields and waves in particular.

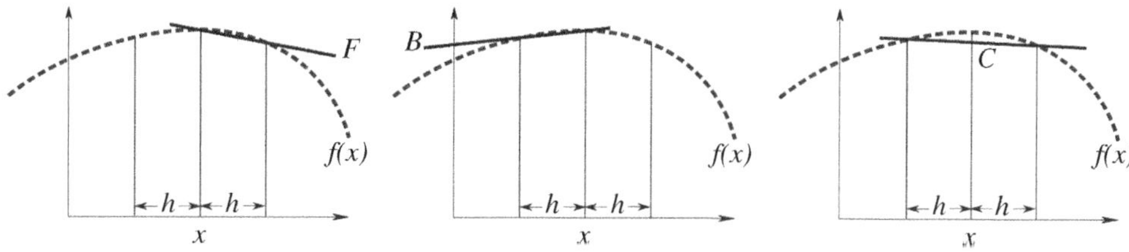

Figure 3.3 Forward (*F*), backward (*B*), and central (*C*) difference approximations each approximate the slope of the curve at *x* in the limit of $h \to 0$.

Now consider $u(x)$, which is a differentiable function of x, and expand it in a Taylor series about x. We write

$$u(x+h) = u(x) + u'(x)h + \frac{u''(x)h^2}{2!} + \frac{u'''(x)h^3}{3!} + \cdots \tag{3.12}$$

and

$$u(x-h) = u(x) - u'(x)h + \frac{u''(x)h^2}{2!} - \frac{u'''(x)h^3}{3!} + \cdots \tag{3.13}$$

Add these two equations to get

$$u(x+h) + u(x-h) = 2u(x) + u''(x)h^2 + O(h^4) \tag{3.14}$$

so we can write, accurate to $O(h^4)$

$$u''(x) = \frac{1}{h^2}[u(x+h) - 2u(x) + u(x-h)] + O(h^4) \tag{3.15}$$

or, with $x = x_i, x \pm h = x_{i\pm1}$, we write

$$u_i'' = \frac{u_{i+1} - 2u_i + u_{i-1}}{h^2} + O(h^4) \tag{3.16}$$

where $u_i \equiv u(x_i)$ and so on. Subtracting the expression for $u(x-h)$ from that for $u(x+h)$ similarly gives

$$u_i' = \frac{u_{i+1} - u_{i-1}}{2h} + O(h^3) \tag{3.17}$$

These two quite simple expressions allow us to replace all the first and second derivatives in our differential equations and boundary conditions with finite differences and then write approximate equations to simulate wave propagation and scattering. Of course, we have to chop up our simulation space into a 3D grid and also take some care to match the approximate boundary conditions at the scatterer surface. The obvious issues with approximating curved surfaces with stair-steps are illustrated in Figure 3.4. For scattering problems, the grid has to be able to reproduce the relevant geometry of the scatterers, but there also has to be a certain number of grid points per wavelength of the highest-frequencies of the waves of interest. Of course, there will be some implementation issues, but conceptually the FDTD method is exactly this straightforward.

Exercise 3.4 Write out the algebra to get the expressions for $u'(x)$ and $u''(x)$ above. Is there any advantage to keeping more terms in the Taylor series to try to get more accurate expressions?

That was all one-dimensional, but it works pretty much the same way in two and three dimensions, except that there might be some advantage to using a coordinate system other than Cartesian. For example, if your primary application was pipeline inspection it might be worth your while to

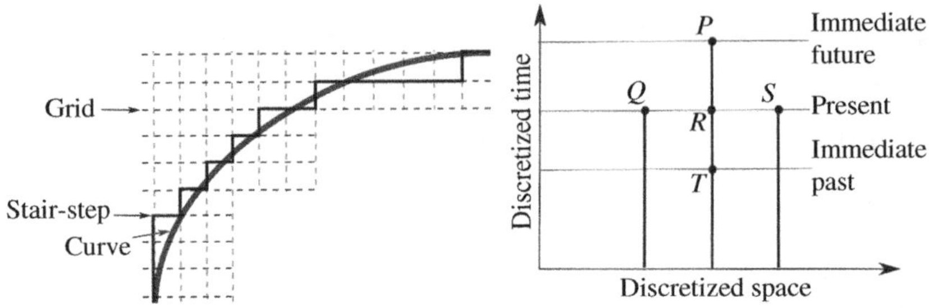

Figure 3.4 The physical space is divided up into a computational grid (left) so any curved surface has to be approximated by a stair-step. Of course, a finer grid will better approximate the curved surface, but that comes at the expense of computational time and necessary data storage. The future value at any node point P depends on the current value at that point R as well as the values of the neighboring points Q and S. It also depends on the prior value at that point T.

discretize things via a cylindrical coordinate system. When you think about it for a moment it's obvious that a cylindrical mesh is going to do a better job of fitting a pipe than will a Cartesian grid, although if your grid gets fine enough, it isn't going to matter much. Again, it's my duty as an old-timer to remind you that, no matter how big a computer you have, you'll want a bigger one, so adapting the mesh to the natural geometry of the problem probably makes good sense. There will be some distortions of the wave propagation that are due to the spatial grid, and minimizing them will come with some computational expense.

Exercise 3.5 What happens to a spherical wave on a rectangular grid?

Now note that the values of the field variables at any point are going to depend only on those adjacent points that are nearest in space and time. The set of nodal points that need to be considered when calculating the value at a particular point is called a stencil. It's often convenient to draw the stencil as a way to describe the particular differencing scheme that is being used.

Consider the 1D wave equation $c^2u'' - \ddot{u} = 0$, where $-\infty < x < \infty$ and $0 \leq t < \infty$ and we have initial conditions $u(x, 0) = f(x)$ and $\dot{u}(x, 0) = g(x)$. For $x = x_i$ and $t = t_j$ we write this as:

$$\frac{c^2}{h^2}\left[u_{i+1,j} - 2u_{i,j} + u_{i-1,j}\right] = \frac{1}{k^2}\left[u_{i,j+1} - 2u_{i,j} + u_{i,j-1}\right] \tag{3.18}$$

which can be rearranged to write, with $\lambda = ck/h$

$$u_{i,j+1} = 2u_{i,j} - u_{i,j-1} + \lambda^2\left[u_{i+1,j} - 2u_{i,j} + u_{i-1,j}\right] \tag{3.19}$$

with the initial condition

$$u_{i,0} = f_i \qquad u_{i,1} = kg_i + f_i \qquad -\infty < i < \infty, \quad j = 0, 1, 2, \ldots \tag{3.20}$$

Note that we have used the forward difference approximation for the second initial condition. With P, Q, R, S, T defined as in Figure 3.4, we write

$$u_P = \lambda^2 u_Q + 2(1 - \lambda^2)u_R + \lambda^2 u_S - u_T \tag{3.21}$$

So, the next value of the wave displacement, u_P, depends on the value at that point at the current time step, u_R, as well as the two neighboring values at the current time step, u_Q and u_S. It also depends on the value at that point one time step ago, u_T, something that strikes most people as strange.

Next consider the traditional continuous version of the scalar wave equation

$$\frac{1}{c^2}\frac{\partial^2 u(\vec{x},t)}{\partial t^2} = \nabla^2 u(\vec{x},t) \tag{3.22}$$

For a plane wave solution of the form

$$u(\vec{x},t) = e^{i(\vec{k}\cdot\vec{x}-\omega t)} \tag{3.23}$$

and we have the dispersion equation $\omega^2 = c^2k^2$ as well as the phase velocity $c_p^2 = \frac{\omega^2}{k^2} = c^2$ where $k^2 = \vec{k}\cdot\vec{k}$. Hence the wave propagation is isotropic and nondispersive, which means that the phase velocity is independent of $\hat{k}$ and $\left|\vec{k}\right|$, respectively. In two Cartesian dimensions, the scalar wave equation is written

$$\frac{\partial^2 u}{\partial t^2} = c^2\left(\frac{\partial^2 u}{\partial x^2} + \frac{\partial^2 u}{\partial y^2}\right) \tag{3.24}$$

If we discretize the space over a square mesh, we can write this as:

$$\frac{d^2 u_{m,n}}{dt^2} = \frac{c^2}{h^2}\left(u_{m-1,n} + u_{m+1,n} + u_{m,n-1} + u_{m,n+1} - 4u_{m,n}\right) = \Diamond u_{m,n} \tag{3.25}$$

where we haven't discretized time at this point. We've used the symbol $\Diamond$ to indicate the stencil in Figure 3.5. Note that this is the obvious way to calculate the differences, but as we'll see shortly there are many other options.

Now let's discretize time as well, and in order to do a better job of keeping track of things, let's get into the excellent habit of using subscripts for the spatial grid points and superscripts for the time step. That seems to be the way most people do it and it's something that I can even keep track of when I'm writing things out by hand. I've said it before but it bears repeatedly repeating that our equations are usually pretty complicated, so it's worth it to think through notations so that they will help you catch inevitable botches.[5]

$$u_{m,n}^{j+1} = -u_{m,n}^{j-2} + 2u_{m,n}^{j} + (c\Delta t/h)^2 \Diamond u_{m,n}^{j} \tag{3.26}$$

Note that the quantity $c\Delta t/h$ is called the Courant number, and $c\Delta t/h \to 0$ returns the continuum limit as expected. In general, the Courant number should be small, something which we'll curse loudly and often. The spatial discretization, h, is sized both by the size of the features we want to model accurately and the Nyquist-like rule of thumb of needing ten or so grid points per wavelength. Hence, complex shapes or even simple shapes that don't match the grid well will be computationally expensive. All other things being equal, higher frequencies will be more computationally expensive because the wavelength shrinks as the frequency increases. Nobody ever really argues with those two points, but now look at the Courant number and note that h is on the denominator. Keeping the Courant number small means that if the spatial grid is very fine then we're going to have to take baby steps in time. That alone will make you long for a faster computer.

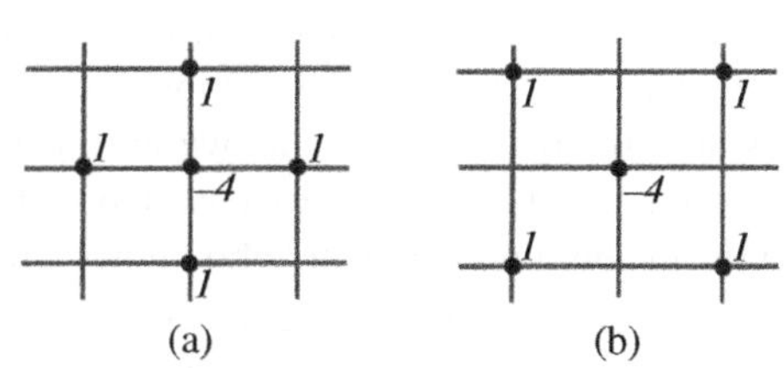

Figure 3.5 The diamond stencil $\Diamond$ (a) uses the grid points to the left/right and top/bottom. The box stencil $\square$ (b) uses the diagonal grid points at the corners of the box.

5 Botches are inevitable. You can only eliminate typos and such asymptotically, so don't be paralyzed by the idea of publishing something with a botch left in it. There might even be one in this next equation, say in the second term.

Next let's write the semidiscrete version of a plane-wave expansion and plug it into the semidiscrete wave equation. We start with

$$u_{m,n}(t) = a(t)e^{\iota\omega(x_m \cos\alpha + y_n \sin\alpha)} \tag{3.27}$$

and get, with $\omega_x = \omega\cos\alpha$ and $\omega_y = \omega\sin\alpha$

$$\begin{aligned}\frac{d^2u_{m,n}}{dt^2} &= c^2 a\left[\frac{(2\cos(\omega_x h) - 2) + (2\cos(\omega_y h) - 2)}{h^2}\right] \\ &= -(\omega c)^2 a\left[\cos^2\alpha\left(\frac{\sin(\omega_x h/2)}{\omega_x h/2}\right)^2 + \sin^2\alpha\left(\frac{\sin(\omega_y h/2)}{\omega_y h/2}\right)^2\right]\end{aligned} \tag{3.28}$$

$$a(t) = a(0)e^{\pm i\omega c^* t} \tag{3.29}$$

where the numerical phase velocity of harmonic plane waves is

$$c^* = c\frac{\sqrt{2}}{\omega h}\left(2 - \cos(\omega_x h) - \cos(\omega_y h)\right)^{1/2} \tag{3.30}$$

and we have

$$u_{m,n} = a(0)e^{i\omega(x_m \cos\alpha + y_n \sin\alpha \pm c^*(\omega,\alpha))} \tag{3.31}$$

Because $c^* = c^*(\omega, \alpha)$, it is clear that the wave propagation is anisotropic. How much the phase velocity depends on direction depends on the wavelength compared to the mesh spacing. Once again, we come back to needing lots of grid points per wavelength in order to do an excellent job of approximating reality.

Now consider the same problem, but with a slightly different stencil (Figure 3.5) using the diagonal neighbors. The semi-discrete wave equation is

$$\frac{d^2u_{m,n}}{dt^2} = \frac{c^2}{2h^2}\left(u_{m+1,n+1} + u_{m+1,n-1} + u_{m-1,n+1} + u_{m-1,n-1} - 4u_{m,n}\right) \tag{3.32}$$

The corresponding phase velocity is given by

$$c^*(\omega, \alpha) = c\frac{\sqrt{2}}{\omega h}\left[1 - \cos(\omega_x h)\cos(\omega_y h)\right]^{1/2} \tag{3.33}$$

which is also anisotropic, but the phase velocity varies with direction in a different way for the □ stencil than it did for the ◇ stencil. For both cases, the waves propagate differently if they happen to be headed in a direction that's aligned with the grid vs. some other direction. That's a bigger problem for us than you might think because we care about scattering problems where things might start out lined up to the mesh, but the scattering is going to happen in pretty much all directions. For most real-world problems that we're going to simulate, the scattered field is going to be pretty subtle and we might be looking for variations that are somehow of the same magnitude as distortions of this sort. Even for strongly scattering situations, the scattered waves are going to tend to decay spherically with distance from the scatterer, so detecting things from "far away" will still be a challenge. In short, we have enough to worry about without numerical artifacts messing up our results. You might suggest that we simply solve the problem via a spherical grid centered about the scatterer, but the incident wave is then going to get distorted by that spherical grid and distortions of the incident wave will do who knows what to the scattered field. It will cost us computer power to fix this, as you might have guessed already.

Rather than inventing yet another stencil, let's just put these two stencils together as follows

$$\frac{d^2u_{m,n}}{dt^2} = c^2\left[\beta\Box u_{m,n} + (1-\beta)\Diamond u_{m,n}\right] \tag{3.34}$$

where β is a parameter between 0 and 1. For $\beta = 1/4$ and $\beta = 3/4$, the results don't look too great, but for $\beta = 1/2$ the results look damn good!

Actually if you do more sophisticated analyses, you find that the best results are obtained when $\beta = 0.4698$, which I'm just a little bit skeptical of because I was expecting a round number or something.

In three dimensions you can do the same thing, but it's a little more complicated. We define three discrete Laplacian operators as:

$$\nabla^2_{100} u^t_{l,m,n} = \left\{ \frac{u^t_{l+1,m,n} - 2u^t_{l,m,n} + u^t_{l-1,m,n}}{h^2} + \frac{u^t_{l,m+1,n} - 2u^t_{l,m,n} + u^t_{l,m-1,n}}{h^2} + \frac{u^t_{l,m,n+1} - 2u^t_{l,m,n} + u^t_{l,m,n-1}}{h^2} \right\} \tag{3.35}$$

and

$$\begin{aligned}\nabla^2_{110} u^t_{l,m,n} = \frac{1}{2} \left\{ \frac{u^t_{l+1,m+1,n} - 2u^t_{l,m,n} + u^t_{l-1,m,n}}{(\sqrt{2}h)^2} + \frac{u^t_{l-1,m+1,n} - 2u^t_{l,m,n} + u^t_{l+1,m-1,n}}{(\sqrt{2}h)^2} \right. \\ + \frac{u^t_{l,m+1,n+1} - 2u^t_{l,m,n} + u^t_{l,m-1,n-1}}{(\sqrt{2}h)^2} + \frac{u^t_{l,m-1,n+1} - 2u^t_{l,m,n} + u^t_{l,m+1,n}}{(\sqrt{2}h)^2} \\ \left. + \frac{u^t_{l+1,m,n+1} - 2u^t_{l,m,n} + u^t_{l-1,m,n-1}}{(\sqrt{2}h)^2} + \frac{u^t_{l-1,m,n+1} - 2u^t_{l,m,n} + u^t_{l+1,m,n-1}}{(\sqrt{2}h)^2} \right\}\end{aligned} \tag{3.36}$$

and

$$\begin{aligned}\nabla^2_{111} u^t_{l,m,n} = \frac{3}{4} \left\{ \frac{u^t_{l+1,m+1,n+1} - 2u^t_{l,m,n} + u^t_{l-1,m-1,n-1}}{(\sqrt{3}h)^2} + \frac{u^t_{l-1,m+1,n+1} - 2u^t_{l,m,n} + u^t_{l+1,m-1,n-1}}{(\sqrt{3}h)^2} \right. \\ \left. + \frac{u^t_{l-1,m-1,n+1} - 2u^t_{l,m,n} + u^t_{l+1,m+1,n-1}}{(\sqrt{2}h)^2} + \frac{u^t_{l+1,m-1,n+1} - 2u^t_{l,m,n} + u^t_{l-1,m+1,n-1}}{(\sqrt{3}h)^2} \right\}\end{aligned} \tag{3.37}$$

If we then write the semi-discrete 3D wave equation as

$$\frac{1}{c^2}\frac{\partial^2 u}{\partial t^2} = \left(\alpha \nabla^2_{100} + \beta \nabla^2_{110} + \gamma \nabla^2_{111}\right) u \tag{3.38}$$

where $\alpha + \beta + \gamma = 1$ so that $h \to 0$ returns the continuum limit, we write the phase velocity as:

$$\begin{aligned}c^* = \frac{c^2}{k^2h^2} \{ 2\alpha \left[3 - \cos(kh \sin\theta \cos\phi) - \cos(kh \sin\theta \sin\phi) - \cos(kh \cos\theta)\right] \\ + \beta \left[3 - \cos(kh \sin\theta \cos\phi)\cos(kh \sin\theta \sin\phi) - \cos(kh \sin\theta \cos\phi)\cos(kh\cos\theta) \right. \\ \left. - \cos(kh \sin\theta \sin\phi)\cos(kh \cos\theta)\right] \\ + 2\gamma \left[1 - \cos(kh \sin\theta \cos\phi)\cos(kh \sin\theta \sin\phi)\cos(kh \cos\phi)\right] \}\end{aligned} \tag{3.39}$$

Minimizing anisotropy then gives $\alpha = 0.299$, $\beta = 0.461$, and $\gamma = 0.240$.

Exercise 3.6 Plot the phase velocity as 3D parametric plots for $\alpha = 1, \beta = \gamma = 0$; $\alpha = 0, \beta = 1, \gamma = 0$; $\alpha = \beta = 0, \gamma = 1$. Then use $\alpha = 0.299$, $\beta = 0.461$, and $\gamma = 0.240$, and you should get a perfect sphere.

So now we've established that discretized wave propagation is dispersive and anisotropic unless you make the computational mesh really small or employ some more complex stencil. Both of those are computationally pretty expensive, so it turns out that getting high-fidelity simulations means no matter how large a computer you're using you're going to want a bigger one. Whatever computer(s) you're using, even if it's the biggest cluster of them on earth, will only be able to deal with a finite number of grid points, which you're keeping track of at every time step in the simulation, and if you've got a complicated stencil the effective number of grid points might be larger than the explicit number of points in your computational space. Remember that we're trying to solve scattering problems where the fields scattered by the target are typically small compared to the incident wave. With a finite-sized computational space, the incident waves will reflect off the edges and if you're not careful you could end up overwhelming the signal you care about with reflections you don't care at all about. A flashlight can be a really bad way to find your way around in a dark house of mirrors. You'll almost certainly end up shining the light back into your eyes and then making faceprints onto a wall before you turn the flashlight off and just feel your way around. Of course, you could imagine having such a very big computer that you can leave some extra room so the echoes from the edges never bother you, but I have to tell you that's pretty unlikely. Instead, what you need is to make the nonphysical walls of your simulation volume not reflect the waves but instead absorb them. Absorbing boundary conditions sometimes work pretty well, but they are hard to make work in 3D for all angles. For elastic waves, it's tricky to deal with both longitudinal and shear waves simultaneously and you have to be careful not to kill off the bulk waves only to generate nonphysical surface waves at the boundary. Nevertheless, it usually pays to spend some of your computational power to damp out outgoing waves that you don't want if you can do that without introducing spurious reflected waves.

That's probably enough introduction and warnings from an old-timer who fondly remembers FORTRAN77. You can worry about most of these issues once you get a basic code up and running. Radar still matters, of course, but the killer application for electromagnetic scattering these days seems to be the new generations of wireless communication. Since electromagnetic wave interaction with the built environment is highly dependent on frequency, changes in the carrier frequency of the signal can have a strong impact on the reach of the network and a device's ability to maintain communications [27–29]. One approach to optimizing the effectiveness and coverage of a wireless network is simulating wave propagation and scattering. You'll be unsurprised to hear that one of the most widely used methods for this is the finite difference time domain (FDTD) method, which was introduced by Kane S. Yee in 1966 [30]. I recommend Taflove's excellent book [31] for details. You should be able to find a used copy quite inexpensively. Obviously, a lot has changed since 2005 and there are lots of more recent books on the subject, but by starting with this classic you will then be in a position to critically evaluate other "more modern" treatments that may include chunks of code in whatever programming language(s) you are comfortable with.

Let's go all old school for a bit and consider Yee's 1966 work.[6] It's good practice to get into the habit of benchmarking your own code(s) with seminal results that have stood the test of time.

---

6 Kane S. Yee was born in Guangzhou, China. He received his BS and MS in electrical engineering from UC Berkeley in 1957 and 1958, and PhD with Bernard Friedman in 1963. His dissertation involved the study of boundary value problems for Maxwell's equations. Dr. Yee first worked at Lockheed Missiles and Space Company, researching diffraction of electromagnetic waves and then was a professor of electrical engineering and mathematics at the University of Florida and later at Kansas State University. He was a consultant to Lawrence Livermore National Laboratory and in 1987, he became a research scientist at Lockheed Palo Alto Research Lab, working on computational electromagnetics problems and retiring in 1996. Surprisingly, his 1966 paper on the use of a finite-difference staggered-grid algorithm in the solution of Maxwell's equations received little attention at the time of its release. Yee was initially motivated by his self-studies in FORTRAN to develop the method.

Maxwell's equations in an isotropic medium are

$$\frac{\partial \vec{B}}{\partial t} + \nabla \times \vec{E} = 0 \qquad \vec{B} = \mu \vec{H}$$

$$\frac{\partial \vec{D}}{\partial t} - \nabla \times \vec{H} = \vec{J} \qquad \vec{D} = \epsilon \vec{E}$$

where $\vec{J}, \mu, \epsilon$ are assumed to be given functions of space and time.

In a rectangular coordinate system, Maxwell's equations are written

$$-\frac{\partial B_x}{\partial t} = \frac{\partial E_z}{\partial y} - \frac{\partial E_y}{\partial z}$$

$$-\frac{\partial B_y}{\partial t} = \frac{\partial E_x}{\partial z} - \frac{\partial E_z}{\partial x}$$

$$\frac{\partial B_z}{\partial t} = \frac{\partial E_x}{\partial y} - \frac{\partial E_y}{\partial x}$$

$$\frac{\partial D_x}{\partial t} = \frac{\partial H_z}{\partial y} - \frac{\partial E_y}{\partial z} - J_x$$

$$\frac{\partial D_y}{\partial t} = \frac{\partial H_x}{\partial z} - \frac{\partial E_z}{\partial x} - J_y$$

$$\frac{\partial D_z}{\partial t} = \frac{\partial H_y}{\partial x} - \frac{\partial E_x}{\partial y} - J_z$$

Now denote a grid point in space as $(i, j, k) = (i\Delta x, j\Delta y, k\Delta z)$ and write, for any function $F(i\Delta x, j\Delta y, k\Delta z, n\Delta t) = F^n_{i,j,k}$.

For many of the most historically significant applications of electromagnetic scattering (e.g. using radar to keep us safe from bad guys trying to sneak into our airspace with bombs and such), it's an excellent approximation to assume that scatterers made from metal are perfect electrical conductors. The boundary conditions are thus $E_{tang} \equiv 0$ and if we assume that conducting surfaces are well approximated by a collection of "stair-step" surfaces parallel to the grid axes, we'll find that plane surfaces perpendicular to the x-axis will be chosen so as to contain points where E_y and E_z are defined.

A set of finite difference equations convenient for 3D simulations are written out fully on the following page. For constant values of ϵ and μ computational stability requires that

$$\sqrt{(\Delta x)^2 + (\Delta y)^2 + (\Delta z)^2} > c\Delta t = \frac{\Delta t}{\sqrt{\mu\epsilon}}$$

If μ, ϵ aren't constant, then the maximum value of the speed of light anywhere in the computational space will determine the limits on the time step relative to the spatial grid size. The mesh size is limited by the need to accurately describe the shapes of the scatterer(s) with a stair-step boundary as well as the requirement that the EM field not change significantly from one grid point to the next, which in practice means that the grid spacing must be a small fraction of a wavelength.

$$H_x\big|^{n+1/2}_{i,j,k} = \left(\frac{1 - \dfrac{\rho'_{i,j,k}\Delta t}{2\mu_{i,j,k}}}{1 + \dfrac{\rho'_{i,j,k}\Delta t}{2\mu_{i,j,k}}}\right) H_x\big|^{n-1/2}_{i,j,k} + \left(\frac{\dfrac{\Delta t}{\mu_{i,j,k}}}{1 + \dfrac{\rho'_{i,j,k}\Delta t}{2\mu_{i,j,k}}}\right)$$
$$\times \left(\frac{E_y\big|^{n}_{i,j,k+1/2} - E_y\big|^{n}_{i,j,k-1/2}}{\Delta z} - \frac{E_z\big|^{n}_{i,j+1/2,k} - E_z\big|^{n}_{i,j-1/2,k}}{\Delta y}\right) \tag{3.40a}$$

$$H_y\Big|_{i,j,k}^{n+1/2} = \left(\frac{1-\frac{\rho'_{i,j,k}\Delta t}{2\mu_{i,j,k}}}{1+\frac{\rho'_{i,j,k}\Delta t}{2\mu_{i,j,k}}}\right) H_y\Big|_{i,j,k}^{n-1/2} + \left(\frac{\frac{\Delta t}{\mu_{i,j,k}}}{1+\frac{\rho'_{i,j,k}\Delta t}{2\mu_{i,j,k}}}\right) \times \left(\frac{E_z\big|_{i+1/2,j,k}^{n} - E_z\big|_{i-1/2,j,k}^{n}}{\Delta x} - \frac{E_x\big|_{i,j,k+1/2}^{n} - E_x\big|_{i,j,k-1/2}^{n}}{\Delta z}\right) \tag{3.40b}$$

$$H_z\Big|_{i,j,k}^{n+1/2} = \left(\frac{1-\frac{\rho'_{i,j,k}\Delta t}{2\mu_{i,j,k}}}{1+\frac{\rho'_{i,j,k}\Delta t}{2\mu_{i,j,k}}}\right) H_z\Big|_{i,j,k}^{n-1/2} + \left(\frac{\frac{\Delta t}{\mu_{i,j,k}}}{1+\frac{\rho'_{i,j,k}\Delta t}{2\mu_{i,j,k}}}\right) \times \left(\frac{E_x\big|_{i,j+1/2,k}^{n} - E_x\big|_{i,j-1/2,k}^{n}}{\Delta y} - \frac{E_y\big|_{i+1/2,j,k}^{n} - E_y\big|_{i-1/2,j,k}^{n}}{\Delta x}\right) \tag{3.40c}$$

$$E_x\Big|_{i,j,k}^{n+1} = \left(\frac{1-\frac{\sigma_{i,j,k}\Delta t}{2\varepsilon_{i,j,k}}}{1+\frac{\sigma_{i,j,k}\Delta t}{2\varepsilon_{i,j,k}}}\right) E_x\Big|_{i,j,k}^{n} + \left(\frac{\frac{\Delta t}{\varepsilon_{i,j,k}}}{1+\frac{\sigma_{i,j,k}\Delta t}{2\varepsilon_{i,j,k}}}\right) \times \left(\frac{H_z\big|_{i,j+1/2,k}^{n+1/2} - H_z\big|_{i,j-1/2,k}^{n+1/2}}{\Delta y} - \frac{H_y\big|_{i,j,k+1/2}^{n+1/2} - H_y\big|_{i,j,k-1/2}^{n+1/2}}{\Delta z}\right) \tag{3.41a}$$

$$E_y\Big|_{i,j,k}^{n+1} = \left(\frac{1-\frac{\sigma_{i,j,k}\Delta t}{2\varepsilon_{i,j,k}}}{1+\frac{\sigma_{i,j,k}\Delta t}{2\varepsilon_{i,j,k}}}\right) E_y\Big|_{i,j,k}^{n} + \left(\frac{\frac{\Delta t}{\varepsilon_{i,j,k}}}{1+\frac{\sigma_{i,j,k}\Delta t}{2\varepsilon_{i,j,k}}}\right) \times \left(\frac{H_x\big|_{i,j,k+1/2}^{n+1/2} - H_x\big|_{i,j,k-1/2}^{n+1/2}}{\Delta z} - \frac{H_z\big|_{i+1/2,j,k}^{n+1/2} - H_z\big|_{i-1/2,j,k}^{n+1/2}}{\Delta x}\right) \tag{3.41b}$$

$$E_z\Big|_{i,j,k}^{n+1} = \left(\frac{1-\frac{\sigma_{i,j,k}\Delta t}{2\varepsilon_{i,j,k}}}{1+\frac{\sigma_{i,j,k}\Delta t}{2\varepsilon_{i,j,k}}}\right) E_z\Big|_{i,j,k}^{n} + \left(\frac{\frac{\Delta t}{\varepsilon_{i,j,k}}}{1+\frac{\sigma_{i,j,k}\Delta t}{2\varepsilon_{i,j,k}}}\right) \times \left(\frac{H_y\big|_{i+1/2,j,k}^{n+1/2} - H_y\big|_{i-1/2,j,k}^{n+1/2}}{\Delta x} - \frac{H_x\big|_{i,j+1/2,k}^{n+1/2} - H_x\big|_{i,j-1/2,k}^{n+1/2}}{\Delta y}\right) \tag{3.41c}$$

For clarity and so, we can make fun of what was possible computationally in 1966, let's restrict ourselves for the moment to 2D. Consider the TM case. When $t = 0$ we specify the initial conditions

$$E_z^0(i,j) \qquad H_y^{1/2}(i+1/2,j) \qquad H_x^{1/2}(i,j-1/2)$$

throughout the entire grid. Boundary conditions for perfect conductors are $E_z^n(i,j) = 0$ for all n for all the grid points corresponding to the scatterer surface. The incident wave is taken to be a half sine wave and the square scatterer has sides of length 4α as shown in Figure 3.6.

We were easily able to reproduce Yee's figures[7] by plotting the value of the E_z component of the wave at various time steps using our 3D FDTD code, the results of which are shown in Figure 3.6.

7 "We" means Dr. Margaret Rooney, who is a Research Mathematician at the Naval Research Laboratory. She holds a BS in Mathematics from St. John's University as well as MS and PhD degrees in Applied Science from William and Mary. FDTD simulations of 5G wireless interaction with the built environment were a key aspect of her dissertation research.

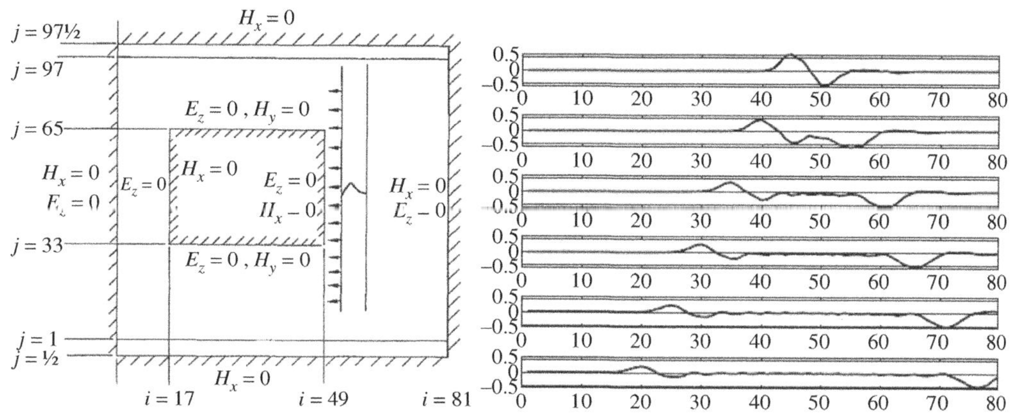

Figure 3.6 Reproduction of Yee's Fig 4 on right. The value of the E_z field in volts/meter is plotted over the 80 horizontal spatial steps in the computational grid. It's good practice to make sure that your new code accurately reproduces classic results in the literature that have stood the test of time.

Due to the appreciable advances in technology in the more than 50 years since Yee's original investigation, these calculations now take mere seconds when run on modern desktop computers.

It bears repeating that the numerical stability of the finite difference method relies on the sizes of the spatial and temporal steps. The choice of a suitable spatial step size depends on the frequencies of the source signal because an accurate simulation of the signal propagation requires spatial steps to be several times smaller than the wavelength of the highest frequency component. For simulations using the higher mmWave frequencies, Δ must be on the order of millimeters to properly characterize signal environment interactions. This change in spatial step proportionally increases the size of the computational grid, often resulting in a much more computationally intensive simulation. That was a bit of an understatement.

So, the size of our computational grid is necessarily finite and rather severely constrained by the reasonable limits imposed by the spatial step to an edge length of several (or even many) meters. Therefore, it is necessary to implement absorbing boundary conditions (ABCs) to prevent reflections from the edges of the grid from reentering the computation space. In the decades since the FDTD method of solving Maxwell's equations was first introduced by Yee, many types of absorbing boundary conditions have been developed to enhance the accuracy of simulations by allowing waves to exit the computational space cleanly, thereby giving the illusion that the waves are propagating through an infinite space.

We performed a comparison between two widely used types of ABCs: Mur's ABCs [32] coupled with Mei-Fang's superabsorption equations [33], and a uniaxial perfectly matched layer (PML) [34] based on Béreneger's PML [35–37]. Our goal was to assess which method would be more practical to use for larger and more complex simulation scenarios. We found [38] there was no significant difference between the reflections allowed by the two types of ABCs, so we decided to proceed using the FDTD version with Mur ABCs. You'll want to revisit this question yourself as you develop your own FDTD code.

Using a parallel computing cluster, we can directly compare measurements and 3D FDTD simulations [38]. Software-defined radio units (SDRs) are used to transmit and receive signals, which then interact with the built environment. The motivation for this work is the newer generations of wireless communication networks at higher frequencies where understanding in some detail the RF interaction with the built environment is key to optimizing network performance.

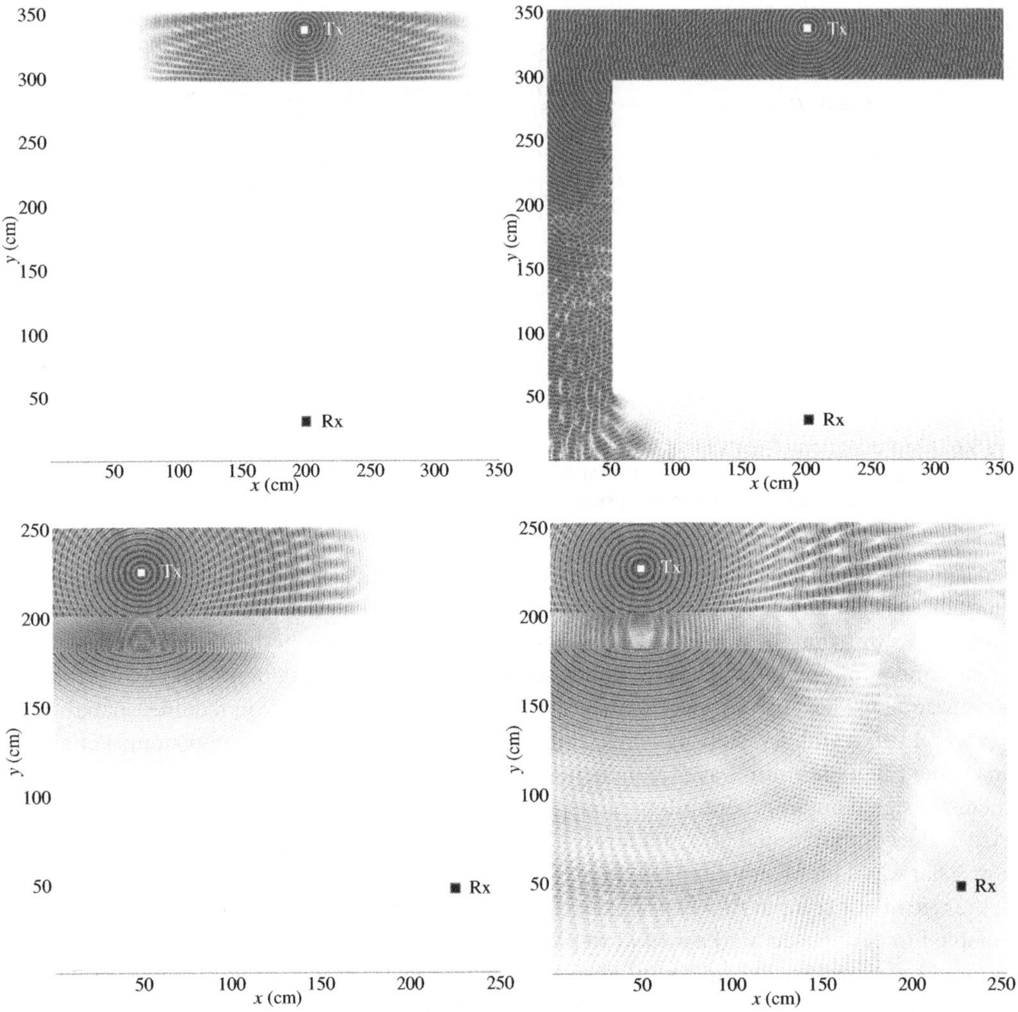

Figure 3.7 Snapshots of simulations where mmWave signals interact with corners in the environment. Top pair is for a food truck with metal skin. Bottom pair is for a corner of a building with a brick facade. Metal is a perfect reflector at mmWave frequencies, whereas brick reflects some of the signal and transmits some, as we will consider in Chapter 4.

In Figure 3.7, we show early and later-time snapshots from simulations where mmWave signals interact with corners. We were able to use these simulations to compare directly to measurements [39].

Our goal is to understand the scattering sufficiently well that we can identify subtle time-domain features in waveforms to distinguish between signal occlusion, by a truck or building corner, from signal termination. Cognitive radio is a way to anticipate and exploit spectrum vacancies in order to make optimal use of the available bandwidth [40–49]. Allowing unlicensed spectrum users to transmit on underpopulated or empty licensed frequency bands encourages efficient use of EM radio spectrum resources. Detecting frequency bands not currently in use by licensed primary users (PUs) gives unlicensed secondary users (SUs) an opportunity to transmit on these channels until the PU resumes its transmission, at which time the SU can switch to another vacant channel if it needs to continue transmission. A key issue in CR and 5G wireless is understanding interactions of RF with the built environment, which means our hard-won knowledge of radar scattering analysis and simulation methods is (again) suddenly important.

3.5 Elastodynamic Simulations

Guided ultrasonic waves are attractive for aircraft structural health monitoring because they are quite sensitive to thinning due to corrosion. In Figure 3.8, Lamb waves in an aluminum plate, with and without a rectangular thinned region, are shown.[8] Most analytic methods only account for the wave speed changes due to thickness changes, but simulations illustrate nicely the complex scattering of the guided waves, which occur even in this rather simple scenario. The elastodynamic finite integration technique (EFIT) evolves from the basic wave equations for elastic solids [50–58]. We start with Hooke's law and Cauchy's equation of motion to give the fundamental equations. The differential form of the equation of motion

$$\rho\dot{v}_x = \frac{\partial\sigma_{xx}}{\partial x} + \frac{\partial\sigma_{xy}}{\partial y} + \frac{\partial\sigma_{xz}}{\partial z} + f_x \tag{3.42}$$

$$\rho\dot{v}_y = \frac{\partial\sigma_{xy}}{\partial x} + \frac{\partial\sigma_{yy}}{\partial y} + \frac{\partial\sigma_{yz}}{\partial z} + f_y \tag{3.43}$$

$$\rho\dot{v}_z = \frac{\partial\sigma_{xz}}{\partial x} + \frac{\partial\sigma_{yz}}{\partial y} + \frac{\partial\sigma_{zz}}{\partial z} + f_z \tag{3.44}$$

and the first time derivative of Hooke's law in differential form

$$\dot{\sigma}_{ij} = \lambda\dot{\epsilon}_{kk}\delta_{ij} + 2\mu\dot{\epsilon}_{ij}, \qquad (i,j = x,y,z) \tag{3.45}$$

where we sum over the repeated index k because that's the tensor index. We can then discretize (3.42)–(3.45) as follows:

$$\rho\dot{v}_x^{(n)}(t) = \frac{\sigma_{xx}^{(n+\hat{x})}(t) - \sigma_{xx}^{(n)}(t)}{\Delta x} + \frac{\sigma_{xy}^{(n)}(t) - \sigma_{xy}^{(n-\hat{y})}(t)}{\Delta y} + \frac{\sigma_{xz}^{(n)}(t) - \sigma_{xz}^{(n-\hat{z})}(t)}{\Delta z} + f_x(t)$$

$$\rho\dot{v}_y^{(n)}(t) = \frac{\sigma_{xy}^{(n)}(t) - \sigma_{xy}^{(n-\hat{x})}(t)}{\Delta x} + \frac{\sigma_{yy}^{(n+\hat{y})}(t) - \sigma_{yy}^{(n)}(t)}{\Delta y} + \frac{\sigma_{yz}^{(n)}(t) - \sigma_{yz}^{(n-\hat{z})}(t)}{\Delta z} + f_y(t)$$

$$\rho\dot{v}_z^{(n)}(t) = \frac{\sigma_{xz}^{(n)}(t) - \sigma_{xz}^{(n-\hat{x})}(t)}{\Delta x} + \frac{\sigma_{yz}^{(n)}(t) - \sigma_{yz}^{(n-\hat{y})}(t)}{\Delta y} + \frac{\sigma_{zz}^{(n+\hat{z})}(t) - \sigma_{zz}^{(n)}(t)}{\Delta z} + f_z(t)$$

$$\dot{\sigma}_{xx}^{(n)}(t) = (\lambda + 2\mu)\frac{v_x^{(n)}(t) - v_x^{(n-\hat{x})}(t)}{\Delta x} + \lambda\left(\frac{v_y^{(n)}(t) - v_y^{(n-\hat{y})}(t)}{\Delta y} + \frac{v_z^{(n)}(t) - v_z^{(n-\hat{z})}(t)}{\Delta z}\right)$$

$$\dot{\sigma}_{yy}^{(n)}(t) = (\lambda + 2\mu)\frac{v_y^{(n)}(t) - v_y^{(n-\hat{y})}(t)}{\Delta y} + \lambda\left(\frac{v_x^{(n)}(t) - v_x^{(n-\hat{x})}(t)}{\Delta x} + \frac{v_z^{(n)}(t) - v_z^{(n-\hat{z})}(t)}{\Delta z}\right)$$

$$\dot{\sigma}_{zz}^{(n)}(t) = (\lambda + 2\mu)\frac{v_z^{(n)}(t) - v_z^{(n-\hat{z})}(t)}{\Delta z} + \lambda\left(\frac{v_x^{(n)}(t) - v_x^{(n-\hat{x})}(t)}{\Delta x} + \frac{v_y^{(n)}(t) - v_y^{(n-\hat{y})}(t)}{\Delta y}\right)$$

8 These simulation snapshots are from the doctoral dissertation of Cara Leckey, who is currently the Branch Head of the Nondestructive Evaluation Sciences Branch at NASA Langley. Cara holds a BS in Physics from Mary Washington as well as MS and PhD in Physics from William and Mary. At LaRC she has served as the Project Lead for the High Performance Computing Incubator as well as the Center Transformation Portfolio Manager and a Detail with NASA Langley Legislative Affairs.

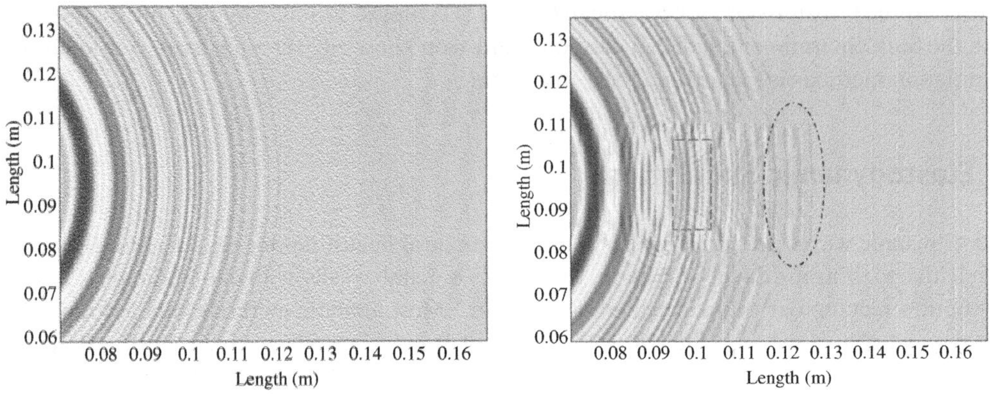

Figure 3.8 Lamb waves in an aluminum plate. Left image is an unflawed plate. Right image is a plate with a rectangular thinning indicated by the dashed rectangle. Note that the wavefront is distorted by interacting with the thinned region, which both scatters the Lamb waves and changes the wave speed.

$$\dot{\sigma}_{xy}^{(n)}(t) = \mu\left(\frac{v_x^{(n+\hat{y})} - v_x^{(n)}}{\Delta y} + \frac{v_y^{(n+\hat{x})} - v_y^{(n)}}{\Delta x}\right)$$

$$\dot{\sigma}_{xz}^{(n)}(t) = \mu\left(\frac{v_x^{(n+\hat{z})} - v_x^{(n)}}{\Delta z} + \frac{v_z^{(n+\hat{x})} - v_z^{(n)}}{\Delta x}\right)$$

$$\dot{\sigma}_{yz}^{(n)}(t) = \mu\left(\frac{v_y^{(n+\hat{z})} - v_y^{(n)}}{\Delta z} + \frac{v_z^{(n+\hat{y})} - v_z^{(n)}}{\Delta y}\right). \tag{3.46}$$

We can invoke stress-free boundary conditions in our simulation boundaries when our interest is in guided elastic waves in structures at MHz frequencies where the solid-air interfaces can be considered traction-free surfaces. That minimizes the need for absorbing boundary conditions, but as always, we must satisfy the Courant–Friedrichs–Levy-criterion by having about 10 grid points per shear wavelength. In practice this means that for a sample that is about 1 mm thick and in the 1 MHz frequency range, we need on the order of 10 grid points per millimeter.

We have also used EFIT simulations to investigate the effects of a limpet mine on a guided waves propagating in ship's hull where the ship hull acts as a thick steel plate and the limpet mine can be estimated as a 3 kg mass adhered to the plate.[9] Figure 3.9 shows the simulation space geometry of the mass loading on the plate as well as the elastic guided wave propagating in the expected manner.

Two more real-world applications of EFIT bear mentioning. The first is for rail-mounted gantry cranes at the Port of Virginia [59]. Depressions in the roadbed cause dips in the rail which lead to flat spots in the wheels which stress the wheel bearings and by the time excess current draw shows up in the electric drive motors the fix can be as much as US 75k. Picking up vibrations propagating down the rail allows imperfections to be detected much earlier when tamping the roadbed costs US 750. EFIT simulations allow us to understand the guided wave propagation in the rails to design machine learning strategies to identify flaw signals as early as possible. The second is an underwater sonar beacon which is deployed beneath Arctic ice floes where oil spills have been detected. The

9 Dr. Jill Bingham grew up in Norfolk, VA and then after earning a BA in Physics from Carleton College (and playing Ultimate) she came back to Virginia and earned a PhD in Applied Science. After a little more than 2 years at the Naval Research Laboratory, Dr. Bingham moved to Seattle where she has been a BR&T NDI Engineer at Boeing since 2011.

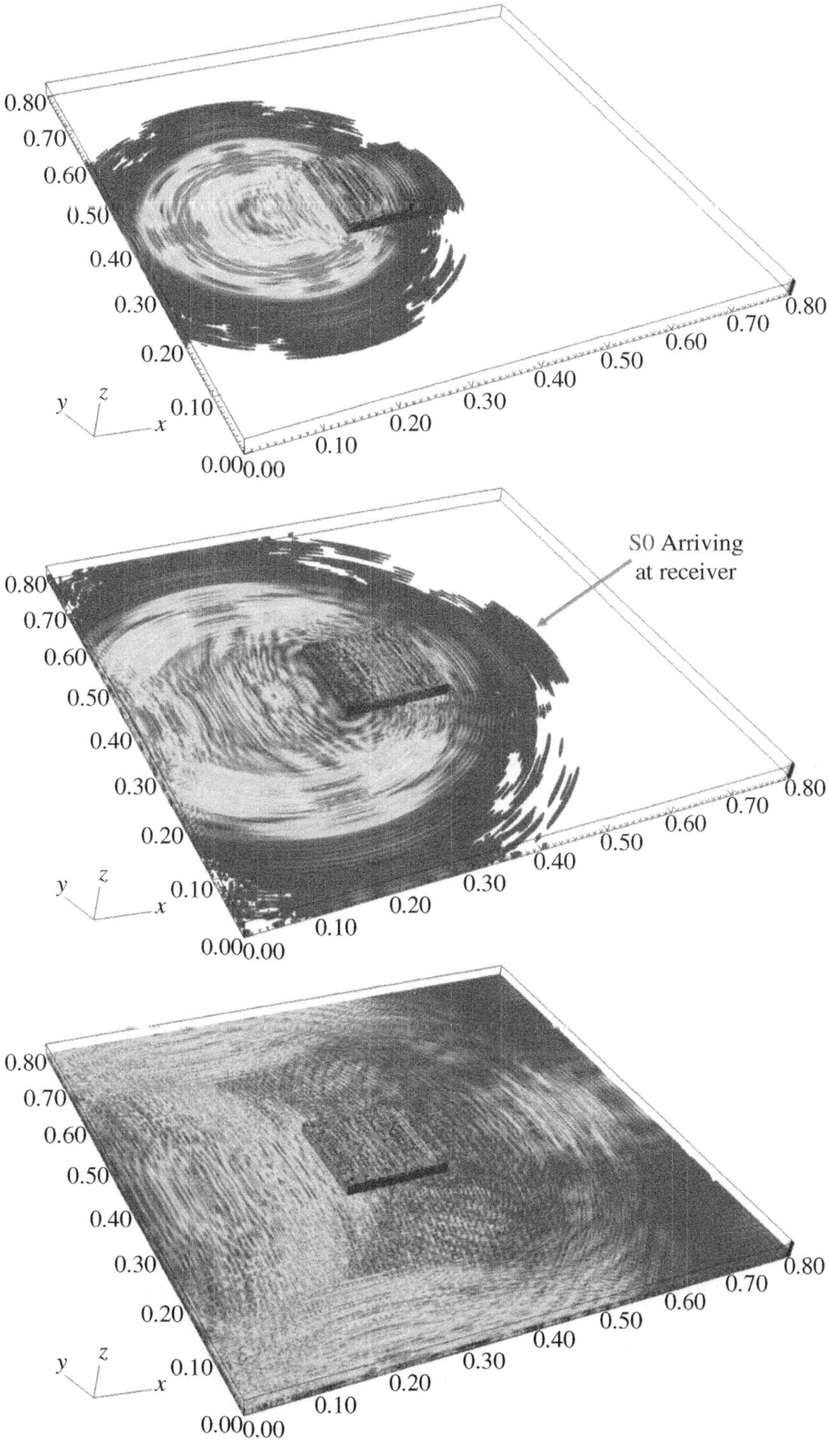

Figure 3.9 As the EFIT simulation steps through time, the guided waves interact with a mass loading on the steel plate and keep on propagating through the plate with distortions due to scattering and then interactions with the edge of the finite plate.

beacon transmits kHz sonar signals up into the ice [60] deflected by an integral reflector cone to preferentially generate Lamb wave modes in the ice, which spread out in all directions to be picked up by air-dropped mesh-radio accelerometer modules that triangulate the source location and then transmit that information to stakeholders via an Iridium satellite link. FDTD simulations were used to optimize the design of the sonar beacon/reflector. As I may have said, it's new things all the time in my lab and we like to solve problems that matter out in the real world.

3.6 The Acoustic Parametric Array

We have been investigating the use of passive infrared and acoustic echolocation sensors for mobile robotics applications [61–65] including classifying oncoming vehicles using acoustic backscattering. We are most interested in the applicability of acoustic echolocation sensors as medium- to long-range sensors, detecting and classifying objects at distances exceeding 50 m. Such long propagation distances limit the upper frequency of the acoustic signal, since the absorption of sound in air is proportional to the frequency squared, requiring an acoustic signal in the audible range.

A directional signal will allow us to focus most of the sound energy on the target in question, increasing the amplitude of the backscattered signal and the probability that the signal contains useful information about the target. The directivity of a speaker depends on the ratio of the wavelength of sound produced to the physical size of the speaker. Sound in the audible range will have wavelengths between 17 m and 17 mm, calculated as $\lambda = \frac{c}{f}$ for frequency f and sound speed $c = 343$ m/s in air. This explains why a normal bookshelf-sized stereo speaker will produce directional high-frequency sound, since the wavelength is much less than the size of the speaker, but nondirectional room-filling low-frequency sound. Creating highly directional, low-frequency sound would require an impractically large array. Even 1–4 kHz acoustic signals contain wavelengths of 8.5–34 cm, requiring a loudspeaker array several meters across in order to create a highly directional signal.[10]

A practical way to create highly directional low-frequency sound is the acoustic parametric array [66–87], which exploits the nonlinearity of air to create a highly directional beam of sound, even at low frequencies. Physically, the device is comprised of many small ultrasound transducers that simultaneously produce waves over a range of ultrasonic frequencies, in pairs. As the large-amplitude ultrasonic signals propagate through a nonlinear medium, nonlinear effects create signals at the sum and difference frequencies. Since attenuation is proportional to the square of frequency, at large propagation distances only the difference (audible) frequency remains.

Because the parametric array is only emitting ultrasonic signals, it can be fairly small yet create directional beams of low-frequency sound. These devices are commercially available and can connect to mp3 players or computers using a standard 1/8″ phone jack, though high-fidelity reproduction of music is difficult due to the narrow bandwidth inherent in the design of the devices.

---

10 Elwood "Woody" G. Norris won the 2005 Lemelson-MIT Prize as Inventor of the Year. He is widely recognized as a prolific inventor and a talented technology integrator. Mr. Norris joined the U.S. Air Force in 1956 and was trained as a Nuclear Weapons Specialist, specializing in electronic fusing systems. He studied electronics at the University of New Mexico. In 1959, he began working at the University of Washington where he continued his practical education, elevating his position from technician to Director of the Engineering Experiment Station within two years. HyperSonic Sound (HSS) is his trade name for a directional loudspeaker that shapes the sound wave to fill only a predetermined area much as a spotlight narrows its beam. Woody founded American Technology Corporation (now LRAD), which is publicly traded and is marketing products to customers around the world.

We began exploring the use of parametric arrays on the day after 9/11. It was clear almost immediately that better airport screening technologies were needed, and focused acoustic waves held promise for both penetrating clothing and reflecting back from secreted, nonmetallic contraband. The parametric array provides a way to do this in a nonimaging, stand-off manner which is key to a variety of civilian and military checkpoint scenarios. Understanding the complex nonlinear acoustic propagation and then the 3D scattering when the difference-frequency beam interacts with the person, led us to do extensive simulations of a variety of scenarios of interest.

After the so-called underwear bomber, in order to make you feel safe while flying, the TSA began the practice of touching your junk at the airport. Complaining that didn't make you feel safe would get you taken into a back room where the TSA would touch up inside your junk and then you'd miss your flight. A noncontact, modesty-preserving method to rapidly screen for underwear bombs seemed like an obvious win-win. Figure 3.10 shows a sequence of snapshots from a simulation of the underwear bomber scenario [88].

To describe this nonlinear sound propagation, we start with the Westervelt equation

$$\nabla^2 p = \frac{1}{c_0^2}\frac{\partial^2 p}{\partial t^2} - \frac{\delta}{c_0^4}\frac{\partial^3 p}{\partial t^3} - \frac{\beta}{\rho_0 c_0^4}\frac{\partial^2 p^2}{\partial t^2} \tag{3.47}$$

Here, δ is the sound diffusivity, described in terms of the shear viscosity μ, bulk viscosity μ_B, thermal conductivity k, and specific heats at constant volume and pressure c_v and c_p

$$\delta = \frac{1}{\rho_0}\left(\frac{4}{3}\mu + \mu_B\right) + \frac{k}{\rho_0}\left(\frac{1}{c_v} - \frac{1}{c_p}\right) \tag{3.48}$$

The coefficient of nonlinearity, β, is described in terms of the measured nonlinearity parameter B/A [89]

$$\beta = 1 + \frac{B}{2A} \tag{3.49}$$

Higher values of β (and likewise, B/A) correspond to greater nonlinear effect. The value of B/A for air at 20 °C is 0.4, compared to a value of 5 in water at the same temperature. This is why propagation of sound in air can be explained as a linear phenomenon in many cases. On the other hand, body fat has a B/A value of 9.9, requiring a nonlinear wave equation to accurately describe the wave propagation in body tissues [90, 91], assuming you want to pre-screen rollercoaster riders for BMI.

A more mathematical description of nonlinear wave propagation comes from the Khokhlov–Zabolotskaya–Kuznetsov (KZK) equation

$$\frac{\partial^2 p}{\partial z \partial \tau} = \frac{c_0}{2}\nabla_\perp^2 p + \frac{\delta}{2c_0^3}\frac{\partial^3 p}{\partial \tau^3} + \frac{\beta}{2\rho_0 c_0^3}\frac{\partial^2 p^2}{\partial \tau^2} \tag{3.50}$$

which describes the propagation of a directional sound beam along spatial dimension z while accounting for diffraction, thermoviscous absorption (δ), and nonlinearity (β). The KZK equation is a parabolic approximation which makes the assumption that effects due to diffraction are much larger than those due to nonlinearity, that is, that transverse changes to the wave are much larger than axial changes. This approximation introduces errors at more than 20° from the axis and within several radii of source. Alternative models exist to more accurately model wave behavior very near the source, but our typical applications do not require us to know the pressure near field.

Numerical simulations of the KZK equation [92, 93] allow us to visualize the nonlinear beam of sound produced by the acoustic parametric array. This gives us some idea of the extent of the beam and its interaction with the environment, allowing us to ensure that most of the sound energy

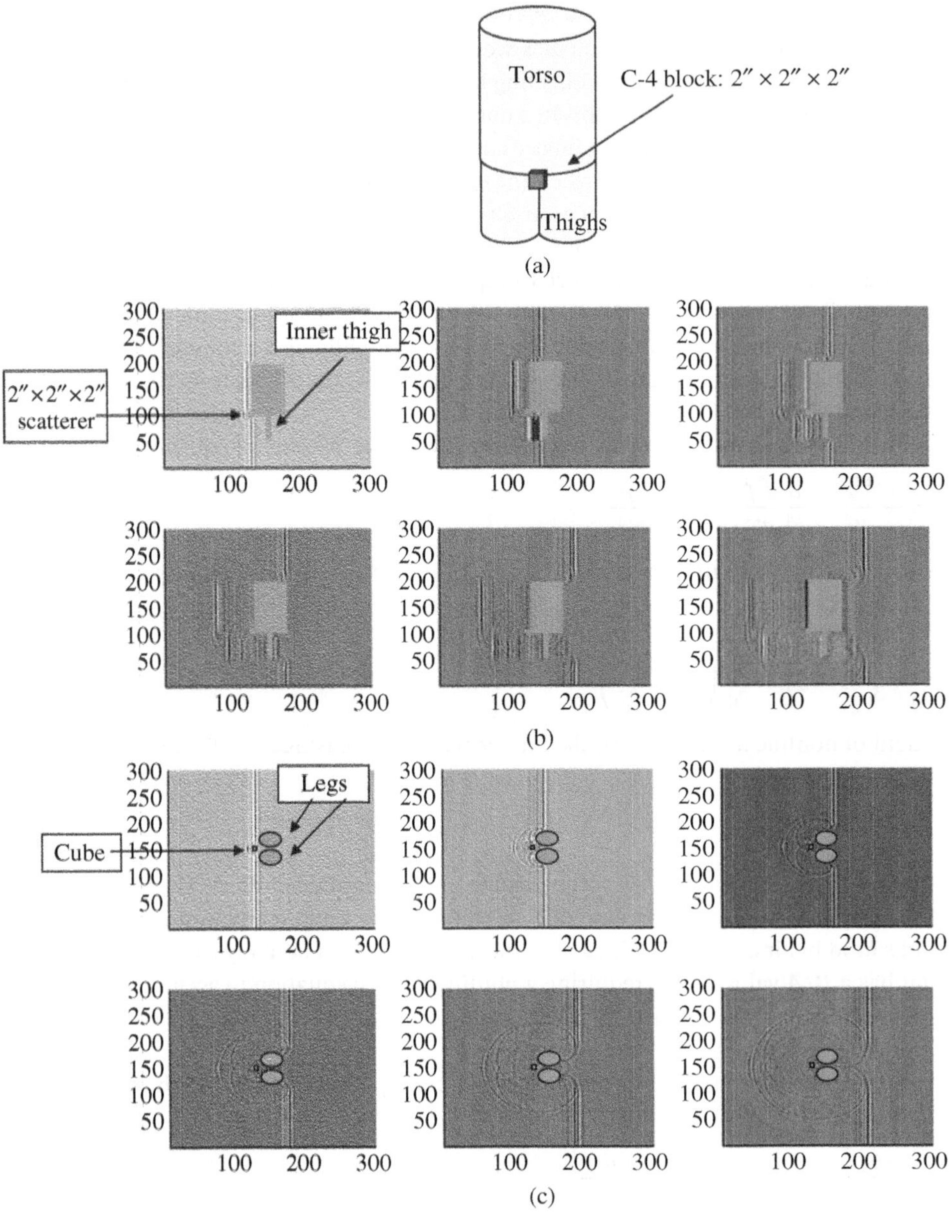

Figure 3.10 Side and top views of acoustic waves scattering from a model torso with a simulated underwear bomb. The goal is to detect concealed explosive devices from beyond the "lethal radius," which I take to mean the distance beyond which the suicide bomber doesn't get his junk all over you when he detonates the bomb. Not shown in this figure are results of the test we performed on backscattering from a simulated underwear bomb. Yes, one of our engineers did agree to wear one and be scanned, but I'm assured that no photos of that day exist. (a) Geometric model of torso and legs with C4, (b) vertical slice, and (c) horizontal slice.

comes from interaction with the target of interest and not from interaction with extraneous objects in the environment. In order to gain a better understanding of the interaction of the acoustic beam with the target, we then switch to a 3D linear, full-wave numerical simulation [94].

We start with the linearized conservation of mass and momentum, generalized to include pressure and velocity source functions M and $\vec{F}$, respectively:

$$\frac{\partial p}{\partial t} + \rho_0 c_0^2 \nabla \cdot \vec{v} = M \tag{3.51}$$

$$\rho_0 \frac{\partial \vec{v}}{\partial t} + \nabla p = \vec{F} \tag{3.52}$$

A finite difference simulation would approximate these derivatives directly, but instead we integrate over a control volume, which is a cube in Cartesian space. For the pressure, we use the standard central-difference

$$\dot{p}^{(t)} = \frac{p^{(t+\frac{\Delta t}{2})} - p^{(t-\frac{\Delta t}{2})}}{\Delta t} \tag{3.53}$$

with time step Δt to arrive at

$$p^{(t+\frac{\Delta t}{2})} = p^{(t-\frac{\Delta t}{2})} - \rho_0 c_0^2 \frac{\Delta t}{\Delta x}\left[\left(v_1^{1+} - v_1^{1-}\right) + \left(v_2^{2+} - v_2^{2-}\right) + \left(v_3^{3+} - v_3^{3-}\right)\right] + M\Delta t \tag{3.54}$$

while for the velocity, we use an integer indexed central difference

$$v^{(t)} = v^{(t-\Delta t)} + \dot{v}^{(t-\frac{\Delta t}{2})}\Delta t \tag{3.55}$$

with spatial step Δx to arrive at

$$\begin{aligned} v_1^{(t)} &= v_1^{(t-\Delta t)} - \frac{\Delta t}{\rho_0 \Delta x} - \left(p^{1+} - p^{1-}\right) + F_1 \frac{\Delta t}{\rho_0} \\ v_2^{(t)} &= v_2^{(t-\Delta t)} - \frac{\Delta t}{\rho_0 \Delta x} - \left(p^{2+} - p^{2-}\right) + F_2 \frac{\Delta t}{\rho_0} \\ v_3^{(t)} &= v_3^{(t-\Delta t)} - \frac{\Delta t}{\rho_0 \Delta x} - \left(p^{3+} - p^{3-}\right) + F_3 \frac{\Delta t}{\rho_0} \end{aligned} \tag{3.56}$$

This discrete set of equations for pressure and velocity given by (3.54) and (3.56) provides updates to the staggered grid in space and time. The size of the spatial step Δx (which is also the size of a single grid cell) is limited by the upper frequency in the simulation. In general, maintaining stability requires at least six grid points per wavelength. For our simulations, we use eight points per wavelength, so the minimum step size is $\Delta x = ds = \frac{\lambda}{8}$. The minimum time step is related to the spatial step size by the Courant condition

$$\Delta t = dt \le \frac{\Delta x}{c\sqrt{3}} \tag{3.57}$$

To study the acoustic scattering from vehicles, we created models of real-world scattering objects and imported them into our computational space. Visualizations of the scattered pressure field allow us to study how a scatterer's shape affects the backscattered reflection, providing specific information that will add to our intuitive knowledge of scattering behavior.

Direct simulations of the 100 and 250 ms chirp signals we use in our measurements are not practical on the distributed computing resources available to us. Instead, we use a delta pulse as input so that the scattered pressure field is effectively the impulse response of the system. Since this is a linear system once the nonlinear conversion to difference frequency is complete, convolution of the impulse response pressure field with any signal will provide the pressure field for that incident signal. Using this technique to create the whole backscattered pressure field would require a convolution at each point in the volume and is not a computationally efficient procedure. However, we

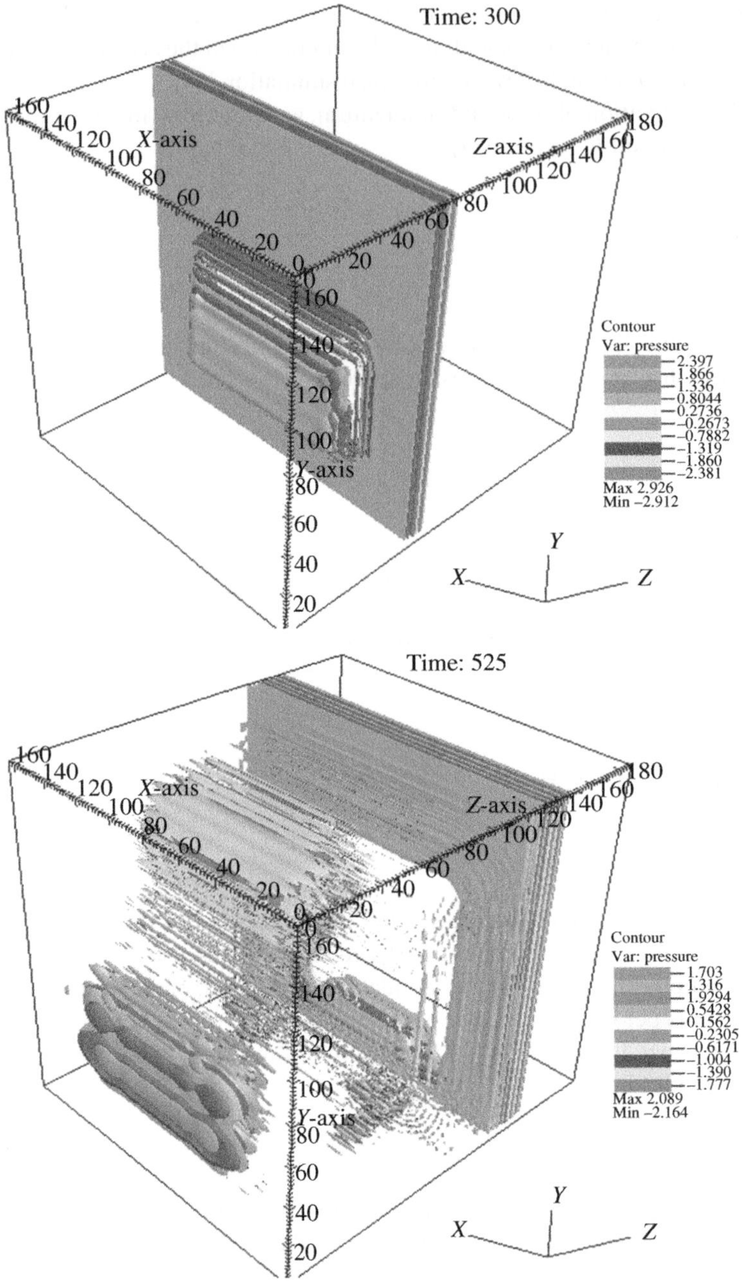

Figure 3.11 A three-dimensional contour plot allows visualization of surfaces of equal pressure at a single time. These visualizations work well for input signals of short duration. Surfaces of equal pressure are shown when the short-pulse signal is first incident on the front of the truck model scatterer and a short time later after the signal has been reflected from the entire truck. Source: [65]/with permission of ProQuest LLC.

can do this when we are looking at pressure data as a function of time at a specific spatial coordinate, for example, the location of the microphone recording the backscattered acoustic signal.

While our simulation output consists of pressure values for each point in the three-dimensional simulation space at a given time step, displaying this data as a three-dimensional volume is often not ideal. Data in the interior of the volume, where the scattering behavior we are most interested in occurs, is obscured by data from the edges of the volume (Figure 3.11). Decreasing the opacity of the pressure data at points on the edges of the space allows observation into the interior at the cost of a direct correlation between color values and pressure values.

However, we have found that the best way to visualize scattering behavior is to take slices through the pseudocolor volume. This also has an advantage over volume visualizations in that it requires less graphical power to create renderings–not a problem for our modest-sized simulation spaces but an important issue in other simulations. The three-dimensional contour plot can also be sliced to create two-dimensional contour lines.

References

1. Stratton, J.A. (1941). *Electromagnetic Theory*. Wiley.
2. Harrington, R.F. (2001). *Time-Harmonic Electromagnetic Fields*. IEEE Press Classic Reissue.
3. Jackson, J.D. (1999). *Classical Electrodynamics*. Wiley.
4. Landau, L.D. (1984). *Electrodynamics of Continuous Media*. Pergamon Press.
5. Jones, D.S. (1979). *Methods in Electromagnetic Wave Propagation*. Oxford.
6. Born, M. and Wolf, E. (1964). *Principles of Optics: Electromagnetic Theory of Propagation, Interference and Diffraction of Light*. Pergamon.
7. Pierce, A.D. (1981). *Acoustics Introduction to Its Physical Principles and Applications*. Acoustical Society of America and American Institute of Physics.
8. Morse, P.M.C. and Ingard, K.U. (1986). *Theoretical Acoustics*. Princeton University Press.
9. Blackstock, D.T. (2000). *Fundamentals of Physical Acoustics*. Wiley.
10. Kinsler, L.E., Frey, A.R., Coppens, A.B., and Sanders, J.V. (2000). *Fundamentals of Acoustics*. Wiley.
11. Crighton, D.G., Dowling, A.P., Williams, J.E.F. et al. (1992). *Modern Methods in Analytical Acoustics Lecture Notes*. Springer-Verlag.
12. Crocker, M.J. (1998). *Handbook of Acoustics*. Wiley.
13. Benade, A.H. (1976). *Fundamentals of Musical Acoustics*. Oxford University Press.
14. Morse, P.M. (1948). *Vibration and Sound*. New York: McGraw-Hill.
15. Graff, K.F. (1975). *Wave Motion in Elastic Solids*. Oxford.
16. Kolsky, H. (1963). *Stress Waves in Solids*. Dover.
17. Brekhovskikh, L.M. and Goncharov, V. (2012). *Mechanics of Continua and Wave Dynamics*, vol. 1. Springer.
18. Achenbach, J. (2012). *Wave Propagation in Elastic Solids*. Elsevier.
19. Lempriere, B.M. (2003). *Ultrasound and Elastic Waves: Frequently Asked Questions*. Elsevier.
20. Hudson, J.A. (1980). *The Excitation and Propagation of Elastic Waves*. Cambridge University Press.
21. Bedford, A. and Drumheller, D.S. (1984). *Introduction to Elastic Wave Propagation*. Wiley.
22. Doyle, J.F. (1989). *Wave Propagation in Structures*. Springer US.
23. Harper, A.H. (1988). *Elastic Waves in Solids*. IOP Publishing.
24. Royer, D. and Dieulesaint, E. (1999). *Elastic Waves in Solids*. Springer.

25 Auld, B.A. (1973). *Acoustic Fields and Waves in Solids*. Krieger.

26 Rose, J.L. (2000). *Ultrasonic Waves in Solid Media*. Cambridge University Press.

27 Swindlehurst, A.L., Ayanoglu, E., Heydari, P., and Capolino, F. (2014). Millimeter-wave massive MIMO: the next wireless revolution? *IEEE Communications Magazine* 52 (9): 56–62.

28 Mamta, A., Abhishek, R., and Navrati, S. (2016). Next generation 5G wireless networks: a comprehensive survey. *IEEE Communications Surveys & Tutorials* 18 (3): 1617–1655.

29 Diago-Mosquera, M.E., Aragón-Zavala, A., and Castañón, G. (2020). Bringing it indoors: a review of narrowband radio propagation modeling for enclosed spaces. *IEEE Access* 8: 103875–103899.

30 Yee, K. (1966). Numerical solution of initial boundary value problems involving Maxwell's equations in isotropic media. *IEEE Transactions on Antennas and Propagation* 14 (3): 302–307.

31 Taflove, A. and Hagness, S.C. (2005). *Computational Electrodynamics: The Finite-Difference Time-Domain Method*. Artech House.

32 Mur, G. (1981). Absorbing boundary conditions for the finite-difference approximation of the time-domain electromagnetic-field equations. *IEEE Transactions on Electromagnetic Compatibility* 4: 377–382.

33 Mei, K.K. and Fang, J. (1992). Superabsorption-a method to improve absorbing boundary conditions (electromagnetic waves). *IEEE Transactions on Antennas and Propagation* 40 (9): 1001–1010.

34 Gedney, S.D. (1996). An anisotropic perfectly matched layer-absorbing medium for the truncation of FDTD lattices. *IEEE Transactions on Antennas and Propagation* 44 (12): 1630–1639.

35 Bérenger, J.-P. (1996). Perfectly matched layer for the FDTD solution of wave-structure interaction problems. *IEEE Transactions on Antennas and Propagation* 44 (1): 110–117.

36 Bérenger, J.-P. (1997). Improved PML for the FDTD solution of wave-structure interaction problems. *IEEE Transactions on Antennas and Propagation* 45 (3): 466–473.

37 Bérenger, J.-P. (2003). Making use of the PML absorbing boundary condition in coupling and scattering FDTD computer codes. *IEEE Transactions on Electromagnetic Compatibility* 45 (2): 189–197.

38 Rooney, M.M. (2022). Characterization of wireless communications networks using machine learning and 3D electromagnetic wave propagation simulations. Doctoral dissertation. William and Mary.

39 Rooney, M.M. and Hinders, M.K. (2023). Identification of transient radio frequency occlusion events in urban environments. *IEEE Access* 11: 68051–68065. https://doi.org/10.1109/ACCESS.2023.3292343.

40 Mitola, J. (2000). *Cognitive Radio: An Integrated Agent Architecture for Software Defined Radio*. Stockholm, Sweden: Royal Institute of Technology Department of Teleinformatics, Royal Inst. Tech. (KTH).

41 Mitola, J. and Maguire, G.Q. (1999). Cognitive radio: making software radios more personal. *IEEE Personal Communications* 6 (4): 13–18.

42 Mitola, J. (1999). Software radio architecture: a mathematical perspective. *IEEE on Selected Areas in Communications* 17 (4): 514–538.

43 Akyildiz, I.F., Lee, W.Y., Vuran, M.C., and Mohanty, S. (2006). NeXt generation/dynamic spectrum access/cognitive radio wireless networks: a survey. *Computer Networks* 50 (13): 2127.

44 Li, H. and Han, Z. (2010). Catch me if you can: an abnormality detection approach for collaborative spectrum sensing in cognitive radio networks. *IEEE Transactions on Wireless Communications* 9 (11): 3554–3565.

45 Wang, J., Ghosh, M., and Challapali, K. (2011). Emerging cognitive radio applications: a survey. *IEEE Communications Magazine* 49 (3): 74–81.

46 Rooney, M.M. and Hinders, M.K. (2021). Machine learning for medium access control protocol recognition in communications networks. *IEEE Access* 9: 110762–110771. https://doi.org/10.1109/ACCESS.2021.3102859.

47 Ullah, H., Nair, N.G., Moore, A. et al. (2019). 5G communication: an overview of vehicle-to-everything, drones, and healthcare use-cases. *IEEE Access* 7: 37251–37268.

48 Bkassiny, M., Li, Y., and Jayaweera, S.K. (2013). A survey on machine-learning techniques in cognitive radios. *IEEE Communications Surveys & Tutorials* 15 (3): 1136–1159.

49 Akyildiz, I.F., Lee, W.-Y., and Chowdhury, K.R. (2009). CRAHNs: Cognitive radio ad hoc networks. *AD Hoc Networks* 7 (5): 810–836.

50 Fellinger, P., Marklein, R., Langenberg, K.J., and Klaholz, S. (1995). Numerical modeling of elastic-wave propagation and scattering with EFIT - elastodynamic finite integration technique. *Wave Motion* 21 (1): 47.

51 Schubert, F., Peiffer, A., Kohler, B., and Sanderson, T. (1998). The elastodynamic finite integration technique for waves in cylindrical geometries. *The Journal of the Acoustical Society of America* 104 (5): 2604–2614.

52 Schubert, F. and Koehler, B. (2001). Three-dimensional time domain modeling of ultrasonic wave propagation in concrete in explicit consideration of aggregates and porosity. *Journal of Computational Acoustics* 9 (4): 1543–1560.

53 Schubert, F. (2004). Numerical time-domain modeling of linear and nonlinear ultrasonic wave propagation using finite integration techniques—theory and applications. *Ultrasonics* 42 (1–9): 221–229.

54 Rudd, K.E., Leonard, K.R., Bingham, J.P., and Hinders, M.K. (2007). Simulation of guided waves in complex piping geometries using the elastodynamic finite integration technique. *The Journal of the Acoustical Society of America* 121 (3): 1449–1458.

55 Rudd, K.E., Bertoncini, C.A., and Hinders, M.K. (2009). Simulations of ultrasonographic periodontal probe using the finite integration technique. *The Open Acoustics Journal* 2: 1–19. http://dx.doi.org/10.2174/1874837600902010001.

56 Bingham, J. and Hinders, M. (2010). 3D elastodynamic finite integration technique simulation of guided waves in extended built-up structures containing flaws. *Journal of Computational Acoustics* 18 (02): 165–192.

57 Leckey, C.A.C., Rogge, M.D., Miller, C.A., and Hinders, M.K. (2012). Multiple-mode Lamb wave scattering simulations using 3D elastodynamic finite integration technique. *Ultrasonics* 52 (2): 193–207.

58 Leckey, C. and Hinders, M. (2012). 3D simulations for the investigation of Lamb wave scattering from flaws. *Review of Progress in Quantitative Nondestructive Evaluation* 31: 111–117.

59 Hendrickson, D.C. and Hinders, M.K. (2023). Monitoring of gantry crane track health with commodity IOT devices. *Journal of Infrastructure Systems* 29 (4): 05023008. https://doi.org/10.1061/JITSE4.ISENG-2292.

60 Skinner, E.D., Kirn, S.L., and Hinders, M.K. (2019). Development of underwater beacon for arctic through-ice communication via satellite. *Cold Regions Science and Technology* 160: 58–79. https://doi.org/10.1016/j.coldregions.2019.01.010.

61 Fehlman, W.L. and Hinders, M. (2009). *Mobile Robot Navigation with Intelligent Infrared Image Interpretation*. London: Springer.

62 Gao, W. (2005). Sonar sensor interpretation for ectogeneous robots. Doctoral dissertation. The College of William and Mary.

63 Gao, W. and Hinders, M.K. (2005). Mobile robot sonar interpretation algorithm for distinguishing trees from poles. *Robotics and Autonomous Systems* 53: 89–98.

64 Hinders, M., Gao, W., and Fehlman, W. (2007). Sonar sensor interpretation and infrared image fusion for mobile robotics. In: *Mobile Robots: Perception & Navigation* (ed. S. Kolski), 69–90. Germany: Pro Literatur Verlag.

65 Dieckman, E. (2013). Use of pattern classification algorithms to interpret passive and active data streams from a walking-speed robotic sensor platform. William and Mary, Department of Applied Science doctoral dissertation.

66 Westervelt, P. (1963). Parametric acoustic array. *The Journal of the Acoustical Society of America* 35: 535–537.

67 Bennett, M.B. and Blackstock, D. (1975). Parametric array in air. *Journal of the Acoustical Society of America* 57: 562–568.

68 Yoneyama, M., Fujimoto, J., Kawamo, Y., and Sasabe, S. (1983). The audio spotlight: an application of nonlinear interaction of sound waves to a new type of loudspeaker design. *Journal of the Acoustical Society of America* 73: 1532.

69 Pompei, F.J. (1999). The use of airborne ultrasonics for generating audible sound beams. *Journal of the Audio Engineering Society* 47: 726–731.

70 Gan, W.-S., Yang, J., and Kamakura, T. (2012). A review of parametric acoustic array in air. *Applied Acoustics* 73: 1211–1219.

71 Achanta, A., McKenna, M., and Heyman, J. (2005). Non-linear acoustic concealed weapons detection. *Proceedings of the 34th Applied Imagery and Pattern Recognition Workshop (AIPR05)*, 727, December 2005.

72 Hinders, M. and Rudd, K. (2010). Acoustic parametric array for identifying standoff targets. In: *31st Review of Progress in Quantitative Nondestructive Evaluation*, vol. 29 (eds. D. Thompson and D. Chimenti), 1757–1764. AIP Press.

73 Calicchia, P., Simone, S.D., Marcoberardino, L.D., and Marchal, J. (2012). Near- to far-field characterization of a parametric loudspeaker and its application in non-destructive detection of detachments in panel paintings. *Applied Acoustics* 73: 1296–1302.

74 Nomura, H., Hedberg, C.M., and Kamakura, T. (2012). Numerical simulation of parametric sound generation and its application to length-limited sound beam. *Applied Acoustics* 73 (12): 1231–1238.

75 Kamakura, T., Nomura, H., Akiyama, M., and Hedberg, C.M. (2011). Parametric sound fields formed by phase-inversion excitation of primary waves. *Acta Acustica United with Acustica* 97 (2): 209–218.

76 Hedberg, C., Haller, K., and Kamakura, T. (2010). A self-silenced sound beam. *Acoustical Physics* 56 (5): 637–639.

77 Tanaka, N. and Tanaka, M. (2010). Active noise control using a steerable parametric array loudspeaker. *The Journal of the Acoustical Society of America* 127 (6): 3526–3537.

78 Komatsuzaki, T. and Iwata, Y. (2011). Active noise control using high-directional parametric loudspeaker. *Journal of Environment and Engineering* 6 (1): 140–149.

79 Lam, B., Gan, W.-S., and Shi, C. (2014). Feasibility of a length-limited parametric source for active noise control applications. International Congress on Sound and Vibration, Beijing, China.

80 Tanaka, K., Shi, C., and Kajikawa, Y. (2017). Binaural active noise control using parametric array loudspeakers. *Applied Acoustics* 116: 170–176. https://doi.org/10.1016/j.apacoust.2016.09.021.

81 van der Rots, R. and Berkhoff, A. (2015). Directional loudspeaker arrays for acoustic warning systems with minimised noise pollution. *Applied Acoustics* 89: 345–354. https://doi.org/10.1016/j.apacoust.2014.09.024.

82 Shi, C., Kajikawa, Y., and Gan, W.-S. (2015). Generating dual beams from a single steerable parametric loudspeaker. *Applied Acoustics* 99: 43–50. https://doi.org/10.1016/j.apacoust.2015.05.004.

83 Aoki, S., Shimizu, K., and Itou, K. (2017). Study of vertical sound image control with parametric loudspeakers. *Applied Acoustics* 116: 164–169. https://doi.org/10.1016/j.apacoust.2016.08.014.

84 Guasch, O. and Sánchez-Martín, P. (2018). Far-field directivity of parametric loudspeaker arrays set on curved surfaces. *Applied Mathematical Modelling* 60: 721–738. https://doi.org/10.1016/j.apm.2018.04.002.

85 Westervelt, P. (1957). Scattering of sound by sound. *The Journal of the Acoustical Society of America* 29: 199–203.

86 Westervelt, P. (1957). Scattering of sound by sound. *The Journal of the Acoustical Society of America* 29: 934–935.

87 Pompei, F.J. (2002). Sound from ultrasound: the parametric array as an audible sound source. Doctoral dissertation. Massachusetts Institute of Technology.

88 Leckey, C.A.C. (2011). Investigation of ultrasonic wave scattering effects using computational methods. Doctoral dissertation. The College of William and Mary.

89 Beyer, R.T. (1960). Parameter of nonlinearity in fluids. *The Journal of the Acoustical Society of America* 32: 719–721.

90 Hamilton, M.F. and Blackstock, D.T. (1998). *Nonlinear Acoustics*. San Diego, CA: Academic Press.

91 Beyer, R.T. (1974). *Nonlinear Acoustics*. Providence, RI: Brown University Department of Physics.

92 Lee, Y.-S. and Hamilton, M. (1995). Time-domain modeling of pulsed finite-amplitude sound beams. *The Journal of the Acoustical Society of America* 97: 906–917.

93 Lee, Y.-S. (1993). Numerical solution of the KZK equation for pulsed finite-amplitude sound beams in the thermoviscous fluids. Doctoral dissertation. Massachusetts Institute of Technology.

94 Rudd, K.E. (2007). Parallel 3D acoustic and elastic wave simulation methods with applications in nondestructive evaluation. Doctoral dissertation. The College of William and Mary.

4

Reflection and Refraction

You might think it's a little funny that there's a whole chapter on reflection and refraction. You might be thinking how much there is to say about Snell's law. Good point. Presumably that was all covered in the optics part of frosh physics, with periodic refreshers as needed over the subsequent semesters. Acoustic reflection and refraction are equally simple, with angle of reflection equal to angle of incidence, and angle of refraction determined by the index of refraction. Easy.

There might be some subtleties involved. From Wikipedia,

> Total internal reflection (TIR) is the optical phenomenon in which waves arriving at the interface (boundary) from one medium to another (e.g. from water to air) are not refracted into the second ("external") medium, but completely reflected back into the first ("internal") medium. It occurs when the second medium has a higher wave speed (i.e. lower refractive index) than the first, and the waves are incident at a sufficiently oblique angle on the interface. For example, the water-to-air surface in a typical fish tank, when viewed obliquely from below, reflects the underwater scene like a mirror with no loss of brightness.

That's a rather poor effort, Wikipedia. How exactly do you expect me to view the surface of a fish tank from below? Somebody should hate-edit that entry, but then it would probably just get edited back. A better way to think about TIR is diamonds, which are neither rare nor inherently valuable, BTW. Everything you think about diamonds is a part of a carefully crafted marketing campaign by the De Beers Corporation who maintained an amazingly effective monopoly over the sale of cut diamonds for a century. Making you think that diamonds are forever is just a way to prevent you from finding out that they have almost no resale value [1]. Lab-grown diamonds are now 80% cheaper, BTW.

The sparkliness of diamonds comes from TIR. Remember we were talking about reflection and refraction? In 1919, Marcel Tolkowsky[1] found that if the diamonds were cut at the correct angle to make use of TIR (Figure 4.1), it would greatly improve their brilliance, fire, and sparkle. This understanding of TIR has been used ever since when cutting diamonds.

One of my graduate students just said to me: "I think the reflection and refraction with water are useful to talk about. A figure of the refraction causing shapes to shift under water can drive home the point. Things like a head a foot away from the body, or the dark ring underwater at about 89° where no light comes through the surface because the full arch was bent into 89° is

1 Marcel Tolkowsky, an engineer by education, was a Belgian member of a Jewish family of diamond cutters from Poland. Tolkowsky, as part of his PhD topic at the University of London, systematically studied the grinding of diamonds. In his book Diamond Design, he published the specifications of what would later be called the American Standard, derived from mathematical calculations that considered both brilliance and fire of the stone.

A Data Engineering Approach to Wave Scattering Analysis: with Applications in Radar, Sonar, Medical Diagnostics, Structural Flaw Detection and Intelligent Robotics, First Edition. Mark K. Hinders.

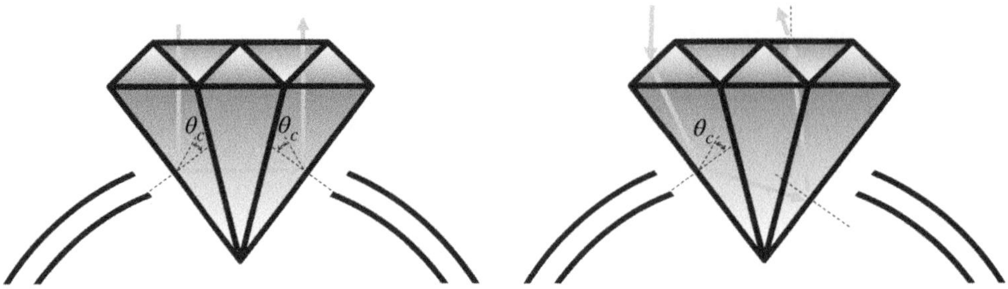

Figure 4.1 Light entering the top of a diamond will be reflected back out the top if it is cut correctly so total internal reflection occurs. Even light entering from the sides of the diamond, if cut correctly, can reflect the light upward when you hold your hand out so everybody can ooh and ahh over your giant rock. My wife used to say that if you can lift your hand, the diamond is too small. De Beers' advertising evil geniuses figured out that, "A woman can easily feel that diamonds are 'vulgar' and still be highly enthusiastic about receiving diamond jewelry" (Caveat emptor: https://youtu.be/N5kWu1ifBGU).

interesting." I don't know his thoughts on Pokémon cards or diamond engagement rings, but he is married and has kids who go to the pool in the summer. He has a fancy camera on his phone. Did he take a picture at the pool to drive home the point? Nope. I guess you'll just have to find some on the internet. If you try to take one yourself, have a container of uncooked white rice handy. To demonstrate the refraction and the limit at 48.5° (half angle for the 89 dark region), Figure 4.2 is included, though. The cut off for water into air reflection can be clearly seen.

When you do get around to searching the internet for interesting pictures that explain TIL, you'll find that examples include mirages and optical fibers. There will be a few underwater photos of fishes and Olympic swimmers, and even some low-angle shots of aquariums of the sort that Wikipedia referred to. There will be photos of glasses of water with straws and that one weird fingerprint picture that seems to show up everywhere. You'll also find a variety of videos purporting to explain this and related optics phenomenon to the general public.

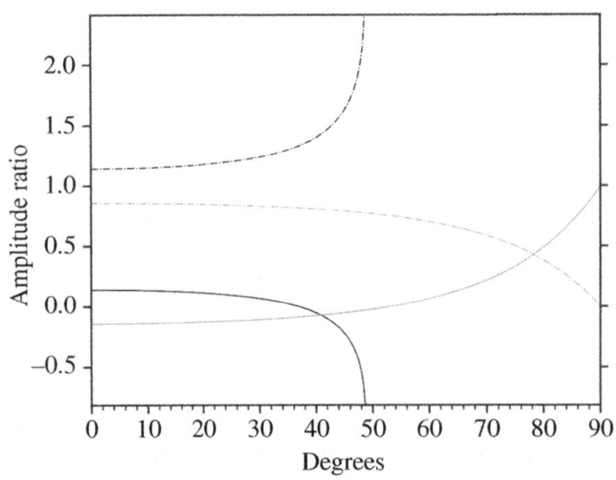

Figure 4.2 The amplitude ratios of the reflected and refracted vertically polarized light traveling through an air–water interface like the surface of a swimming pool. The dark dash-dot line is reflection from water into air. The solid dark line is refraction from water into air. The gray dash-dot line is reflection from air into water. The solid gray line is refraction from water into air. Note the clear cutoff, where no light is refracted, at 48.5° or beyond.

Exercise 4.1 Are there any TIR-related phenomena that would make it feasible to fuse underwater and overhead video streams to make swimming competitions more interesting? Obviously, the surface of the water gets disturbed, but that might be a relatively small effect for the leaders in some cases. I have a water-proof GoPro around here somewhere and a very large ultrasound immersion tank for some proof-of-concepts experiments. You'll want to roll up your sleeves, of course.

We'll come back to optics using the full machinery of electromagnetics, but first, I want to discuss reflection and refraction of elastic waves, which have the additional complication that we call mode conversion. That turns out to make it worth having a whole chapter about reflection and refraction.[2]

In an infinite homogeneous isotropic elastic medium, the longitudinal and transverse waves propagate independently, that is, they do not interact with each other. If the material parameters vary in space, then a propagating L wave generates T waves and vice versa. The same takes place at a boundary between two media. This "mode coupling" is included implicitly in the boundary conditions. We'll consider four different types of interfaces.

1) **Welded contact between elastic half-spaces:** The displacement as well as the force per unit area of the interface S must be continuous.

$$\left(u_j^{(1)} - u_j^{(2)}\right)_S = 0 \qquad \left(\sigma_{nj}^{(1)} - \sigma_{nj}^{(2)}\right)_S = 0$$

for $j = 1, 2, 3$, and where n is the index of the axis normal to S. The superscripts 1 or 2 denote the half-spaces.

2) **Contact between half-spaces with slip:** The normal displacement and the normal component of stress are continuous, the tangential component of force being zero at S.

$$\left(u_n^{(1)} - u_n^{(2)}\right)_S = 0 \qquad \left(\sigma_{nn}^{(1)} - \sigma_{nn}^{(2)}\right)_S = 0$$

$$\sigma_{nj}^{(1)}\Big|_S = \sigma_{nj}^{(2)}\Big|_S = 0 \qquad (j \neq n)$$

Note that there is no summation on the n despite the repeated index. Also note that the common situation we'll face is a fluid in contact with a solid, in which case the normal stress in the solid is balanced by the pressure in the fluid at the boundary and the shear stress in the solid is zero at the interface.

3) **Welded contact with an infinitely rigid wall:** All displacements will be zero at the boundary.

$$u_j\Big|_S = 0 \qquad (j = 1, 2, 3)$$

---

2 Prof. Robert E. Green Jr., who came to Johns Hopkins in 1960, established the Center for Nondestructive Evaluation, an interdisciplinary group of academic and industrial partners. Green was raised near the naval shipyards in Virginia, where his father worked as a machinist. At a young age, he knew he wanted to be a scientist when he grew up, but he didn't consider physics until he attended William & Mary where he received his bachelor's degree in physics in 1953. He went on to study under Robert Bruce Lindsay, an acoustic physicist, at Brown University, where he received his master's degree in 1956 and his PhD in 1959. Green considered himself to be an "applied physicist." In an interview with the American Institute of Physics in 2000, he recounted his invitation to join the faculty at Johns Hopkins 40 years earlier: "I received a letter from Johns Hopkins asking me if I wanted to apply for [an] assistant professor position in their mechanical engineering department. And I told them, 'No, I was a physicist, I didn't want to be an engineer, I didn't like engineers,' and they wrote back, 'Wonderful. That's the kind of guy we are looking for."' As Department Chair he viewed his primary job as beating up on the Dean. Green felt blessed to have worked with so many "outstanding students" who went on to do so well. Bob died on 21 November 2017 at age 85.

When Prof. Green retired, he passed the leadership of CNDE to his deputy, Dr. Boro Djordjevic, but Boro was less skilled at beating up on deans (having spent most of his career in industry) and CNDE folded. Born in Yugoslavia, Boro was a William and Mary alum, where his mother was a Chemistry professor, and he played on the W&M soccer team some years before John Stewart. Boro died 26 November 2023.

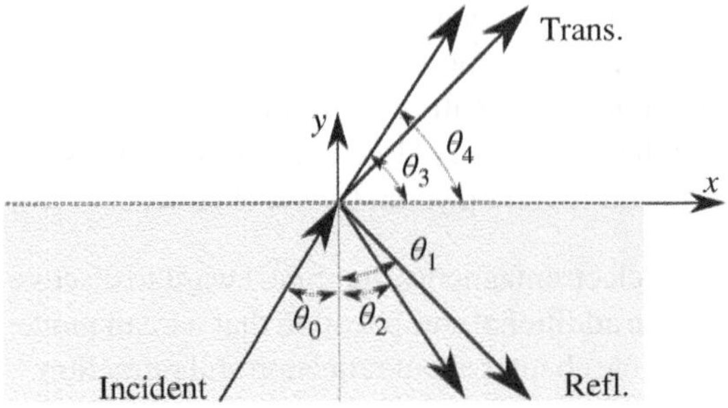

Figure 4.3 Incident, reflected, and transmitted waves at an interface between two half-spaces.

4) **Boundary of elastic solid with vacuum:** Often referred to as a stress-free surface or just a free surface, the normal surface tractions are zero.

$$\sigma_{nj}\Big|_S = 0 \qquad (j = 1, 2, 3)$$

Air has a low-enough density compared to solids that elastic waves at MHz frequencies reflect from a solid-air boundary like a free surface. Water is dense enough that there is significant transmission through solid-water boundaries. In that case, use condition 2 above with shear stress zero in the fluid and normal stress equal to pressure.

Now let's consider the reflection and refraction of elastic waves from the interface between two elastic half-spaces, illustrated in Figure 4.3. This is something that's straightforward to simulate using the EFIT technique from the Chapter 3. Indeed, I recommend that students do exactly this as soon as they get their EFIT codes up and running. I was going to apologize for Figure 4.4 being upside down, but you shouldn't get too hung up on the particular way I drew Figure 4.3. Rotate the page if it's a problem.

I should note that in this chapter, I'm mostly following the notation of K. Graff's excellent book [2], which you want for your personal library. While you're at it, pick up a copy of [3] and keep a look out for the book by Viktorov [4] which I checked out from my university library more than a decade ago and plan to renew every year until I retire because I'm afraid that it will get discarded

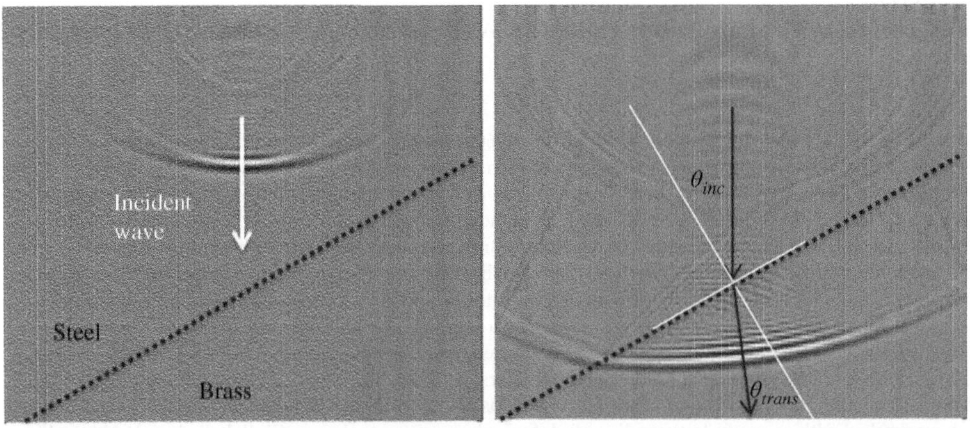

Figure 4.4 Elastic wave in a steel block incident upon a brass wedge angled at 30°. A dotted black line shows the location of the steel-brass boundary. The left image shows a 2D vertical slice through the 3D simulation space of the wave at time $t = 127$ μs; the image on the right shows the behavior of the scattered and transmitted longitudinal waves at $t = 282$ μs. The white lines in the right image show the surface normal at the steel-brass boundary. The calculated longitudinal transmitted (refracted) angle is shown.

along with other old books to make room for more massage chairs or 3D printers or whatever. (Look back to the Chapter 3 references for a bunch more books that are relevant.) You probably prefer eBooks, so keep a look out for PDF versions that you can download and archive locally or in a cloud service that you control. Things that are freely available online today could end up behind a paywall tomorrow. "Who controls the past controls the future; who controls the present controls the past." I remember a time several years ago when legally purchased copies of 1984 got yanked back off of everybody's Kindles overnight. Because reasons. The incident, reflected, and refracted waves (refer to Figure 4.3 as needed) are written

$$\vec{u}^{(n)} = A_n d^{(n)} e^{i\eta_n} \tag{4.1}$$

where A_n is the amplitude, $d^{(n)}$ is the polarization, and the phase term is

$$\eta_n = k_n \left(x_1 p_1^{(n)} + x_2 p_2^{(n)} - c_n t) \right) \tag{4.2}$$

where $n = inc, refl, trans$ are for incident, reflected, and transmitted waves. Yes, I know this notation is kind of cumbersome and I don't particularly care for it, but the idea is to show how all the various cases are alike and different.

For the incident wave, the propagation direction is

$$\vec{p}^{(inc)} = \sin\theta_i \hat{\imath}_1 + \cos\theta_i \hat{\imath}_2 \tag{4.3}$$

If the incident wave is a longitudinal wave, we have

$$\vec{d}^{(inc)} = \vec{p}^{(inc)} \qquad c_i = c_L$$

and for an incident transverse wave we have

$$\vec{d}^{(inc)} \cdot \vec{p}^{(inc)} = 0 \qquad c_i = c_T$$

Note that this allows for two polarization, SV-waves and SH-waves. That terminology is from seismology, where the transverse, or shear, waves are the second to arrive because they are slower than the longitudinal, or primary, waves. With regard to the surface of the earth, the SH-waves are horizontally polarized, which means that the physical displacement is back and forth parallel to the surface. The SV-waves are vertically polarized for a wave that is traveling horizontally. For incident transverse waves, we have

$$SV: \quad \vec{d}^{(inc)} = \hat{\imath}_3 \times \vec{p}^{(i)} \qquad SH: \quad \vec{d}^{(inc)} = \hat{\imath}_3 \qquad c_i = c_T$$

We will find that the L and SV waves are coupled together at the boundary while the SH wave is not coupled to either one. Reflected and transmitted (refracted) waves will be both L and SV waves if the incident wave is either L or SV. To summarize, incident, reflected, and transmitted L and SV waves are:

$$\begin{aligned}
\vec{u}^{L}_{inc} &= A_0 \left(\sin\theta_0 \hat{\imath}_1 + \cos\theta_0 \hat{\imath}_2\right) \exp ik_0 \left(x_1 \sin\theta_0 + x_2 \cos\theta_0 - c_L t\right) \\
\vec{u}^{L}_{refl} &= A_1 \left(\sin\theta_1 \hat{\imath}_1 - \cos\theta_1 \hat{\imath}_2\right) \exp ik_1 \left(x_1 \sin\theta_1 - x_2 \cos\theta_1 - c_L t\right) \\
\vec{u}^{SV}_{refl} &= A_2 \left(\cos\theta_2 \hat{\imath}_1 + \sin\theta_2 \hat{\imath}_2\right) \exp ik_2 \left(x_1 \sin\theta_2 - x_2 \cos\theta_2 - c_T t\right) \\
\vec{u}^{L}_{trans} &= A_3 \left(\sin\theta_3 \hat{\imath}_1 + \cos\theta_3 \hat{\imath}_2\right) \exp ik_3 \left(x_1 \sin\theta_3 + x_2 \cos\theta_3 - c'_L t\right) \\
\vec{u}^{SV}_{trans} &= A_4 \left(-\cos\theta_4 \hat{\imath}_1 + \sin\theta_4 \hat{\imath}_2\right) \exp ik_4 \left(x_1 \sin\theta_4 + x_2 \cos\theta_4 - c'_T t\right)
\end{aligned}$$

Here A_0 is the amplitude of the incident L-wave and A_1–A_4 are unknown amplitudes. These coefficients, as well as the angles θ_1–θ_4 and wavenumbers k_1–k_4 are to be determined from the boundary conditions. Recall welded contact

$$u_1^{inc} + u_1^{refl} = u_1^{trans} \qquad u_2^{inc} + u_2^{refl} = u_2^{trans}$$

$$\sigma_{12}^{inc} + \sigma_{12}^{refl} = \sigma_{12}^{trans} \qquad \sigma_{22}^{inc} + \sigma_{22}^{refl} = \sigma_{22}^{trans}$$

all of which are applied at $x_2 = 0$. Note that we haven't bothered to write continuity of the other tangential displacements or of the other shear stresses, since four equations are enough to solve for the four unknowns A_1–A_4.

Now recall Hooke's law and the strain–displacement relation

$$\sigma_{ij} = \lambda\epsilon_{kk}\delta_{ij} + 2\mu\epsilon_{ij} \qquad \epsilon_{ij} = \frac{1}{2}\left(\partial_j u_i + \partial_i u_j\right) \tag{4.4}$$

so we can write the stress components that we need

$$\begin{aligned}\sigma_{22} &= \lambda(\epsilon_{11} + \epsilon_{22}) + 2\mu\epsilon_{22} \\ &= (\lambda + 2\mu)\frac{\partial u_2}{\partial x_2} - 2\mu\frac{\partial u_1}{\partial x_1} \\ \sigma_{12} &= 2\mu\epsilon_{12} = \mu\left(\frac{\partial u_1}{\partial x_2} + \frac{\partial u_2}{\partial x_1}\right)\end{aligned}$$

Hence

$$\sigma_{22}^{inc} = ik_0 A_0 \left\{(\lambda + 2\mu)(\cos^2\theta_0 + \sin^2\theta_0) - 2\mu\, \sin^2\theta_0\right\} e^{ik_0(x_1 \sin\theta_0 + x_2 \cos\theta_0 - c_L t)} \tag{4.5}$$

or

$$\sigma_{22}^{inc} = ik_0 A_0 \left\{(\lambda + 2\mu) - 2\mu\, \sin^2\theta_0\right\} e^{ik_0 x_1 \sin\theta_0} e^{ik_0 x_2 \cos\theta_0} e^{-i\omega t} \tag{4.6}$$

and

$$\sigma_{12}^{inc} = ik_0\mu A_0 \left\{2\sin\theta_0 \cos\theta_0\right\} e^{ik_0 x_1 \sin\theta_0} e^{ik_0 x_2 \cos\theta_0} e^{-i\omega t} \tag{4.7}$$

and so on for the reflected and transmitted stress components. Now consider the first displacement boundary condition equation

$$u_1^{inc} + (u_1^L + u_1^{SV})^{refl} = (u_1^L + u_1^{SV})^{trans}$$

where we've explicitly written that the reflected and transmitted waves have both L and SV parts. We have

$$\begin{aligned}u_1^{inc} &= A_0 \sin\theta_0 e^{ik_0 x_1 \sin\theta_0} e^{ik_0 x_2 \cos\theta_0} e^{-i\omega_0 t} \\ u_1^{refl} &= A_1 \sin\theta_1 e^{ik_1 x_1 \sin\theta_1} e^{-ik_1 x_2 \cos\theta_1} e^{-i\omega_1 t} + A_2 \cos\theta_2 e^{ik_2 x_1 \sin\theta_2} e^{-ik_2 x_2 \cos\theta_2} e^{-i\omega_2 t} \\ u_1^{trans} &= A_3 \sin\theta_3 e^{ik_3 x_1 \sin\theta_3} e^{ik_3 x_2 \cos\theta_3} e^{-i\omega_3 t} + A_4 \cos\theta_4 e^{ik_4 x_1 \sin\theta_4} e^{ik_4 x_2 \cos\theta_4} e^{-i\omega_4 t}\end{aligned}$$

so we write, at the boundary $x_2 \equiv 0$

$$\begin{aligned}&A_0 \sin\theta_0 e^{ik_0 x_1 \sin\theta_0} e^{-i\omega_0 t} + A_1 \sin\theta_1 e^{ik_1 x_1 \sin\theta_1} e^{-i\omega_1 t} + A_2 \cos\theta_2 e^{ik_2 x_1 \sin\theta_2} e^{-i\omega_2 t} \\ &\quad = A_3 \sin\theta_3 e^{ik_3 x_1 \sin\theta_3} e^{-i\omega_3 t} + A_4 \cos\theta_4 e^{ik_4 x_1 \sin\theta_4} e^{-i\omega_4 t}\end{aligned} \tag{4.8}$$

This boundary condition equation must hold for all time, so we can conclude that $\omega_0 = \omega_1 = \omega_2 = \omega_3 = \omega_4 \equiv \omega$, which isn't particularly surprising. It follows then that

$$k_0 = k_1 = \frac{\omega}{c_L} \equiv k \qquad k_2 = \frac{\omega}{c_T} \equiv K$$

$$k_3 = \frac{\omega}{c_L'} \equiv k' \qquad k_4 = \frac{\omega}{c_T'} \equiv K'$$

Similarly, because the boundary conditions must hold for all values of x_1 we get Snell's law which relates the angles to each other

$$k \sin\theta_0 = k \sin\theta_1 = K \sin\theta_2 = k' \sin\theta_3 = K' \sin\theta_4 \tag{4.9}$$

Hence, the boundary condition equation for continuity of u_1 simplifies to

$$A_0 \sin\theta_0 + A_1 \sin\theta_1 + A_2 \cos\theta_2 = A_3 \sin\theta_3 - A_4 \cos\theta_4 \tag{4.10}$$

Similarly for the u_2 boundary condition equation, we get

$$A_0 \cos\theta_0 - A_1 \cos\theta_1 + A_2 \sin\theta_2 = A_3 \cos\theta_3 + A_4 \sin\theta_4 \tag{4.11}$$

The algebra for stresses is similar. We can then write the four boundary condition equations in a matrix equation

$$\begin{bmatrix} -\sin\theta_1 & -\cos\theta_2 & \sin\theta_3 & -\cos\theta_4 \\ \cos\theta_1 & \sin\theta_2 & \cos\theta_3 & \sin\theta_4 \\ \sin 2\theta_1 & \frac{c_L}{c_T}\cos 2\theta_2 & \frac{\mu'}{\mu}\frac{c_L}{c'_L}\sin 2\theta_3 & -\frac{\mu'}{\mu}\frac{c_L}{c'_T}\cos 2\theta_4 \\ -\left(\frac{c_L}{c_T}\right)^2\cos 2\theta_1 & \frac{c_L}{c_T}\sin 2\theta_2 & \frac{\mu'}{\mu}\frac{c_L}{c'_L}\left(\frac{c'_L}{c'_T}\right)^2\cos 2\theta_3 & \frac{\mu'}{\mu}\frac{c_L}{c'_T}\sin 2\theta_4 \end{bmatrix} \begin{pmatrix} A_1 \\ A_2 \\ A_3 \\ A_4 \end{pmatrix} = A_0 \begin{pmatrix} \sin\theta_0 \\ \cos\theta_0 \\ \sin 2\theta_0 \\ \left(\frac{c_L}{c_T}\right)^2\cos 2\theta_0 \end{pmatrix} \tag{4.12}$$

Exercise 4.2 Do the algebra to verify this expression and implement it on a computer to solve for and plot A_1/A_0 and A_2/A_0 as functions of θ_0.

For incident SV waves, the incident plane wave is

$$\vec{u}_{inc}^{SV} = B_0\left(-\cos\theta_0\hat{i}_1 + \sin\theta_0\hat{i}_2\right) e^{iKx_1 \sin\theta_0} e^{iKx_2\cos\theta_0} e^{-i\omega t} \tag{4.13}$$

with B_1–B_4, the unknown coefficients of the reflected and transmitted waves, which are of the same form as before. The boundary condition equations are written

$$\begin{bmatrix} -\sin\theta_1 & -\cos\theta_2 & \sin\theta_3 & -\cos\theta_4 \\ \cos\theta_1 & -\sin\theta_2 & \cos\theta_3 & \sin\theta_4 \\ \sin 2\theta_1 & \frac{c_L}{c_T}\cos 2\theta_2 & \frac{\mu'}{\mu}\frac{c_L}{c'_L}\sin 2\theta_3 & -\frac{\mu'}{\mu}\frac{c_L}{c'_T}\cos 2\theta_4 \\ -\left(\frac{c_L}{c_T}\right)^2\cos 2\theta_1 & \frac{c_L}{c_T}\sin 2\theta_2 & \frac{\mu'}{\mu}\frac{c_L}{c'_L}\left(\frac{c'_L}{c'_T}\right)^2\cos 2\theta_3 & \frac{\mu'}{\mu}\frac{c_L}{c'_T}\sin 2\theta_4 \end{bmatrix} \begin{pmatrix} B_1 \\ B_2 \\ B_3 \\ B_4 \end{pmatrix} = B_0 \begin{pmatrix} -\cos\theta_0 \\ \sin\theta_0 \\ -\frac{c_L}{c_T}\cos 2\theta_0 \\ \frac{c_L}{c_T}\sin 2\theta_0 \end{pmatrix} \tag{4.14}$$

Exercise 4.3 Do the algebra to verify this expression and implement it on a computer to solve for and plot B_1/B_0 and B_2/B_0 vs. θ_0.

Note from Snell's law that when $\theta_0 = 0$ (normal incidence), the other angles are also zero. Consider the L-wave case

$$\begin{bmatrix} 0 & -1 & 0 & -1 \\ 1 & 0 & 1 & 0 \\ 0 & \frac{c_L}{c_T} & 0 & -\frac{\mu'}{\mu}\frac{c_L}{c'_T} \\ -\left(\frac{c_L}{c_T}\right)^2 & 0 & \frac{\mu'}{\mu}\frac{c_L}{c'_L}\left(\frac{c'_L}{c'_T}\right)^2 & 0 \end{bmatrix} \begin{pmatrix} A_1 \\ A_2 \\ A_3 \\ A_4 \end{pmatrix} = A_0 \begin{pmatrix} 0 \\ 1 \\ 0 \\ \left(\frac{c_L}{c_T}\right)^2 \end{pmatrix} \tag{4.15}$$

where A_2 and A_4 are decoupled from A_0, A_1, and A_3. We conclude that $A_2 = A_4 = 0$ and find that

$$\frac{A_1}{A_0} = \frac{\rho' c'_L - \rho c_L}{\rho' c'_L + \rho c_L} \equiv \frac{Z' - Z}{Z' + Z} \tag{4.16}$$

$$\frac{A_3}{A_0} = \frac{2\rho c_L}{\rho' c_L' + \rho c_L} \equiv \frac{2Z}{Z' + Z} \tag{4.17}$$

where $Z = \rho c_L$ and $Z' = \rho' c_L'$ are the acoustic impedances of the two materials.

By eliminating the A_2 and A_4 columns, and the u_2 and σ_{12} rows we could have just written the following 2×2 system for normal incidence

$$\begin{bmatrix} -1 & -1 \\ -\left(\frac{c_L}{c_T}\right)^2 & \frac{\mu'}{\mu}\frac{c_L}{c_L'}\left(\frac{c_L'}{c_T'}\right)^2 \end{bmatrix} \begin{pmatrix} A_1 \\ A_3 \end{pmatrix} = A_0 \begin{pmatrix} 1 \\ \left(\frac{c_L}{c_T}\right)^2 \end{pmatrix} \tag{4.18}$$

Two other simple special cases are for reflection from a rigid wall and reflection from a free surface. In both cases, A_3 and A_4 are zero, by definition because there's no transmitted waves. For the rigid wall, the displacements are zero at $x_2 = 0$ which gives

$$\begin{bmatrix} -\sin\theta_1 & -\cos\theta_2 \\ \cos\theta_1 & \sin\theta_2 \end{bmatrix} \begin{pmatrix} A_1 \\ A_2 \end{pmatrix} = A_0 \begin{pmatrix} \sin\theta_0 \\ \cos\theta_0 \end{pmatrix} \tag{4.19}$$

which you can solve by hand using Cramer's rule to get expressions for A_1/A_0 and A_2/A_0 vs. θ_0.

For reflection from a stress-free surface at $x_2 = 0$, we similarly have

$$\begin{bmatrix} \sin 2\theta_1 & \frac{c_L}{c_T}\cos 2\theta_2 \\ -\frac{c_L^2}{c_T^2}\cos 2\theta_1 & \frac{c_L}{c_T}\sin 2\theta_2 \end{bmatrix} \begin{pmatrix} A_1 \\ A_2 \end{pmatrix} = A_0 \begin{pmatrix} \sin 2\theta_0 \\ \frac{c_L^2}{c_T^2}\cos 2\theta_0 \end{pmatrix} \tag{4.20}$$

The explicit solution to this system gives the reflection coefficients

$$\frac{A_1}{A_0} = \frac{\sin 2\theta_0 \sin 2\theta_2 - \frac{c_L^2}{c_T^2}\cos^2 2\theta_2}{\sin 2\theta_0 \sin 2\theta_2 + \frac{c_L^2}{c_T^2}\cos^2 2\theta_2} \tag{4.21}$$

$$\frac{A_2}{A_0} = \frac{2\frac{c_L^2}{c_T^2}\sin 2\theta_0 \cos 2\theta_2}{\sin 2\theta_0 \sin 2\theta_2 + \frac{c_L^2}{c_T^2}\cos^2 2\theta_2} \tag{4.22}$$

Next consider an L-wave in water incident upon a solid elastic half-space, where we can assume that there are no reflected shear waves $A_2 \equiv 0$, but we must allow for transmitted shear waves in the solid. We could get to this result by considering the fluid to be a solid with a shear rigidity that has gone to zero ($\mu \to 0$) giving the 3×3 system

$$\begin{bmatrix} -\sin\theta_1 & \sin\theta_3 & -\cos\theta_4 \\ 0 & \mu'\frac{c_L}{c_L'}\sin 2\theta_3 & -\mu'\frac{c_L}{c_T'}\cos 2\theta_4 \\ -\rho c_L^2\cos 2\theta_1 & \mu'\frac{c_L}{c_L'}\left(\frac{c_L'}{c_T'}\right)^2\cos 2\theta_3 & \mu'\frac{c_L}{c_T'}\sin 2\theta_4 \end{bmatrix} \begin{pmatrix} A_1 \\ A_3 \\ A_4 \end{pmatrix} = A_0 \begin{pmatrix} \sin\theta_0 \\ 0 \\ \rho c_L^2\cos 2\theta_0 \end{pmatrix} \tag{4.23}$$

Exercise 4.4 Check this result by performing the limiting process from the full 4×4 system. You'll need to perform some row operations before taking the limits, of course.

Exercise 4.5 Derive the 3×3 systems for both L- and SV-waves in an elastic solid half-space obliquely incident on a fluid boundary. Recall that the coefficient matrix is the same for both cases. Plot the L and SV reflection coefficients and the L-wave transmission coefficients vs. angle for a bunch of different materials for both L- and SV-wave incidence. When you run your code you may find some angles beyond which things don't behave "nicely" and that those angles are different for different materials.

4.1 Reflection from a Free Surface

Consider the reflection of an oblique L wave at a free surface as shown in Figure 4.5. Figure 4.6 includes several snapshots from a simulation for a steel block with an angled free surface. The reflected mode converted waves are noted, as are the angles of reflection for both the L and SV

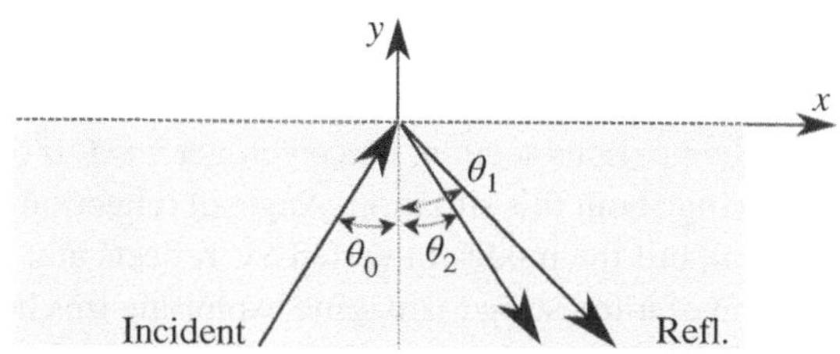

Figure 4.5 Incident and reflected elastic waves at a free surface of a half-space.

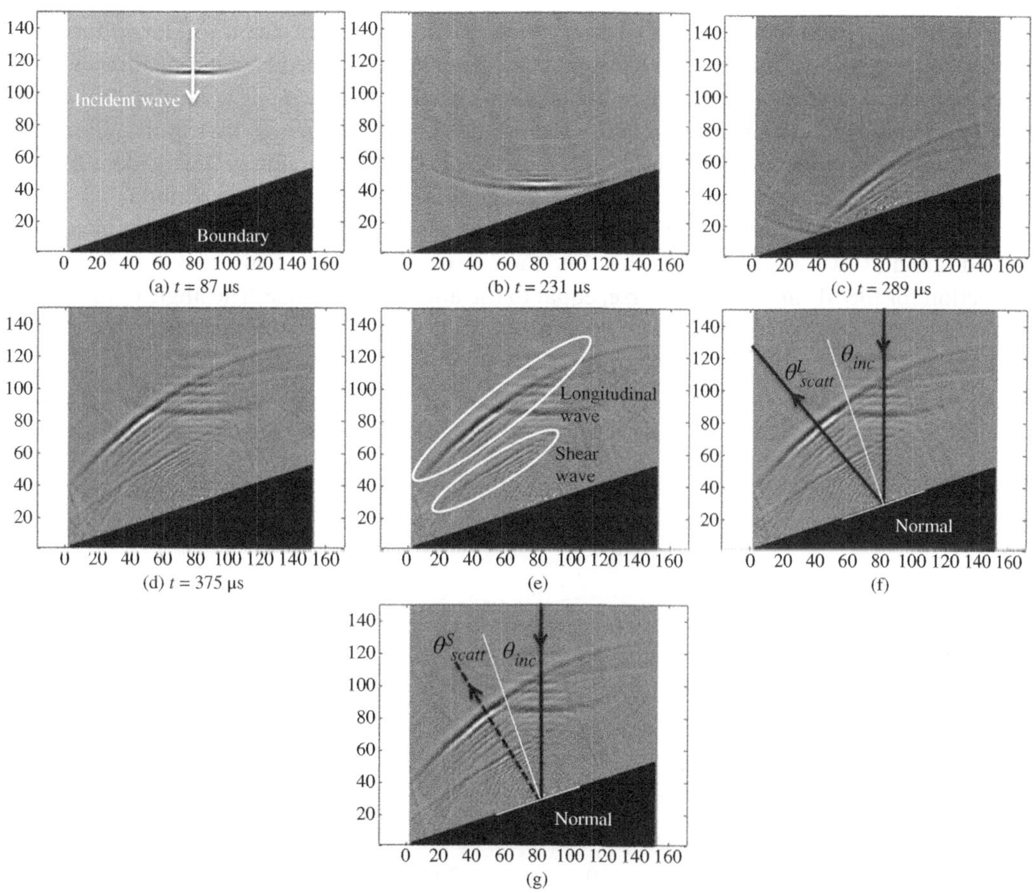

Figure 4.6 Elastic wave in a steel block scattering from an angled free surface at 20°. Both axes represent the number of spatial steps. The circular transducer is on the top surface of the block. Image (a) shows a 2D vertical slice through the 3D simulation space of the wave at time $t = 87$ μs, (b)–(d) show additional snapshots of wave propagation, (e) longitudinal and shear waves after reflection and mode conversion, (f) calculated angle of reflection for longitudinal wave, and (g) calculated angle of reflection for the shear wave.

incidence cases. Again, rotate the page if you need to, since it's a different orientation from the previous figure. Sorry, not sorry.

The amplitude ratios are given by

$$\frac{A_1}{A_0} = \frac{\sin 2\theta_1 \sin 2\theta_2 - (c_L/c_T)^2 \cos^2 2\theta_2}{\sin 2\theta_1 \sin 2\theta_2 + (c_L/c_T)^2 \cos^2 2\theta_2} \tag{4.24}$$

$$\frac{A_2}{A_0} = \frac{2 \sin 2\theta_1 \cos 2\theta_2}{\sin 2\theta_1 \sin 2\theta_2 + (c_L/c_T)^2 \cos^2 2\theta_2} \tag{4.25}$$

as in the previous section. Once you get used to the idea of mode conversion, there's nothing too surprising about this situation. Angle of reflection for the L wave is equal to the angle of incidence, as usual, but the mode-converted SV reflects at a different angle, which depends on the material. It's kind of interesting to imagine exploiting this behavior to design measurement schemes, where the SV wave(s) can be steered into directions that would otherwise be difficult to reach. You can also imagine all sorts of confusion arising from echoes returned by mode-converted wave modes scattering from features in places that weren't expected. This is doubly true if you forget for a hot minute that the L and SV modes have different velocities. We get so used to mapping time delay of echoes to physical distance that, if you don't know for sure which wave mode is causing the echoes you're looking at, you can misinterpret both the direction and distance of the feature. The other thing that becomes apparent, for example, in Figure 4.7, is that the fraction of the incident L wave energy that is mode converted into the SV wave depends both on angle and on material.

Next let's consider the same case (Figure 4.5), but for an incident SV wave. We'll again have both reflected L and SV waves, except that this time the L wave is the mode-converted one. The angle of reflection for the SV mode is, of course, equal to the angle of incidence. The angle of reflection

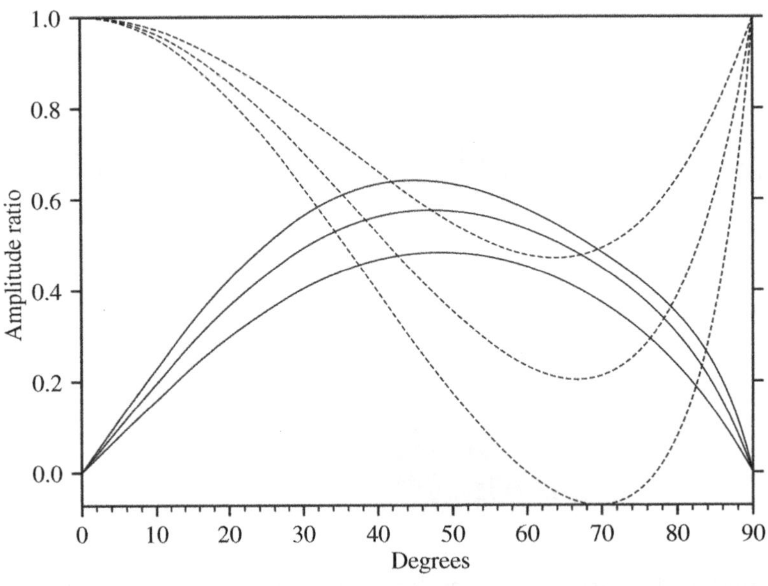

Figure 4.7 Incident and reflected elastic waves at a free surface of a half-space. Dashed lines are for A_1/A_0 for various Poisson's ratios. Solid lines are for the mode-converted waves A_2/A_0. Note that for normal incidence, $\theta_1 = 0$, there is no mode conversion and, of course, for $\theta_1 = 90\,°$, there is no reflection, per se. Note that the angle of maximum mode conversion into reflected shear wave depends somewhat on the material.

for the L mode is different, and depends on the material according to Snell's Law. The amplitude ratios are given by

$$\frac{B_1}{B_0} = \frac{\sin 2\theta_1 \sin 2\theta_2 - (c_L/c_T)^2 \cos^2 2\theta_2}{\sin 2\theta_1 \sin 2\theta_2 + (c_L/c_T)^2 \cos^2 2\theta_2} \tag{4.26}$$

$$\frac{B_2}{B_0} = \frac{-2(c_L/c_T)^2 \sin 2\theta_1 \cos 2\theta_2}{\sin 2\theta_1 \sin 2\theta_2 + (c_L/c_T)^2 \cos^2 2\theta_2} \tag{4.27}$$

Note that there is some value of θ_2 such that $\sin\theta_1 = (c_L/c_T)\sin\theta_2 = 1$. If you code up these simple equations and set about making some plots, you'll be faced with the question: What about θ_2 greater than the "critical" angle where the plot cuts off? Figures 4.8 and 4.9 illustrate this situation.

Since $c_L > c_T$ we have $c_L/c_T > 1$ and for θ_2 beyond the critical angle Snell's law predicts that $\sin\theta_1 > 1$, which is clearly impossible. It turns out that we'll have to go back and re-examine our governing equations, since the simple reflection model we've set up here doesn't seem to be working beyond the critical angle. Note that in Figure 4.5 the reflected L wave is between the reflected SV wave and the surface. This is always the case because the L waves are faster than the SV waves. Please page back and stare at Snell's Law for a moment. So, as the angle of incidence gets more and more oblique, the reflected L wave will start to get squeezed up between the reflected SV wave and the surface, eventually resulting in the situation shown in Figure 4.8. More on this situation in just a bit. Take a look at Figure 4.9 in the meantime.

But first we need to talk about earthquakes. And lifequakes. Tectonic plates move about as fast as your toenails grow. Some people think getting a mani-pedi is self-care and some people don't want strangers touching their feet. Some people who have to hold their breath while they cut their

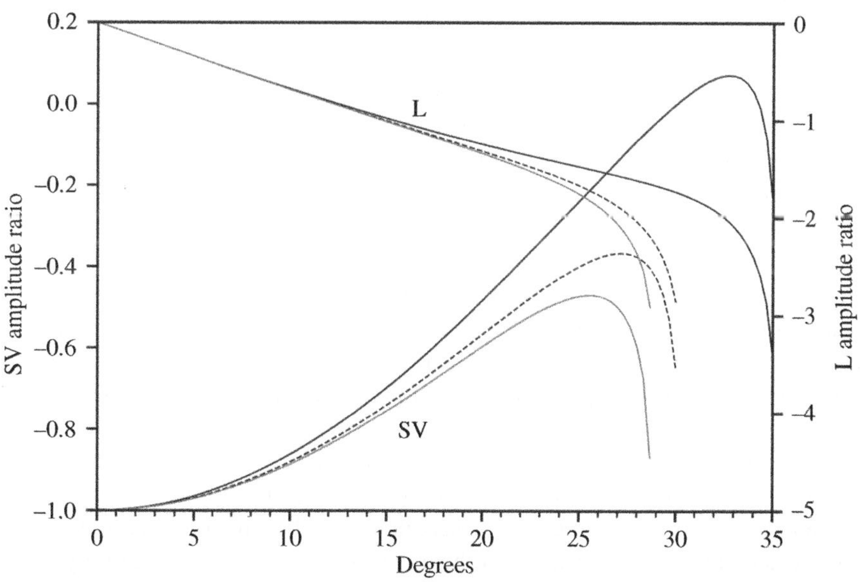

Figure 4.8 Incident and reflected SV elastic waves at a free surface of a half-space beyond the critical angle generates a wave mode that travels along the surface instead of reflecting from it. The critical angle depends on Poisson's ratio and so is different for different materials. The phenomenon is analogous to total internal reflection in optics and turns out to be both quite interesting and quite useful. The wave mode that travels along the surface is called a Rayleigh wave.

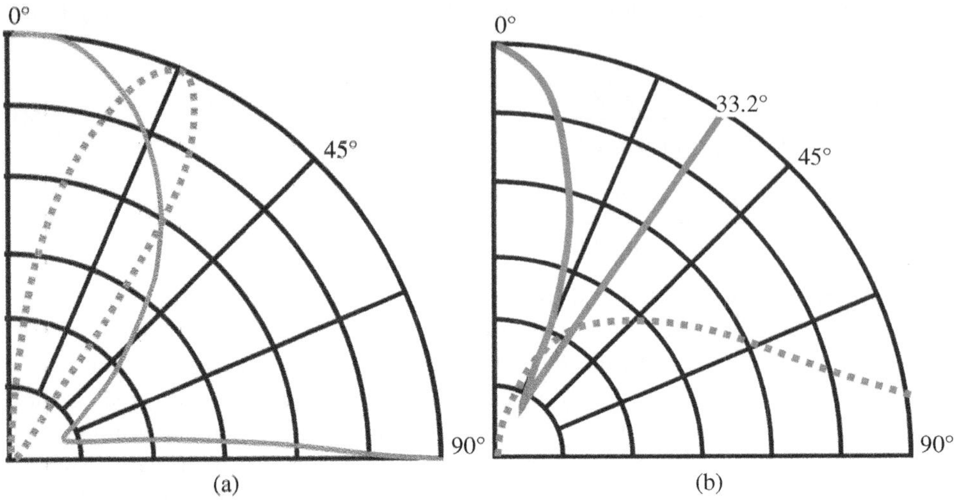

Figure 4.9 Reflection at steel/air interface for incident L-wave (a) and incident SV-wave (b). Dashed line is the mode-converted reflected wave, normalized to highlight the angular scattering behavior. Note for the incident SV-wave, there is a critical angle of 33.2° beyond which there is total internal reflection.

own toenails should probably cut down on the carbs a bit. As tectonic plates try to slide against each other, like is happening in California, they generate shear stress which is periodically released when some local stress limit is reached and part of the one plate jerks forward. There's no way to know when "the big one" is going to happen except that it will happen eventually. Earthquakes radiate out in all directions from the point(s) of release. It's important to keep in mind that the Richter Scale is base-ten logarithmic, so an additional point on the Richter Scale is ten times as bad. Recently there was a 5.1 earthquake in southern California, which is no big deal. It only made the news because it happened during the first tropical storm to hit southern California in about a century and reporters on the scene were having trouble getting people excited about four inches of rain in the desert even though that was record rainfall for them. The new word, "hurriquake" briefly trended and then we were right back to wildfires and whatnot.

I mentioned earthquakes because there are four wave modes to be concerned about, and the Rayleigh waves are the most worrisome. Assuming that the point of origin of the earthquake is down under the ground somewhere, which seems to be a safe assumption, then the seismic waves will propagate outward from that point in three dimensions. Some of that energy will be noted by seismologists in other countries, perhaps quite far away, who will be all excited because they live for that. They will call their friends to see if they got the signals too. The game then becomes analyzing the signals to figure out where the earthquake was and how severe it was.

Nearer to the source, the seismic waves will head up toward the surface. The fastest wave modes are longitudinal waves, which seismologists call P waves because they arrive first and so are Primary. There are also transverse waves, which seismologists call S waves because they are slower and arrive second so they are Secondary and they are also shear waves so the S seems to fit pretty well. The horizontal and vertical polarizations of the shear waves are called SH and SV for the obvious reason that the earth is locally flat-ish and side-to-side swaying feels different from up-and-down shaking. Look back at Figure 4.8 for just a second, and note that the SV waves will convert to Rayleigh waves for those surface locations far enough away from the underground source that the SV waves arrive at the surface beyond the critical angle.

One more bit of geometry. The L, SV, and SH waves are radiating out in three dimensions, so the fall-off in amplitude with distance goes like $1/r^3$. But the Rayleigh waves are bound to the surface, which is why we call them surface waves, and that means that their fall-off in amplitude with distance goes like $1/r^2$. Thus, they pack a punch farther away from the source. In addition, they shake your house both up-and-down and front-to-back. They actually make your house do a little elliptical dance, which the structural engineers may or may not have allowed for in the design. Certainly, they didn't design your bungalo to get shaken that way while at the same time, it was being buffeted by a tropical storm and simultaneously hit with a wall of mud from that hillside that just burned and is now ineffectively absorbing a year's worth of rain in an afternoon.

I may have mentioned that I did radar scattering while in the Air Force. I almost did seismology. After the Challenger exploded and the space program stopped, instead of going on active duty, I stayed at the university and got a master's degree. The thesis I wrote was on attenuation of elastic waves due to scattering from inhomogeneities, with application to modeling seismic wave attenuation. Then the Air Force insisted that I put on my uniform and get a haircut and report for duty in California because I owed four years of service in exchange for a scholarship. Somehow my advisor got my orders changed and had the Air Force instead assign me to Hanscom AFB near Boston so I could continue in graduate school while I was on active duty. It almost worked out perfectly, because that base was home to the Air Force Geophysics Laboratory where they did seismology research. I said, "almost."

One more word for you, since I assume you're old enough to remember the COVID-19 pandemic. Lifequake is a word that describes some significant disruption in your life. It could be mostly just you or it could be everybody. Lifequakes seem to happen every 6–10 years or so. I've been through a bunch of them. You can never tell when they're going to happen, and you can't just sit around waiting for the sky to fall. You will almost certainly find as you go through your career that something you were becoming expert in and was the best thing going will suddenly become irrelevant. Redirect your efforts. Reinvent yourself. Get back to work.

That's enough philosophizing from an almost-seismologist who unintentionally became an expert in radar scattering just before the Cold War ended. Rayleigh waves are useful for lots of things besides earthquakes, so let's take another crack at the equations and derive some solutions that model them on purpose.

4.2 Surface Waves

Recall that a Helmholtz decomposition of the displacement vector allows us to write $\vec{u} = \nabla\Phi + \nabla \times \vec{H}$ and so in two dimensions we have

$$u_x = \frac{\partial \Phi}{\partial x} + \frac{\partial H_z}{\partial y} \qquad u_y = \frac{\partial \Phi}{\partial y} - \frac{\partial H_z}{\partial x}$$

$$u_z = -\frac{\partial H_x}{\partial y} + \frac{\partial H_y}{\partial x} \qquad \frac{\partial H_x}{\partial x} + \frac{\partial H_y}{\partial y} = 0$$

so that Φ and the components of $\vec{H}$ satisfy

$$\nabla^2\Phi = \frac{1}{c_L^2}\frac{\partial^2\Phi}{\partial t^2} \tag{4.28}$$

$$\nabla^2 H_p = \frac{1}{c_T^2}\frac{\partial^2 H_p}{\partial t^2} \qquad (p = x, y, z) \tag{4.29}$$

Note that there is no z-dependence for any of the above quantities because that's the way we've set up the geometry. We can thus consider the plane strain case where solutions are of the form

$$\Phi = f(y)e^{i\xi x-\omega t} \qquad H_z = h_z(y)e^{i\zeta x-\omega t} \tag{4.30}$$

for the time-harmonic case. These give

$$\frac{d^2f}{dy^2} + \alpha^2 f = 0 \qquad \frac{d^2h_z}{dy^2} + \beta^2 h_z = 0 \tag{4.31}$$

where

$$\alpha^2 = \frac{\omega^2}{c_L^2} - \xi^2 \qquad \beta^2 = \frac{\omega^2}{c_T^2} - \zeta^2 \tag{4.32}$$

The plane wave solutions for Φ and H_z are then given by

$$\Phi = A_1 e^{i(\xi x-\alpha y-\omega t)} + A_2 e^{i(\xi x+\alpha y-\omega t)} \tag{4.33}$$

$$H_z = B_1 e^{i(\zeta x-\beta y-\omega t)} + B_2 e^{i(\zeta x+\beta y-\omega t)} \tag{4.34}$$

where

$$\xi = k\sin\theta_1 \qquad \alpha = k\cos\theta_1 \qquad k = \omega/c_L$$

$$\zeta = K\sin\theta_1 \qquad \beta = K\cos\theta_1 \qquad K = \omega/c_L$$

Note in this solution that we have assumed $\alpha^2 > 0$ in deriving the results. Snell's law allows us to write

$$\alpha^2 = k^2\left(1 - \sin^2\theta_1\right) = k^2\left(1 - \frac{K^2}{k^2}\sin^2\theta_2\right) \tag{4.35}$$

Hence, at the critical angle, where $\frac{K^2}{k^2}\sin^2\theta_2 = 1$ we have $\alpha^2 = 0$ and beyond the critical angle where $\frac{K^2}{k^2}\sin^2\theta_2 > 1$ we have $\alpha^2 < 0$.

We are, of course, correct in saying that $\sin\theta_1$ cannot be greater than unity, but that just means that beyond the critical angle α is no longer real. Let's go back and reevaluate, by considering the governing equations for $\alpha^2 = 0$ and $\alpha^2 < 0$. Recall that

$$\Phi = f(y)e^{i(\xi x-\omega t)} \qquad \frac{d^2f}{dy^2} + \alpha^2 f = 0 \tag{4.36}$$

For the $\alpha^2 = 0$, case we have

$$\frac{d^2f}{dy^2} = 0 \Longrightarrow f = A_1 y + A_2 \tag{4.37}$$

and we conclude that

$$\Phi = (A_1 y + A_2)e^{i(\xi x-\omega t)} \tag{4.38}$$

The integration coefficient A_1 must be zero on physical grounds because non-zero A_1 would make Φ tend to infinity at large y. Yes, I know that finite A_1 is a perfectly acceptable mathematical solution, but this is exactly the sort of place where the applied mathematician gets to pull rank. We have a particular sort of physical problem that we're trying to model here. A plane elastic wave is reflecting from a free surface and there's some angle where the math tells us some stranger thing is going on in the upside down. However, it's not strange enough for a finite-amplitude disturbance to give rise to some sort of weird reflected wave at a critical angle that ends up having infinite amplitude. Sorry. We're thus left with

$$\Phi = A_2 e^{i(\xi x - \omega t)} \tag{4.39}$$

which is a plane wave of constant amplitude traveling parallel to the free surface. Nothing wrong with that. It even gets an acronym: SSLW stands for surface-skimming longitudinal wave. It shows up in ultrasonics fairly often, although real transducers are finite in size and are typically excited by a spike or a tone burst rather than a continuous signal. Nevertheless, this math does a pretty good job of describing what is seen experimentally. We could try to mess with the mathematicians at this point and argue that this analysis only works exactly at the critical angle and no surface is exactly flat even if you could point your transducer in exactly the right direction, etc., but SSLWs do show up in practice not just in the upside down.

When $\alpha^2 < 0$, we have

$$\frac{d^2 f}{dy^2} = -\bar{\alpha}^2 f = 0 \qquad \bar{\alpha}^2 = -\alpha^2 \tag{4.40}$$

and hence

$$\Phi = A_1 e^{+\bar{\alpha}y} e^{i(\xi x - \omega t)} + A_2 e^{-\bar{\alpha}y} e^{i(\xi x - \omega t)} \tag{4.41}$$

Again we discard the A_1 term because that part of the field grows exponentially as y increases. The result is a wave propagating in the x-direction whose amplitude falls off exponentially in the y-direction.

Let's step back for just a minute and see if this math makes physical sense. We start with an obliquely incident shear wave, and up to a certain angle we get both a reflected shear wave and a reflected longitudinal wave. That may still sound weird to you, but mode conversion is a fact of life with elastic waves. Some people think that's what makes them fun. Others were very happy to unceremoniously dump the elastic solid theory of the aether in favor of Maxwell's equations because mode conversion makes elastic waves kind of hard. Plus you don't need tensor notation to model electromagnetics.

Now note that the mode-converted P-wave is more shallow than the reflected SV-wave, although the particular angle the P-wave reflects at depends on the material constants, and so there will be some incidence angle where the reflected P-wave gets sort of mashed up against the surface and "reflects" parallel to it. That's the SSLW and it happens at the critical angle. Beyond the critical angle, we don't have an SSLW anymore, but because the displacement falls off exponentially with depth it's obvious that most of the wave energy is near the surface. We call this a "surface wave" and we'll find them to be useful enough that we're now going back to the field equations again and see about deliberately modeling such things. In addition, I recently purchased a *Stranger Things* Ouija board at Target and was disappointed to find that there was nothing printed on the underside. It's not the first time I've wasted twenty bucks at Target. Back to what happens beyond the critical angle.

We have

$$\nabla^2\Phi + \frac{1}{c_L^2}\frac{\partial^2\Phi}{\partial t^2} \qquad \nabla^2 H_z + \frac{1}{c_T^2}\frac{\partial^2 H_z}{\partial t^2} \tag{4.42}$$

and

$$\Phi = f(y)e^{i(\xi x - \omega t)} \qquad H_z = h_z(y)e^{i(\xi x - \omega t)} \tag{4.43}$$

but with $f(y)$ and $h_z(y)$ given by solutions to

$$\frac{d^2f}{dy^2} - \overline{\alpha}^2 f = 0 \qquad \frac{d^2h_z}{dy^2} - \overline{\beta}^2 h_z = 0 \tag{4.44}$$

where $\overline{\alpha}^2 = -\alpha^2$ and $\overline{\beta}^2 = -\beta^2$. The physical parts of the solution are

$$\Phi = Ae^{-\overline{\alpha}y}e^{i\xi(x-c_R t)} \qquad H_z = Be^{-\overline{\beta}y}e^{i\xi(x-c_R t)} \tag{4.45}$$

Boundary conditions are $\tau_{yy} = \tau_{xy} = 0$ at $y = 0$, which give two equations in the two unknown coefficients A and B

$$\left(\overline{\beta}^2 + \xi^2\right)A + 2i\overline{\beta}\xi B = 0$$

$$-2i\overline{\alpha}A + \left(\overline{\beta}^2 + \xi^2\right)B = 0$$

so that

$$\frac{A}{B} = \frac{-2i\overline{\beta}\xi}{\overline{\beta}^2 + \xi^2} = \frac{\overline{\beta}^2 + \xi^2}{2i\overline{\alpha}\xi} \tag{4.46}$$

and

$$\left(\overline{\beta}^2 + \xi^2\right)^2 - 4\overline{\alpha}\overline{\beta}\xi^2 = 0 \tag{4.47}$$

where

$$\overline{\alpha}^2 = \xi^2 - \omega^2/c_L^2 \qquad \overline{\beta}^2 = \xi^2 - \omega^2/c_T^2$$

Noting that $\omega = \xi/c_R$ gives

$$\left(2 - \frac{c_R^2}{c_T^2}\right)^2 = 4\left(1 - \frac{c_R^2}{c_T^2}\right)^{\frac{1}{2}}\left(1 - \frac{c_R^2}{c_T^2}\right)^{\frac{1}{2}} \tag{4.48}$$

This can be rewritten as:

$$\frac{c_R^2}{c_T^2}\left\{\frac{c_R^6}{c_T^6} - 8\frac{c_R^4}{c_T^4} + \left(24 - 16/\kappa^2\right)\frac{c_R^2}{c_T^2} - 16\left(1 - 1/\kappa^2\right)\right\} = 0 \tag{4.49}$$

which is a reduced cubic equation in $\frac{c_R^2}{c_T^2}$ whose roots are dependent on Poisson's ratio because

$$\kappa^2 = \frac{2(1-\nu)}{1-2\nu}$$

Note that

$$\frac{c_R}{c_T} \approx \frac{0.87 + 1.12\nu}{1+\nu}$$

The particle motion for the Rayleigh wave is elliptical and retrograde with respect to the direction of propagation, that is, counterclockwise for a wave traveling to the right, which is analogous to but opposite from water waves. The vertical component of displacement is about 1.5 times greater than the horizontal component, typically. The motion decreases in amplitude away from the surface, but at a depth of 0.192 wavelengths, the direction of particle motion reverses.

Exercise 4.6 Consider the reflection of an obliquely-incident *SH*-wave from a free surface. Recall that *SH*-waves are not coupled to *P*-waves like *SV*-waves are. Set up and solve this simpler boundary condition problem, and if any critical angles show up determine whether surface waves result.

One of the reasons I don't get too freaked out about lifequakes is that I know math, and so when the world changes and something that had been very important to me suddenly isn't, I can take the math I know and apply it to something else. To wit, the math for reflection and refraction and surface waves applies equally well to ultrasonic waves as it does to seismic waves. Rayleigh waves can destroy a structure, as we have discussed, but they can also be important for nondestructive evaluation of structures. The math is identical, and as long as we appropriately scale frequency with some characteristic size, the physical behavior is the same as well. It works for kilo-Hertz and mega-Hertz frequencies, all the way up to about 2 GHz or so at which point we will call the measurement scheme acoustic microscopy. The basic idea is that we exploit refraction beyond the critical angle to deliberately generate Rayleigh waves, which interact with surface features in our samples in particular ways that highlight topographical or material discontinuities of interest.

Another reason I don't get too freaked out about lifequakes is knowing stories like Laszlo Adler, Taine McDougal Professor Emeritus in the Department of Integrated Systems Engineering at the Ohio State University. He is known for his work in Ultrasonics, Acousto-optics, and Nondestructive Evaluation of Materials. He has been active in scientific research for over 60 years, and was a participant from the beginning of the nationwide interdisciplinary program Quantitative Nondestructive Evaluation - QNDE. Adler had just started elementary school when Nazi Germany invaded Poland in 1939. He lived with his family in Jewish-hostile Hungary until the German invasion in 1944. In about three weeks of occupation, he was moved into the Ghetto of Debrecen. A few weeks later, he was sent to a concentration camp called "the brick factory" and, from there, to a labor camp, Strasshof, near Vienna. In April 1945, the Germans sent the family to an extermination camp, but the railway station was bombed by Russian planes, which stopped the deportation. They were liberated a few weeks later by the Russian Army.

4.3 Acoustic Microscopy

Rayleigh waves turn out to be quite useful for very high-frequency inspection of small, near-surface features in objects. Many people casually throw about the term "acoustic microscopy" using it for any high-frequency ultrasound inspection method. Here, we're going to mean something quite specific. In particular, we'll refer to a scanning acoustic microscope (SAM) as a pulse-echo ultrasound system similar to a traditional ultrasound immersion tank, but with a highly-focused lens. The reason for that will become clear momentarily. A common acoustic microscope lens geometry is shown in Figure 4.10.

Successful development of the SAM began with the realization that it is not possible to make a high-resolution acoustic lens that can image more than one point of an object at a time, but it is possible to make an acoustic lens that has excellent focusing properties on its axis [5]. At moderate frequencies, the focusing lens is often made from quartz and with single-crystal sapphire lenses, it is possible to image at frequencies as high as 2 GHz. Rather than a large immersion tank, acoustic microscopes typically have a small dish of water for coupling or just some water on the surface of the sample. The very high index of refraction at the lens-water interface allows a single lens to be used even when the numerical aperture is large. A lens for use at 2 GHz would typically have a cavity of radius about 40 μm. Because the lens has a large opening angle, Rayleigh waves are generated that travel along the surface of the sample while "leaking" acoustic energy back into the coupling fluid, as long as the lens is close enough to the surface that the geometric focal point is beneath the surface, as in Figure 4.10. There is also a direct pulse-echo reflection from the surface,

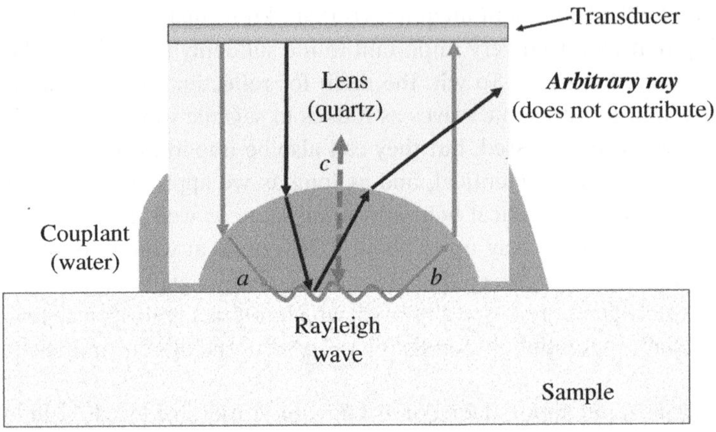

Figure 4.10 The transducer is at the top of a delay line with a lens that focuses rays beyond the critical angle. At very high frequencies, a small amount of water couplant is held between the delay line and the specimen's surface by fluid tension. The ray marked *a* is oblique enough to the specimen surface that Rayleigh waves are generated, and as they propagate along the surface they leak energy back up into the lens, as indicated by the ray marked *b*. In addition, there will be direct rays, marked *c*, which are perpendicular to the specimen surface and reflect bulk waves from the surface. Interference between the reflected bulk waves and the ray paths containing the Rayleigh waves changes as the distance between the lens and the specimen is changed. If the geometric focus is at the surface, there will be no Rayleigh waves, but as the lens is defocused to various depths below the surface, the length of the Rayleigh wave path changes. An arbitrary ray is shown, which reflects from the surface specularly and those reflections refract back into the lens at angles such that this ray doesn't contribute to the signal recorded at the transducer.

which interferes with the leaky Rayleigh wave signal, since both are recorded by the transducer at the back end of the lens/delay line. By changing the defocus distance, z, the amount of constructive or deconstructive interference between the bulk wave and the leaky Rayleigh wave can be controlled.

Consider the relatively simple case of oblique plane waves in water incident upon a elastic (solid) half-space, for example, a block of aluminum. At small angles of incidence, that is, close to perpendicular, a refracted longitudinal wave enters the solid, as expected, and of course there's also a refracted transverse wave in the solid, but close to normal incidence that mode-converted wave is pretty weak. As the angle of incidence gets to 13.56°, something odd happens. The refracted longitudinal wave disappears. When the angle of incidence gets to 29.2° the transmitted transverse wave disappears as well. If the elastic half-space was PMMA or fused silica, the same thing would happen, except that the two critical angles would be different.

In order to understand what's happening at the critical angles, let's look in detail at the mathematics of this reflection–refraction problem. We'll make the interface the plane $z = 0$ and assume without loss of generality that the xz-plane is the plane of the incident, reflected, and transmitted waves. There's no variation in y so the problem is 2D. In the fluid we have incident and reflected longitudinal waves while in the solid, we have transmitted longitudinal and transverse waves. The z-component of the four-wave vectors is denoted by α, with the appropriate subscript. The x-component of all the waves is the same, by Snell's law, and is denoted by β.

We thus write

$$\phi_0 = \phi_i e^{i\alpha_0 z} + \phi_r e^{-i\alpha_0 z} \qquad (\alpha_0 = \alpha_i = -\alpha_r) \tag{4.50}$$

for $z \geq 0$ and

$$\phi_1 = \phi_l e^{i\alpha_l z} \qquad \psi_1 = \psi_s e^{i\alpha_s z} \tag{4.51}$$

where the total displacement field is

$$\vec{u} = \nabla\phi = \nabla \times \vec{\psi} \tag{4.52}$$

and we have suppressed the common factor of $e^{i(\beta x-\omega t)}$. Boundary conditions at the liquid–solid interface are

1) Balance of normal stress and pressure
2) Vanishing shear stress in fluid
3) Continuity of normal surface displacement

In the fluid at $z = 0$, we have

$$\begin{aligned} u_z &= i\alpha_0\phi_i + i\alpha_0\phi_r \\ \sigma_{zz} &= -Bk_0^2(\phi_i + \phi_r) \\ \sigma_{xz} &- 0 \end{aligned} \tag{4.53}$$

where B is modulus. In the solid at $z = 0$ we have

$$\begin{aligned} u_z &= i\alpha_l\phi_l - i\beta\psi_s \\ \sigma_{zz} &= -2C_{44}\left(\alpha_l^2\psi_l - \alpha_s\psi_s\right) - \left(C_{11} - C_{44}\right)k_l^2\phi_l \\ &= -2C_{44}\left[\left(\beta^2 - \alpha_s^2\right)\psi_l - 2\alpha_s\beta\psi_s\right] \\ \sigma_{xz} &= C_{44}\left[-2\alpha_l\beta\phi_l - \left(\beta^2\alpha_s^2\right)\psi_s\right] \end{aligned} \tag{4.54}$$

Next we substitute

$$\rho_0\omega^2/k_0^2 = B \qquad \rho_1\omega^2/k_s^2 = C_{44}$$

$$p \equiv \left(\beta^2 - \frac{1}{2}k_s^2\right)/\beta = (\beta^2 - \alpha_s^2)/2\beta$$

Putting everything together, canceling ω^2, and using the third BC equation in the first gives

$$\alpha_0(\phi_i - \psi_r) - (\alpha_l\phi_l - \beta\psi_s) = \alpha_0(\phi_i - \phi_r) - (k_s^2/2\beta)\psi_s = 0 \tag{4.55}$$

$$\rho_0(\phi_i + \phi_r) + \frac{\rho_l}{k_s^2}\left[(\beta^2 - \alpha_s^2)\phi_l + 2\alpha_s\beta\psi_s\right] = \rho_0(\phi_i + \phi_r) + \frac{2\rho_l\beta}{k_s^2}(p\phi_l + \alpha_s\psi_s) = 0 \tag{4.56}$$

Writing the reflection coefficient as $R = \phi_r/\phi_i$ and the two stress amplitude transmission coefficients as $T_l = (\rho_1/\rho_0)\phi_l/\phi_i$ and $T_s = (\rho_1/\rho_0)\psi_s/\phi_i$, we find

$$\begin{aligned} R &= \frac{4\alpha_0\beta^2(\alpha_l\alpha_s + p^2)\rho_1/\rho_0 - \alpha_l k_s^4}{4\alpha_0\beta^2(\alpha_l\alpha_s + p^2)\rho_1/\rho_0 + \alpha_l k_s^4} \\ T_l &= \frac{-4\alpha_0 p\beta k_l^2}{4\alpha_0\beta^2(\alpha_l\alpha_s + p^2)\rho_1/\rho_0 + \alpha_l k_s^4} \\ T_s &= \frac{\alpha_1}{p}T_l \end{aligned} \tag{4.57}$$

If we recall Snell's law $\beta = k\sin\theta = k_1\sin\theta_1 = k_s\sin\theta_s$, we have $\alpha_0 = k\cos\theta$ $\alpha_1 = k_1\cos\theta_1$ $\alpha_s = k_s\cos\theta_s$, where we have used

$$p = -k_s\frac{\cos 2\theta_s}{2\sin\theta_s}$$

We can define impedances as:

$$Z = \frac{\rho_0 c_0}{\cos\theta} \qquad Z_1 = \frac{\rho_1 c_1}{\cos\theta_1} \qquad Z_s = \frac{\rho_s c_s}{\cos\theta_s} \tag{4.58}$$

and we then have

$$\begin{aligned} R(\theta) &= \frac{Z_l \cos^2 2\theta_s + Z_s \sin^2 2\theta_s - Z}{Z_l \cos^2 2\theta_s + Z_s \sin^2 2\theta_s + Z} \\ T_l(\theta) &= \frac{2Z_l \cos 2\theta_s}{Z_l \cos^2 2\theta_s + Z_s \sin^2 2\theta_s + Z} \\ T_s(\theta) &= \frac{-2Z_l \sin 2\theta_s}{Z_l \cos^2 2\theta_s + Z_s \sin^2 2\theta_s + Z} \end{aligned} \tag{4.59}$$

If we define $Z_{tot} = Z_l \cos^2\theta_s + Z_s \sin^2 2\theta_s$, we have the familiar form

$$R(\theta) = \frac{Z_{tot} - Z}{Z_{tot} + Z} \tag{4.60}$$

4.3.1 *V*(*z*) Curves

Contrast in acoustic microscopy varies very sensitively with the distance between the lens and the surface of the specimen. The first, and strongest, effect is that the signal is greatest when the specimen is at the focus of the lens. The second effect is more subtle. The contrast varies as the specimen is moved away from focus. It doesn't change monotonically, but can undergo a series of oscillations. There can even be contrast reversal as the "defocus" is changed.

In the $V(z)$ curve sketched in Figure 4.11, V is the signal intensity and z is the amount by which the specimen surface is displaced from the focal plane of the lens. Note that getting closer to the surface is "negative defocus" and that is where the most interesting phenomena occur because of the interplay of the bulk and surface waves.

Exercise 4.7 Verify the above algebra starting from first principles. Implement the final formula and make some interesting plots of $R(\theta)$ etc.

Note that at the critical angle, the reflection coefficient is unity, which means that all of the energy is reflected. Also note that up to the critical angle the phase of the reflection coefficient is zero, indicating that R is real. Beyond the critical angle, the phase is nonzero and the reflection coefficient is complex, which indicates the presence of surface waves. When there's a second critical angle, beyond which $|R|$ is unity, so there can be no excitation of longitudinal or shear waves away from the surface, there can be excitation of waves that decay exponentially into the bulk. This turns out to be the leaky Rayleigh wave that we care about in acoustic microscopy.

Consider again the diagram of an acoustic microscope transducer/lens, Figure 4.10. As we have discussed, in the ray model of an acoustic microscope lens, only two rays are of interest. Arbitrary rays are reflected specularly from the specimen and pass back through the lens with an inappropriate angle to contribute significantly to the excitation of the transducer. The axial ray c represents bulk waves, which reflect at normal incidence from the specimen. For normal incidence we have, as usual,

$$R = \frac{Z - Z_0}{Z + Z_0} \tag{4.61}$$

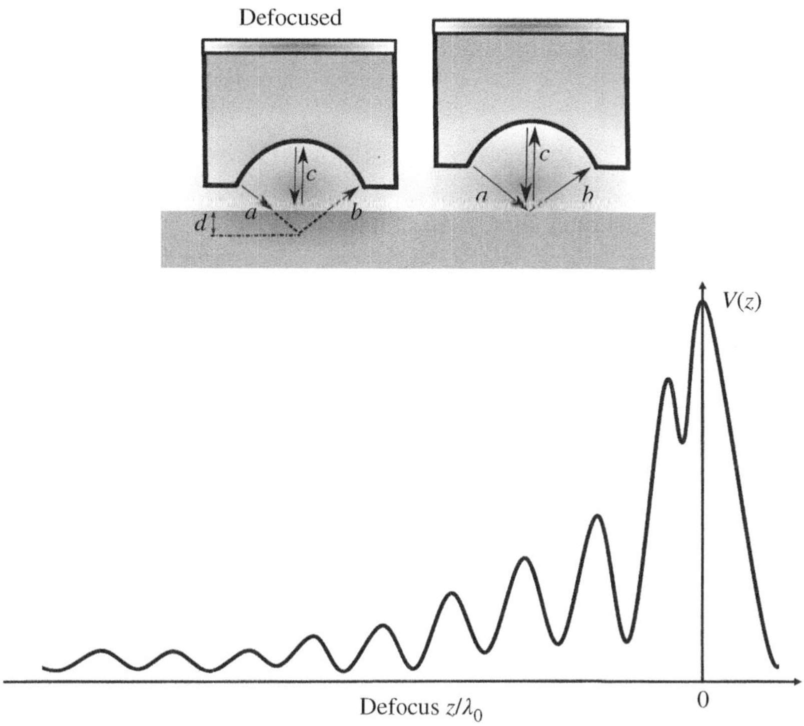

Figure 4.11 The left lens at top is defocused by an amount *d* as indicated by the dashed lines. The right lens is focused at the surface of the specimen so it will not generate Rayleigh waves. Bottom is a sketch of a *V*(*z*) curve, with the vertical axis voltage and the horizontal axis defocus. The largest signal is at zero defocus, that is, when the geometric focus is at the surface. Note that as the defocus increases (to the left of 0 on the z/λ_0-axis) the geometric focus is beneath the surface so there is constructive and destructive interference between the reflected *L* wave and the signal that includes the Rayleigh wave. Because the Rayleigh wave velocity and critical angle depend on material, the spacing of the peaks and nulls in the *V*(*z*) curve depends on material. Topography also affects the *V*(*z*) curve because changes in topography are effective changes in defocus. As an acoustic microscope lens is raster-scanned across a sample, small changes in material or topography can give a reversal of contrast, for example, going from peak to adjacent null in the *V*(*z*) response. Abrupt changes in topography, materials, cracks, etc. will reflect, refract and/or diffract the Rayleigh waves. This combination of effects, along with potentially very high frequencies, is what gives acoustic microscopy such excellent sensitivity to small features of interest.

where $Z_0 = \rho_{water} c_{water}$ and $Z = \rho_{solid} c_{solid}$, so this ray is sensitive to the material properties of the specimen, as well as to the surface topography as the time delay of the reflection is recorded.

The other important ray *a* is incident upon the specimen at the Rayleigh angle $\theta_r \equiv \sin^{-1}(v_0/v_r)$ and excites a surface wave in the specimen. The surface wave leaks energy (radiates a wave at θ_R) back into the fluid as it propagates. Because the “leaked” energy is a wave at the Rayleigh angle, it contributes to the signal measured by the transducer (*b*).

Both the axial ray and the Rayleigh ray contribute to the signal at the transducer. Although they are incident at different places on the transducer, the piezoelectric voltages they excite are summed (with respect to amplitude and phase) so that their complex-valued sum is detected, and therefore interference effects between them are observed.

As *z* changes, the phases of the two rays change at different rates so the combined effect alternates between constructive and destructive interference, viz the *V*(*z*) curve.

The phase of the geometrically reflected normal ray is $\phi_G = -2kz$. Here k is the wavenumber in the fluid. The phase ϕ_R of the Rayleigh ray is a little more complicated. We write

$$\begin{aligned}\phi_R &= -2(k \sec \theta_R - k_R \tan \theta_R)Z - \pi \\ &= -2kz\left(\frac{1 - \sin^2\theta_R}{\cos \theta_R}\right) - \pi\end{aligned} \tag{4.62}$$

since by Snell's law $k_R = k \sin \theta_R$. Trig identities allow us to simplify this to

$$\phi_R = -2kz \cos \theta_R - \pi \qquad (z < 0) \tag{4.63}$$

If the output of the transducer is detected by a phase-insensitive circuit – as is usual – then we care about

$$\phi_G - \phi_R = -2kz(1 - \cos \phi_R) + \pi \tag{4.64}$$

As the specimen is moved toward the lens, the two rays will alternate between being in phase and out of phase. The period of the resulting oscillations in $V(z)$ is the movement in the z-direction needed for a change of 2π in the relative phase

$$\Delta z = \frac{2\pi}{2k(1 - \cos \theta_R)} \tag{4.65}$$

Expressed in terms of the wavelength in water this gives

$$\Delta z = \frac{\lambda_0}{2(1 - \cos \theta_R)} \tag{4.66}$$

where $\sin \theta_R = c_0/c_R$.

Exercise 4.8 Go back through this derivation to understand the details. By an analogous derivation, show that the change in the total attenuation suffered by the Rayleigh ray is $\Delta\alpha = 2z(\alpha_0 \sec \theta_R - \alpha_R \tan \theta_R)$.

In the late 1990s, I had a postdoc who had done acoustic microscopy for her doctoral work and, although her project with me was primarily Lamb waves, I was happy to have her do some acoustic microscopy as well. The basic components necessary to do acoustic microscopy are: (i) a high-precision *xyz*-scanner (check), (ii) a fast-enough A/D board to capture very high-frequency signals (check), and (iii) a high-frequency, highly-focused transducer/lens. I told her to spec out the transducer/lens she wanted at 100 MHz because that would make use of our existing scanner and electronics, and we placed the order. The custom transducer/lens was expensive, but that's only a problem if you don't specify carefully enough the opening angle and it then won't generate the Rayleigh waves needed to do acoustic microscopy. Remember, without the Rayleigh waves at defocus you're not doing acoustic microscopy you're just doing very high-frequency pulse-echo ultrasound. Oops.

This was actually our fallback position. At NASA Langley near us was one of the handful of 2 GHz SAMs in the world (www.ksisam.com) and our friends and collaborators there didn't have any existing expertise to make good use of it because they had been through a significant reduction in force a few years prior. I volunteered to have my postdoc come there and put it to good use, but we were told no. "She might break it," we were told. "She's quite expert at this particular make and model," we responded. "What if she damages the 2 GHz transducer/lens?" we were asked. "What if we buy one of those and put in on the shelf in case it's needed?" was our offer. We eventually concluded that the civil servant in charge of that apparatus didn't actually know how to use it since the contractor who had been operating it was let go in the recent budget cuts. He seemed to be concerned that if someone who really knew how to use it came there and used it that would make it obvious that he didn't know how to use it, which is so basic. We let the issue drop, but it kind of stuck in our craw because the other owners of similar ELSAMs all seemed quite lonely and were eager to have someone come there and collaborate. I got the sense that acoustic microscopy was a dormant field in those days, a cool niche technology but without a killer application.

In the intervening years, the electronics necessary to digitize 2 GHz signals has become widely available and quite inexpensive, which means that acoustic microscopy is a technique that's worth reconsidering. From our perspective, the currently open technical questions are no longer hardware. It's scattering. In particular, the key technical challenge now is understanding and modeling/simulating the interaction of Rayleigh waves with near-surface flaws such as cracks. The ray model of acoustic microscopy that we've just been doing won't work for that, so we're going to have to plow ahead with a wave model. Fair warning, it's mathematically a bit extra.

I'm following Andrew Briggs,[3] who developed the application of acoustic microscopy to image and measure the elastic structure of a wide range of materials, developing the theory to explain how cracks and other defects give rise to surface wave contrast. His excellent 1992 monograph "Acoustic Microscopy" remains in print, with a new chapter on acoustically excited probe microscopy in the second edition [5]. There aren't many other books on the subject except [6] which I take as an indication that it's still a niche subject.

4.3.2 Wave Model of Acoustic Microscopy

Now consider the same transducer/lens/specimen geometry, but with wave theory instead of rays.

- The waves radiated by the transducer are refracted by the lens to form a spherical wavefront centered on the focal point of the lens.
- Each point on the wavefront can be described by its angular coordinates from the focus (θ, ϕ).

Thus, the spherical wave emerging from the lens can be described by a function $L_1(\theta, \phi)$. The wave is reflected from the specimen surface according to $R(\theta, \phi)$. The reflected wave returns through the lens to the transducer, where it is detected with a sensitivity $L_2(\theta, \phi)$. The total signal at focus is therefore

$$V = \int_0^{\pi/2} \int_{-\pi}^{\pi} L_1(\theta, \phi) R(\theta, \phi) L_2(\theta, \phi) \sin\theta d\phi d\theta \qquad (z = 0) \tag{4.67}$$

Since L_1 and L_2 depend on geometry it is convenient to define a pupil function

$$P(\theta, \phi) = L_1(\theta, \phi) L_2(\theta, \phi) / \cos\theta$$

to get

$$V = \int_0^{\pi/2} \int_{-\pi}^{\pi} P(\theta, \phi) R(\theta, \phi) \sin\theta \cos\theta d\phi d\theta \qquad (z = 0) \tag{4.68}$$

3 Andrew Briggs is the inaugural holder of the Chair of Nanomaterials at Oxford. "His research interests focus on materials and techniques for quantum technologies and their incorporation into practical devices. Current hot topics include vibrational states of nanotubes and charge transport through single molecules in graphene nanogaps, and machine learning for measuring and tuning quantum devices. He has more than 650 publications, with over 28,000 citations." (https://andrewbriggs.org).

If the specimen is isotropic and the lens has axial symmetry then the ϕ integration may be absorbed in P. We then get

$$V = \int_0^{\pi/2} P(\theta)R(\theta) \sin\theta \cos\theta d\theta \qquad (z = 0) \tag{4.69}$$

If the specimen is moved away from the focal position, then there will be a phase shift that depends on θ. If the wavenumber in the coupling medium is $k = 2\pi/\lambda_0$ then the z component of the wavevector is $k_z = k\cos\theta$. Defocusing the specimen by an amount z causes a phase delay of $2zk_z$ or $2kz\cos\theta$. Expressing this phase delay as the complex exponential of a phase angle gives

$$V(z) = \int_0^{\pi/2} P(\theta)R(\theta)e^{-2izk\cos\theta} \sin\theta \cos\theta d\theta \tag{4.70}$$

The hard part in this calculation is the pupil function. (You did $R(\theta)$ as an exercise recently.) However, since it's a function of the lens you're using, it's a matter of finding it once. Of course, $P(\theta)$ might be something that's supplied by the lens manufacturer.

Exercise 4.9 How would you simulate or measure $P(\theta)$? For a spherical lens it would be 2D axi-symmetric. Line-focus (cylindrical) lenses could be considered 2D, of course.

So, the ray model gives us a useful account of what's going on when Rayleigh waves dominate the contrast, but even with the additional complexity of the wave model, all we've really got so far is a way to measure material property variations and topography as we scan across a surface. That begs the obvious question: What about defects? This is a book about scattering.

One sort of defect that acoustic microscopes are quite good at finding are surface-breaking cracks. Finding them can be made a bit easier by using a cylindrical lens instead of a spherical one because it's the Rayleigh waves that hit the crack broadside which are going to interact with it most strongly. Scans will have to be a bit more complex using a cylindrical lens, of course, but then it's not like we'd be doing the measurement by hand anyway. The scanning in SAM means we set it up and press go. About the most we have to do while the SAM is scanning is keeping one eye on it to make sure that the measurement is progressing as expected. It's all motorized and automatic; we don't even have to pedal it. Of course, we do have to have a scanning system that can both move in the xyz directions and rotate 180°, and do it all very precisely if we're going to use very high frequencies.

Real cracks won't be straight and there's no particular reason to expect them to be perpendicular to the surface, but let's consider the simplest 2D cases for now. Because the Rayleigh wave will reflect from the crack, we'll get an interference pattern related to the distance the crack is from the focal line of the cylindrical lens. SAM images often have "Rayleigh fringes" which are the hallmark of such cracks. Instead of simply considering $V(z)$ as before, we now need the *much* more complicated $V(x, z)$. We define a Fourier domain scattering function $S(k_x, k_x')$, which describes how the waves incident on a 2D crack at normal incidence are scattered. The prime refers to the incident wave and the unprime to the scattered wave. We write

$$V(x, z) = \int_{-k'}^{k'} \int_{-k}^{k} e^{i(k_z'-k_z)z} L_1(k_x')L_2(k_x)S(k_x, k_x')e^{i(k_x'-k_x)x} dk_x dk_x' \tag{4.71}$$

In this, $L_1(k_x')$ and $L_2(k_x)$ are the lens functions for outgoing and incoming waves, and the crack is taken to be at the origin. The first term in the integrand allows for phase shifts due to defocus, and the last term allows for phase shifts due to lateral displacement of the lens from directly over the crack.

If we grind through the Fourier transforms (and other details that I typed out and then deleted because you really should get yourself a copy of Brigg's book anyway) and put the geometric term back in (it's unaffected by the crack) we have

$$S(k_x,k_x')=\left[R_0(k_x)+\frac{i4\alpha_R k_p}{k_x^2-k_p^2}\right]\delta(k_x-k_x')+\frac{2\alpha_R}{\pi}\left[\frac{(T_R-R_R-1)k_xk_x'+(T_R+R_R-1)k_p^2}{(k_x^2-k_p^2)(k_x'^2-k_p^2)}\right] \tag{4.72}$$

where α_R describes the attenuation of the Rayleigh wave due to radiation (leaking) into the fluid. Note carefully the assumptions explicit in this derivation:

- Rayleigh wave scattering dominates the contrast.
- The fluid loading is light.
- The effect of the crack on the geometrical reflection can be neglected.

Also note that we haven't yet done the hard part of the scattering problem since we've simply used the symbols R_R and T_R for the 2D Rayleigh wave reflection and transmission coefficients of the crack. Because the energy of the Rayleigh wave falls off exponentially with depth, it's pretty obvious that crack depth is going to have to show up in the reflection and transmission coefficients. At this stage, it's pretty much anybody's guess what else about the crack might matter in calculating how a Rayleigh wave scatters from it. Don't be too disappointed if that turns out to be something that we're forced to address numerically.

We can consider two joined quarter spaces using the same equations, except that $S(k_x,k_x')$ is now the delightful expression

$$\begin{aligned}S(k_x,k_x')&=R_0(k_x)\delta(k_x-k_x')+2i\delta(k_x-k_x')\left[\frac{\alpha_1k_{p_1}}{k_x^2-k_{p_1}^2}+\frac{\alpha_2k_{p_2}}{k_x^2-k_{p_2}^2}\right]\\&+\frac{1}{2\pi}\left\{2\sqrt{\alpha_1\alpha_2}\left[\frac{T_{R_2}}{(k_x+k_{p_1})(k_x'+k_{p_2}^2)}+\frac{T_{R_1}}{(k_x+k_{p_2})(k_x'+k_{p_1}^2)}\right]\right.\\&-\frac{2\alpha_1}{k_x+k_{p_1}}\left[\frac{1}{k_x'+k_{p_1}}+\frac{R_{R_1}}{k_x'-k_{p_1}}\right]-\frac{2\alpha_2}{k_x+k_{p_2}}\left[\frac{1}{k_x'-k_{p_2}}+\frac{R_{R_2}}{k_x'+k_{p_2}}\right]\\&\left.+\frac{4}{k_x'-k_x}\left[\frac{\alpha_1k_{p_1}}{k_x^2-k_{p_1}^2}-\frac{\alpha_2k_{p_2}}{k_x'^2-k_{p_2}^2}\right]\right\}\end{aligned} \tag{4.73}$$

where T_{R_1}, T_{R_2} and R_{R_1}, R_{R_2} are the transmission and reflection coefficients incident from sides 1 and 2, respectively. Again, the reflection and transmission coefficients are the missing pieces.

Forgetting about the fluid loading for the moment, let's consider the simplest possible 2D Rayleigh wave scattering problems. Although we may be inclinded to set up the quarterspace problem in terms of reflection and refraction of Rayleigh waves alone, it turns out that the actual situation is rather a lot more complex, as sketched in Figure 4.12. When the surface waves interact with even quite simple features, reflected and transmitted Rayleigh waves will be generated, of course, but bulk waves will also be generated due to mode conversion. What sort and how much mode conversion happens depends on the details of the feature. Whether they matter for acoustic microscopy is a whole other question. Oh, and that was for the simplest possible 2D case. If either the wave field or the crack (or both) aren't fully 2D then we'll have to talk about the scattering pattern of the Rayleigh waves on the surface as well as the 3D bulk waves, which are scattering in an even more complex pattern down into the bulk of the sample. If the sample is thick enough, the bulk waves might just "go away" but if the sample is thin or there are any near-surface scattering features then there could easily be bulk wave reflections that make their way back up through the

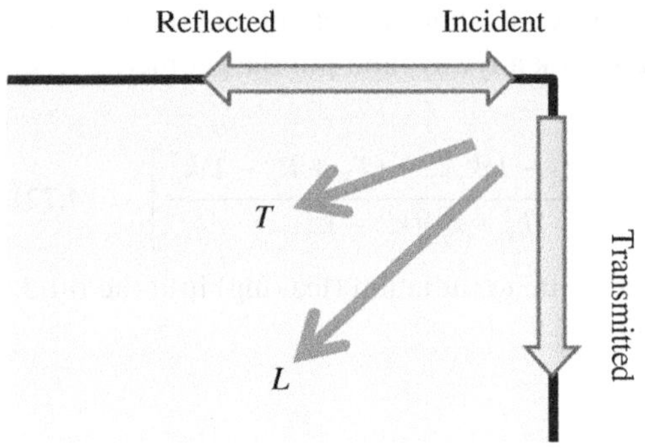

Figure 4.12 Rayleigh waves incident on the end of a quarterspace will reflect both Rayleigh and bulk waves, in addition to "transmitting" Rayleigh waves down the edge. It's not as simple as assuming some form for these wave modes and then applying the boundary conditions. The good news is that this is a rather simple geometry to simulate with EFIT.

coupling fluid and into our microscope lens. It's frightfully/delightfully complicated, depending on your perspective.

4.3.3 Detecting Cracks in Teeth

Once upon a time a sailor in a ballistic missile submarine had a toothache. A bad one. So bad that he couldn't be treated on the boat and they had to surface so he could be taken off by a helicopter for treatment at a hospital on shore. It was probably a bit more than a toothache. Whatever, but this incident prompted the Navy to conclude that better ways to screen all the sailors for potential dental issues prior to each six-month deployment were mission critical. I don't know quite why, but Cmdr Charlie Richardson of the Naval Dental School contacted Dr. Joseph Heyman[4] at the Nondestructive Evaluation Sciences Branch at the NASA Langley Research Center to see if any of the techniques NASA was developing for aircraft structural health monitoring could be adapted to monitoring dental health. Cmdr Richardson provided dental expertise, and also periodically fished around in his head vat to find samples that might be interesting for NASA to scan. Fortunately, there was a fearless gadgeteer at NESB, named John Companion, who wasn't grossed out at all. He happily scanned whatever samples the Navy provided, and patented a notional technique to use diagnostic ultrasound pulse-echo measurements to painlessly diagnose periodontal disease. It turns out that the reason you really should floss more often is that if left undisturbed, the naturally occurring bacterial plaque attacks the ligaments that hold your teeth in place. It's the crest of the periodontal ligament that dental hygienists imagine they're measuring when they poke you with

4 Dr. Joseph Heyman retired as NASA Langley Research Center's Senior Technologist in 2001. Joe was Director of Langley's Commercialization Program Office for five years, after founding and building the Nondestructive Evaluation Sciences Branch to a staff of 80 researchers over a span of more than 13 years. He joined Nascent Technology Solutions in 2001, which was acquired by Luna Innovations and operated as their Hampton, Virginia location. Joe was the Luna Chief Scientific Officer through their IPO, leading a variety of technology development and commercialization efforts funded primarily via the SBIR/STTR program.

Joe holds a BA from Northeastern and MA and PhD from Washington University, with concentrations in physics and ultrasonics. He has been an adjunct professor of physics and applied science at William and Mary since 1979, and founded the NDE program there. He holds 26 US Patents and has authored over 100 scientific publications and presentations. Among his numerous honors include being the first person to win 4 IR&D 100 Awards Blood Microemboli Monitor, Ultrasonic Bolt Monitor, Acoustic Power Detector, Geodynamic Stress Monitor. He's also in NASA Langley's Hall of Fame: https://news.northeastern.edu/2023/04/28/magazine/nasa-hall-of-honor-northeastern-coop.

Joe convinced me to come to Williamsburg in 1993 and take over the NDE graduate program at W&M. His branch continued to fund a number of research scientists through W&M and we charged NASA the off-campus overhead rate because they worked at LaRC, which allowed the savings to be used to fund several graduate students to do their dissertation research on problems of interest to NASA's mission. When Joe retired from NASA, I convinced him to join a startup near NASA that I had been shepherding in the development and commercialization of NDE technologies.

that probe before exhorting you to do a better job of flossing regularly before leaving you in the chair to contemplate your life choices while you wait for the dentist to drop by briefly, mispronounce your name, and then look under your tongue for a tumor or whatever.

The particular room at NESB, where John Companion was scanning cadaver jaws in one corner, also happened to be where some of my PhD students and one of my research scientists worked most days. I was only kind of aware of what John was doing though, until NASA LaRC reorganized and NESB got moved from Instrument Division to Materials Division. Dr. Heyman got promoted from the NESB Branch Chief to Deputy of Technology Transfer during that reorganization, which took him out of the chain of command. Normally that wouldn't be a problem, except that the new Materials Division Head, Charlie Harris, was Joe's frenemy and he promptly said something to the effect of, "Get this dental stuff out of my division. We do aerospace materials." I got a call from my colleagues at NASA asking me if I would be willing to take on the dental ultrasound project. They would provide funding for John (and me and a graduate student) and would transfer to the College all of John's equipment, materials, and supplies. I immediately agreed. It was two pickup-trucks worth of stuff, including a small wooden box with a human skull in it. NASA was very interested in highlighting non-aerospace Spinoff technologies.[5]

So that's how I ended up being a leading expert in dental ultrasonography [7]. It's a much harder problem than you might think because it's not an imaging application of medical ultrasound. Instead, the individual A-lines need to be interpreted, automatically and in real time, to identify the distance from the gum line to the crest of the periodontal ligament. At three places on each side of each tooth. In addition, the equipment needs to be foot-pedal operated and inexpensive enough that it can compete with that simple metal probe that the hygienist pokes you with repeatedly every year or so. We worked with both startups and giant dental companies, and got very close to commercializing the technology, except that our clinical partners never seemed to be able to come through with the patients needed to provide training data for machine learning. If you don't have sufficient training data, your machine learning system isn't learning, it's just memorizing your training data. I get kind of tired of trying to tell people that. Oh, and since the periodontal ligament, gums, etc. are all soft tissue, it makes no sense to do tests on cadaver jaws because the ultrasonic properties are totally different.

Another problem that diagnostic ultrasound can be useful for in the dental office is detecting cracks [8]. If you foolishly ate Captain Crunch cereal as a bedtime snack when you were a kid in the 1970s, you probably have large fillings in several of your molars and now that you're in your (late) 50s, those teeth are in some danger of cracking. X-rays do a terrible job of detecting cracks, but diagnostic ultrasound detects cracks quite easily. Acoustic microscopy can detect very, very small cracks because the surface waves scatter from even superficial cracks. The question we faced was whether acoustic microscopy could be implemented in the dental office to detect cracks in teeth. We came up with a potential solution, happily knowing that we could develop the method in the laboratory on extracted teeth, and then subsequently demonstrating clinical viability with human patients in the dental office.

As you're well aware, acoustic microscopy uses a lens with an opening angle that's greater than the critical angle to launch Rayleigh waves, which are then leaked back into the coupling fluid and picked up by the other side of the lens. Our idea was to turn the lens inside out, so that Rayleigh waves travel away from the lens in both directions. We demonstrated this in the laboratory by machining a triangular delay line as shown in Figure 4.13. We showed that in flat samples, the surface waves traveling to the left and the right away from the transducer were reflected from

5 See https://spinoff.nasa.gov/Spinoff2008/pdf/spinoff2008.pdf, pp. 60–61.

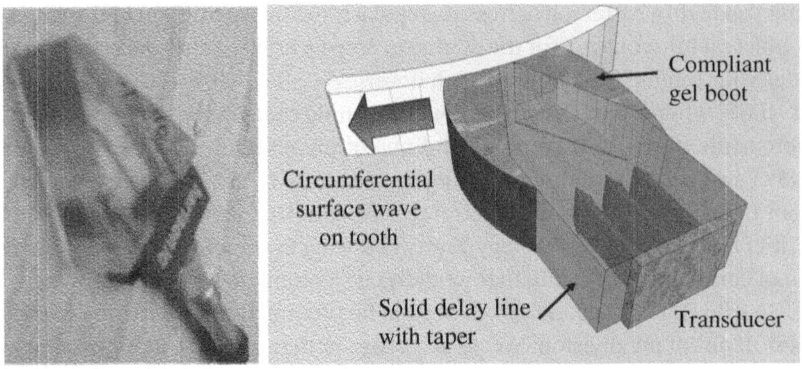

Figure 4.13 The transducer generates longitudinal waves in the delay line, which will refract across the tapered boundary and be coupled to the tooth surface by the compliant gel boot, so as to generate circumferential surface waves in the tooth enamel.

the left and right edges of the sample, respectively, and those reflections were picked up as leaky Rayleigh waves by the transducer. Since Rayleigh waves follow the curvature of a surface, we were able to generate counterpropagating circumferential surface waves that can ring all the way around a cylindrical sample and be recorded by the other side of the transducer. The concept is illustrated in Figure 4.13 where a compliant boot is used to contain couplant so that a handpiece can be used to make the measurement in teeth in a clinical setting. The circumferential Rayleigh waves will interact with any cracks in the tooth, even if the cracks are between the teeth, and time delay of the crack signals will indicate where the flaw is located.

We thought this was a clever adaptation of acoustic microscopy concepts to solve an important clinical need in dentistry, and we had done enough preliminary (phantom) experiments in the laboratory to be confident of the way forward. Two things happened, which were each horrifying in their own way but bear relating because it's important to understand that having a clever idea is only one part of developing new technologies. The first thing that happened was with an NIH proposal we wrote for ultrasound detection of cracks in teeth. It was a beautiful proposal and we had worked on it for almost a year. We were just ready to submit it via the normal mechanism when the 2009 Stimulus funding flooded the system, and so we added some additional work to the plan in order to fit the million-dollar budget cap of those Challenge Grants. You may recall that there was nearly a trillion dollars in play and our project was, to use the phrase of the day, Shovel Ready. I review quite a lot of proposals for both NIH and NSF, and during that flood of money through the system I reviewed way more than usual. All the experienced reviewers were overloaded, so they seem to have had to call on quite a lot of n00bs in order to process all the proposals swamping the system. Our proposal went to a n00b who got the scoring system backward and meant to rate it highly but gave it high numbers when lower numbers are better. I recall a blast email that went out to all reviewers reminding everybody that at NIH, low numbers are good and high numbers are bad, and thought to myself, "Who would be dumb enough to get the scoring system backward?" Also because of the flood of proposals in the system, the usual error correction mechanisms like panel discussions weren't used that cycle and so a single high score got our proposal streamlined.

The second thing that happened was with a patent application which my technology transfer office submitted (US20120040312A1). Patent examiners know how to google, and the first thing they always do is google a few terms and then generate a preliminary office action based on what they find. There's typically a back-and-forth, which is how patent agents generate billable hours and patent examiners demonstrate they're working. If there is prior art in the public domain, typically via a publication, that means the concept isn't patentable. Research proposals don't count because they are confidential and only used for review purposes, unless one of your sponsors posts your student's excellent fellowship proposal to the open internet as an example of a proposal for other students to see. The patent examiner will find that via Google and will deny the patent based on that. A nonlazy, noninept technology transfer director would respond to that office action appropriately. Mine didn't. I'm still a little salty about it.

4.3.4 Inspection of V22 Hydraulic Lines

I have one more related story. For tube inspection, this same acoustic microscopy scheme can be employed because the guided surface waves follow the curvature of the inner surface of the tube. Even for regions of the tube containing bends, the surface waves follow the compound curvature of the inner surface just as easily as they do any flat surface. Of course, it's more difficult to envision (or draw) the ray paths, but as long as the lens has an opening angle large enough to reach the Rayleigh critical angle for all orientations then the important interference effects discussed above are still present. For small tubes, the primary difficulty lies with making the transducer/delay line/lens system small enough to both fit and fish around corners. Fortunately, we had developed expertise at building small endoscopic scanning probes for urology applications. In this project, we proposed to design and fabricate such an endoscopic acoustic microscopy probe for detecting small cracks in Ti tubing. We envisioned the transducer being placed axially with the ray path bent 90° via reflection. In one implementation, shown conceptually in Figure 4.14, the transducer (dark gray) is attached to a solid delay line which has a triangular section that bends the ultrasonic beam 90° before the curved lens section projects a focused ultrasound beam at the inner surface of the tube. An alternative implementation uses the same geometry, but instead of a solid delay line to turn and focus the beam, it uses a curved reflector. In our previous urological work, we have found this scheme particularly useful for the very small probes used for scanning the prostate from inside the urethra. We built a number of these probes with diameters less than 3 mm using ultrasonic frequencies of 10 MHz. The probes have long and flexible drive shafts in order to accommodate the bend just prior to entering the prostate. For the Ti tubes, we expected to use somewhat higher frequencies and larger transducers, although a primary goal was to optimize these ultrasonic parameters for best sensitivity to the crack sizes of interest.

Kind of a neat concept, and my friends (from graduate school) and I wrote up an SBIR proposal describing it, in response to a call for inspecting hydraulic lines on the V22 Tiltrotor. As the university subcontractor, I provided my input ahead of time because the small-business prime does the submission. We got everything ready to go the day before. Everything was uploaded and ready to submit, but my colleague wanted to read everything over once more in the morning and then click submit well before the five o'clock deadline. Brahm called me mid-morning to tell me that he just realized that the deadline was 5 am, not 5 pm. They had changed the deadline in order to prevent the system from crashing when everybody clicked submit just before five. I've known Brahm[6] since

6 Dr. Brahm Rhodes is General Partner and co-founder of Malaika Ventures. He invests in early-stage climate tech startups through a climate justice lens to enable a just transition and open the door to a sustainable future for everyone. Founders can apply at: https://www.malaikaventures.com/startup-application although I assume the deadlines are firm. Dr. Rhodes is also a Fellow in On Deck Founders (ODF8) and On Deck Angels (ODA4), exploring responsible AI/ML and building entrepreneurial ecosystems in underserved communities. After earning a PhD in Engineering at Boston University, Dr. Rhodes did a postdoc in the Radiology Department at Harvard Medical School focusing on high-performance computing.

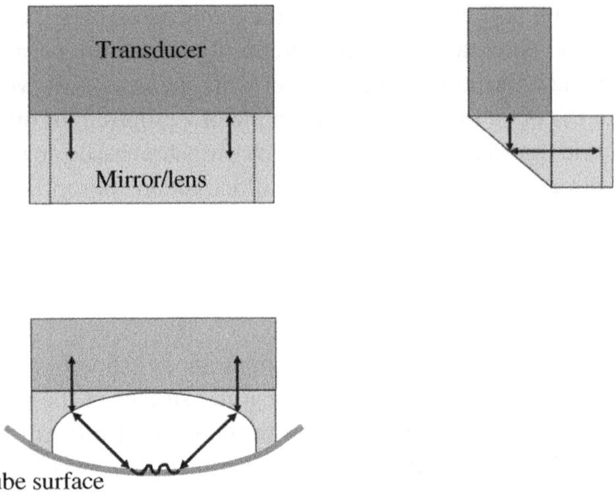

Figure 4.14 A right-angle delay line with a curved opening allows acoustic microscopy to be done on the inner surface of metal tubing. Ultrasonic ray paths shown by arrows in top, front, and side views. The solid mirror/lens both turns the rays and then focuses them sharply enough to generate the Rayleigh waves (wiggly line) that propagate along the inner surface of the tube.

we shared an office in graduate school. Back in the day, he figured out that you could drive your proposal to the FedEx office at the airport if you didn't get it done before the pickup deadline.

I've written hundreds of proposals over the years, and have reviewed many thousands. I have long since come to grips with proposals not being funded because the competition is often quite fierce. That's part of what makes it so fun to get proposals funded. You both get money to develop the solution you envisioned, and you get the satisfaction of winning the brutal competition for that money. It kind of stings when a colleague you know well or an anonymous reviewer makes a dumb mistake and your hard work goes in the dumpster. Sometimes it takes some years before you're ready to dust things off and try that concept again.

References

1 Epstein, E.J. (1982). Have you ever tried to sell a diamond? In: *The Atlantic*. Atlantic Media Company.

2 Graff, K.F. (1975). *Wave Motion in Elastic Solids*. Clarendon Press. You might find this in a used book store, but it's available at Google Books or you can get an excellent, inexpensive version at https://store.doverpublications.com/0486667456.html (accessed 16 September 2024).

3 Kolsky, H. (1953). *Stress Waves in Solids, Monographs on the Physics and Chemistry of Materials*. London: Oxford University Press. You can find it on-line as a pdf or from Dover at https://store.doverpublications.com/0486610985.html (accessed 16 September 2024).

4 Viktorov, I.A. (1967). *Rayleigh and Lamb Waves: Physical Theory and Applications*. New York: Plenum Press.

5 Briggs, A. and Kolosov, O. (2009). *Acoustic Microscopy, Monographs on the Physics and Chemistry of Materials*, 2e. Oxford. Online edn, Oxford Academic, 1 February 2010. https://doi.org/10.1093/acprof:oso/9780199232734.001.0001 (accessed 24 August 2023).

6 Maev, R.G. (2008). *Acoustic Microscopy: Fundamentals and Applications*. Wiley.

7 Ghorayeb, S.R., Bertoncini, C.A., and Hinders, M.K. (2008). Ultrasonography in dentistry. *IEEE Transactions on Ultrasonics, Ferroelectrics, and Frequency Control* 55 (6): 1256–1266. https://doi.org/10.1109/TUFFC.2008.788.

8 Bertoncini, C.A., Hinders, M.K., and Ghorayeb, S.R. (2010). Ultrasonographic detection of tooth flaws. *AIP Conference Proceedings* 1211 (1): 1559–1565.

5

Guided Waves

Guided elastodynamic waves, capable of propagating relatively long distances in plates, pipes, and shells, are quite useful for structural health monitoring. Their propagation properties depend on the frequency as well as on the thickness and material properties of the structural components. Flaws such as disbonds, corrosion, and cracks represent changes in effective thickness and/or local material properties, and therefore measurement of variations in guided wave propagation can be used to assess the integrity of these structures. Measurements can be made for a number of relative positions of small embedded or attached transducers in order to cover large areas with only a few transducers. The physics is complicated enough that it's not realistic to expect the NDT technician to directly interpret guided wave signals, or even to reproducibly set up in-the-field measurements that hinge on using angle blocks and Snell's law to select particular wave modes in order to maximize sensitivity to particular flaw types. Fortunately, data engineering can be used to automatically extract features from measurements for classification via machine learning. The unimaginably rapid increase in computational power in embeddable devices in recent years has made possible this type of sophisticated on-the-fly analyses of complex signals, meaning that the necessary "artificial intelligence" can now be implemented in compact measurement systems of the type desired for structural health monitoring. Tiny computerized instruments now contain ultrasonic and A/D function, as well as sufficient processing and memory/storage to perform all of the expert system interpretation, functions in real time.

One important characteristic of guided waves is that they can have multiple modes propagating simultaneously, each with distinctly different dispersion properties. However, most research on guided waves has used only a single mode due to the inability to robustly and automatically interpret multimode behavior. Different modes have different wave structures, with the velocities of the different modes depending on frequency-thickness in characteristic manners. Hence, for a specific defect present in the structure, such as surface corrosion, some modes may be more sensitive than other modes. In our previous work developing the technique of Lamb wave tomography, three decades ago we started actively down the road to developing data engineering algorithms that could interpret multimode guided wave signals. It turns out that we were doing machine learning before that was even fashionable.

At ultrasonic frequencies, these guided waves are confined to the plate-/shell-/pipe-like structure itself and so follow its shape and curvature, with sensitivity to material discontinuities at either surface as well as in the interior. These various guided wave modes are typically referred to as Lamb waves, a terminology we will tend to use here. Plotted vs. a combined frequency-thickness parameter in Figure 5.1, the phase and group velocities of the symmetric and anti-symmetric families of modes are as shown for aluminum plates, although other structural materials have similar behavior. With the exception of the zeroth-order modes, all Lamb wave modes have a

A Data Engineering Approach to Wave Scattering Analysis: with Applications in Radar, Sonar, Medical Diagnostics, Structural Flaw Detection and Intelligent Robotics, First Edition. Mark K. Hinders.

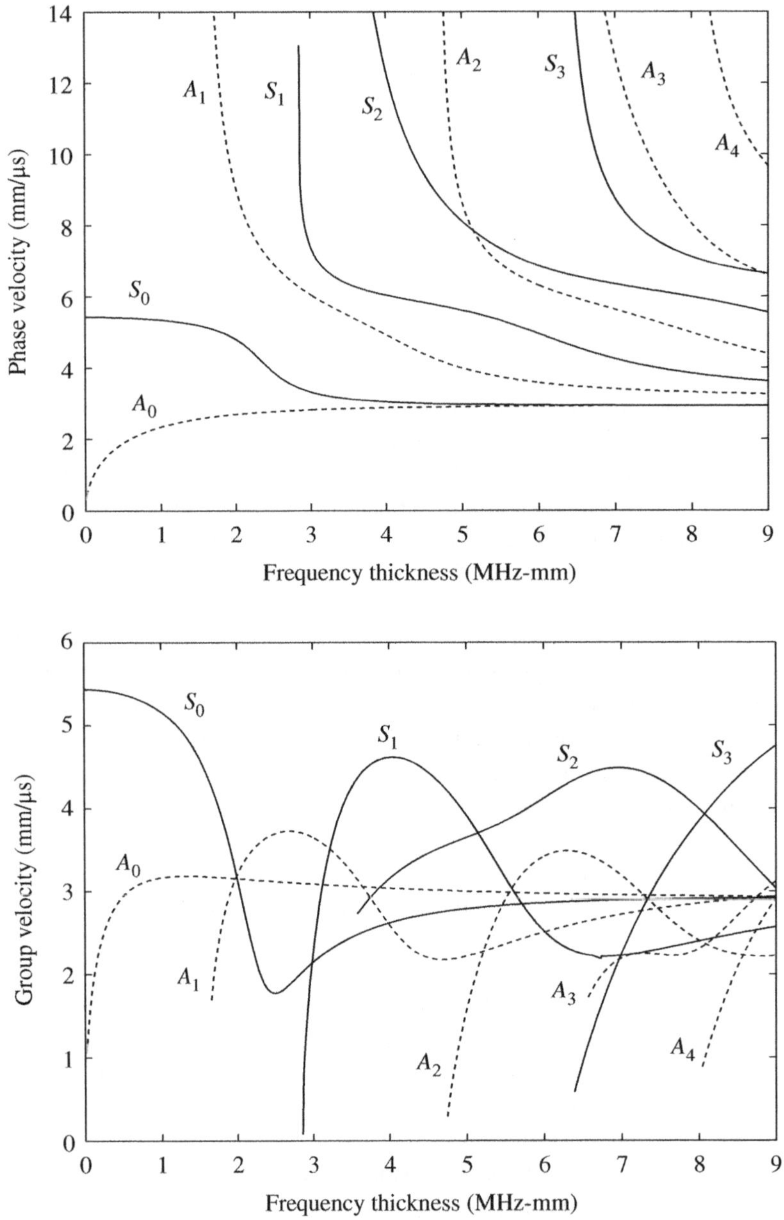

Figure 5.1 Dispersion curves for an aluminum plate. Solutions to the Rayleigh–Lamb wave equations are plotted here for both symmetric (solid lines) and antisymmetric (dashed lines) for both phase and group velocity.

cutoff frequency-thickness value where their phase and group velocities tend to infinity and zero, respectively, and hence below that value of frequency-thickness those modes do not propagate.

Characteristic change of group velocity with thickness changes is what makes Lamb waves so useful for detecting flaws, such as corrosion and disbonds, which represent effective changes in thickness. Sometimes called plate waves, the antisymmetric modes are flexural, while the symmetric modes are dilatational.

A typical Lamb wave signal will have multiple modes, and the parts of the signal that are of interest may not even be the highest in amplitude so that traditional peak-detection sorts of approaches will fail rather badly. Angle blocks, comb transducers, etc. can be employed to select purer modes, or perhaps a wise choice of frequency-thickness product can give easier to interpret signals. For example, $fd = 4$ MHz-mm could give a single fast S1 mode with all other modes much slower. Of course, most researchers simply choose a value below $fd = 2$ MHz-mm, where all but the two fundamental modes are cutoff. Some even go much lower in frequency where the A0 mode is not very significant and the S0 mode is not very dispersive, although that tends to minimize the most useful aspects of Lamb waves for flaw detection, which is that the different modes each have different through-thickness displacement profiles. Optimal detection of a particular flaw type depends on choice of modes with displacement profiles that will interact strongly with it, that is, scatter from it. Moreover, this scattering interaction will cause mode mixing to occur which can be exploited to better identify, locate, and size flaws. Lamb wave scattering from flaws is inherently a three-dimensional process.

There is a large literature on the use of Lamb waves for nondestructive evaluation and structural health monitoring.[1] Mathematically, they were first described by Lamb in 1917 [1] and demonstrated at ultrasonic frequencies by Worlton in the 1950s [2]. The main mathematical approaches to the solution of Lamb wave problems together with the most general results and conclusions are summarized in several well-known treatises on wave propagation theory written by Viktorov [3], Achenbach [4], Graff [5], Brekhovskikh and Goncharov [6], Rose [7], Auld [8], and others. A decade or so ago, I did a Lamb wave "state-of-the-art" literature review for a multinational passenger aircraft manufacturing company, which ran to more than 650 entries. A popular area of applied research, there are 100s of papers published worldwide each year. A more manageable Lamb wave literature review can be found in [9], which focuses on methods to identify and exploit signal features of interest for machine learning.

5.1 Guided Waves in Plates

For homogeneous, isotropic solids, SH-waves are not coupled to L- and SV-waves during reflection from a stress-free plane. That makes a half-space rather uninteresting because the angle of reflection is equal to the angle of incidence and that's about it, but if we instead consider an infinite plate of finite thickness some interesting effects do show up. Let's consider a plate of thickness $2b$ so that the planes $y = \pm b$ are the top and bottom stress-free surfaces, as shown in Figure 5.2.

1 Joseph L. Rose began his academic career at Drexel University as an assistant professor in 1970, after earning master's and doctorate degrees in Applied Mechanics there. In 1988, he was named Albert and Harriet Soffa Professor in Mechanical Engineering, a position he held for three years. While at Drexel, he advised 20 doctoral students. Rose also worked in industry for several years at Hale Fire Pump (known today as Hale Products) and SKF Group. He then moved to Penn State where, during 27 years, he was the principal adviser to 40 doctoral students and more than 60 master's students, retiring in 2019. Prof. Rose taught a class titled Business Opportunities in Engineering, where he alerted students to the many entrepreneurship and intrapreneurship paths to success. One of his most famous quotations for both engineering and business students that he's to be remembered for is "Failure is on the path to success. If you've never failed, it means that you are not doing anything." He holds 30 patents, has authored five text books and published more than 600 articles on such topics as ultrasonic NDE, wave mechanics, medical ultrasound, adhesive bonding, pipe and tubing inspection, bridge and rail inspection, composite material inspection, ice detection, structural health monitoring, signal processing, and pattern recognition. His publication work has received more than 18,000 citations.

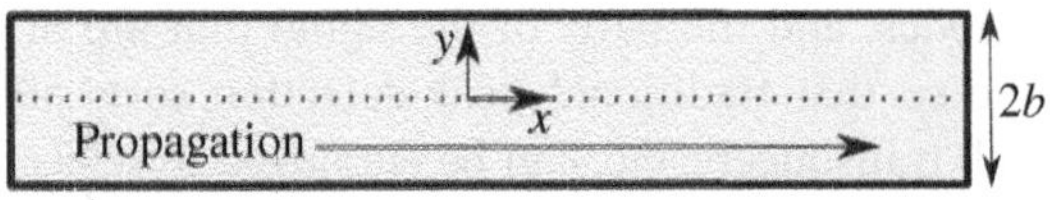

Figure 5.2 Plate of thickness $y = \pm b$ with propagation in the x-direction.

For SH-waves, the governing equation is

$$\nabla^2 u_z = \frac{1}{c_T^2}\frac{\partial^2 u_z}{\partial t^2} \tag{5.1}$$

where $u_z = u_z(x, y, t)$ and $c_T^2 = \mu/\rho$ is the shear wave speed. We'll consider solutions given by

$$u_z = h(y)e^{i(\xi x - \omega t)} \tag{5.2}$$

so that we are sure to be modeling waves that are propagating in the x-direction. Plugging (5.2) into (5.1), we conclude that the function $h(y)$ satisfies

$$\frac{d^2h}{dy^2} + \beta^2 h = 0 \qquad \beta^2 = \frac{\omega^2}{c_T^2} - \xi^2 \tag{5.3}$$

which leads to the general solution

$$u_z = \left(A_1 \sin \beta y + A_2 \cos \beta y\right) e^{i(\xi x - \omega t)} \tag{5.4}$$

The free surface boundary conditions are $\tau_{yy} = \tau_{xy} = \tau_{zy} = 0$ at $y = \pm b$, but we only need the last of these for this SH-case. This gives

$$\frac{\partial u_z}{\partial y} = 0 \,\text{at}\, y = \pm b$$

and hence the two equations

$$A_1 \cos \beta b - A_2 \sin \beta b = 0 \qquad A_1 \cos \beta b + A_2 \sin \beta b = 0$$

from which we conclude that $\cos \beta b \sin \beta b = 0$, which is satisfied by $\beta b = n\pi/2$ for $(n = 0, 1, 2, \dots)$. Thus, given a frequency, ω, the resulting wavenumber is

$$\xi = \left(\omega^2/c_T^2 - n^2\pi^2/4\right)^{\frac{1}{2}}$$

Actually, it's wavenumbers (plural) because of the integer, $n = 0, 1, 2, \dots$, which means that there are an infinite number of guided wave modes in the plate. For $n = 0$ things are relatively simple because then $\xi = \omega/c_T$ and it's just a plane shear wave. For any nonzero n, the through-thickness displacement profile is not uniform and gets more complicated the larger n is. In addition, note importantly that for nonzero n, the wavenumber is no longer proportional to frequency, which means that all those modes will be dispersive. That may not seem like a big deal, but any real signal will almost always be a pulse or toneburst in time and will thus contain at least some frequency spread about the center frequency. Dispersion will cause an initially well-formed pulse or toneburst to distort as it propagates because the different frequency components will propagate at different velocities. That turns out to make interpretation of guided wave signals delightfully complicated.

Now note that for $n = 1, 3, 5, \dots$ we have $\cos \beta b = 0 \Rightarrow A_2 = 0$ and hence

$$u_z = A_1 \sin \beta y e^{i(\xi x - \omega t)}$$

which is antisymmetric with respect to the mid-plane of the plate, $y = 0$. Similarly, if $n = 0, 2, 4, \dots$ we have $\sin \beta b = 0 \Rightarrow A_1 = 0$ and hence

$$u_z = A_2 \cos \beta y e^{i(\xi x - \omega t)}$$

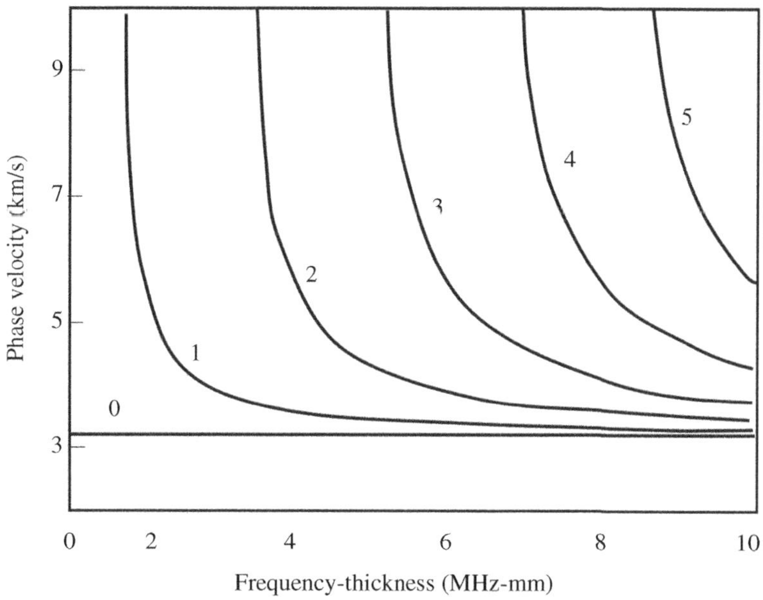

Figure 5.3 Plate wave dispersion curves for SH modes. Note that the lowest order (0) SH mode is dispersionless, so the curve is a horizontal line. The other modes are only dispersionless at the high-frequency limit, and each has a cutoff value below which they don't propagate. The horizontal axis is frequency-thickness, so we talk about thin and thick plates for small and large values of *fd*, respectively.

which is symmetric with respect to the mid-plane of the plate, $y = 0$. In discussing guided waves, it will often be useful to describe the various modes as symmetric or antisymmetric, and then to number them in order to keep track of which particular mode we're dealing with. Plots of phase velocity vs. frequency-thickness are called dispersion curves, as sketched in Figure 5.3.

Exercise 5.1 Repeat the aforementioned analysis for the case where the plate is in welded contact with an elastic half-space on the lower surface. These guided waves are called Love waves.[2]

Now let's go back and consider guided waves in a plate more generally. With

$$\vec{u} = \nabla\Phi + \nabla \times \vec{H} \qquad (\nabla \cdot \vec{H} = 0) \tag{5.5}$$

---

2 Professor A. E. H. Love, Sedleian Professor of Natural Philosophy at Oxford and Fellow of the Queen's College, died on 5 June 1940, following an operation. He was aged 77. Up to a very short time before his death, he was fulfilling the full duties of his Chair, lecturing and attending meetings of the Sub-Faculty of Mathematics. For the last few years, his health had been frail, but only to the extent that he took a taxi to go into Oxford for his lectures from St Margaret's Road where he resided. To the end, he retained full use of all his faculties, and there was never any apparent dimming of the acuteness with which he would deal with a piece of university business, the precision of his lecturing, or the wisdom and judgment which he contributed to matters of current policy. Under the present statutes, he was, at his age, ineligible for service on the Board of Faculty of the Physical Sciences, or the Board of Visitors of the University Observatory, but he never on that account forsook the society of his colleagues as they gathered at their informal lunch club before meetings of the Sub-Faculty. From: https://royalsocietypublishing.org/doi/10.1098/rsbm.1941.0015.

I currently drive oldster carpool to the Methodist church on Sunday because my mother and her friends don't drive much anymore. I get caught up in all the gossip and afterward we talk smack about the sermon. A recent sermon invoked Love waves, which the minister had read about on the internet. Since the church is directly across the street from the College, he was a little concerned that there might be someone in the sanctuary who was an expert in seismology and would call him out on some technical details. I made a point of telling him on the way out that his description was pretty much spot on. Seismologically, that is. I offered no theological opinion, although everybody in the carpool had them.

we have

$$\nabla^2\Phi = \frac{1}{c_L}\frac{\partial^2\Phi}{\partial t^2} \tag{5.6}$$

$$\nabla^2 H_p = \frac{1}{c_T}\frac{\partial^2 H_p}{\partial t^2} \qquad (p = x, y, z) \tag{5.7}$$

If variations with respect to z are excluded, we write

$$u_x = \frac{\partial\Phi}{\partial x} + \frac{\partial H_z}{\partial y} \tag{5.8}$$

$$u_y = \frac{\partial\Phi}{\partial y} - \frac{\partial H_z}{\partial x} \tag{5.9}$$

$$u_z = \frac{\partial H_y}{\partial x} - \frac{\partial H_x}{\partial y} \tag{5.10}$$

and for plane waves, we have

$$\Phi = f(y)e^{i(\xi x-\omega t)} \tag{5.11}$$

$$H_x = h_x(y)e^{i(\xi x-\omega t)} \tag{5.12}$$

$$H_y = h_y(y)e^{i(\xi x-\omega t)} \tag{5.13}$$

$$H_z = h_z(y)e^{i(\xi x-\omega t)} \tag{5.14}$$

where we have assumed without loss of generality that the waves are propagating in the xy-plane. We have, as we've done before

$$\Phi = (A\cos\alpha y + B\sin\alpha y)e^{i(\xi x-\omega t)} \tag{5.15}$$

$$H_x = (C\cos\beta y + D\sin\beta y)e^{i(\xi x-\omega t)} \tag{5.16}$$

$$H_y = (E\cos\beta y + F\sin\beta y)e^{i(\xi x-\omega t)} \tag{5.17}$$

$$H_z = (G\cos\beta y + H\sin\beta y)e^{i(\xi x-\omega t)} \tag{5.18}$$

where

$$\alpha^2 = \omega^2/c_L^2 - \xi^2 \qquad\qquad \beta^2 = \omega^2/c_T^2 - \xi^2 \tag{5.19}$$

The displacements are then

$$u_x = \{i\xi\,(A\cos\alpha y + B\sin\alpha y) + \beta\,(-G\sin\beta y + H\cos\beta y)\}\, e^{i(\xi x-\omega t)}$$

$$u_y = \{\alpha\,(-A\sin\alpha y + B\cos\alpha y) - i\xi\,(G\cos\beta y + H\sin\beta y)\}\, e^{i(\xi x-\omega t)}$$

$$u_z = \{-\beta\,(-C\sin\beta y + D\cos\beta y) + i\xi\,(E\cos\beta y + F\sin\beta y)\}\, e^{i(\xi x-\omega t)}$$

and the stresses are

$$\tau_{yy} = (\lambda + 2\mu)\frac{\partial u_y}{\partial y} + \lambda\frac{\partial u_x}{\partial x}$$

$$\tau_{xy} = \mu\left(\frac{\partial u_y}{\partial x} + \frac{\partial u_x}{\partial y}\right) \qquad\qquad \tau_{yz} = \mu\frac{\partial u_z}{\partial y}$$

Note that we have eight unknowns $A \dots H$ but the stress free boundary conditions and the top and bottom surfaces only give six equations. We can get two more from $\nabla \cdot \vec{H} = 0$, which is here

$$\frac{\partial H_x}{\partial x} + \frac{\partial H_y}{\partial y} = 0 \text{ at } y = \pm b$$

The resulting system of equations is

$$\left\{(\lambda + 2\mu)\alpha^2 + \lambda\xi^2\right\}(A\cos\alpha b + B\sin\alpha b) + 2i\mu\xi\beta\,(-G\sin\beta b + H\cos\beta b) = 0$$

$$\left\{(\lambda + 2\mu)\alpha^2 + \lambda\xi^2\right\}(A\cos\alpha b - B\sin\alpha b) + 2i\mu\xi\beta\,(G\sin\beta b + H\cos\beta b) = 0$$

$$\beta^2\,(C\cos\beta b + D\sin\beta b) + i\xi\beta\,(-E\sin\beta b + F\cos\beta b) = 0$$

$$\beta^2\,(C\cos\beta b - D\sin\beta b) + i\xi\beta\,(E\sin\beta b + F\cos\beta b) = 0$$

$$2i\xi\alpha\,(-A\sin\alpha b + B\cos\alpha b) + (\xi^2 - \beta^2)\,(G\cos\beta b + H\sin\beta b) = 0$$

$$2i\xi\alpha\,(A\sin\alpha b + B\cos\alpha b) + (\xi^2 - \beta^2)\,(G\cos\beta b - H\sin\beta b) = 0$$

$$\beta\,(-E\sin\beta b + F\cos\beta b) + i\xi\,(C\cos\beta b + D\sin\beta b) = 0$$

$$\beta\,(E\sin\beta b + F\cos\beta b) + i\xi\,(C\cos\beta b - D\sin\beta b) = 0$$

The structure becomes a bit clearer if we write this as a matrix equation. To get the dispersion relation, we need to set the determinant of the coefficient matrix to zero, but the determinant can be expanded as the product of four subdeterminants, each of which may be separately set to zero in order to recover four different families of guided wave modes.

$$\begin{vmatrix}
\left\{(\lambda+2\mu)\alpha^2+\lambda\xi^2\right\}\cos\alpha b & \left\{(\lambda+2\mu)\alpha^2+\lambda\xi^2\right\}\sin\alpha b & 0 & 0 & -2i\mu\xi\beta\sin\beta b & 2i\mu\xi\beta\cos\beta b & 0 & 0 \\
\left\{(\lambda+2\mu)\alpha^2+\lambda\xi^2\right\}\cos\alpha b & -\left\{(\lambda+2\mu)\alpha^2+\lambda\xi^2\right\}\sin\alpha b & 0 & 0 & 2i\mu\xi\beta\sin\beta b & 2i\mu\xi\beta\cos\beta b & 0 & 0 \\
0 & 0 & -i\xi\beta\sin\beta b & i\xi\beta\cos\beta b & 0 & 0 & \beta^2\cos\beta b & \beta^2\sin\beta b \\
0 & 0 & i\xi\beta\sin\beta b & i\xi\beta\cos\beta b & 0 & 0 & \beta^2\cos\beta b & -\beta^2\sin\beta b \\
-2i\xi\alpha\sin\alpha b & 2i\xi\alpha\cos\alpha b & 0 & 0 & (\xi^2-\beta^2)\cos\beta b & (\xi^2-\beta^2)\sin\beta b & 0 & 0 \\
2i\xi\alpha\sin\alpha b & 2i\xi\alpha\cos\alpha b & 0 & 0 & (\xi^2-\beta^2)\cos\beta b & -(\xi^2-\beta^2)\sin\beta b & 0 & 0 \\
0 & 0 & -\beta\sin\beta b & \beta\cos\beta b & 0 & 0 & i\xi\cos\beta b & i\xi\sin\beta b \\
0 & 0 & \beta\sin\beta b & \beta\cos\beta b & 0 & 0 & i\xi\cos\beta b & -i\xi\sin\beta b
\end{vmatrix} = 0 \tag{5.20}$$

Solution I: The solution with only the subdeterminant corresponding to nonzero C, F is

$$\beta(\xi^2 + \beta^2)\cos^2\beta b = 0 \tag{5.21}$$

and we have

$$u_x = u_y = 0 \qquad u_z = (\beta C + i\xi F)\sin \beta y e^{i(\xi x - \omega t)} \tag{5.22}$$

Solution II: The solution with only the subdeterminant corresponding to nonzero D, E is

$$\beta(\xi^2 + \beta^2)\sin^2 \beta b = 0 \tag{5.23}$$

and we have

$$u_x = u_y = 0 \qquad u_z = (-\beta D + i\xi E)\cos \beta y e^{i(\xi x - \omega t)} \tag{5.24}$$

Solution III: The solution with only the subdeterminant corresponding to nonzero A, H is

$$\frac{\tan \beta b}{\tan \alpha b} = -\frac{4\alpha\beta\xi^2}{(\xi^2 - \beta^2)^2} \tag{5.25}$$

and we have

$$\begin{aligned} u_x &= (i\xi A \cos \alpha y + \beta H \cos \beta y)e^{i(\xi x - \omega t)} \\ u_y &= -(\alpha A \sin \alpha y + \xi H \sin \beta y)e^{i(\xi x - \omega t)} \\ u_z &= 0 \end{aligned} \tag{5.26}$$

Solution IV: The solution with only the subdeterminant corresponding to nonzero B, G is

$$\frac{\tan \beta b}{\tan \alpha b} = -\frac{(\xi^2 - \beta^2)^2}{4\alpha\beta\xi^2} \tag{5.27}$$

and we have

$$\begin{aligned} u_x &= (i\xi B \sin \alpha y - \beta G \sin \beta y)e^{i(\xi x - \omega t)} \\ u_y &= (\alpha B \cos \alpha y - i\xi G \cos \beta y)e^{i(\xi x - \omega t)} \\ u_z &= 0 \end{aligned} \tag{5.28}$$

The frequency spectra for *I* and *II* are just the *SH* plate waves we discussed previously. *III* and *IV* are coupled *L*- and *SV*-waves in a plate, which are usually called Lamb waves. Lamb waves turn out to be incredibly useful for inspecting large plate-like structures, which we'll discuss shortly. First, however, notice the obvious point that Lamb waves are going to be dispersive because the frequency spectra are given by the solution of transcendental equations. For both *III* and *IV* there will be an infinite number of antisymmetric and symmetric Lamb wave modes, respectively. This was illustrated for an aluminum plate in Figure 5.1. Corresponding dispersion curves for other metals look pretty similar, although there are subtle differences.

You never quite know when esoteric knowledge of wave propagation will be useful. In the 80 years since a trucker named Malcolm McLean first conceived the modern-day shipping container, shipping has exploded into a US$ 400 billion a year industry [10]. The shipping container, however, has revolutionized much more than the shipping industry. It has radically transformed supply chains, fundamentally changed domestic and international economies across the world, and changed societies in the process. In a recent survey by the World Shipping Council, whose members operate approximately 90% of the global liner ship capacity, the international liner shipping industry reported transporting approximately 130 million containers packed with cargo, with an estimated value of more than US$ 4 trillion [11]. These metal boxes will remain a staple in the future for an industry that relies upon their ease of use for transporting all types of cargo.

Despite their versatility, shipping containers are easy to break into, and difficult to keep track of, as they are transported around the globe. Cargo theft and cargo loss are estimated to cost the industry at least US$ 50 billion annually, according to The National Cargo Security Council [12]. Losses can occur in a myriad of ways, from containers being mislaid, mislabeled, or simply failing to arrive at their destination, to instances involving premeditated criminal intent, such as breaking into ports to steal goods, or pirates attacking crews at gunpoint for their valuable cargo. In the last several decades, security and visibility of the global supply chain has largely been addressed through improved locking mechanisms and radio tracking devices that are attached to shipping containers, and it has been estimated that there are now over one million remote tracking systems in containers worldwide [13]. The latest adaptations to the existing technology include GPS-capable devices that use satellite, cellular, or Wi-Fi connectivity.

These methods have come at a high monetary cost due to the expensive infrastructure needed to track devices that use short-range radio frequency identification (RFID) and in some cases, high bandwidth costs to transmit the data wirelessly. Vulnerabilities and shortfalls still exist in current RFID tracking methods, especially during disaster relief when wireless networks are destroyed, in emerging markets that exist outside wireless coverage areas, and in providing real-time information to distant stakeholders.

Technology now offers lower cost and higher-efficiency sensors connected to the internet that can better inform decision-making. Recent developments in this technology have improved sensing capability with lower power consumption. Advances in computational abilities, to handle the large amount of data produced by these sensors, have made it possible to uncover information. Ubiquitous devices still have bandwidth and power limitations, but the challenge of working in austere environments has been improved via on-board processors with the ability to do edge computing. By processing data near the sensor, bandwidth is dramatically reduced between the sensors and any central data center. Thus, it is now possible to equip containers with a device that can talk to stakeholders during their entire voyage in the transportation network. Better yet, it is possible to put sensors on the inside of the shipping container, without drilling holes to feed signal wires, by transmitting data through the container wall using ultrasonic guided waves, as illustrated in Figure 5.4. The first antitheft sensor that we explored systematically was long-wave infrared [14]. IR cameras used to cost as much as my house, but now they are snap-on attachments to smartphones that cost at most a few hundred dollars. They are particularly sensitive to body heat, which makes them ideally suited to detection of thieves breaking into shipping containers. Even very quiet ninjas can't

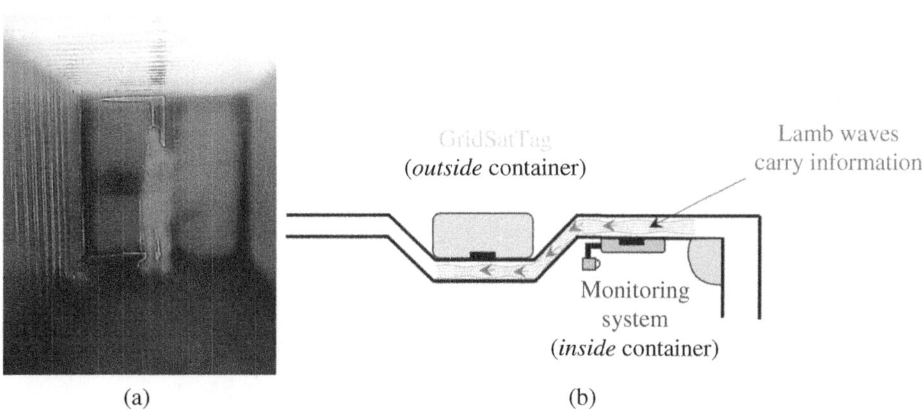

Figure 5.4 Infrared cameras can detect a person's body heat entering a shipping container (a). The Lamb waves are able to propagate across the corrugation (b) and so both the internal and external units can be located to minimize damage as cargo is being loaded/unloaded and containers are being stacked/unstacked.

hide from IR, and we've demonstrated an approach to use cadiotropic reflectors to cover the interior of a 40-ft container, and then use machine learning to automatically detect the presence of humans. A GRIDSAT Tag Architecture utilizes interconnected GRID tags configured in a self-healing mesh network, which communicate with a GRIDSAT tag. Messages are compiled and transmitted via satellite communications to stakeholders through a GIS Software Application Package [15].

Lamb waves are used to communicate sensor/imager information gathered inside a sealed shipping container to the GRID tag attached to the outside of the container. One subtlety is that encoding information into the Lamb wave signals via frequency modulation must take into account the dispersion of the various Lamb wave modes, so the choice of frequency relative to container wall thickness is rather important. I feel a little bad that we haven't really done any scattering yet, so Figure 5.5 is a teaser from many years ago where scattering of guided waves showed up. Sun and Johnston [16] experimentally studied interaction of Lamb waves with adhesively bonded lap splice and doubler joints in aluminum sheets with and without rivets. They monitored the amplitude of the S_0 mode while scanning a pair of contact water-coupled piezoelectric transducers along and across the joint. For the doublers, disbonds showed up as the amplitude maxima, for the lap joints, they corresponded to the amplitude minima. The rivet rows produce amplitude minima due to scattering. Interaction of guided waves with defects is a very complicated mechanism involving scattering [17] and mode conversion. One goal of our research has been the development of Lamb wave tomography as a fully automated inspection technique that is robust enough to rapidly assess large areas, and represent the defect information in a form, convenient for visual interpretation. Key to this concept is methods for automatically extracting features from many, many measurements in order to provide inputs for tomographic reconstruction. That could be used as our working definition of data engineering.

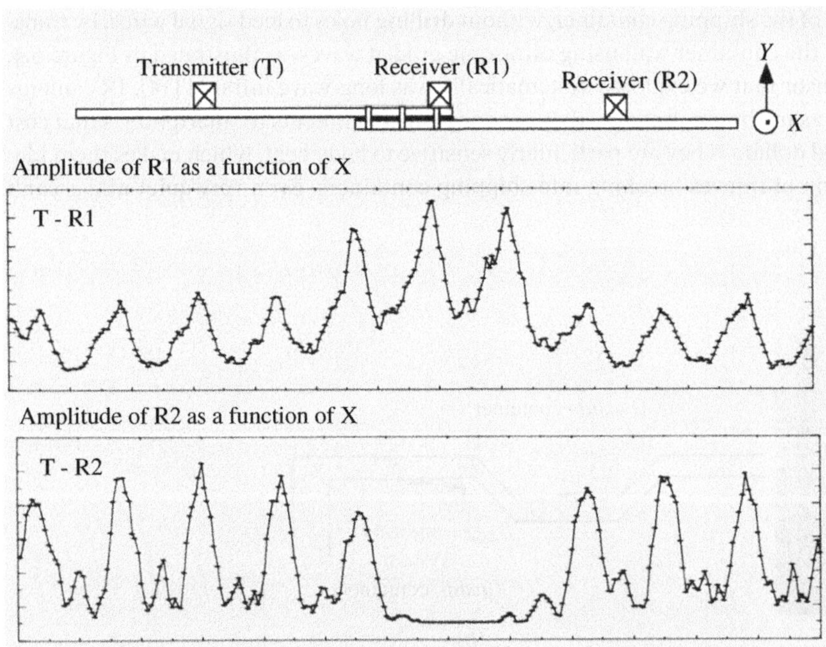

Figure 5.5 Lamb wave detection of disbonds in a riveted lap joint. As the transducers are scanned along the joint, the rivet-scattering signal is enhanced/suppressed depending on whether the receiver is on the top/bottom plate.

5.2 Cylindrical Guided Waves

Guided waves in plates and shells are incredibly useful for a wide range of applications, but as you may have noticed there is some significant amount of math involved. That's a feature, not a bug, BTW. We worked hard for many years to learn lots of math. We like it when high-level math provides a barrier to competition. With that in mind, let's level up.

There are analogous guided waves in rods and pipes, which are likewise incredibly useful for a wide range of applications. Mechanical vibrations can be induced to propagate from one place to another, accumulating information about the structure as they travel along. The vibrations recorded at some other location can then be interpreted to infer things about structural health. For example, consider a section of piping that goes under a road or is covered by lagging or transits a compartment that is inaccessible. Guided waves, which are sensitive to pipe wall thickness, can be used to detect corrosion-thinning in places where it would be expensive, inconvenient, or unpleasant to expose the pipe and check directly. On aircraft carriers there are sewage tanks tucked here and there, because when one flushes a toilet that doesn't get dumped into the ocean. The same is true for airliners, BTW.[3] Periodically, when a carrier is in for maintenance, some sailor has to go into the sewage tanks and inspect for corrosion that might cause a leak. You may agree with me that Lamb waves can be a better way to do that. In that case, I think you'd also agree that guided waves might be quite useful to inspect piping that for whatever reason has been routed through sewage tanks. I happen to know that this is in issue because aircraft carriers are built near me in Newport News, Virginia. We had the apprentice shipbuilders make us up some representative samples of piping going through bulkheads so that we could demonstrate in the laboratory that such a guided wave inspection scheme could be made to work. We presented our preliminary results at an industry conference with the excellent name MegaRust. Corrosion is a significant cost driver for the Navy.

Back to math. Way back in Chapter 2, I suggested an exercise where you go through separation of variables for the Helmholtz equation in cylindrical and spherical coordinates. Here's one place where you'll need that in cylindrical coordinates. There will be others. If you've never come across Bessel, Neumann, and Hankel functions before, take a bit of time to read up on them. Not just on Wikipedia. Ideally, you would have picked up used copies of "Methods of Theoretical Physics" by Herman Morse and Philip M. Feshbach which was published in 1953.[4] It's two volumes and over a thousand pages. You can probably find downloadable versions as well. I walked across the Charles River on an annually Smoothed bridge[5] and paid real money for copies at the MIT bookstore when I was a graduate student at BU in the 1980s. We'll talk more about these functions in the chapter on scattering from cylinders, so you can also read ahead in this book, but do keep an eye out for both volumes of Morse and Feshbach. Having a reference like that at hand and knowing how to find answers to high-level math questions in it can be highly lucrative.

Now consider guided wave propagation in a solid, cylindrical rod as shown in Figure 5.6. As in the treatment of plate problems, we formulate the stresses and displacements in terms of potential functions Φ and $\vec{H}$:

$$\vec{u} = \nabla\Phi + \nabla \times \vec{H} \qquad \nabla \cdot \vec{H} = F(r,t) \tag{5.29}$$

3 See MythBusters S9.E2 https://www.imdb.com/title/tt1887371 where Grant, Kari and Tory look into the urban legend of the Blue Ice – frozen toilet waste falling from the sky.

4 When one of my graduate advisors was a graduate student at MIT in the 1950s, he got assigned the task of working through all the exercises in both volumes of Morse and Feshbach to proof them prior to publication. After MIT, Guido Sandri was a postdoc for Oppenheimer at Princeton.

5 https://www.atlasobscura.com/places/harvard-bridge-smoot-measurements.

Figure 5.6 Cylindrical rod of radius $r = a$ with propagation in the z-direction.

where $F(r,t)$ is an arbitrary function. In cylindrical coordinates, we can write the components of displacement as:

$$u_r = \frac{\partial \Phi}{\partial r} + \frac{1}{r}\frac{\partial H_z}{\partial \theta} - \frac{\partial H_\theta}{\partial z} \tag{5.30}$$

$$u_\theta = \frac{1}{r}\frac{\partial \Phi}{\partial \theta} + \frac{\partial H_r}{\partial z} - \frac{\partial H_z}{\partial r} \tag{5.31}$$

$$u_z = \frac{\partial \Phi}{\partial z} + \frac{1}{r}\frac{\partial}{\partial r}(rH_\theta) - \frac{1}{r}\frac{\partial H_r}{\partial \theta} \tag{5.32}$$

Just to reinforce the point about having the right reference materials at hand, how often in life does one need to know gradient and curl in cylindrical coordinates? It's not something you remember; it's something you look up when you need it. Knowing where to look is the key skill. I know it seems old fashioned to look things like that up in musty old books, but remember that something like Morse and Feshbach was thoroughly checked for accuracy 70 years ago and has been used by literal generations of researchers ever since. Who knows whether the most recent edits of Wikipedia are correct?

The potentials Φ and $\vec{H}$ satisfy the scalar and vector wave equations

$$\nabla^2\Phi + \frac{1}{c_L^2}\frac{\partial^2\Phi}{\partial t^2} \qquad \nabla^2\vec{H} + \frac{1}{c_L^2}\frac{\partial^2\vec{H}}{\partial t^2} \tag{5.33}$$

where

$$\nabla^2\Phi = \frac{\partial^2\Phi}{\partial r^2} + \frac{1}{r}\frac{1}{r}\frac{\partial \Phi}{\partial r} + \frac{1}{r^2}\frac{\partial^2\Phi}{\partial \theta^2} + \frac{\partial^2\Phi}{\partial z^2} \tag{5.34}$$

$$\nabla^2\vec{H} = \left(\nabla^2 H_r - \frac{H_r}{r} - \frac{2}{r^2}\frac{\partial H_\theta}{\partial \theta}\right)\hat{e}_r + \left(\nabla^2 H_\theta - \frac{H_\theta}{r} + \frac{2}{r^2}\frac{\partial H_r}{\partial \theta}\right)\hat{e}_\theta + \nabla^2 H_z \hat{e}_z \tag{5.35}$$

Sorry, but I can't help mentioning again that Laplacian in cylindrical coordinates is one of those things that you might just need some time. We just needed the Laplacian of a scalar and then a vector in cylindrical coordinates. I'm pretty sure I typed those equations correctly, but how sure are you that there isn't some typo? Authoritative reference sources on your bookshelf can make it easy to check things like that to defend against propagating a simple error in your own work.

The stresses are given by Hooke's law and the strain–displacement relations, which was simple to do in Cartesian coordinates, but they are somewhat more complicated in cylindrical coordinates

$$\tau_{rr} = \lambda\left(\frac{\partial u_r}{\partial r} + \frac{u_r}{r} + \frac{1}{r}\frac{\partial u_\theta}{\partial \theta} + \frac{\partial u_z}{\partial z}\right) + 2\mu\frac{\partial u_r}{\partial r} \tag{5.36}$$

$$\tau_{\theta\theta} = \lambda\left(\frac{\partial u_r}{\partial r} + \frac{u_r}{r} + \frac{1}{r}\frac{\partial u_\theta}{\partial \theta} + \frac{\partial u_z}{\partial z}\right) + 2\mu\left(\frac{u_r}{r} + \frac{1}{r}\frac{\partial u_\theta}{\partial \theta}\right) \tag{5.37}$$

$$\tau_{zz} = \lambda\left(\frac{\partial u_r}{\partial r} + \frac{u_r}{r} + \frac{1}{r}\frac{\partial u_\theta}{\partial \theta} + \frac{\partial u_z}{\partial z}\right) + 2\mu\frac{\partial u_z}{\partial z} \tag{5.38}$$

$$\tau_{r\theta} = \mu\left(\frac{\partial u_\theta}{\partial r} - \frac{u_\theta}{r} + \frac{1}{r}\frac{\partial u_r}{\partial \theta}\right) \tag{5.39}$$

$$\tau_{\theta z} = \mu\left(\frac{1}{r}\frac{\partial u_z}{\partial \theta} + \frac{\partial u_\theta}{\partial z}\right) \tag{5.40}$$

$$\tau_{zr} = \mu\left(\frac{\partial u_r}{\partial z} + \frac{\partial u_z}{\partial r}\right) \tag{5.41}$$

The boundary conditions for the problem will be

$$\tau_{rr} = \tau_{r\theta} = \tau_{zr} = 0 \text{ at } r = a \tag{5.42}$$

which is only three equations instead of the six that we had for the plate. Because Φ and $\vec{H}$ satisfy the wave equation, we can write the general form of these potential functions for a z-propagating harmonic wave as:

$$\Phi = f(r)\Theta_\phi(\theta)e^{i(\xi z - \omega t)} \tag{5.43}$$

$$H_r = h_r(r)\Theta_r(\theta)e^{i(\xi z - \omega t)} \tag{5.44}$$

and so on. Plugging back into the wave equation, we find for Φ:

$$f''\Theta_\phi + \frac{1}{r}f'\Theta_\phi + \frac{1}{r^2}f\Theta_\phi'' - \xi^2 f\Theta_\phi = -\frac{\omega^2}{c_L^2}f\Theta_\phi \tag{5.45}$$

which can be rewritten as:

$$r^2\frac{f''}{f} + r\frac{r'}{r} - \left(\xi^2 - \frac{\omega^2}{c_L^2}\right)r^2 = -\frac{\Theta_\phi''}{\Theta_\phi} = k^2 \tag{5.46}$$

Thus we have the solution

$$\Theta_\phi = A\sin k\theta + B\cos k\theta \tag{5.47}$$

and single-valuedness on Θ_ϕ makes $k = n$ an integer. Later considerations on the nature of θ-dependence for the longitudinal, torsional, or flexural modes would lead us to discard either the sin or cos terms in each. Hence we write

$$\Phi = f(r)\cos n\theta e^{i(\xi z - \omega t)} \tag{5.48}$$

$$H_r = h_r(r)\sin n\theta e^{i(\xi z - \omega t)} \tag{5.49}$$

$$H_\theta = h_\theta(r)\cos n\theta e^{i(\xi z - \omega t)} \tag{5.50}$$

$$H_z = h_z(r)\sin n\theta e^{i(\xi z - \omega t)} \tag{5.51}$$

Now consider the radial dependence. Recall that for Φ we have

$$\frac{d^2 f}{dr^2} + \frac{1}{r}\frac{df}{dr} + \left(\alpha^2 - \frac{n^2}{r^2}\right)f = 0 \tag{5.52}$$

where we have written $\alpha^2 = \frac{\omega^2}{c_L^2} - \xi^2$. This is Bessel's equation of order n, which has the solution

$$f(r) = AJ_n(\alpha r) + BN_n(\alpha r) \tag{5.53}$$

For $n = 0, 1, 2$, these radial functions are certainly available at Wikipedia, so take a look and note that the Neumann functions are all singular at the origin. Since $f(r)$ must be finite at $r = 0$, we here set $B = 0$. We get a similar result for $h_z(r)$ but with α^2 replaced by $\beta^2 = \frac{\omega^2}{c_T^2} - \xi^2$

$$h_z(r) = BJ_n(\beta r) \tag{5.54}$$

The remaining two equations for h_r and h_θ will be coupled

$$\frac{d^2h_r}{dr^2} + \frac{1}{r}\frac{dh_r}{dr} + \frac{1}{r^2}\left(-n^2h_r + 2nh_\theta - h_r\right) - \xi h_r + \frac{\omega^2}{c_T^2}h_r = 0$$

$$\frac{d^2h_\theta}{dr^2} + \frac{1}{r}\frac{dh_\theta}{dr} + \frac{1}{r^2}\left(-n^2h_\theta + 2nh_r - h_\theta\right) - \xi h_\theta + \frac{\omega^2}{c_T^2}h_\theta = 0$$

Here comes a nice trick. It's one of those clever things that someone figured out many years ago and wrote down in a musty book. Simply add and subtract these two equations to get

$$\left\{\frac{d^2}{dr^2} + \frac{1}{r}\frac{d}{dr} + \beta^2 - \frac{(n+1)^2}{r^2}\right\}(h_r - h_\theta) = 0 \tag{5.55}$$

$$\left\{\frac{d^2}{dr^2} + \frac{1}{r}\frac{d}{dr} + \beta^2 - \frac{(n-1)^2}{r^2}\right\}(h_r + h_\theta) = 0 \tag{5.56}$$

which have the solutions

$$h_r - h_\theta = 2B_2J_{n+1}(\beta r) \tag{5.57}$$

$$h_r + h_\theta = 2B_1J_{n-1}(\beta r) \tag{5.58}$$

where we haven't bothered to write the Neumann part of the solutions. From these, we can then write

$$h_r = B_1J_{n-1}(\beta r) + B_2J_{n+1}(\beta r) \tag{5.59}$$

$$h_\theta = B_1J_{n-1}(\beta r) - B_2J_{n+1}(\beta r) \tag{5.60}$$

We thus have four coefficients A, B, B_1, B_2 but only three boundary conditions. Recall that in the plate case, we used $\nabla \cdot \vec{H} = 0$ to get additional equations. Here we will use the gauge invariance to eliminate one of the constants. Setting $B_1 = 0$ gives $h_r(r) = -h_\theta(r)$, which results in no loss of generality. We then have

$$\begin{aligned}
u_r &= \left\{f' + \left(\frac{n}{r}\right)h_z + \xi h_r\right\}\cos n\theta e^{i(\xi z - \omega t)} \\
u_\theta &= \left\{-\left(\frac{n}{r}\right)f\xi h_r - h_z'\right\}\sin n\theta e^{i(\xi z - \omega t)} \\
u_z &= \left\{-\xi f - h_z' - (n+1)h_r/r\right\}\cos n\theta e^{i(\xi z - \omega t)}
\end{aligned} \tag{5.61}$$

and the stress components we need are

$$\begin{aligned}
\tau_{rr} &= \left[-\lambda(\alpha^2 + \xi^2)f + 2\mu\left\{f'' + \frac{n}{r}\left(h_z' - \frac{h_z}{r}\right) + \xi h_r'\right\}\right]\cos n\theta e^{i(\xi z - \omega t)} \\
\tau_{r\theta} &= \mu\left[-\frac{2n}{r}\left(f' - \frac{f}{r}\right) - (2h_z'' - \beta^2h_z) - \xi\left(\frac{n+1}{r}h_r - h_r'\right)\right]\sin n\theta e^{i(\xi z - \omega t)} \\
\tau_{rz} &= \mu\left[-2\xi f' - \frac{n}{r}\left\{h_r' + \left(\frac{n+1}{r} - \beta^2 + \xi^2\right)h_r\right\} - \frac{n\xi}{r}h_z\right]\cos n\theta e^{i(\xi z - \omega t)}
\end{aligned} \tag{5.62}$$

Dispersion relations are calculated by plugging f, h_r, h_z into these, setting $r = a$ and then setting the determinant of the 3×3 characteristic matrix to zero.

$$0 = \begin{vmatrix} \left\{ \frac{\lambda(\alpha^2+\xi^2)(\alpha a)^2}{2\mu\alpha^2} + (\alpha a)^2 - n^2 \right\} J_n(\alpha a) + \alpha a J_n'(\alpha a) & \left\{ n^2 - (\beta a)^2 \right\} J_n(\beta a) - \beta a J_n'(\beta a) & 2n \left\{ \beta a J_n'(\alpha a) - J_n(\alpha a) \right\} \\ n \left\{ \alpha a J_n'(\alpha a) - J_n(\alpha a) \right\} & -n \left\{ \beta a J_n'(\beta a) - J_n(\beta a) \right\} & -\left\{ 2n^2 - (\beta a)^2 \right\} J_n(\beta a) + 2\beta a J_n'(\beta a) \\ -\alpha a J_n'(\alpha a) & \frac{\xi^2-\beta^2}{2\xi^2} \beta a J_n'(\beta a) & n J_n(\beta a) \end{vmatrix} \tag{5.63}$$

Note that the θ-dependence as well as the factor $e^{i(\xi z - \omega t)}$ are not in the coefficient matrix. As we found in the plate case, there will be an infinite number of solutions and most of these guided wave modes will have dispersion because wave number and frequency will not be proportional. Fortunately, some of the modes do have fairly simple physical interpretation, so we can at least give names to the various families of guided wave modes in rods.

5.2.1 Torsional Modes in a Rod

A family of torsional modes results when only the u_θ displacement is assumed to exist. Wikipedia tells me that, "An Indian burn … is a pain-inducing prank, where the prankster grabs onto the victim's forearm or wrist, and starts turning the skin away from themselves with one hand, and with another hand towards themselves, causing an unpleasant burning sensation to the skin." While I grew up using that term, that was the 1970s and it's now rather offensive and I would never use such a term. Moreover, what we now call the Americas were fully populated in 1492 when Columbus sailed the ocean blue with his guns, germs, and steel. That the New World was depopulated over the course of a century or two is profoundly sad. It's a shame that the Aztecs didn't have the technology and foresight to tell the conquistadors to, "Get off my lawn!"

Back to alternating torsional deformations in a metal rod, not your arm. Since only $H_z \neq 0$ and we have

$$H_z = B_3 J_0(\beta r) e^{i(\xi z - \omega t)} \tag{5.64}$$

Using the $\tau_{r\theta} = 0$ boundary condition

$$r \frac{\partial}{\partial r} \left(\frac{u_\theta}{r} \right) = 0 \text{ at } r = a$$

gives

$$\beta a J_0(\beta a) = 2 J_1(\beta a) = 0 \tag{5.65}$$

The dispersion curves for these torsional modes are sketched later (Figure 5.7). Note that the lowest-order mode is dispersionless just as for *SH*-waves in a plate and that all other modes are dispersive.

5.2.2 Longitudinal Waves in a Rod

Next consider the case when $u_\theta = 0$ and recall that

$$u_r = \frac{\partial \Phi}{\partial r} - \frac{\partial H_\theta}{\partial z} \tag{5.66}$$

$$u_z = \frac{\partial \Phi}{\partial z} + \frac{1}{r} \frac{\partial}{\partial r} (r H_\theta) \tag{5.67}$$

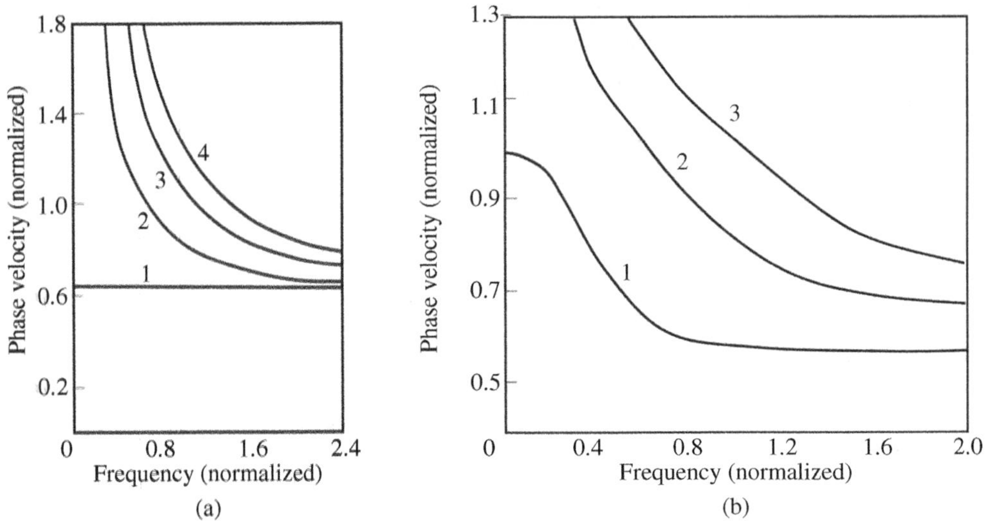

Figure 5.7 Torsional (a) and longitudinal (b) wave mode dispersion curves for a rod with a Poisson's ratio of 0.29. Note that the lowest-order torsional mode is dispersionless. The other modes are not. Where their phase velocities tend to infinity indicates that they are cut off, and below those values of frequency normalized by radius they do not propagate.

where, for $n = 0$

$$\Phi = AJ_0(\alpha r)e^{i(\xi z-\omega t)} \tag{5.68}$$

$$H_\theta = -B_2 J_1(\beta r)e^{i(\xi z-\omega t)} \tag{5.69}$$

The dispersion relation is then

$$\frac{2\alpha}{a}(\beta^2+\xi^2)J_1(\alpha a)J_1(\beta a)-(\beta^2-\xi^2)^2 J_0(\alpha a)J_1(\beta a)-4\xi^2\alpha\beta J_1(\alpha a)J_0(\beta a)=0 \tag{5.70}$$

which is called the Pochhammer frequency equation for longitudinal modes in a rod. The displacements for this mode are

$$u_r = B_2\left\{-\frac{A}{B_2}\alpha J_1(\alpha r)+\xi J_1(\beta r)\right\}e^{i(\xi z-\omega t)} \tag{5.71}$$

$$u_z = B_2\left\{\frac{A}{B_2}i\xi J_0(\alpha r)-\beta J_0(\beta r)\right\}e^{i(\xi z-\omega t)} \tag{5.72}$$

where

$$\frac{A}{B_2} = -\left(\frac{\beta}{\alpha}\right)^2\frac{\beta^2-\xi^2}{2\xi^2}\frac{J_1(\beta a)}{J_1(\alpha a)} \tag{5.73}$$

so we see that these are analogous to the coupled L- and SV-modes in a plate. Dispersion curves for the first few modes are also sketched in Figure 5.7.

5.2.3 Flexural Waves in a Rod

The torsional and longitudinal modes were for $n = 0$. The case of $n = 1$ corresponds to the lowest-order family of flexural modes. Higher-order flexural modes arise for $n \geq 2$. The frequency spectra for $n = 2$ are shown in Figure 5.8. The spectrum looks similar for $n > 2$.

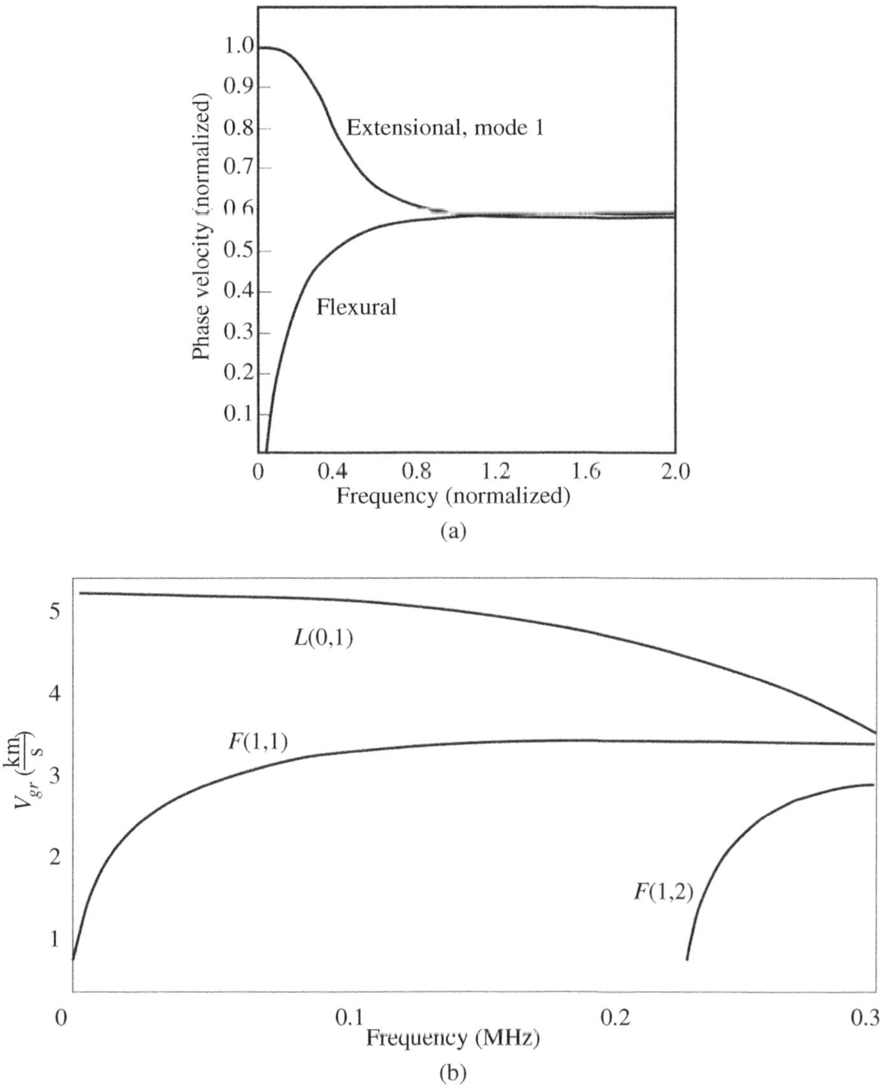

Figure 5.8 Phase velocities of extensional and flexural modes in a cylindrical bar with Poisson's ratio of 0.29 (a). The normalization is a little different, but what's called the extensional mode here should look familiar from Figure 5.7. Different authors normalize things in different ways, so I haven't tried to present dispersion curves in a uniform manner. I'm hoping that you'll code up the various dispersion relations and plot phase- and group-velocity in different ways until things make sense to you. You'll then have a tool at the ready to help understand complex guided wave physics in a variety of scenarios. Plot (b) shows group velocities for guided wave modes in an aluminum rod of 8 mm diameter. All existing modes (except fundamental torsional mode) up to 300 kHz are shown. Note that for some frequencies some of the longitudinal (L) and flexural (F) modes are reasonably nondispersive, that is, the group velocity is relatively constant. For other frequencies, the group velocities are not even close to horizontal flat lines, so those modes are dispersive at those frequencies. I hope you noticed that the axes of this plot aren't normalized. Most people normalize the *x*-axis with a characteristic thickness (for plates/pipes) or radius (rods) because the results are more general, but then if you're plotting your own dispersion curves, do it in whatever way makes the most sense to you.

Exercise 5.2 Consider in some detail how the various special cases follow from the general solution, and how they can be derived independently.

5.3 Guided Waves in Pipes

Now consider the same geometry as before, but this time the elastic medium is an infinite cylindrical shell with inner radius a and outer radius b. The problem geometry and a simple inspection scenario are shown in Figure 5.9. Boundary conditions are

$$\tau_{rr} = \tau_{r\theta} = \tau_{rz} = 0 \text{ at } r = a, b \tag{5.74}$$

The potentials are as before

$$\Phi = f(r) \cos n\phi \cos(\omega t + \xi z) \tag{5.75}$$

$$H_r = h_r(r) \sin n\phi \sin(\omega t + \xi z) \tag{5.76}$$

$$H_\theta = h_\theta(r) \cos n\phi \sin(\omega t + \xi z) \tag{5.77}$$

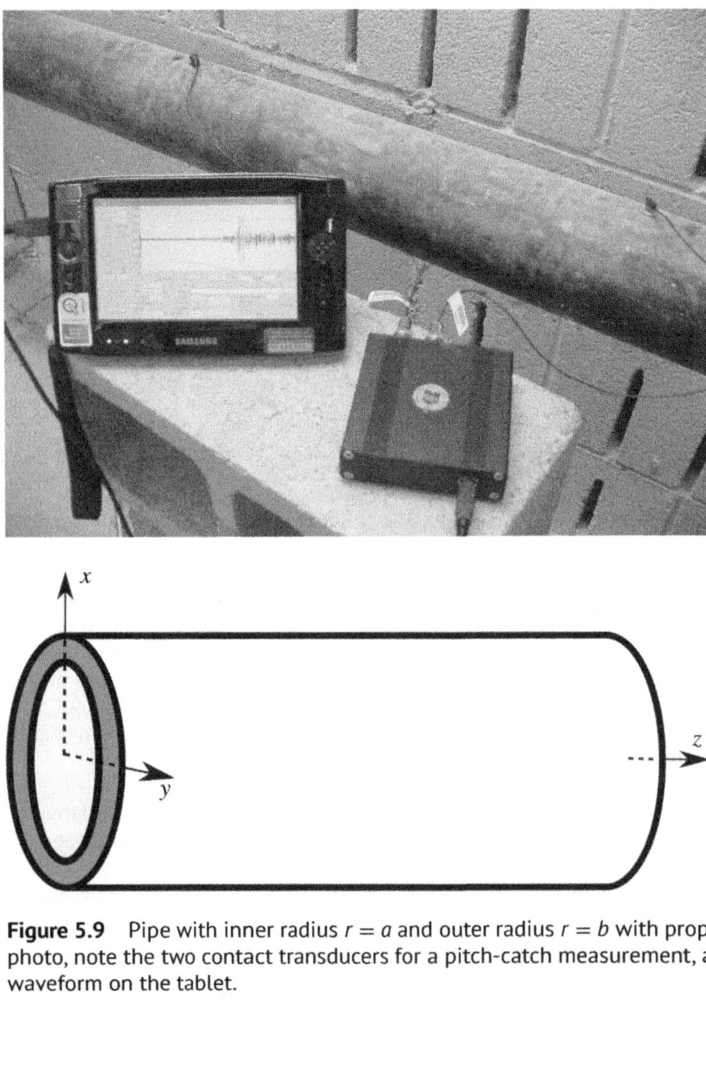

Figure 5.9 Pipe with inner radius $r = a$ and outer radius $r = b$ with propagation in the z-direction. In the photo, note the two contact transducers for a pitch-catch measurement, and then the rather complex waveform on the tablet.

$$H_z = h_z(r)\sin n\phi \cos(\omega t + \xi z) \tag{5.78}$$

with the radial functions

$$f = AJ_n(\alpha r) + BN_n(\alpha r) \tag{5.79}$$

$$h_z = CJ_n(\beta r) + DN_n(\beta r) \tag{5.80}$$

Similarly to the rod, we write

$$2h_1 = h_r - h_\theta = 2A_1J_{n+1}(\beta r) + 2B_1N_{n+1}(\beta r) \tag{5.81}$$

$$2h_2 = h_r + h_\theta = 2C_1J_{n-1}(\beta r) + 2D_1N_{n-1}(\beta r) \tag{5.82}$$

Note that there is no concern about the singularity of the Neumann functions because $a > 0$. Gauge invariance allows us to set $h_2 = 0$ without loss of generality. Displacements and stresses are then

$$u_r = \{f' + (n/r)h_z + \xi h_1\}\cos n\theta \cos(\omega t + \xi z)$$

$$u_\theta = \{-(n/r)f + \xi h_1 - h_z'\}\sin n\theta \cos(\omega t + \xi z)$$

$$u_z = \{-\xi f - h_1' - (n+1)h_1/r\}\cos n\theta \sin(\omega t + \xi z)$$

$$\tau_{rr} = \left[-\lambda(\alpha^2 + \xi^2)f + 2\mu\left\{f'' + \frac{n}{r}\left(h_z' - \frac{h_z}{r}\right) + \xi h_1'\right]\cos n\theta \cos(\omega t + \xi z)$$

$$\tau_{r\theta} = \mu\left[-\frac{2n}{r}\left(f' - \frac{f}{r}\right) - (2h_z'' - \beta^2 h_z) - \xi\left(\frac{n+1}{r}h_1 - h_1'\right)\right] \times \sin n\theta \cos(\omega t + \xi z)$$

$$\tau_{rz} = \mu\left[-2\xi f' - \frac{n}{r}\left\{h_1' + \left(\frac{n+1}{r} - \beta^2 + \xi^2\right)h_1\right\} - \frac{n\xi}{r}h_z\right] \times \cos n\theta \sin(\omega t + \xi z)$$

Applying the boundary conditions at $r = a$ and $r = b$ gives six equations, which can be solved for a dispersion relation by setting the determinant of the coefficient matrix equal to zero.

Exercise 5.3 Write down that determinant.

In general, the frequency spectrum is very complicated, but the overall structure is pretty similar to Figure 5.1. Next, consider a slight generalization of the problem that turns out to make it quite a lot more complicated. If the cylindrical elastic tube is filled with a liquid, then guided waves can propagate in either the elastic solid or in the liquid itself, although the liquid only supports compressional waves, of course. Actually, it's perhaps even a bit more complex than that, since there could even be cases where modes propagate at the interface between the solid and liquid, somewhat analogous to leaky Rayleigh waves at the interface between solid and liquid halfspaces.

Boundary conditions are

$$\tau_{rr} = \tau_{r\theta} = \tau_{rz} = 0 \text{ at } r = b \tag{5.83}$$

and, at the inner diameter

$$\tau_{rr} = p \text{ and } \tau_{r\theta} = \tau_{rz} = 0 \text{ at } r = a \tag{5.84}$$

along with an additional boundary condition that the u_r be continuous at the inner diameter. That will give a 7×7 coefficient matrix whose determinant is set to zero to solve for a frequency spectrum. But why might I care about that? Good question. Here's two answers. The obvious application is that there are all manner of piping systems that should be inspected, but they are filled with fluids. Without looking at the dispersion curves, it would be tough to design an inspection scenario. The ultrasonic energy might travel mostly in the fluid and flaws in the pipe wouldn't show up. There might be some particularly interesting effects whereby particular wave modes give quite good sensitivity to particular flaws of interest. A less obvious answer is that long-bones filled with marrow might be well approximated by this admittedly simple geometry, and ultrasound propagating in long bones could measure fracture healing or bone thinning without the use of ionizing radiation. A potential application I learned about recently isn't obvious at all. Spies could attach a MHz-frequency guided wave transmitter to a water pipe inside a building they want to spy on. Their adversaries might have devices to check for radiowave transmissions, but they wouldn't think to make sure that pretend plumbers haven't attached a device that transmits vibrations out to the street where the spies have plumbed a receiver which senses the guided wave coming from inside the building. Megahertz frequencies are inaudible and the vibrations can't be felt by touching the pipe. Sneaky, huh?

Exercise 5.4 Write down the 7×7 determinant and see if you can solve it to get a sense of the dispersion curve behavior.

5.4 Data Engineering for Tomography

Detection of hidden corrosion in plates and pipes is a very important industrial problem because very often only one side of the object is open for the inspection. The pulse-echo thickness measurement technique with bulk waves is tedious and can be inaccurate, especially in corroded areas with rough surfaces. The frequency-thickness dependence of the velocity of ultrasonic guided waves means their arrival time will change accordingly, making it possible to judge the size and degree of the corrosion damage. Corrosion also affects the amplitudes of reflected and transmitted signals, and filters out some modes which may have been initially present in the signal, but have an $f \cdot d$ product in the corroded region below their cut-off frequency.

In science things can go in and out of fashion.[6] Take radar, for instance. It was highly fashionable during and after World War II, even though the atomic bomb got all the publicity. At the height of the Cold War big aerospace companies were paying highly inflated salaries to anybody who could

6 I know quite a lot about fashion because my wife used to buy sportswear for a chain of 132 department stores. At social events, her friends didn't want to talk about the mathematics of scattering, so I learned to talk about fashion. I never ascribed to "fashion before comfort" although I do understand the function of an irrationally expensive handbag and I appreciate that the women in my life have strong shoe games. I wear pretty much the same thing most of the time, and as fashions fluctuate, I'm sometimes inadvertently on trend. Or not. I don't really care all that much about fashion.

A few years ago, I switched over from danglie ties to bow ties. Each time when I took off my tie I would throw it away. Initially, my wife said things like, "Good, I've always hated that tie." Since I don't wear ties all that often, this process took almost a year, and as I got down to the last several ties, my wife said things like, "You aren't going to throw that tie away are you?" to which I responded something like, "Haven't you been paying attention for the last year?" and she said, "But I gave you that tie." Inside my head, and only inside my head, I said, "You also gave me those other ties that you 'always hated' and were glad when I threw them in the trash." I know enough about fashion to understand that a foundational principle is that people need to hate all their clothes four times a year and go buy new ones or the department stores would all go out of business. So far none of my bowties have been declared ugly.

Ludwig Prandtl gets credit for boundary layer theory in fluid mechanics, which accounts for viscosity of the fluid flow near the surface of the wing or whatever. The Nazis often used Prandtl's international reputation as a scientist to promote Germany's scientific agenda. The Prandtl number was named after him. It's quite disappointing to read that he supported the Nazis because I remembered a funny anecdote that I heard in a fluid mechanics class in the 1980s. Guests were coming over for dinner and Prandtl came downstairs dressed in an ugly tie. His wife told him to go back upstairs and change his tie because the dinner guests were starting to arrive. When he didn't come back down promptly, his wife excused herself and went upstairs to fetch him, only to find him asleep in his bed. Since taking off his tie was the first part of his nightly routine, after he took off his tie he continued to undress and put on his jammies and went to sleep. He was probably thinking about the mathematics of fluid mechanics rather than small talk about fashion with his dinner guests.

recognize Maxwell's equations in neon lights, no matter what ridiculous outfit they showed up for work wearing. My daily wardrobe was the uniform of an Air Force Captain, and I had developed some expertise in using Maxwell's equations for understanding radar scattering, but by the time I traded in my dress blues for blue pinstripes, the Berlin Wall had come down and radar scattering had gone out of fashion. Most of those big aerospace companies merged with, or were gobbled up, by competitors. I went and did other things that were in fashion, like medical imaging and structural health monitoring, sometimes using Maxwell's equations and sometimes not.

It turns out that the thing that I've been doing professionally for four decades now is suddenly on trend. The currently fashionable name for it is *data engineering*, which encompasses many concepts in disciplines such as signal processing and computer engineering, representing the critical data acquisition and processing steps required to prepare massive datasets for study with modern machine learning algorithms. For example, biomedical datasets are measured in petabytes and comprise data types ranging from DNA sequences to text-based medical histories to wearable sensor-generated outputs like heartrate. NIH envisions a biomedical enterprise in which data and information generated in the field, laboratory, and clinic are processed and analyzed in real time and readily shared. In an increasingly data-rich world, these advances are essential to enhancing health, lengthening life, and reducing illness and disability.

As you may have noticed, this book purports to be about the mathematics of the scattering of acoustic, electromagnetic, and elastic waves from various objects, discontinuities, or inclusions. What we're doing is called *forward scattering*, meaning that the incident wave is specified as is the size, shape, orientation, and composition of the scatterer. The mathematics is used to predict the resulting scattered waves in order to gain insight into the behavior. Analytic solutions are often used in conjunction with experimental measurement and numerical simulations in order to understand enough of the scattering behavior that strategies can be developed to deduce the properties of an unknown scatterer from features in the scattered field(s). The so-called *inverse scattering* problem is, in general, unsolvable. Or so we used to think.

If you have fed it a sufficient training data set, your fashionable new machine learning system can output "answers" for any new, unknown input. Some people get a little overexcited about all of this, and then are disappointed when it doesn't work very well in the real world. That's usually because they've underestimated the amount of training data they'll need by an order of magnitude or three. The answer, happily for us, is data engineering. You have to put some effort into figuring out what inputs to present to your machine learning system or it will GINGO–GINGO–GINGO all over your fancy new danglie necktie [18].

So, we use our hard-won knowledge of the scattering (forward) behavior to make intelligent decisions about what features to use from our input data streams to feed to our classifier. In my research group, we first ran into this family of issues as we were developing a structural health monitoring technique called Lamb Wave Tomography. Accurate tomographic reconstructions require that features from rather a lot of waveforms need to be reliably extracted as quickly as possible. There are geometric constraints, but for our work, we were able to overcome many of those to figure out what was likely to be practical. I'll tell you the story in the following text, with the requisite equations and diagrams, of course, but flagging along the way the key issue we were facing, which we now know to call data engineering. But first, I have to tell you a story about structural engineering.

On 28 April 1988, a Boeing 737-200, operated by Aloha Airlines, experienced an explosive decompression and structural failure at 24,000 feet, while en route from Hilo to Honolulu, Hawaii [19]. Approximately 18 feet of the cabin skin and structure aft of the cabin entrance door and above the passenger floor line separated from the airplane during flight. One flight attendant was sucked out during the decompression; seven passengers and one flight attendant received serious injuries. As a result of the accident, the airplane was determined to be damaged beyond repair. It was dismantled and sold for parts and scrap. There's a terrible made-for-TV movie about the incident called "Miracle Landing" starring Connie Sellecca, Wayne Rogers, Ana Alicia, and Nancy Kwan. The special effects are really hokey, so I'm quite surprised to see that the movie was awarded the 1990 Primetime Emmy Award for Outstanding Individual Achievement in Special Visual Effects. The National Transportation Safety Board determined that the probable cause of this accident was the failure of the Aloha Airlines maintenance program to detect the presence of significant disbonding and fatigue damage which ultimately led to the failure of the lap joint at S-10L and the separation of the fuselage upper lobe. They emphasized the great importance of encouraging further research into improved corrosion detection and prevention methods, and to employ methods that minimize human performance inadequacies. I take that to mean that the goal is automated interpretation of NDT measurements.

Military and civil aircraft, ships, waste-storage reservoirs, and railway tank cars are subjected to in-service stresses and harsh environmental conditions, which eventually leads to corrosion, cracks, delaminations, fatigue, etc. Different defects grow with different speeds and reach their critical states at different times, so periodic inspections are scheduled, tailored for the particular application.

The most common ultrasonic inspection routine uses point measurements in a pulse-echo or through-transmission mode to estimate variations of material properties with depth at a given location [20]. This uses longitudinal and/or shear bulk waves and can achieve very high resolution at high frequencies. For large areas, however, the ultrasonic transducer must be scanned over the entire region to map the structure. This conventional method, while very effective for small parts becomes rather time consuming and mechanically cumbersome for field inspection of large structures. To save time, inspection is typically carried out at selected locations, thus increasing the probability of missing a flaw. There often isn't much data engineering involved because the back wall or any delamination in a doubler will give a sharp reflection and that peak is easy to identify. Time delay then gives the thickness or depth of delamination.

Our work has included the development of an alternative ultrasonic inspection technique that is optimized for large area plates, pipes, and shells and is capable of assessing structural integrity quickly and efficiently. We make use of ultrasonic Lamb waves propagating along the plate, pipe, or shell for large distances, exploiting wave properties that are very different from those of bulk waves. Data engineering is key to interpreting these waves because a Lamb wave is a combination of coupled longitudinal and vertical-shear motions in the xz plane, and there are symmetric and antisymmetric families of Lamb waves with corresponding dispersion behavior. At any given frequency, in general, several modes can propagate, each having a different phase velocity. The higher the frequency, the more modes coexist in the plate, and their phase velocities converge to the common limit, which is the velocity of the Rayleigh surface wave (Figure 5.1). All of this means that the key barrier to practical use of Lamb waves for structural health monitoring is being able to automatically identify subtle changes in arrival times of complex, multimode signals, which are each dispersive in particular ways. The wave packets distort as they propagate, and the different parts of the wave packets have different group velocities and the various Lamb wave modes interact with different types of structural flaws in different ways. It's a delightfully complex set of data engineering issues.

An often underappreciated aspect of data engineering is the data acquisition step(s), which usually includes choices about transducers. This strikes me as strange because such choices affect the data quality and character in such profound ways. Feel free to skip ahead and come back to this section later, though. There hasn't really been much in the way of new transducer/technique development in the last 25 years[7] except that the electronics cost/size has plummeted precipitously. We all have lots of expensive equipment that we keep in our labs as decoration or memorabilia. The property control folks still come around every so often to scan the equipment tags, though, because they have everything in their databases according to the original procurement price.

There are a great variety of techniques to generate Lamb waves. The simplest and the most widespread method uses normal or shear contact transducers. If the excitation pulse is broadband enough, it can simultaneously generate several distinct modes at different frequencies, although the energy of the initial pulse is distributed between these modes unequally. It may happen that the mode of interest receives only a small fraction of the input energy, while undesirable modes dominate and complicate the measurement process.

Variable incidence-angle methods require either a contact wedge or an immersion tank to couple ultrasonic bulk waves into the plate at a chosen angle. Variable incidence angle creates preferable conditions for a particular mode and suppresses the others. This mechanism is based on Snell's law for refraction because the phase velocity V_{ph} of the expected Lamb waves propagating in the x direction satisfies the relation $V_{ph} = V_L / \sin\theta$ with V_L the longitudinal velocity in the wedge material and θ the incidence angle in the wedge. Since each mode has a different phase velocity V_{ph} at a given frequency ω, one can select the desired mode by tuning the angle θ. It is worth noting that the above is true only for plane wave incidence [7]. Real wavefronts emanating from transducers are not planar and have some variation of incidence angles.

Comb transducers can create a specified phase displacement pattern on the surface of the plate to excite the desired mode with a given phase velocity at a given frequency. When all elements in the array are equally spaced and connected to the same source (in phase), the excited guided wave will have the same wave length as the transducer spacing [3]. If the elements are connected to different sources with appropriate phase shifts, it is possible not only to control the wavelength of the resulting guided wave but also to make it propagate in a preferred direction [21–23]. The last effect is achieved due to the interference between waves, created by different parts of the array. As a result, the excitation energy is used more efficiently and the wave can propagate further, which is very important in many large-scale pipe and plate inspection applications.

Air-coupled transducers are capacitive devices vibrating under alternating voltage and emitting ultrasound waves into air. Due to the large acoustic impedance mismatch between gases and solid materials, air-coupled transducers require specialized electronics with very high voltage pulsers and narrow-band, high-gain amplifiers. The total reflection losses can exceed 120 dB [24] and manufacturers must use sophisticated combinations of matching layers in order to receive signals with acceptable signal-to-noise ratio. Current advances in technology have made it possible to overcome these limitations and manufacture air-coupled transducers capable of performing most general inspection tasks including generation and detection of Lamb waves.

Electromagnetic-acoustic transducers (EMATs) exploit the principle that an electromagnetic wave incident on the surface of an electrical conductor induces eddy currents within the skin of the conductor [25]. A typical EMAT consists of a large magnet to produce the external field

---

7 Dr. Alfred L. Broz's 40 year career in nondestructive evaluation was with the US Army and the FAA. For the last decade of service as Chief Scientist, he was proud to note that there were no fatalities in transport aircraft in the United States. Dr. Broz attended St. Thomas University in St. Paul, MN, the South Dakota School of Mines for his Masters, and earned his PhD in Physics at Notre Dame. I remember, in particular, a conversation with him at the QNDE Conference in Golden, CO, where he articulated that there was nothing new anymore, but he agreed with my assertion that automated interpretation of results was a new thing. We didn't know to call it machine learning back in 2004.

and a high-frequency coil to induce eddy currents in the sample skin. By varying the mutual orientation of the coil, magnet and the sample, it is possible to excite both normal and shear displacements in the sample. No couplant is required for the regular operation of EMATs, but the low signal-to-noise ratio and need for large magnets to excite useful levels of ultrasonic signals limits their practical application. In addition, EMATs work only with ferrous metal samples. They have been used extensively, however, for pipe and rod inspection, where the geometry allows for the efficient excitation of guided waves.

Laser ultrasound is a fully noncontact method of Lamb wave excitation [26–28]. It uses periodic heating of the sample surface with a chopped, or Q-switched laser beam, thus generating thermal transients. Harmonic heat expansion and shrinking occurs in the target region and gives rise to ultrasonic waves. Using laser beams, it is possible to select both the desired mode and the direction of the resulting Lamb wave in a way similar to contact periodic arrays. Although very elegant and noncontact, the laser generation method has several significant drawbacks. One of these is the low excitation efficiency; high-intensity laser radiation can cause permanent damage to the sample. Laser-based methods are attractive because a single scanned laser transmitter can be substituted for a whole array of piezoelectric transducers, but the receiving part of the equipment is costly to implement.

5.4.1 Tomography Overview

Our early work on Lamb wave tomography explored both reconstruction geometries and associated tomographic algorithms in order to find a balance between reconstruction fidelity and practical measurements [29–35]. In the parallel scheme, employed in the first generation of medical CAT-scanners [36, 37] a single transmitter and a single receiver step in tandem to complete one projection. The sample, or apparatus, then rotates to take another projection and the whole process is repeated until projection angles cover the interval $(0, 2\pi)$. A more efficient fan beam geometry, implemented in third-generation CAT scanners, contains one transmitter and an array of receivers distributed uniformly along a circular arc. This scheme is much faster since rotation of the sample is the only mechanical motion involved. The so-called crosshole geometry [38, 39], widely employed in geophysics and seismology, uses two linear arrays of transmitters and receivers which are placed into boreholes and then all possible combinations of rays are recorded by the principle "one send all receive." Employing a four-sided perimeter array of transmitters and receivers is what we call the double-crosshole scheme.

For all scanning geometries, the accuracy of any reconstruction method strongly depends on the spatial density of the "rays" penetrating the object. In general, higher image quality is observed in well-covered central regions, while poorly illuminated peripheral areas suffer from noise and reconstruction artifacts. In addition, the task of tomography can be reduced to the reconstruction of the object from its Fourier spectrum, and the latter should not be undersampled. In other words, the directions of wave vectors should cover the range $(0, 2\pi)$ as densely and as uniformly as possible. However, these two requirements hold together only for the parallel projection geometry, where the ray density and the wave vector directivity pattern are uniform. In addition, while the wave vector coverage for the fan beam scheme is complete in a global sense, both the local directivity pattern and the ray density are nonuniform throughout the region. For the single crosshole geometry, even the global wave vector coverage is rather poor and limited by the minimum and maximum possible ray angles. Moreover, even well-covered regions [38] may have rays with a very limited range of angles, resulting in very poor resolution of structures that are oriented perpendicular to the raypaths for X-rays and parallel to the raypaths for ultrasound, respectively. This lack of local angular coverage qualitatively changes the situation from a well-constrained imaging problem to one where there is no unique solution even if an infinite amount of perfectly noise-free data with an infinitely small sampling interval were available, which it isn't.

Originally, all tomographic reconstruction algorithms were based on the assumption of straight rays connecting the transmitter and receiver positions. This assumption is valid for X-rays propagating in biological tissues and can, in principle, hold for ultrasound in nonrefractive media. However, the majority of interesting defects do scatter ultrasound and reconstruction is more properly done via diffraction tomography. Weak spatial variations in the index of refraction result in bending of ray paths, thus calling for ray tracing algorithms to describe their complex behavior. Reconstruction of strong scatterers requires inversion of the wave equation that can, in general, only be approximated. Lamb waves introduce additional complications because they are multimode and dispersive. Simple peak-detecting schemes are ineffective.

5.4.2 Fan Beam Tomography

The convolution-backprojection algorithm proved to be successful in reconstructing plate defects from the projections acquired with the Parallel Projection scanning technique. A source-receiver pair linearly scans over the length of a projection, then rotates and scans the next projection until the entire circle is covered. The method requires rotation of either the sample or the transmitter–receiver assembly, which is slow and mechanically cumbersome. This worked beautifully in the laboratory for small samples, but is impractical to realize in field conditions when large objects are being scanned, with near real-time requirements for the data acquisition process.

In the "fan-beam" scheme, widely used in medical imaging [37], the projections are generated by a single transducer emanating a fan-like beam which is recorded by a bank of receivers. Usually, the receivers are located either along the arc centered at the transmitter at equiangular intervals, or are equally spaced along a straight line. The transmitter–detector assembly is rotated to measure projections at corresponding angles. The data-collection procedure is much faster than in the parallel case, but mechanically moving parts are still present in the system unless a complete ring of transducers is used.

Image reconstruction from fan beam projections is very similar to that from parallel projections, except for some small changes in the algebra. The whole reconstruction algorithm is described by Kak and Slaney [40]. You might also like the book by Devaney [41]. They recommend convolving the result with an additional smoothing filter to enhance the reconstruction quality.

We started working on fan beam tomography because of the possibility of using a circular perimeter array of transducers capable of both transmitting and receiving ultrasound. In this arrangement, transducers are equally spaced along a circle, which would be free from mechanically moving parts and could be implemented for scanning large areas with Lamb waves. In the proof-of-concept laboratory setup (Figure 5.10), the receiving transducer R moves along the circular arc of radius r centered at O. The line OR is actually a rigid arm geared to a stepper motor. The scanning angle $\gamma = \angle ROK$ is the angle between the central ray TK of the fan beam and the current direction of the arm OR. It is desirable to make the opening angle of the fan beam 90° having the holder arm OR sweep the whole 180°. In this case, the best-covered area, or the circle formed as an intersection of all fan beams, will be exactly embedded into the image square with side $a = |OR| \cdot \sqrt{2}$. This reconstruction image will also have the biggest possible size because it is, in turn, embedded into the transducer array circle.

If we ignore both ray bending and diffraction effects and use a straight ray assumption for image reconstruction, the ray length can be determined as the distance between transmitting

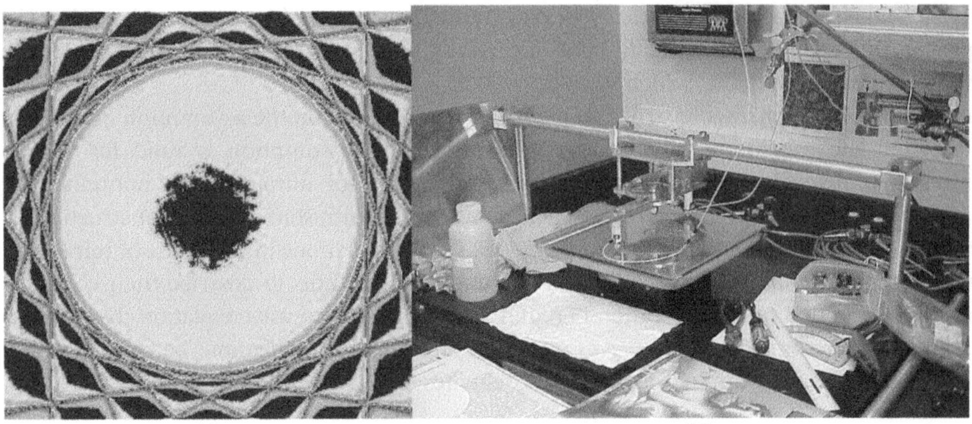

(a) (b)

Figure 5.10 Fan beam tomography reconstruction for a 25-mm, 50% thickness flat-bottom hole in an aluminum plate (a) and photo of the fan beam scanner in W&M NDE lab (b). The radius of the artifact-free circular area is 56 mm due to the geometric limitations of the fan beam method, which makes the technique rather impractical.

and receiving transducers. We use this to correct experimental data for changing ray lengths, something that's not an issue for X-ray CT of humans because there's essentially no attenuation of the X-rays in the air surrounding the patient.

Before committing to the expense of a circular array of transducers, we mimicked its behavior using a single transmitter–receiver pair, implemented in a custom fan beam scanner (Figure 5.10). The scanner is mounted on a support bar and consists of a static arm and a motor-controlled swing arm rotating about the geometrical center. The source and receive transducers are mounted on both arms at equal distances from the center of rotation. The transducers are equipped with small footprint delay lines and are spring-loaded to ensure good coupling with the sample plate. The sample is placed on the rotary table under the arm assembly. The swing arm is worm-geared to a stepper motor and can sweep through an arc of specified opening angle to complete a single fan beam projection. Additional projections are acquired by incrementing the rotary table holding the sample.

The most important parameter to be extracted from the digitized waveforms is the arrival time of the fastest mode. This time depends upon the presence of any defects and, obviously, upon the distance between transmitter and receiver. We therefore need to detect the rather small variations in arrival time which carry the information about defects. The reconstruction in Figure 5.10 shows very good quantitative correspondence with the defect size, but the usable fill factor is limited by the area of the overlapping fans.

We were unable to obtain a satisfactory reconstruction image for an Al plate with five through holes. According to the single projection data, we should have been able to get a satisfactory result at least for the filtered data. But instead of three separate holes, we got a large dark spot covering almost the entire image area. The reason could be an inability of that version of the reconstruction program to resolve more than one spatially separated defect, that is, the assumption of straight rays breaks down. It also seems obvious that there's some scattering effects coming into play here.

5.4.3 Double Crosshole Tomography

Parallel projection and fan beam algorithms both belong to the convolution-backprojection family, which requires strictly determined scanning configurations and is very sensitive to any

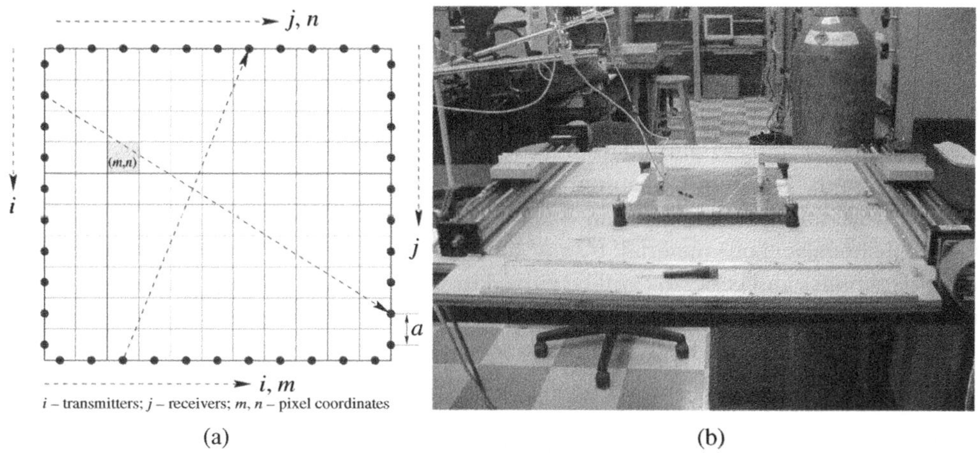

Figure 5.11 Explanation of the ART algorithm for the double crosshole geometry (a) a – distance between transducers; pixels (m, n) are indexed as shown; ray enumeration order (i,j) is different for different projections. Photo of the crosshole scanner in our lab is at (b).

incompleteness or noise in the experimental data. The algebraic reconstruction technique (ART) provides a solution for the problem of recovering an object from its projections, without many of the geometric constraints inherent in the convolution-backprojection methods. What makes it useful for Lamb wave tomography is its iterative nature and its great flexibility which allows practically any scanning geometry and incomplete data sets.

For the sake of clarity, we will briefly describe the four-legged crosshole geometry ART algorithm for our particular scanning setup as shown in Figure 5.11. Circles are the transducer positions, which can all be occupied by transducers in the case of a perimeter array of transducers. In our laboratory implementation, only two positions are occupied at a time since we use a single transmitter–receiver pair. Both transducers are attached to linear slider screws and can be moved back and forth using two stepper motors controlled by the computer. The transmitting transducer steps along the lower edge from left to right incrementing i from 0 to $N - 1$. The receiving transducer slides along the upper edge sweeping all N available j-positions for a fixed i. For each relative position, the whole wave train is recorded and stacked into a data file. After N^2 measurements, the first crosshole projection is complete. We then move the transmitter down from the upper left corner and move the receiver up and down along the right side. Again, N^2 measurements complete the second projection. The increased number of crosshole projections is the only difference between the double- and the single-projection crosshole geometries.

The ray density is critical to the reconstruction quality. It can be increased by increasing the number of transducer locations per side, which could be rather expensive in the case of an array. Alternatively, one can measure additional data using the available setup. For example, two more crosshole projections can easily be obtained by swapping transmitting and receiving transducers. Furthermore, transmitting from one and receiving on the other three sides of the rectangle would provide additional valuable information and also increase ray density. The optimal solution is to be able to take data in both directions along the rays connecting any two points on the square (Figure 5.11). This could be accomplished by having a perimeter array of multiplexed transducers, but is much harder to realize when the perimeter is covered by moving transducers mechanically. We, therefore, typically limited ourselves to acquiring only two mutually orthogonal crosshole projections.

As we have seen (Figure 5.11), each ray can be assigned two coordinates i, j. For each projection, $i, j \in [0, N)$. The resulting square image can be divided into cells (pixels) in an arbitrary fashion, but the simplest way is to make the square pixel side length equal to the transducer step size a. This will preserve all the information contained in the data. Each pixel in the resulting matrix has unique coordinates $m, n \in [0, N)$ starting from the lower left corner. Each $ray[i, j]$ crosses a certain number of pixels on its way from transmitter to receiver. The term "ray" actually stands for the path the Lamb wave travels in the medium (plate). These paths are straight lines if the plate is made of isotropic homogeneous material. If some anisotropy or defects are present the paths will differ from straight lines and generally will be curved in some complex fashion. It is obvious that in the latter case, the arrival time of a particular Lamb wave mode and the pixels it travels through will differ from those in the straight ray assumption. To take ray bending effects into account and thus to improve the resolution of reconstruction methods, Lamb wave diffraction tomography is needed.

For the chosen scanning geometry, the total number of equations in a single-crosshole projection equals N^2, where N is the number of transmitter positions. In our experiments, for example, we often use $N = 100$. To avoid the direct inversion of such a large matrix, the less computationally intensive iterative ART is commonly used. The ART algorithm updates pixel velocities ray-by-ray leading to the so-called "salt-and-pepper" noise [40] in the resulting image. To eliminate it, at each iteration one can first calculate the updates for all the rays and only then update all the pixel velocities simultaneously. The modified method is called simultaneous iterative reconstruction technique (SIRT).

After measuring the first set of crosshole projections as shown in Figure 5.11, we place the sliders along the other two opposing sides and record the second set of projections, which we call double-crosshole tomography. Each set of crosshole projections results in a stack of N^2 5000-point waveforms, which needs to be processed in order to extract one or several values of interest from every waveform. Although fairly robust, our time extraction algorithm produced various estimation errors ranging from fractions of the quarter-wavelength to complete failure to detect signal. Fortunately, the percentage of severe errors is small enough (1%), so that the entire data set is not fatally distorted. In addition, we correct the aforementioned failures by substituting out-of-range data points with values of their closest neighbors. However, the algorithm runs well on experimental data and produces satisfactory images.

It was surprising to us that the straight ray assumption, used in both reconstruction algorithms, worked well even for the through holes. Lamb waves definitely diffract around the hole, and their arrival time increases. Assuming straight paths, the reconstruction algorithm interprets that increase as a wave slowdown and assigns smaller velocity values to some (often wrong) pixels. As a result, the shape and the exact size of the defect may be distorted upon reconstruction and the image cannot always serve as a velocity map. A method is needed that naturally incorporates ray bending effects into the reconstruction process. Diffraction tomography is one such method.

Visual comparison of the reconstruction quality of the ART algorithm vs. SIRT shows (Figure 5.12) that the latter always yields much smoother picture and shows fewer artifacts than the former. The "salt-and-pepper" noise, specific for the ART, is completely absent on the SIRT images. The run time and memory usage are approximately the same for both algorithms. Therefore, it seemed sensible to use SIRT and its derivatives as a primary reconstruction algorithm in subsequent imaging. The ART algorithm can also be utilized for revealing reconstruction artifacts and using this information, improve the images, obtained with SIRT. The spatial resolution of both methods is good enough to ensure subjective visual separation of all the defects studied. In the case of two flat bottom holes, the imaging artifact slightly complicates this.

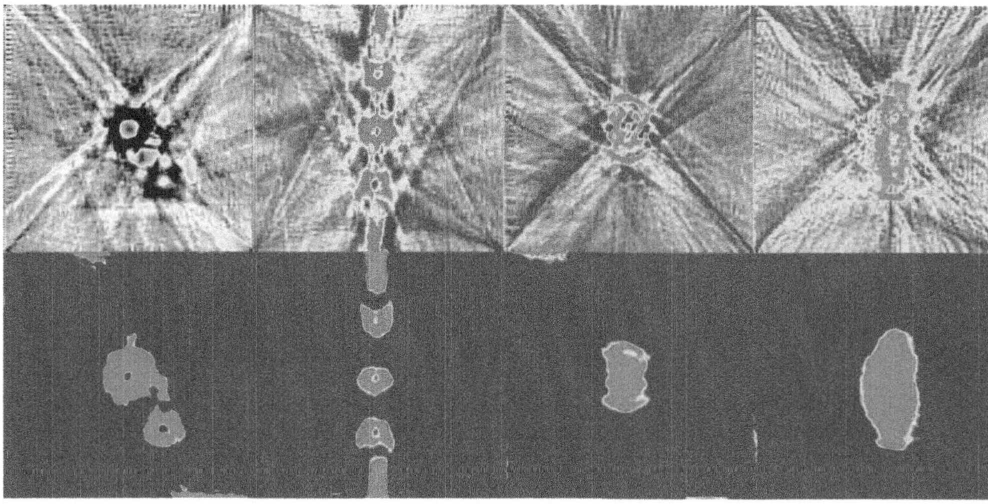

Figure 5.12 Sequential ART reconstruction of two circular flat-bottom holes in aluminum plate (top left). Defect diameters: 29 and 25 mm. Distance between centers: 53 mm. Image size: 200 × 200 mm. $f = 1.2$ MHz. Note the level of "salt-and-pepper noise" throughout the image. Next is for sequential ART reconstruction of five equally spaced 15 mm-diameter through-holes in aluminum plate. Note the level of "salt-and-pepper noise" and distorted shape of the holes near the edges. Third image is for a rectangular thinning in aluminum plate and the right image is for an oblong thinning region in aluminum plate. SIRT reconstruction of the same four samples (bottom). Image size: 200 × 200 mm. In all cases, we used a thin layer of water along the transducer scan paths for coupling purposes. The transducers were equipped with cone-shaped delay lines made of acrylic resin. The footprint diameter of the delay line was 2 mm.

5.4.4 Arrival Time Determination

Buckle in please. When you do research for a living you never quite know what combination of events might kick off a new and fruitful line of inquiry, causing you to be inadvertently on-trend many seasons hence. First, I'm going to tell you a story or three and then I'll tell you in some detail how we began to systematically solve a new class of problems. It will also provide an opportunity for us to remind ourselves that the whole point of understanding the mathematics and physics of wave propagation and scattering is so that we can design more effective measurement schemes and better interpret measured signals of various types.

The ultrasonographic periodontal probe development project, which I took on as a favor for my colleagues at NASA, presupposed that ultrasonic A-lines could be automatically interpreted because in the dental office, they do not do ultrasound imaging at all. Preclinical work on preserved cadaver jaws turned out to be misleading because those are all rubbery (and gross), whereas for living tissues, the backscattering from the bottom of the periodontal pocket is far too subtle for simple peak-detection approaches to work at all. The economics and workflow of a typical dental office meant that these complex backscattered signals had to be automatically interpreted in real time. This spurred us to develop time-frequency and time-scale representations of time-domain signals so that image processing approaches could be brought to bear and then image features could be used for machine-learning determination of periodontal pocket depth. See Figure 5.13. At the same time, we were working on a novel concept to diagnose and localize treatments for prostate cancer using 3D ultrasound. Actually, the system used two 3D ultrasound probes, one was a trans-rectal probe and the other was a trans-urethral probe. Two probes meant that we could go to higher frequency in order to get better resolution, and both probes were scanned under computer

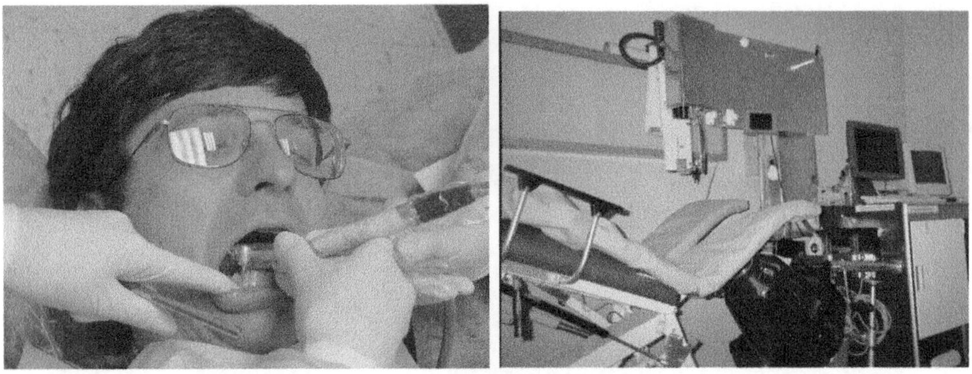

Figure 5.13 Two investigational ultrasound devices we developed based on concepts patented by John Companion. Left is an ultrasonographic periodontal probe (US5755571A), which uses pulse-echo to measure the distance from the gum line to the crest of the periodontal ligament. Right is a dual-probe prostate cancer detection prototype (US5282472A) in use at the Walter Reed Army Medical Center urology department, which performs computer-controlled 3D scanning with both trans-rectal and trans-urethral ultrasound probes. For both of these, the key technical challenge is automatic interpretation of the ultrasound signals to identify the anatomical variations of interest. See [9] for details, or [42] if you prefer the fictionalized version.

control, so when we found something suspicious in the ultrasound signals, we would know with certainty the 3D location, and then the robotic biopsy subsystem could accurately sample the tissue there. Yes, there was a special chair involved. Although it was the 1990s and we didn't use the term machine learning, it was a straight machine learning play. As it turned out with the periodontal probe project, our clinical partners never quite got us the training data necessary to do machine learning. That's a common problem and it's always very frustrating.

When Dr. Keun Jenn Sun[8] was doing Lamb wave scanning at NASA, he had a clever piece of analog hardware called a pulsed phase-lock loop (P2L2) which accurately determined subtle arrival-time changes in Lamb waves that indicated thickness variations, disbonds, scattering, etc. As we were writing a sequence of SBIR proposals to get non-NASA funding so he could keep his NASA funding, one of our collaborators put in the concept of tomography while revising a previous

8 Keun Jenn Sun was born in Keelung, Taiwan, but immigrated to the United States to complete his master's degree at the University of Ohio-Akron in 1979 and Doctoral studies in physics at the University of Wisconsin-Milwaukee in 1986. He then moved to Virginia to take a position at the NASA Langley Research Center, where he continued to work on and off for 21 years. Dr. Sun specialized in Lamb wave detection of flaws in aging aircraft structures. In his spare time, he enjoyed tending to his large fish tanks and cultivating the backyard of his home, where he raised a family and lived for 19 years. He was modest by nature but consistently insightful, reliable, sincere, and just as both a father and a husband. Keun passed away on 8 August 2007.

Dr. Sun was among a group of NDE scientists at NASA Langley who were funded via annually-renewed contracts at W&M, so when I arrived here in 1993, we co-mentored students and collaborated on Lamb wave research. The contractors' cheese all got moved when the NASA administrator decreed that procuring scientific research staff via contracts was bad, despite there not being enough civil-service slots to deliver the needed workscope. Contractors could only be funded by NASA if they had a tiny bit of other funding, and they could only come to work at NASA every day if they had an office at their home institution. We allocated shared desk space for them all on campus, and set about writing proposals to other agencies to fund a portion of their time. After a few tries, an SBIR proposal was funded based on some of Dr. Sun's Lamb wave research and we moved the equipment to our new building on campus a day before the government shut down and access to NASA facilities was barred for several weeks. NASA did cut the funding for all the on-site contractors.

proposal. Sometimes you have to revise and resubmit a few (or several) times before a proposal gets funded, and the tomography was my collaborator's part of the project, so I didn't pay close attention to those tasks. When he called to tell me that the proposal had been funded, he also told me that he was moving to a new company so it turned out that I was on the hook to figure out how to make Lamb wave tomography work.

We figured out that the crosshole tomography methods developed for seismology could be made practical with parallel arrays of transducers, multiplexed to get the necessary crisscross pattern of rays. We proposed that for Phase II of the project, which was funded. We mimicked the measurement scheme in the laboratory with two linear slides and began collecting crosshole tomography Lamb wave data. We immediately found that the P2L2 wasn't going to work because the path length straight across was too different from the diagonal paths, and the P2L2 would lose lock. We began systematically collecting data and working to figure out how to automatically extract Lamb wave arrival times from thousands of waveforms in near-real time. The modern term for this is data engineering, which is very fashionable right now.

I assume that you're still buckled in, so here we go. Lamb waves are inherently dispersive and multimode, so finding their arrival time is more of a problem than in the more common bulk wave case. In nondispersive situations the shape of the excitation pulse is preserved and finding the arrival time of a particular echo is simply a matter of finding the position of its sharp front edge or detecting the arrival time of a "peak" in the signal. Dispersion, on the other hand, means that the propagation velocity of an elastic disturbance depends on its frequency. Because of the finite duration of the excitation pulse, its spectrum always contains frequencies within some band of varying width. The traveling elastic disturbance can be treated as a superposition of harmonic waves of different frequencies. Due to dispersion these waves propagate with different velocities, and the resulting wave packet gets distorted with time. Its shape and length evolve, and determining the arrival time is no longer a straightforward problem, especially for practical tomographic geometries where the ray lengths vary considerably. It's also often the case that the Lamb wave mode that's most of interest is neither the first-arriving nor the one with the largest amplitude.

Figure 5.14 shows theoretical arrival times computed as a ratio of the transmitter–receiver distance to the velocity of the S_0 mode in the defect-free plate, along with the first part of a typical experimental waveform. To compare the output of various time-delay estimation methods, we performed two typical measurement sets on two aluminum plates. One plate was defect-free and the other had a 25.4-mm-diameter 50% thickness reduction flat-bottom-hole in the middle. The data were acquired in tomographic crosshole experiments with 40 transmitter and receiver positions. For each plate, we recorded total of 1600 waveforms resulting from all possible transmitter–receiver positions on a 2.23-mm-thick aluminum plate at a frequency of 1.2 MHz. The transducer step size was 5 mm, and the distance between transmitter and receiver lines was 200 mm.

A variety of time-delay estimation algorithms were applied to these experimental data sets. We then computed mean square errors between 1600-point sequences of experimental and theoretical arrival times in a defect-free aluminum plate. The value of the mean square error served as an independent measure of the performance of each algorithm.

One of the powerful time-domain detection methods is pattern matching. Given some signal-like pattern, one can match it point by point against parts of the signal until a specified accuracy (or minimum error) is reached. Using this technique some acceptable results were obtained [29–34]. Another popular way of finding similarities in different traces is to search for peaks of their cross-correlation functions [43].

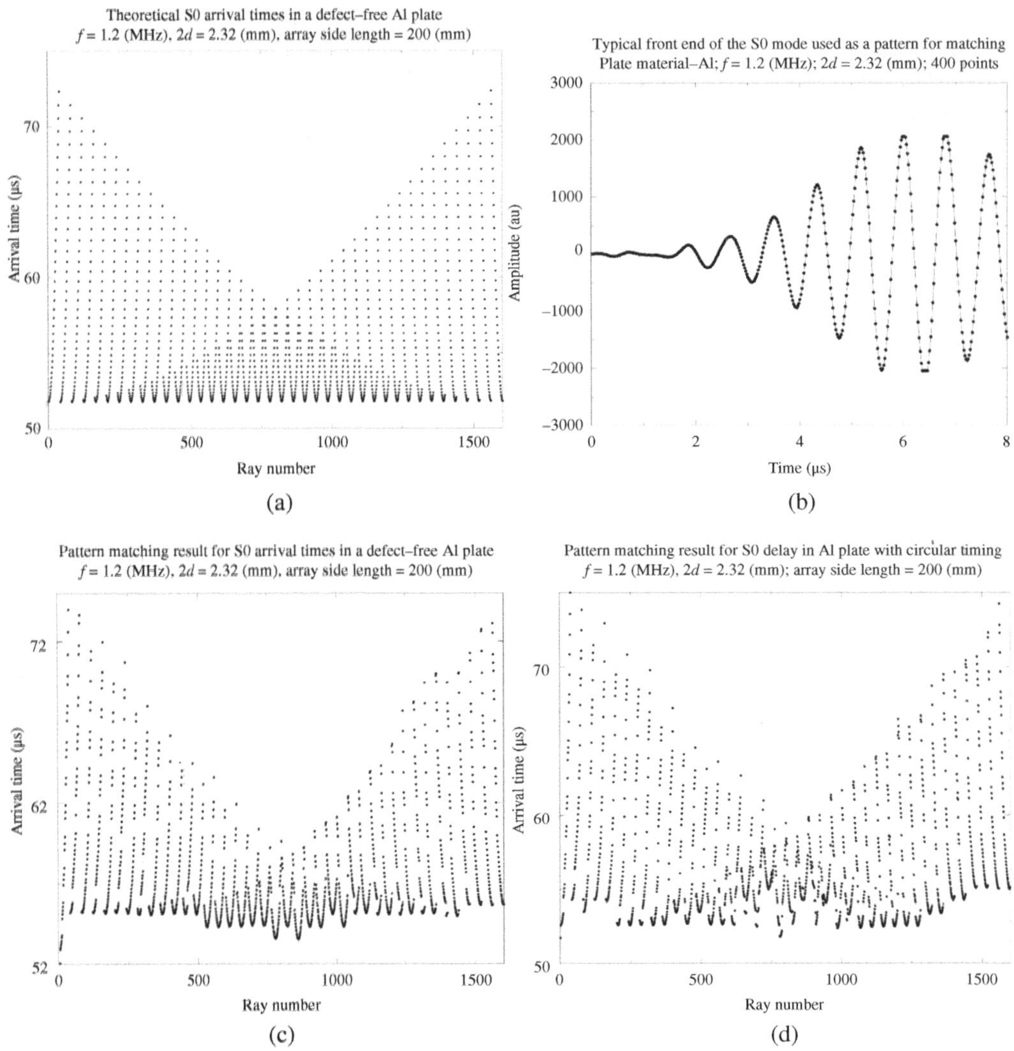

Figure 5.14 Theoretical arrival times of the S_0 mode in a defect-free aluminum plate (a). To generate this pattern the source-receiver distance was changed the same way as in our crosshole scanner. Typical shape of the leading edge of the fastest mode is shown at (b). It is often difficult to accurately find the beginning of the wave packet. Bottom plots show arrival times in a defect-free aluminum plate (c) obtained with adaptive pattern matching method. Comparison with theoretical times yields MS error $\delta_{ms} = 2.74$. At (d) are arrival times in aluminum plate with 25.4-mm-diameter flat-bottom hole in the center, obtained with an adaptive pattern matching method.

Figure 5.14 also shows the result of application of an adaptive pattern matching algorithm to the test tomographic data set acquired on a defect-free aluminum plate and one with a flat-bottom hole. The pattern length N in this case was chosen to be 400 points. Comparison with theoretical arrival times for unflawed plates shows an MS error of 2.74. Although the method shows good potential in tracking the arrival time of successive signals, it can be seen that its results often differ from their expected values by several wavelengths.

The time domain pattern matching approach requires a great deal of empirical knowledge about the signal of interest. For instance, to start the adaptive pattern matching algorithm, the operator must supply the hand-picked arrival time of the first signal in the set and use some theoretically calculated arrival times to outline the search zone. We next considered a time-frequency approach as a promising substitute to the pure time domain methods.

The idea behind time-frequency analysis is that the energy density spectrum tells us which frequencies existed during the whole duration of the signal and no indication as to when they existed [44]. A great variety of more or less successful methods have been developed to combine the time and frequency domains of the signal into a more powerful and informative representation. The advantage of a joint time-frequency representation would be the simultaneous knowledge about the presence of particular frequencies in the signal and times at which those frequencies occur. So far, no universal distribution, capable of handling different signals with equally good resolution, has been constructed. All existing successful transforms are tailored to particular signals and our purpose was to find one yielding enough resolution both in time and frequency domains for accurate extraction of the features of interest from the recorded Lamb wave signals. Such a distribution would combine the most useful information in both domains and simplify detection of the arrival time of the event at a particular frequency.

In our tomographic experiments, we excited Lamb waves with a narrow-band tone burst and the first received arrival is a wave packet composed of the amplitude-modulated central carrier frequency. The spectrum of such a packet has a sharp maximum at the carrier frequency. The part of the signal before the first arrival would not contain that frequency. Using an appropriate time-frequency algorithm, it is possible to determine the time when the carrier frequency enters the signal spectrum – that is, the signal arrival time.

The simplest time-frequency representation can be constructed by means of a moving-window Fourier transform. The window of a fixed size moves along the signal and at each step, its content is Fourier transformed to determine which frequencies existed in that time interval. Narrowing the window to a certain extent increases the resolution in time, but for very short windows the spectrum becomes meaningless because of the increasing bandwidth of the short signals.

The window size in the moving-window Fourier transform cannot be made arbitrarily small without affecting the bandwidth of the corresponding part of the signal. The time resolution limit of a spectrogram will therefore be equal to the size of the smallest window capable of reliably detecting the carrier frequency. Such a detection is possible only when the part of signal within the window includes at least one complete period at a carrier frequency. At 1 MHz such a period T will be $T = \frac{1}{f} = 1\mu s$. A typical value for the phase velocity of the S_0 mode in aluminum in our experiments is $v_{ph} = 3.8\ mm/\mu s$. Hence, if the distance between transmitter and receiver changes by 0.26 mm, the phase of the wave shifts 2π or one period. The best temporal resolution of the spectrogram in our case is equivalent to that when the transmitter–receiver distance randomly changes by $\pm$0.26 mm. Of course, for better frequency resolution, the real window must be wider than just one temporal period.

The low temporal resolution of a spectrogram forced us to look for alternative distributions, which would be capable of handling our signals. The Wigner distribution [44] is the prototype of distributions that are qualitatively different from the spectrogram.

The Wigner distribution does not belong to a class of positive distributions because it is not positive throughout the time-frequency plane. In addition, it is not necessarily zero when the signal is zero and at frequencies that do not exist in the signal spectrum. These phenomena are called interference or "cross terms" and lead to numerous artifacts in the resulting two-dimensional representation. Figure 5.15 demonstrates the signal typical for our experiments, its Fourier power spectrum and the corresponding Wigner transform image. The cross-term artifacts show up at frequencies higher than the peak frequency on the signal power spectrum. These artifacts

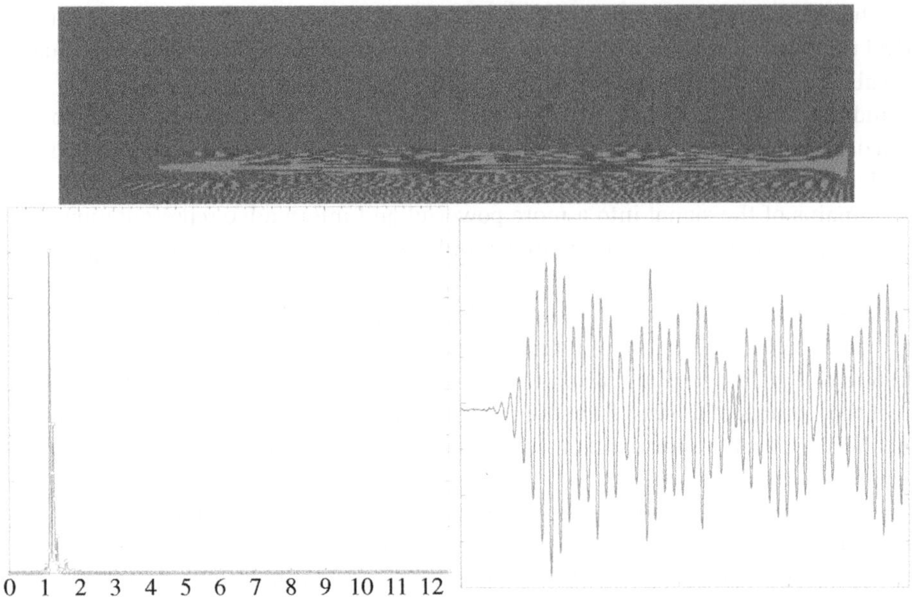

Figure 5.15 Wigner transform of the signal (top). The power spectrum and original signal are shown below. Note severe artifacts at frequencies above the carrier frequency, that is, the peak frequency on the power spectrum.

complicate traversing the time frequency plane while looking for the signal arrival time. It would be desirable, however, to reduce the level of cross-term artifacts and to speed up the computation process. Straightforward numerical computation of the Wigner distribution of an N-point signal requires N Fourier transforms. Although rather fast for one signal, this process becomes the major bottleneck if many (tens of thousands) of signals are involved. In addition, the Wigner transform needs higher sampling rate than the conventional DFT to avoid aliasing; its frequency resolution is reported to be only a quarter of that obtained by DFT [45].

A successful attempt to adapt Wigner distribution to study the dispersion of Lamb waves is described in [45]. This approach involves calculation of so called Wigner–Ville distribution, which is a Wigner distribution of the analytic signal. The analytic signal is a complex signal, where the real component is the original signal and the imaginary component is its Hilbert transform. The Wigner–Ville transform reduces the sampling requirement to that of the Nyquist criteria. To minimize the cross-term interference, the Wigner–Ville distribution is convolved with a Gaussian window function. The resulting pseudo Wigner–Ville distribution is positive and much less noisy than its predecessors. It was effectively employed in [45] for arrival time estimation and building Lamb wave dispersion curves.

A very general and accurate delay estimation method for dispersive media was proposed by Herman [46], tested by Goudswaard et al. [47], and further developed by Ernst and Herman [48]. The method called generalized traveltimes is based on an error norm, related to the phase of the wavefield. To avoid the cumbersome step of interpreting arrivals, it computes phase "traveltime" as a logarithm of the Fourier transform of the signal part cut around a theoretically guessed time of flight, simplifying the task and making it almost automatic.

The authors of [48] comment that no single traveltime can be associated with the direct guided wave due to its dispersive nature. Since each frequency travels with a different phase velocity, the traveltime of a certain event also varies with frequency, leading to a dispersive $\tau(\omega)$. The definition of generalized traveltimes is both an intuitive and physically correct approach to group velocity in dispersive media. We adapted it to Lamb wave tomography and studied its performance on the two test data sets.

After testing the performance of these and various other techniques,[9] we reached maximum accuracy and speed with a time domain group delay estimation method. The method determines the time delay of the leading edge of the envelope of the fastest mode and requires well-filtered signals with distinctive first arrivals. The arrival times extracted with the method yield the lowest MS error when compared to the theoretical ones. Another success criterion is the quality of the tomographic reconstruction. Most of the results of this work were obtained with time domain group delay estimation methods, very efficiently handling Lamb wave signals typical to our experiments.

Time-frequency analysis based on a custom-built positive distribution is a robust autonomous method, although less accurate than the group delay method. It cannot compute time delay with enough precision because of the uncertainty principle. However, it provides a reliable first estimate of the arrival time that can be used as an input to more accurate algorithms if theoretical prediction is a problem. The method is fast, highly insensitive to signal quality, and works practically without human intervention. Using wavelet transform instead of conventional DFT may improve time localization accuracy, as we'll see later, after we've done some actual scattering analysis. Nevertheless, the method of generalized traveltime is automatic, fast, and based on the deep insight into the nature of guided waves and the concept of group velocity. Preliminary results obtained with the method were promising, although not free from the DFT-specific uncertainty. Both pattern matching and neural network approaches have been explored extensively within our work. Although the results for some samples were quite satisfactory, we were unable to reach the reliability and accuracy of our more successful methods.

I realize that was rather a long discussion, which you may or may not have cared all that much about. I thought it was important to include at this point in a book about scattering because the whole point of what we're doing is utterly reliant on making sense of complex time-domain signals out in the real world. As we're about to see, acoustic, electromagnetic, and elastic wave scattering is horrifyingly (delightfully) complex and making sense of complex signals is key to exploiting our knowledge of subtleties in the scattering behavior. What used to be called signal conditioning or filtering or preprocessing isn't going to cut it anymore. We're going to want to bring to bear much more sophisticated data engineering approaches so that we can get the most out of the signals that we have. It's no longer good enough to have detected something-something. The charge these days is to automatically process the recorded signals and various other ancillary data streams to determine in near-real time exactly what, where, how big, and who that signal just scattered from. Ideally, the data-engineering preprocessing steps will continue to work as the algorithms they're feeding into are improved. In our work, the next step was to relax the assumption that the Lamb waves travel in straight lines, because we know full well that they scatter and diffract and reflect and refract. It's delightfully complex.

---

9 See [49] for many of the details. Drs. Eugene and Daryia Malyarenko both did their PhDs at W&M, and after the birth of their second child naturally planned a trip home to Kiev to visit grandparents in September of 2002. They had only planned on being gone for two weeks, but got stranded for several months because American consular officials were being *extra* careful processing visas in the year(s) after 9/11. I remember my daughter bringing home school assignments for their oldest so I could email them to Kiev. Science has been fully internationalized for decades, but every so often distances and boundaries suddenly matter quite a lot.

5.4.5 Curvilinear SIRT

As I may have mentioned, this is a book about scattering. We'll get to that for real, I promise. The aforementioned discussion about Lamb wave tomography assumed that the guided waves travel in straight lines from source to receiver, but of course, that is a simplifying assumption. Lamb waves will reflect and refract and diffract at any material discontinuity. It turns out that we had enough to deal with accounting for distortions to the wave multimode packets via dispersion because the path lengths change so much in the practical tomography geometries. I should mention again that this was a surprise to us three decades ago. In our early Lamb wave scanners and our parallel-projection implementation of tomography, the separation between the transducers was fixed and we had a clever piece of analog hardware called a Pulsed Phase-Lock Loop (P2L2) that very accurately tracked small changes in arrival time. When we tried to implement crosshole tomography, the path-length changes were large enough the P2L2 would lose lock. That pushed us to investigate software methods to track arrival times, which we have been doing ever since. We eventually got to the point that we could automatically track arrivals of multiple modes in quite complex Lamb waveforms. Feel free to peek ahead at Chapter 10 to see where our data engineering journey has led us. TL;DR: time-scale representations work even better than time-frequency approaches.

Ray bending is a common feature of traditional crosshole tomography using seismic waves, particularly for layered media. The same thing happens for underwater sound applications where temperature and salinity variations cause refraction. Since the result of our Lamb wave tomographic reconstruction is a slowness grid (which relates to thickness), the question is whether this can be used as a starting point to improve reconstructions by accounting for refraction across grid boundaries and hence ray bending. The answer turns out to be yes, and the reconstructions are improved somewhat (Figure 5.16).

The curved ray SIRT is conceptually very similar to the straight ray one [35] except that it stores all the ray data in memory instead of computing them analytically. It is indeed impossible to know in advance the number of pixels the bent ray will cross as well as the lengths of the pixel segments it will cut. We therefore have to compute these quantities for each traced ray and accumulate them into a huge array. After the tracing of all the rays is completed, we invert the matrix using the same ART or SIRT algorithm as for the straight rays. This means that reconstruction from curved rays preserves all tomographic artifacts specific for a given data acquisition geometry and reconstruction algorithm. In addition, refraction at defects often creates severe focusing or defocusing depending on the difference in the wave speed inside and outside the defect. Due to these effects, the resulting ray density and wave vector coverage can become highly nonuniform throughout the scanned area making the resolution of the reconstruction algorithm depend on the size and location of the defects.

The degree of ray bending is determined by the size and properties of the smoothing mask. This approach is not as artificial as it seems because the smoothed image can be treated as an initial guess for the slowness map. Lower degrees of smoothing leave artifacts that strongly distort ray paths and make accurate tracing impossible. This is in some sense equivalent to the presence of strong scatterers for which the whole geometrical acoustics approach fails. On the other hand, very high degrees of smoothing blur out defects and ray paths more closely resemble straight lines.

As we developed an iterative tomographic procedure using both straight and curved ray SIRT to reconstruct Lamb wave velocity maps for isotropic plates, we found that success of the ray tracing approach strongly depends on the degree of smoothing of the image, reconstructed with a straight

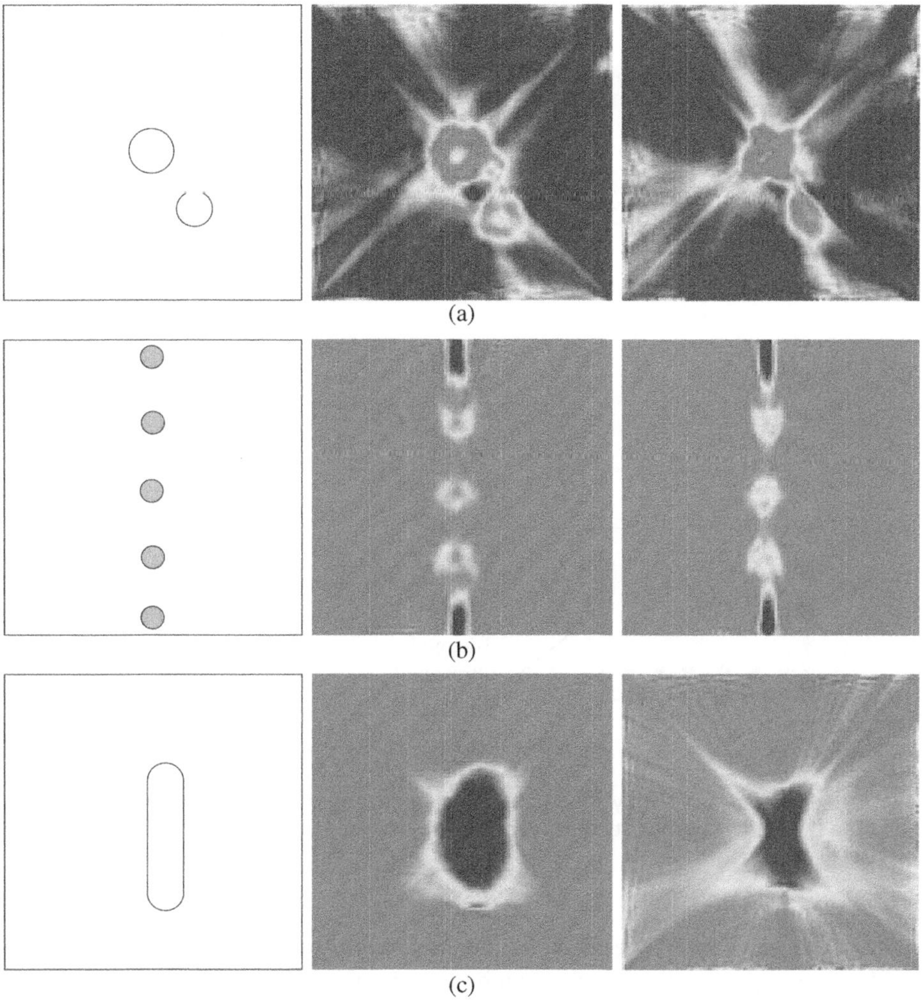

Figure 5.16 Aluminum plate with two circular thinnings (a), five through holes (b) oblong thinning (c). In each, the center image is straight ray SIRT and the right image is curved ray SIRT. Image size: 200 × 200 mm.

ray SIRT. Changing this degree affects the size of the reconstructed defects, but it is possible to find optimal smoothing conditions where defect dimensions will be reconstructed correctly. At the same time, ray bending always distorts the uniformity of the ray density and introduces additional artifacts into the final image. This distortion, however, is proportional to the size and severity of the defect. It does not occur for smaller defects and can be minimized even for larger ones by sufficient smoothing of the input image.

The double-crosshole reconstructions are improved somewhat if we also include all the diagonal raypaths from adjacent legs of a square array. The six projection crosshole geometry has high and uniform ray density plus broad local wave vector directivity pattern throughout the scanned area to insure satisfactory resolution over the entire image.

Once we automated the arrival-time extraction algorithms, we were able to further double-down on the amount of data via a "walking toneburst" scheme where we transmit at each source location

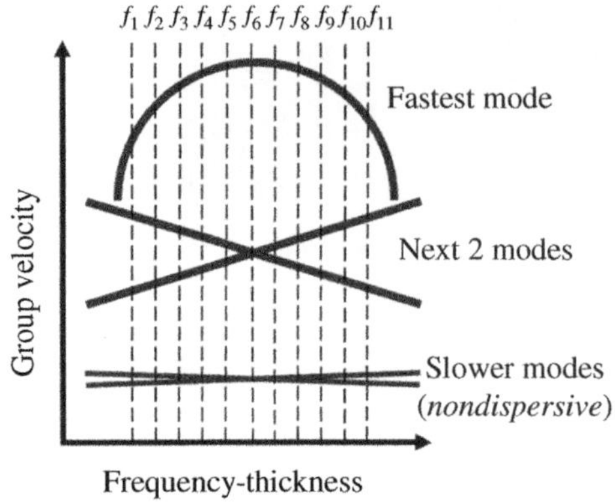

Figure 5.17 A sequence of tonebursts at increasing frequencies are transmitted, and the characteristic changes in arrival times with frequency-thickness for the fastest, middle, and slower modes are used to identify multiple modes in a complex signal. Source: [50]/with permission of Leonard, Kevin Raymond.

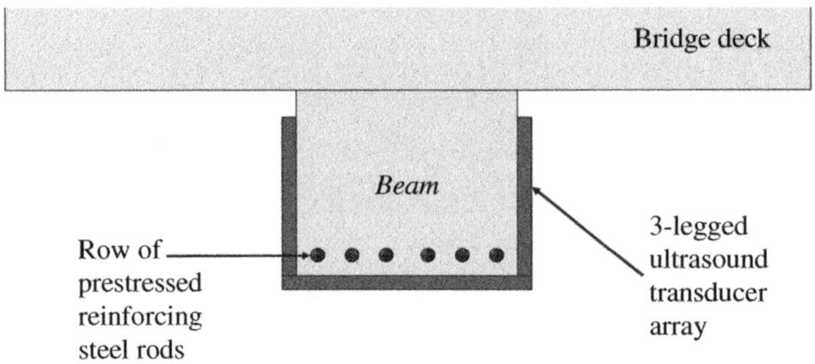

Figure 5.18 A three-legged transducer array allows for tomographic reconstructions, with the many crisscross paths having the highest ray density precisely where it is needed to inspect the steel rebar for corrosion.

a sequence of toneburst signals where the center frequency covers a group velocity hump as shown in Figure 5.17. Having recognized that because Lamb waves in pipes and tanks follow the curvature, we can consider them to be rolled-up plates and thus the crosshole tomography geometry can be implemented with either circumferential belts of transducers or meridional (linear) arrays of transducers. Criss-cross patterns then become helical rays, but the reconstruction algorithms map over directly. Here are two other applications that we have considered. Figure 5.18 shows a concept to inspect the rebar in a commonly occurring steel-reinforced concrete bridge structure. The steel takes up the tensile bending stresses because concrete only withstands compression well, so the cylindrical steel scatterers are toward the bottom of the beam. This three-sided crosshole tomography geometry has been well developed in seismology, although here the rebar is likely to scatter quite strongly and that will need to be accounted for.

When a new "acoustic camera" technology became available,[10] this caused us to think about how it could be exploited for tomographic imaging. Figure 5.19 shows a concept for breast cancer screening. A single large transducer emits an ultrasonic fan beam, which is recorded by the 2D array of sensors in the acoustic camera. The mechanism is rotated similar to a CT scanner to make 3D reconstructions of the uncompressed breast without ionizing radiation.

10 https://www.imperiuminc.com/the-acoustocam.

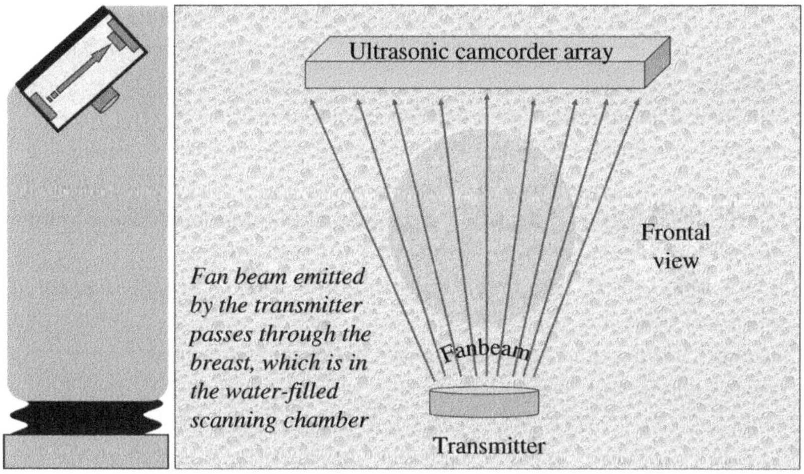

Figure 5.19 Woman leans over scanner (left) with breast suspended into scanning chamber. Water fills chamber and provides medium for ultrasound propagation in a through-transmission fanbeam tomography geometry. All data recording and computer processing of signal are automatic.

References

1 Lamb, H. (1917). On waves in an elastic plate. *Proceedings of the Royal Society of London Series A*, Containing papers of a mathematical and physical character 93 (648): 114–128.

2 Worlton, D.C. (1956). Ultrasonic testing With Lamb Waves. United States. https://doi.org/10.2172/4356069; https://www.osti.gov/servlets/purl/4356069. You can either download the report with DECLASSIFIED stamped on it, or if you prefer it's available in paperback from Amazon: https://www.amazon.com/Lamb-waves-at-ultrasonic-frequencies/dp/B003Z6PQQE.

3 Viktorov, I.A. (1967). *Rayleigh and Lamb Waves: Physical Theory and Applications*. New York: Plenum Press.

4 Achenbach, J.D. (1984). *Wave Propagation in Elastic Solids*. Netherlands: North-Holland.

5 Graff, K.E. (1991). *Wave Motion in Elastic Solids*. New York: Dover.

6 Brekhovskikh, L. and Goncharov, V. (1994). *Mechanics of Continua and Wave Dynamics*. New York: Springer.

7 Rose, J.L. (1999). *Ultrasonic Waves in Solid Media*. Cambridge: Cambridge University Press.

8 Auld, B.A. (1990). *Acoustic Fields and Waves in Solids*. Malabar, FL: Kreiger.

9 Hinders, M.K. (2020). *Intelligent Feature Selection for Machine Learning using the Dynamic Wavelet Fingerprint*. Springer.

10 World Shipping Council (2017). http://www.worldshipping.org (accessed 26 October 2017).

11 Grey, E. (2017). *Cargo Theft: A Billion Dollar Problem*. Ship Technology. https://www.ship-technology.com/features/featurecargo-theft-a-billion-dollar-problem-5882653/ (accessed 1 January 2019).

12 MFAME Team (2017). *Cargo Theft: A Billion-dollar Problem*. MFAME. http://mfame.guru/cargo-theft-billion-dollar-problem/ (accessed 26 October 2017).

13 Edmonds, J. (2016). *The History of the Shipping Container*. Freightos. https://www.freightos.com/the-history-of-the-shipping-container/ (accessed 1 January 2019).

14 Trujillo, V.E. II (2019). Global shipping container monitoring using machine learning with multi-sensor hubs and catadioptric imaging. College of William & Mary, Department of Applied Science Doctoral Dissertation.

15 Schreib, B., Bostic, A., Mitchell, J. et al. (2015). Geo-Referencing Identification (GRID) Tag Final Report. Study funded by Bureau of Safety and Environmental Enforcement. Contract # E14PC00027.

16 Sun, K.J. and Johnston, P.H. (1994). Disbond detection in bonded aluminum joints using Lamb wave amplitude and time-of-flight. *Review of Progress in QNDE* 13: 1507–1514.

17 McKeon, J.C.P. and Hinders, M.K. (1999). Lamb wave scattering from a through hole. *Journal of Sound and Vibration* 224 (5): 843–862.

18 Nowitall, I.B. (2020). *Scientific Adulting and BS Detection.* W&N Edutainment Press.

19 DC National Transportation Safety Board, Washington (1989). Aircraft Accident Report: Aloha Airlines, Flight 243, Boeing 737-200, N73711, Near Maui, Hawaii, April 28, 1988. U.S. Department of Commerce, NTTS. https://www.ntsb.gov/investigations/AccidentReports/Reports/AAR8903.pdf (accessed 16 September 2024).

20 Krautkramer, J. and Krautkramer, H. (1990). *Ultrasonic Testing of Materials*, 4e. Springer-Verlag.

21 Zhu, W. and Rose, J.L. (1999). Lamb wave generation and reception with time-delay periodic linear arrays: a BEM simulation and experimental study. *IEEE Transactions on Ultrasonics, Ferroelectrics, and Frequency Control* 46 (3): 654–664.

22 Rose, J.L. (2000). Guided wave nuances for ultrasonic nondestructive evaluation. *IEEE Transactions on Ultrasonics, Ferroelectrics, and Frequency Control* 47 (3): 575–583.

23 Demol, T., Blanquet, P., and Delebarre, C. (1995). Lamb waves generation using a flat multi-element array device. *IEEE Ultrasonics Symposium*, 791–794. IEEE.

24 Strycek, J.O., Grandia, W.A., and Loertscher, H. (1997). Wave modes produced by air coupled ultrasound. *NDT.net* 2 (5).

25 Bray, D.E. and Stanley, R.K. (1997). *Nondestructive Evaluation. A Tool in Design, Manufacturing and Service.* Revised edition. CRC Press.

26 Scruby, C.B. and Drain, L.E. (1990). *Laser Ultrasonics: Techniques and Applications.* Adam Hilger.

27 Friedman, A. (2000). Theoretical and experimental study of generation mechanisms for laser ultrasound in woven graphite/epoxy composites with translaminar stitching. PhD in Applied Science Thesis. Williamsburg, VA, USA: College of William and Mary.

28 Tretout, H. (1998). Review of advanced ultrasonic techniques for aerospace structures. *NDT.net* 3 (9).

29 Hinders, M.K. and McKeon, J.C.P. (1998). Lamb wave tomography for corrosion mapping. 2nd joint NASA/FAA/DoD Conference on Aging Aircraft.

30 Pei, J., Yousuf, M.I., Degertekin, F.L. et al. (1995). Lamb wave tomography and its application in pipe erosion/corrosion monitoring. *IEEE Ultrasonics Symposium*, 795–798.

31 Pei, J., Yousuf, M.I., Degertekin, F.L. et al. (1996). Plate tomography with dry contact Lamb wave trasducers. *Acoustical Imaging* 22: 725–730.

32 Hildebrand, B.P., Davis, T.J., Posakony, G.J., and Spanner, J.C. (1999). Lamb wave tomography for imaging erosion/corrosion in piping. *Review of Progress in QNDE* 18: 967.

33 Hinders, M.K., Malyarenko, E.V., and McKeon, J.C.P. (1998). Contact scanning Lamb wave tomography. *Journal of the Acoustical Society of America* 104 (3, Part 2): 1790.

34 McKeon, J.C.P. and Hinders, M.K. (1999). Parallel projection and crosshole Lamb wave contact scanning tomography. *Journal of the Acoustical Society of America* 106 (5): 2568–2577.

35 Malyarenko, E.V. and Hinders, M.K. (2000). Fan beam and double crosshole Lamb wave tomography for mapping flaws in aging aircraft structures. *Journal of the Acoustical Society of America* 108 (4): 1631–1639.

36 Natterer, F. (2001). *The Mathematics of Computerized Tomography*. New York: Wiley.

37 Webb, S. (1988). *The Physics of Medical Imaging*. IOP Publishing Ltd.

38 Bregman, N.D., Bailey, R.C., and Chapman, C.H. (1989). Crosshole seismic tomography. *Geophysics* 54 (2): 200–215.

39 Pratt, R.G. and Goulty, N.R. (1991). Combining wave-equation imaging with traveltime tomography to form high resolution images from crosshole data. *Geophysics* 56 (2): 208–224.

40 Kak, A.C. and Slaney, M. (1988). *Principles of Computerized Tomographic Imaging*. New York: IEEE Inc.

41 Devaney, A.J. (2012). *Mathematical Foundations of Imaging, Tomography and Wavefield Inversion*. Cambridge University Press.

42 Wogglebug, M.H. (2024). *Incubator*. W&N Edutainment Press.

43 Taner, M.T. and Koehler, F. (1998). Estimation of unbiased delays. *Geophysics* 63 (2): 738–742.

44 Cohen, L. (1995). *Time-Frequency Analysis*. Englewood Cliffs, NJ: Prentice Hall.

45 Prosser, W.H. and Seale, M.D. (1999). Time-frequency analysis of the dispersion of Lamb modes. *Journal of the Acoustical Society of America* 105 (5): 2669–2675.

46 Herman, G. (1992). Generalization of traveltime inversion. *Geophysics* 57 (1): 9–14.

47 Goudswaard, J.C.M., ten Kroode, F.P.E., Snieder, R.K., and Verdel, A.R. (1998). Detection of lateral velocity contrasts by crosswell traveltime tomography. *Geophysics* 63 (2): 523–533.

48 Ernst, F. and Herman, G. (2000). Tomography of dispersive media. *Journal of the Acoustical Society of America* 108 (1): 105–116.

49 Malyarenko, E.V. (2000). Lamb wave diffraction tomography. William and Mary Doctoral Dissertation.

50 Leonard, K.R. (2004). Ultrasonic guided wave tomography of pipes. William and Mary Doctoral Dissertation.

6

Scattering from Spheres

The same approach that we took to analyze the reflection and refraction/transmission of waves at planar boundaries can be done for a variety of geometries. The "algebra" will be more complicated, of course, but we'll be fine as long as we do that algebra carefully and don't panic about the so-called special functions that will show up. That's not to minimize the complexity of what we're about to do, though. We will end up with really complicated mathematical expressions that we won't be able to just look at and deduce the *wave scattering* behavior. Instead, we'll have to rely on computers of some sort to evaluate the expressions and show us the behavior. I'm old enough to remember when that was really hard, but now the trickiest part is usually typing the expressions correctly.

6.1 Clebsch–Mie Scattering

When I was a graduate student, theoreticians and experimentalists and those who used computers were distinct personages. The senior scientist who had the office next to me at the Air Force was developing new integral equation methods for electromagnetic scattering, which is a numerical approach to scattering, but he didn't have a computer of any sort on his desk. He had an assistant who did the numerical work for him. My advisor had a collaborator in another state who did the numerical aspects of their research, and they collaborated primarily by mail and phone. Thirty years ago, it would have been rather unusual for an experimental apparatus to be computer controlled in the sense we think of it today, so experimentalists didn't really use computers either. Needing a computer-literate collaborator in order to make a plot from your data or equations strikes all my students these days as quite funny, even though I sometimes ask them to make a plot for me. Scientists also had people to type for them back in those days, and the best typists could take your literal cut-and-paste and scribble-and-whiteout draft and turn it into a perfectly typed technical manuscript because the IBM Selectric typewriter had interchangeable balls to do both text and symbols. Personally, I've never had a secretary who typed as well as me and I have always written my own equations and grunted my own code to make my own plots. Turning output into a 3D plot took quite a long time on computers in the 1980s, BTW.

It's best if you can derive your own equations and program your own computers and collect your own data. You spend a lot less time waiting on collaborators to get back to you with their stuff. Often it turns out that some or all of the data you might need is readily available, and experimentalists will be happy to collaborate with you so they don't have to try to slog through all the math. When you're choosing problems to attack, it's also a good idea to consider questions that have practical and even commercial importance.

A Data Engineering Approach to Wave Scattering Analysis: with Applications in Radar, Sonar, Medical Diagnostics, Structural Flaw Detection and Intelligent Robotics, First Edition. Mark K. Hinders.

In 1908, Gustav Mie[1] investigated the scattering of light from spherical particles in an attempt to explain the colors produced by colloidal gold suspensions, specifically the colors of stained glass [1]. Mie theory, as it's now called, gives the exact solution for the optical effects of spherical particles and provides a first-order approximation of the effects from nonspherical particles. Explaining the colors of stained glass was of practical interest, and there was ample experimental data for Mie to compare his results to. It was a problem that a number of great minds had attacked, but without satisfactory results. The new electromagnetic theory of light turned out to be much better suited than the rather unwieldy elastodynamic theory of the æther, which didn't quite work. The mathematical tools necessary had recently been worked out as well so Mie sucked it up and did his own computations, which must have been quite onerous. Once you've got the problem set up, though, it's a rather straightforward boundary value problem to solve, as we'll see later. L.V. Lorenz did that algebra well before Mie, but he published it in Danish more than a century before the internet, so Mie wouldn't have known about his paper.

Doing a literature review before *Google scholar* was a tedious process, which I kind of miss. I had mad photocopy skills back in the day. I still have three-ring binders full of photocopies of the early Mie scattering papers all the way back to the 1860s. When I traced things back to Clebsch's foundational paper and gingerly photocopied it, I noticed on the card in the pocket of that volume that the last person to check it out was Nelson Logan, so I'm going to quote the introductory paragraphs of his review paper, "Survey of Some Early Studies of the Scattering of Plane Waves by a Sphere," which he published in 1965 [2]:

> On October 30, 1861, A. Clebsch (1833–1872) completed a 68-page memoir [3] in which he developed the mathematical theory required to solve (by the method of separation of variables) the class of boundary-value problems in which a wave propagating in an elastic medium impinges upon a spherical surface. This paper could have become the cornerstone upon which future generations of scientists could base their theoretical studies. Scores of writers since Clebsch have sought to increase our understanding of this class of problems. However, the mathematical ingenuity of this master craftsman was doomed to lie buried within the pages of one of the leading mathematical journals of the middle of the 19th century while later writers rediscovered the results which were to be found in Clebsch's paper.
>
> The same fate was to befall the equally great memoir [4], upon the reflection and refraction of light by a transparent sphere, which was published in 1890 by L. Lorenz (1829–1891).

My read of the historical situation is that A. Clebsch set up the problem and L. Lorenz solved it, but G. Mie (1868–1957) got credit for it. Note that Clebsch and Lorenz were at the end of their

1 Gustav Mie was a professor of physics with a strong background in mathematics. After moving to the University of Greifswald in North-Eastern Germany, he became acquainted with colloids, and one of his PhD students investigated the scattering and attenuation of light by gold colloids experimentally. Mie used his previously acquired knowledge of Maxwell's equations and solutions of very similar problems in the literature to concisely treat the theoretical problem of scattering and absorption of light by a small absorbing sphere. He also presented many numerical examples which explained all the effects that had been observed until then. Since all calculations were done by hand, Mie had to limit his theoretical results to three terms in infinite expansions; thus, he only could treat particles smaller than 200 nm at visible wavelengths. Mie's paper had remained hardly noticed for the next 50 years, most likely because of the lack of computers. It experienced a revival later and up to now it has been referenced more than 4000 times, owing to the widespread use of Mie's approach in sciences such as astronomy, meteorology, fluid dynamics, and many others.

Gustav Mie did not consider his work on scattering of light by small particles as very important, since he just tried to explain the effects which his students had observed. He concentrated on hot topics in theoretical physics, for example, the theory of matter. He wrote several textbooks on relativity, gravitation theory, and electromagnetism.

Figure 6.1 Problem geometry for scattering from a sphere.

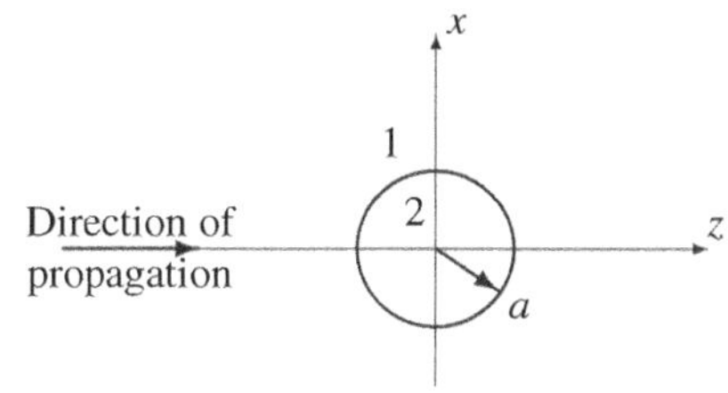

careers while Mie still had a long life ahead of him during which he could, as needed, remind people that the answers they wanted were available in his 1908 paper and here let me send you a reprint. Logan also points out that the essential mathematics of what we now call Rayleigh scattering was published by Clebsch in 1869, but Rayleigh perhaps didn't read German....

There are innumerable treatments of the Mie scattering problem, that is, plane electromagnetic scattering from a sphere, but I found the treatments by Born and Wolf [5] and Bohren[2] and Huffman [6] especially clear and concise so that's more or less what I'm including here.

We consider a plane, monochromatic, electromagnetic wave propagating in a homogeneous, isotropic medium (like, maybe glass) that is incident upon a spherical inclusion with finite conductivity (like, maybe a gold particle). Figure 6.1 is a drawing of the problem geometry that I programmed in LaTeX in 1988. At the origin of our coordinate system, we have a sphere of radius a which has different material properties from the surrounding medium. Call the region $r > a$ medium 1 and $r < a$ medium 2.

Since we're going to solve the problem in frequency domain, we will suppress the $e^{-i\omega t}$ time dependence, and write Maxwell's equations as:

$$\nabla \times \vec{H} = -i\omega\epsilon\vec{E} \qquad \nabla \times \vec{E} = i\omega\mu\vec{H} \tag{6.1}$$

along with $\nabla \cdot \vec{E} = 0$ and $\nabla \cdot \vec{H} = 0$, where the wave numbers are

$$k_1 = \omega\sqrt{\epsilon_1\mu_1} \qquad k_2 = \omega\sqrt{\left(\epsilon_2 + i\frac{\sigma_2}{\omega}\right)\mu_2} \tag{6.2}$$

and we note that the finite conductivity of the scatterer turns out to be really important because without accurate values for σ_2, the model won't predict the colors of the stained glass. Now take the curl of the first two of Maxwell's equations and use my favorite vector identity

$$\nabla \times (\nabla \times \vec{A}) = \nabla(\nabla \cdot \vec{A}) - \nabla^2\vec{A} \tag{6.3}$$

to get the vector wave equation for both $\vec{E}$ and $\vec{H}$

$$\nabla^2\vec{E} + k^2\vec{E} = 0 \qquad \nabla^2\vec{H} + k^2\vec{H} = 0 \tag{6.4}$$

Since we're going to want a plane wave propagating in the z-direction, the incident wave solution is

$$E_x^{(i)} = e^{ik_1 z} \qquad E_y^{(i)} = E_z^{(i)} = 0 \tag{6.5}$$

with the magnetic field components computed with Maxwell's equation.

We'll then assume appropriate functional forms for the scattered electric and magnetic fields in region 1 ($r > a$) as well as in region 2 ($r < a$). Those solutions are going to be in spherical

---

2 Craig F. Bohren, Distinguished Professor Emeritus of Meteorology at Pennsylvania State University, is an atmospheric scientist and physicist who wrote fundamental books on light scattering, atmospheric thermodynamics, and radiative transfer, as well as popular science books on atmospheric optics such as "Clouds in a Glass of Beer: Simple Experiments in Atmospheric Physics" in 1987.

coordinates, of course. Don't freak out. I'll try to include the details here (and throughout) so you can follow along without looking up esoterica in books you'd have to go borrow from some old-timer like me because then you're likely to end up having to sit through some long, boring story before being allowed to leave with the book you need.

When my advisor died, I had the somber task of sorting through his three offices and deciding which things to save and which to discard. I had earned my stipend as a first-year graduate student by photocopying papers for him. He would make a check mark by the references he wanted in a paper he was reading and I would go find them in one of the science libraries at Harvard and photocopy them. He would make check marks by the references he wanted from those, etc. I wasn't even a student at Harvard, BTW, but I had a master key to the physics department there, so I could get into the various libraries and copy rooms and such after hours. There was no swiping in with ID cards back in those days. It was many dozens of bankers, boxes worth of papers. It took quite a while.[3]

The boundary conditions for electromagnetic scattering are continuity of tangential electric and magnetic fields, applied at the surface of the scatterer ($r = a$). Since I made you read through an anecdote, I owe you all the mathematical details. Spherical coordinates are related to Cartesian coordinates by

$$x = r\sin\theta\cos\phi \qquad y = r\sin\theta\sin\phi \qquad z = r\cos\theta \tag{6.6}$$

and the components of any vector $\vec{A}$ are transformed according to

$$\begin{aligned} A_r &= A_x \sin\theta\cos\phi + A_y \sin\theta\sin\phi + A_z\cos\theta \\ A_\theta &= A_x\cos\theta\cos\phi + A_y\cos\theta\sin\phi - A_z\sin\theta \\ A_\phi &= -A_x\sin\phi + A_y\cos\phi \end{aligned} \tag{6.7}$$

We're going to need the curl of a vector in spherical coordinates, which has the r, θ, ϕ components

$$\begin{aligned} (\nabla\times\vec{A})_r &= \frac{1}{r^2\sin\theta}\left[\frac{\partial(rA_\phi\sin\theta)}{\partial\theta} - \frac{\partial(rA_\theta)}{\partial\phi}\right] \\ (\nabla\times\vec{A})_\theta &= \frac{1}{r\sin\theta}\left[\frac{\partial A_r}{\partial\phi} - \frac{\partial(rA_\phi\sin\theta}{\partial r}\right] \\ (\nabla\times\vec{A})_\phi &= \frac{1}{r}\left[\frac{\partial(rA_\theta)}{\partial r} - \frac{\partial A_r}{\partial\theta}\right] \end{aligned} \tag{6.8}$$

so that Maxwell's curl equations can be written out as:

$$\begin{aligned} -i\omega\epsilon E_r &= \frac{1}{r^2\sin\theta}\left[\frac{\partial(rH_\phi\sin\theta)}{\partial\theta} - \frac{\partial(rH_\theta)}{\partial\phi}\right] \\ -i\omega\epsilon E_\theta &= \frac{1}{r\sin\theta}\left[\frac{\partial H_r}{\partial\phi} - \frac{\partial(rH_\phi\sin\theta)}{\partial r}\right] \\ -i\omega\epsilon E_\phi &= \frac{1}{r}\left[\frac{\partial(rH_\theta)}{\partial r} - \frac{\partial H_r}{\partial\theta}\right] \end{aligned} \tag{6.9}$$

---

3 I did get to keep all the books I wanted, though, and now they're available to my own students to make good use of. I recently realized that the copy of Born and Wolf [5] that I've had all these years was borrowed by my advisor from Don Chodrow, who wrote his name in ink on the inside cover. I Googled him and found that he earned a PhD in physics from Harvard in 1973, which explains how my advisor happened to have the book that has been here in Virginia since 1993. It turns out that Dr. Chodrow was a professor at James Madison University, also here in Virginia, for all these years, but he died in a car crash in 2012, so I've lost my chance to give him his book back.

and

$$
\begin{aligned}
i\omega\mu H_r &= \frac{1}{r^2 \sin\theta}\left[\frac{\partial(rE_\phi \sin\theta)}{\partial\theta} - \frac{\partial(rE_\theta)}{\partial\phi}\right] \\
i\omega\mu H_\theta &= \frac{1}{r\sin\theta}\left[\frac{\partial E_r}{\partial\phi} - \frac{\partial(rE_\phi \sin\theta)}{\partial r}\right] \\
i\omega\mu H_\phi &= \frac{1}{r}\left[\frac{\partial(rE_\theta)}{\partial r} - \frac{\partial E_r}{\partial\theta}\right]
\end{aligned}
\tag{6.10}
$$

These expressions will allow us to write boundary condition equations that ensure continuity of the tangential (θ and ϕ) electric and magnetic fields at the constant coordinate surface $r = a$. But first some math. It's pretty mathy. Again. Don't. Freak. Out. I'm going to include the details.

Since we know that $\nabla \cdot \vec{E} = 0$ and $\nabla \cdot \vec{H} = 0$, we can write both $\vec{E}$ and $\vec{H}$ as the curl of some potential function. Since we're making some effort to revere Clebsch, let's invoke Clebsch's theorem, which says that any vector can be written as the sum of two parts, one of which has no curl and the other of which has no divergence. For this next step, the various books and papers are all over the map on notation and such, but I'll try to be consistent throughout this chapter. Kerker [7] included a handy table in his book translating the notations used by various authors.

Let's write things as:

$$\vec{E} = \nabla \times (r\Pi_{TM}) \qquad \vec{H} = \frac{1}{k}\nabla \times \nabla \times (r\Pi_{TE}) \tag{6.11}$$

where the (scalar) potential functions Π_{TM} and Π_{TE} each satisfy the (scalar) Helmholtz equation

$$\nabla^2\Pi + k^2\Pi = 0 \tag{6.12}$$

This is important because we know the solution to this equation in various coordinate systems, including spherical. Separation of variables allows us to write

$$\Pi = R(r)\Theta(\theta)\Phi(\phi) \tag{6.13}$$

which says that the general solution for Π is a product of three functions, which each only depend on r, θ, ϕ, respectively. Those three functions each then have to obey the following ODEs:

$$
\begin{aligned}
\frac{d^2(rR)}{dr^2} + \left(k^2 - \frac{\alpha}{r^2}\right) rR &= 0 \\
\frac{1}{\sin\theta}\frac{d}{d\theta}\left(\sin\theta\frac{d\Theta}{d\theta}\right) + \left(\alpha - \frac{\beta}{\sin^2\theta}\right)\Theta &= 0 \\
\frac{d^2\Phi}{d\phi^2} + \beta\Phi &= 0
\end{aligned}
\tag{6.14}
$$

where α and β are separation constants. The third of these should look familiar to you. If not, make the variable substitution of $\beta = m^2$, which we're going to need to enforce single-valuedness anyway, and I hope you'll agree that the general solution for $\Phi(\phi)$ is

$$\Phi(\phi) = a_m \cos(m\phi) + b_m \sin(m\phi) \tag{6.15}$$

where m is an integer. At this point, Born and Wolf say that the second ODE "is the well-known equation for spherical harmonics." I'll expect you to nod knowingly about things like that and maybe say, "Yes, of course." Confidence is the key to bullshitting your way through such things. You can Google shit later. In that vein, I'm sure you'll also agree that a necessary and sufficient condition for a single-valued solution is that $\alpha = l(l+1)$ where $l > |m|$ is an integer.

Most books will make a variable substitution at this point, such as $\xi = \cos\theta$, so that the second ODE can be written

$$\frac{d}{d\xi}\left[(1-\xi^2)\frac{d\Theta}{d\xi}\right] + \left[l(l+1) - \frac{m^2}{1-\xi^2}\right]\Phi = 0 \tag{6.16}$$

at which point you can say, "Oh, now I recognize it. It's Legendre's equation. My favorite!" if you didn't bullshit confidently enough at the previous step.

I didn't do quite as well in my first differential equations class as I might have liked. It was taught in a movie theater. The university had worked out a deal with a new multiplex on campus that showed artsy movies at night, so that the auditoria could be used for classes during the day. The snack bar wasn't open during the day, though. There were chalkboards down front, of course, but they were covered with curtains in the evening. The primary issue, was that the theater seats didn't have the fold-away writing surfaces for taking notes, which you certainly should do in something like differential equations. Even if your brain is only half engaged, you need to have your pencil trace out in your notebook most or all of what the professor's chalk is doing on the board. There was a box of little boards in the back that you could use if you wanted, but geez. Class attendance was kind of sparse, except for the midterm when the professor walked in, plopped the pile of exam papers down on the floor at the front of the single-aisle auditorium and said, "Come get your exams and get started." I don't think he was thrilled at teaching in the movie theater across campus from the math department, or maybe he wanted some *Jujubes* to chew on while we did the test. I sat right down front for the rest of the exams, including the final where I didn't do as well as I might have because I was sick all finals week. I had to rest twice on my way to the differential equations final. The student health center was way at the far end of campus, so it didn't really occur to me to go there instead.

Differential equations often have names. They're named for mathematicians who studied them to death and worked out their solutions. One of the things I remember from movieplex math is that in some cases you simply guess the solution to an ODE. If that doesn't work, you try a series solution. For differential equations that are special enough to get named, their series solutions often get the corresponding name and a symbol and Wikipedia pages. Hence, the solutions to Legendre's equation are the special functions named Legendre functions. Legendre and others have worked out all the properties of Legendre functions. I recommend getting a copy of *Abramowitz and Stegun: Handbook of Mathematical Functions*, which you can download or buy inexpensively in paperback or swipe from the office of an oldster. For all the special functions you're likely to come across in your work, Abramowitz and Stegun will have the grown-up versions of the trig identities you're going to need. I buy copies for my incoming graduate students so they won't steal mine. Most of the pages of that book are tables of output, the idea being that you could check your computations with the special functions against standards. Milton Abramowitz and Irene Stegun worked for the National Bureau of Standards, which is now the National Institute of Standards and Technology, so we can trust their numbers. In addition, this was one of the very few scientific activities of the 1950s led by a #girlboss.

Associated Legendre functions are what we're going to use as our guess for $\Phi = P_l^{(m)}(\cos\theta)$ and when $m = 0$, the so-called Legendre polynomial will be written as $P_l(\cos\theta)$. Abromowitz and Stegun have plots of them for various values of l, m but *Matlab* et al. know all about special functions these days so computing with them isn't usually any more difficult than sinusoids.

We have one ODE yet to deal with. Born and Wolf suggest that we set

$$kr = \rho \qquad R(r) = \frac{1}{\sqrt{r}}Z(r) \tag{6.17}$$

so that we can all instantly recognize our old friend, the Bessel equation

$$\frac{d^2Z}{d\rho^2} + \frac{1}{\rho}\frac{dZ}{d\rho} + \left[1 - \frac{(l+\frac{1}{2})^2}{\rho^2}\right] Z = 0 \tag{6.18}$$

which of course has Bessel functions for solutions. Duh. In particular, we want the half-order cylindrical Bessel functions $R = \frac{1}{\sqrt{kr}} Z_{1+\frac{1}{2}}(\rho)$, which are the spherical Bessel functions. There's going to be several Bessel functions, but let's start with Bessel, Neumann, and Hankel which are related by

$$H_l(\rho) = J_l(\rho) \pm iN_l(\rho) \tag{6.19}$$

although some books insist on using $Y_l(\rho)$ for the Neumann functions. The cylindrical Bessel functions are typically written with capital letters, while the spherical Bessel functions are written with the corresponding lowercase letters.

$$j_l(\rho) = \sqrt{\frac{\pi}{2\rho}} J_{1+\frac{1}{2}}(\rho) \qquad n_l(\rho) = \sqrt{\frac{\pi}{2\rho}} N_{1+\frac{1}{2}}(\rho) \qquad h_l(\rho) = \sqrt{\frac{\pi}{2\rho}} H_{1+\frac{1}{2}}(\rho)$$

There are also the Ricatti versions of each of these, $\psi_l(\rho) = \rho j_l(\rho)$ and so on. That book you didn't think you needed has all their properties and relationships spelled out for you in a convenient place.

So the general solution to the scalar wave equation in spherical coordinates is a linear combination of sinusoids, Associated Legendre functions, and spherical Bessel functions. Recall that l and m were integers, and we're going to use the Ricatti Bessel functions because that will give us the most compact answer at the end. There's no way to know that at this stage of course, so I imagine that Lorenz or Mie or whoever slogged all the way to the end and realized that if they were to go back and make that simple substitution, their final answer would be much more elegant. In any event, here's the functional form of Ψ

$$r\Psi = \sum_{l=0}^{\infty} \sum_{m=-l}^{l} \left[a_m \cos(m\phi) + b_m \sin(m\phi)\right] P_l^{(m)}(\cos\theta) \left[c_l \psi_l(kr) + d_l \chi_l(kr)\right] \tag{6.20}$$

where there's an r tacked on in order to make the equations work out better and $\chi_l(kr) = krn_l(kr)$. In addition, a_m, b_m and c_l, d_l are arbitrary constants. This general solution will be important because the boundary value problem we're about to set up will be constructed with functional forms that follow this.

Recall that we are considering the case of a plane electromagnetic wave propagating in the z-direction, polarized in the transverse direction. In spherical coordinates, this is going to be

$$\begin{aligned}
E_r^{inc} &= e^{ikr\cos\theta} \sin\theta\cos\phi & H_r^{inc} &= \sqrt{\frac{\epsilon}{\mu}} e^{ikr\cos\theta} \sin\theta\sin\phi \\
E_\theta^{inc} &= e^{ikr\cos\theta} \cos\theta\cos\phi & H_\theta^{inc} &= \sqrt{\frac{\epsilon}{\mu}} e^{ikr\cos\theta} \cos\theta\sin\phi \\
E_\phi^{inc} &= -e^{ikr\cos\theta} \sin\phi & H_\phi^{inc} &= \sqrt{\frac{\epsilon}{\mu}} e^{ikr\cos\theta} \cos\phi
\end{aligned} \tag{6.21}$$

And then like magic we're going to use Bauer's formula

$$e^{ikr\cos\theta} = \sum_{l=0}^{\infty} i_l(2l+1) j_l(kr) P_l(\cos\theta) \tag{6.22}$$

although Born and Wolf write $\psi_l(kr)/kr$ instead of $j_l(kr)$, which is equivalent. Hence, we know how to write our incident plane wave in terms of spherical harmonics, except without referring to that book you still don't really think you need your own copy of, there's no chance in hell you'd know to use the identities

$$e^{ikr\cos\theta}\sin\theta \equiv -\frac{1}{kr}\frac{\partial}{\partial\theta}\left(e^{ikr\cos\theta}\right)$$

$$\frac{\partial}{\partial\theta}P_l(\cos\theta) \equiv -P_l^{(1)}(\cos\theta) \qquad P_0^{(1)}(\cos\theta) \equiv 0$$

in order to get the incident transverse magnetic and transverse electric plane wave potentials

$$r\Psi_{TM}^{inc} = \frac{1}{k_1^2}\sum_{l=1}^{\infty} i^{l-1}\frac{2l+1}{l(l+1)}\psi_l(k_1 r)P_l^{(1)}(\cos\theta)\cos\phi \tag{6.23}$$

$$r\Psi_{TE}^{inc} = \frac{1}{k_1^2}\sqrt{\frac{\epsilon_1}{\mu_1}}\sum_{l=1}^{\infty} i^{l}\frac{2l+1}{l(l+1)}\psi_l(k_1 r)P_l^{(1)}(\cos\theta)\sin\phi \tag{6.24}$$

I assume you're OK with sticking in factors of i^l or whatever inside the summation, or at least are (convincingly) nodding knowingly that it will make things simpler down the road. I also assume you're saying loudly WTF? inside your head about the term $\frac{2l+1}{l(l+1)}$, which just showed up out of the blue. It shows up because when you're doing separation of variables, there are good and less good choices for the separation constants. Let's just go with it so we can get on to solving something.

When an incident plane wave impinges on the scatterer, it will give rise to a scattered field for $r > a$ and a transmitted field for $r < a$, which will all have to balance at $r = a$ in order to satisfy the boundary conditions of continuity of tangential electric and magnetic fields. In terms of the potential functions, we write these four equations as:

$$\frac{\partial}{\partial r}r\Psi_{TM}^{inc} + \frac{\partial}{\partial r}r\Psi_{TM}^{scat} = \frac{\partial}{\partial r}r\Psi_{TM}^{trans} \tag{6.25}$$

$$\frac{\partial}{\partial r}r\Psi_{TE}^{inc} + \frac{\partial}{\partial r}r\Psi_{TE}^{scat} = \frac{\partial}{\partial r}r\Psi_{TE}^{trans} \tag{6.26}$$

$$k_1 r\Psi_{TM}^{inc} + k_1 r\Psi_{TM}^{scat} = k_2 r\Psi_{TM}^{trans} \tag{6.27}$$

$$k_1 r\Psi_{TE}^{inc} + k_1 r\Psi_{TE}^{scat} = k_2 r\Psi_{TE}^{trans} \tag{6.28}$$

which are evaluated at $r = a$. Obviously, the TM and TE problems can be dealt with separately. First, however, we have to invoke some common sense. I hope you agree that the scattered field has to spread out as it gets farther away from the scatterer, and furthermore that it has to decay away to nothingness as $r \to \infty$. That's easy enough to accomplish if we *choose* spherical Hankel functions for the scattered wave. In addition, the transmitted field inside the scatterer must be finite, so we can't use spherical Neumann functions for that because they are singular as $r \to 0$. We'll put some unknown modal coefficients in each of the scattered and transmitted fields, so that we'll have something to solve for when we enforce the boundary conditions.

The scattered fields are written, with $\zeta_l = \psi_l \pm i\chi_l$, as:

$$r\Psi_{TM}^{scat} = \frac{1}{k_1^2}\sum_{l=1}^{\infty}\left(\frac{\Delta_1}{\Delta_0}\right) i^{l-1}\frac{2l+1}{l(l+1)}\zeta_l^{(1)}(k_1 r)P_l^{(1)}(\cos\theta)\cos\phi \tag{6.29}$$

$$r\Psi_{TE}^{scat} = \frac{1}{k_1^2}\sqrt{\frac{\epsilon_1}{\mu_1}}\sum_{l=1}^{\infty}\left(\frac{\Delta_2}{\Delta_0}\right) i^{l}\frac{2l+1}{l(l+1)}\zeta_l^{(1)}(k_1 r)P_l^{(1)}(\cos\theta)\sin\phi$$

and the transmitted fields are

$$r\Psi_{TM}^{trans} = \frac{1}{k_2^2}\sum_{l=1}^{\infty}\left(\frac{\Delta_3}{\Delta_0}\right) i^{l-1}\frac{2l+1}{l(l+1)}\psi_l(k_2 r)P_l^{(1)}(\cos\theta)\cos\phi \tag{6.30}$$

$$r\Psi_{TE}^{trans} = \frac{1}{k_2^2}\sqrt{\frac{\epsilon_2 - i\sigma_2/\omega}{\mu_2}}\sum_{l-1}^{\infty}\left(\frac{\Delta_4}{\Delta_0}\right) i^{l}\frac{2l+1}{l(l+1)}\psi_l(k_2 r)P_l^{(1)}(\cos\theta)\sin\phi$$

This looks worse than it is. We simply have to solve for the unknown modal coefficients, Δ_0–Δ_4, but it's two pairs of equations, so it's just two equations with two unknowns, twice. In addition, there's this magical thing called orthogonality, which lets us ignore the summation and solve the equations for each value of l. The quick-and-dirty explanation is that you multiply each of the equations by $P_{l'}(\cos\theta)$ and then integrate over θ to find that the answer is zero for all values of $l' \neq l$ which kills off the summation. It's in that book. You should get a copy.

The grown-up way to solve a system of equations is via Cramer's Rule, which is why I wrote the unknown modal coefficients as Δ's. In case you kind of zoned out with all the math I just inflicted on you, here's where we are. We used Bauer's formula to write the incident plane wave in terms of spherical Bessel functions and associated Legendre functions and sines and cosines. We used common sense to pick functional forms for the scattered and transmitted fields, which were zero at infinity and finite at the origin, respectively. We were careful to use k_1, etc. for $r > a$ and k_2 etc. for $r < a$ because the scatterer is a different material (like a metal) than the surrounding glass or whatever. We apply the boundary conditions of tangential electric and magnetic fields, but the TM and TE cases are orthogonal, which is obvious if you notice that the former has $\cos\phi$ while the latter has $\sin\phi$ and those are orthogonal in sort of the same way as the Legendre functions because if you integrate $\cos\phi\sin\phi$ over 2π you'll get hard zero.

So then we do some algebra and find to our delight that the seemingly random factors that we added to things along the way did, in fact, make the answers work out elegantly:

$$\begin{aligned}
\Delta_0 &= \zeta'^{(1)}_l(k_1 a)\psi_l(k_2 a) - \zeta_l^{(1)}(k_1)\psi_l'(k_2 a)\\
\Delta_1 &= \psi_l'(k_1 a)\psi_l(k_2 a) - \psi_l(k_1 a)\psi_l'(k_2 a)\\
\Delta_3 &= \psi_l(k_1 a)\psi_l'(k_2 a) - \psi_l'(k_1 a)\psi_l(k_2 a)
\end{aligned} \tag{6.31}$$

and so on for the TE case. Note that prime indicates differentiation with respect to argument for the Bessel functions. The Δ's give exact expressions for the scattered and transmitted fields, so you can, if you want to, calculate the scattered and transmitted electric and magnetic fields. It's a whole lot easier than it was in 1908 because you don't have to do it by hand. You also don't have to find a computer person to collaborate with because you can just carefully type the expressions into *Matlab* or whatever and make some plots. What to plot is the question, of course.

Note that the size parameter of the scatterer $k_1 a$ is a dimensionless measure of the size of the scatterer compared to wavelength, and that matters a lot. If you plot things as a function of size parameter, you'll find quite a lot of variation. If you plot things as a function of direction, you'll also find quite often that small changes in size parameter or relative properties of the scatterer and surroundings change things dramatically.

When you type in equations like this in order to make plots, it's excellent practice to reproduce a bunch of the standard results in textbooks and such that have stood the test of time. It's so easy to have one little typo and then have all your output be garbage. Generally speaking, the behavior is too complex to look at it and tell whether it looks right or not.

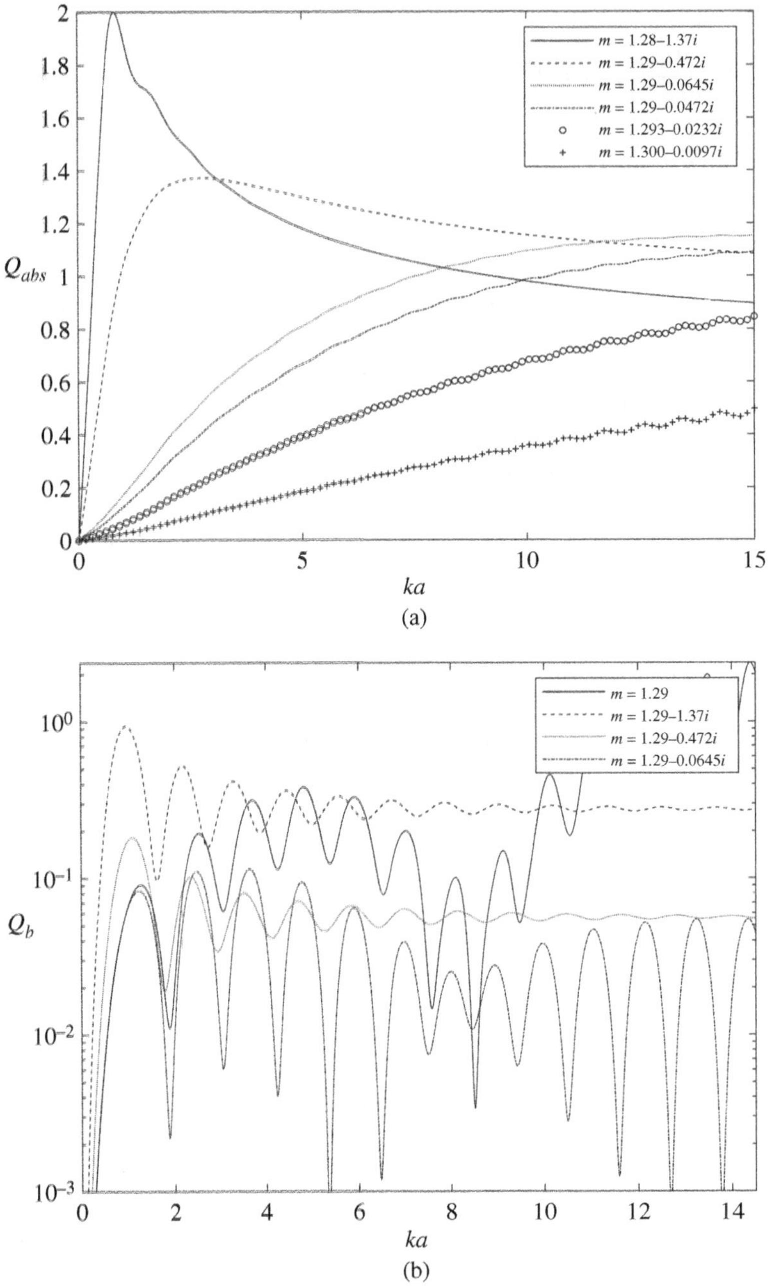

Figure 6.2 Absorption efficiency for various refractive indices vs. *ka* (a). Effect of refractive index on backscatter gain as a function of *ka* (b).

Figures 6.2–6.5 are recreations of historic Mie scattering plots of various scattering parameters such as absorption efficiency, backscatter, extinction efficiency, and total scattering for various refractive indices m as a function of the size parameter, ka, where k is the wavenumber of the incoming wave and a is the radius, for spheres [8]. The extinction efficiency is defined as the

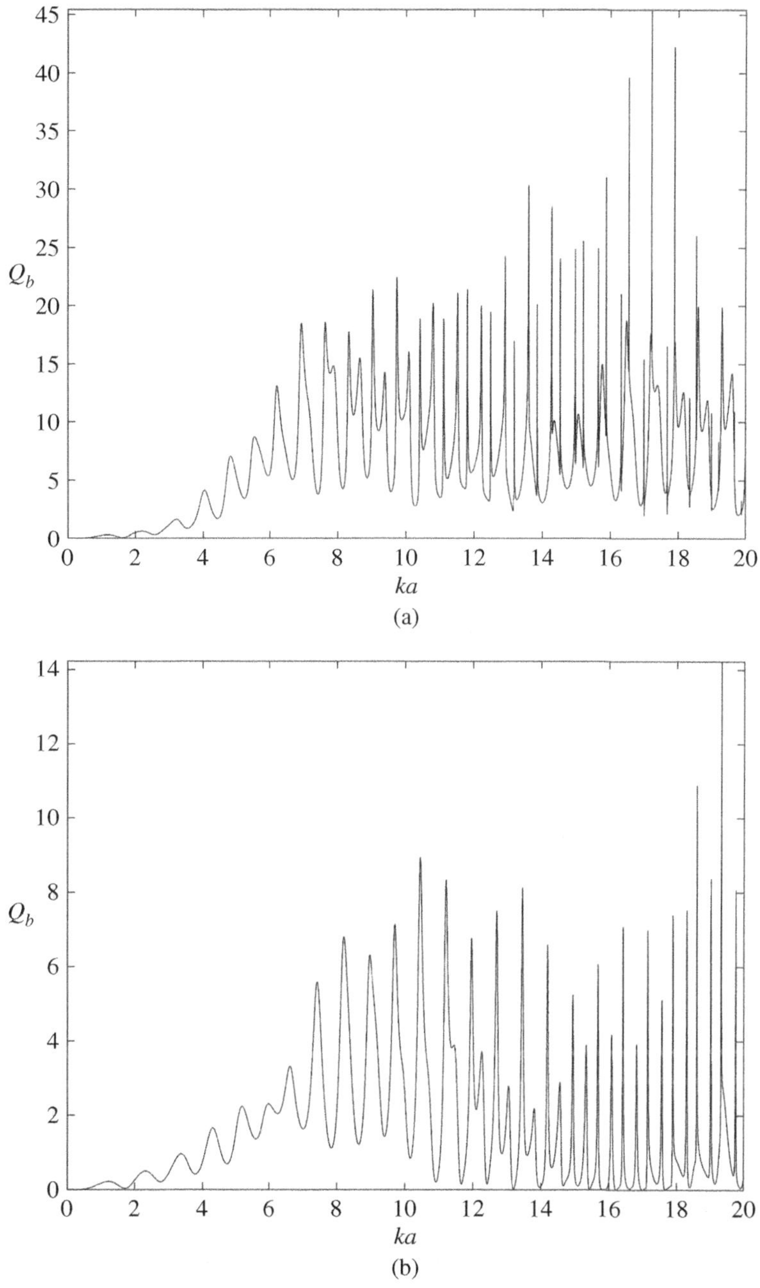

Figure 6.3 Backscatter gain for spheres with refractive index 1.61 (a) and 1.4821 (b).

sum of the absorbed and total scattering efficiencies. These scattering functions for a homogeneous sphere are dependent on its physical parameters such as optical size, complex refractive index, and angle of observation. Generally speaking, the scattering efficiencies, Q, can be written as $Q = \frac{\sigma}{\pi a^2}$, where σ is the cross section for the corresponding process. This cross section is the ratio of the energy flux absorbed, backscattered, extinguished, or scattered in total by the particle to the

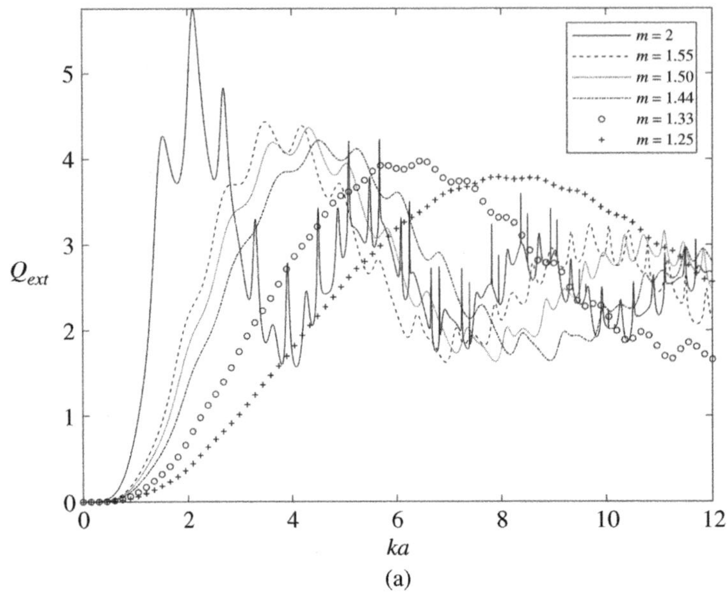

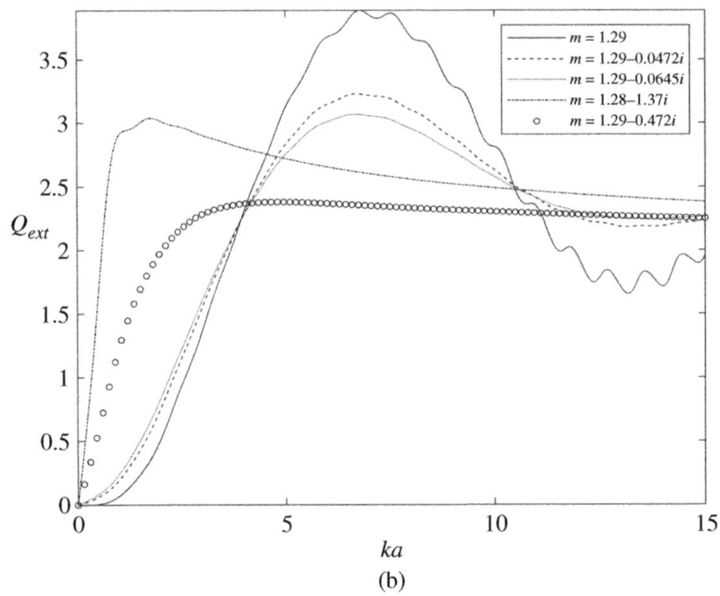

Figure 6.4 Extinction efficiency of spheres for various values of refractive index as a function of *ka* (a). The extinction efficiency is the ratio of the sum of the total scattering and absorption cross sections and the geometric cross section (b).

incident energy flux density, that is, to the energy of incident electromagnetic wave per unit area oriented normal to the wave front. The cross section is an area, while the efficiency coefficients are dimensionless. Scattering cross sections are calculated by integrating the Pointing vector with respect to angle.

As can be seen, the scattering behavior is highly size parameter dependent. Backscatter can vary as much as a factor of 4 or so within tenths of a change in size parameter as depicted by

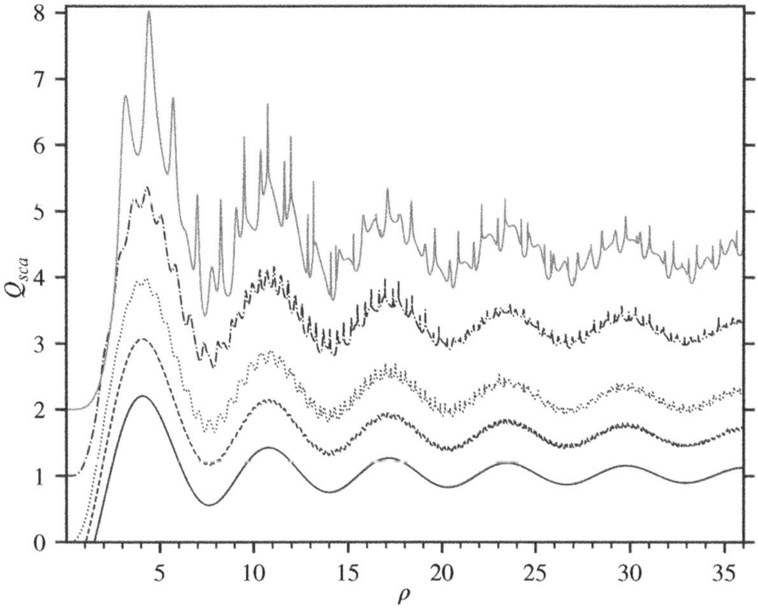

Figure 6.5 The total scattering from selected real refractive indexes (m), showing that the further that m gets from 1 the more fluctuation there is. The plots are vertically displaced to highlight the differences. Replicates Kerker [7] Fig. 4.8. The black line is $m = 1.01$ displaced −1. The dashed line is $m = 1.15$, displaced by −0.5. The dotted line is $m = 1.33$, and is not displaced. The dash-dot-dash-dot line is $m = 1.5$ and is displaced by +1. The gray line is for $m = 2.105$, is displaced by +2.0.

the spheres in Figure 6.3. Visually, this MATLAB implementation of Mie's influential scattering solutions agrees well with the historic plots computed by H. C. Van De Hulst and Milton Kerker.[4]

For most people these days, their instinct would be to simply make plots of the scattered field surrounding the scatterer. Perfectly understandable, except for the question of what exactly it is that you're going to plot and then the follow-on question of what that means. Recall that Mie's motivation was explaining the colors of stained glass, and that it required finite conductivity of the metallic particles in the glass to explain it.

The first step to explaining the colors of stained glass is to assume that the scattering by any single sphere is representative of the collective behavior of multiple spheres because multiple-scattering is really, really complicated. The good news here is that if the concentration of scatterers is dilute enough, this single-scattering approximation works pretty well. We're going to just go with that. The second step is to understand that as the light travels through the glass, some energy is removed due to scattering and some is removed due to absorption via the finite conductivity of the scatterers. So what we have to calculate is energy lost due to scattering and absorption.

The radial component of the complex Poynting vector is

$$S_R^* = \frac{1}{2}\left(E_\theta \tilde{H}_\phi - E_\phi \tilde{H}_\theta\right) \tag{6.32}$$

4 Milton Kerker was a world-renowned expert in aerosol and colloid science, a leading researcher in the use of light scattering to study aerosols and colloids. Author of over 200 papers, he is probably best remembered for coauthoring the book "The Scattering of Light and Other Electromagnetic Radiation." He was also a pioneer in surface-enhanced Raman spectroscopy and an excellent teacher and graduate student and postdoctoral mentor. He was a United States Army Veteran serving from 1942 to 1945 and was awarded the Bronze Star. See Hopke et al. [9].

and because outside the sphere we have

$$\vec{E} = \vec{E}^{inc} + \vec{E}^{scat} \qquad \vec{H} = \vec{H}^{inc} + \vec{H}^{scat} \tag{6.33}$$

we can write

$$\begin{aligned} S_R^* &= \frac{1}{2}\left(E_\theta^{inc}\tilde{H}_\phi^{inc} - E_\phi^{inc}\tilde{H}_\theta^{inc}\right) + \frac{1}{2}\left(E_\theta^{scat}\tilde{H}_\phi^{scat} - E_\phi^{scat}\tilde{H}_\theta^{scat}\right) \\ &+ \frac{1}{2}\left(E_\theta^{inc}\tilde{H}_\phi^{scat} + E_\theta^{scat}\tilde{H}_\phi^{inc} - E_\phi^{inc}\tilde{H}_\theta^{scat} - E_\phi^{scat}\tilde{H}_\theta^{inc}\right) \end{aligned} \tag{6.34}$$

Now draw about the scattering sphere, a concentric spherical surface of radius $b > a$. The real part of S_R^* integrated over the surface of this sphere is equal to the net flow of energy across its surface. The total energy absorbed by the sphere is then

$$W_a = -\Re \int_0^\pi \int_0^{2\pi} S_R^* b^2 \sin\theta d\phi d\theta \tag{6.35}$$

and therefore we write

$$\begin{aligned} W_a &= -\frac{1}{2}\Re \int_0^\pi \int_0^{2\pi} \left(E_\theta^{inc}\tilde{H}_\phi^{inc} - E_\phi^{inc}\tilde{H}_\theta^{inc}\right) b^2 \sin\theta d\phi d\theta \\ &-\frac{1}{2}\Re \int_0^\pi \int_0^{2\pi} \left(E_\theta^{scat}\tilde{H}_\phi^{scat} - E_\phi^{scat}\tilde{H}_\theta^{scat}\right) b^2 \sin\theta d\phi d\theta \\ &-\frac{1}{2}\Re \int_0^\pi \int_0^{2\pi} \left(E_\theta^{inc}\tilde{H}_\phi^{scat} + E_\theta^{scat}\tilde{H}_\phi^{inc} - E_\phi^{inc}\tilde{H}_\theta^{scat} - E_\phi^{scat}\tilde{H}_\theta^{inc}\right) b^2 \sin\theta d\phi d\theta \end{aligned} \tag{6.36}$$

The first integral measures the flow of energy in the incident wave and the second measures the outward scattered energy

$$W_{scat} = \frac{1}{2}\Re \int_0^\pi \int_0^{2\pi} \left(E_\theta^{scat}\tilde{H}_\phi^{scat} - E_\phi^{scat}\tilde{H}_\theta^{scat}\right) b^2 \sin\theta d\phi d\theta \tag{6.37}$$

The sum of the absorbed and scattered energy removed from the incident wave, called extinction, is given by the third integral

$$W_{ext} = -\frac{1}{2}\Re \int_0^\pi \int_0^{2\pi} \left(E_\theta^{inc}\tilde{H}_\phi^{scat} + E_\theta^{scat}\tilde{H}_\phi^{inc} - E_\phi^{inc}\tilde{H}_\theta^{scat} - E_\phi^{scat}\tilde{H}_\theta^{inc}\right) b^2 \sin\theta d\phi d\theta \tag{6.38}$$

Doing these integrals isn't so terrible if you happen to know that

$$\int_0^{2\pi} \sin^2\phi d\phi = \int_0^{2\pi} \sin^2\phi d\phi = \pi \tag{6.39}$$

$$\int_0^{2\pi} \sin\phi\cos\phi d\phi = 0 \tag{6.40}$$

$$\int_0^\pi \left[\left(\frac{P_l^{(1)}(\cos\theta)}{\sin\theta}\right)^2 + \frac{\partial}{\partial\theta}\left(P^{(1)}(\cos\theta)\right)\right] \sin\theta d\theta = \frac{2[l(l+1)]^2}{2l+1} \tag{6.41}$$

$$\int_0^\pi \frac{P_l^{(1)}(\cos\theta)}{\sin\theta}\frac{\partial}{\partial\theta}\left(P^{(1)}(\cos\theta)\right)\sin\theta d\theta = 0 \tag{6.42}$$

which are magic orthogonality relations conveniently found in a book I might have mentioned a time or three. Doing the integrations and normalizing, we end up with the scattering and extinction cross sections[5]

5 Part of the reason it's important to benchmark your code against well-established results is to make sure that you've gotten the equations correct. Every author seems to use a slightly different notation and you're never going to be absolutely sure that you've gotten all the little things correct. I just walked over to my lab and found Don Chodrow's copy of Born and Wolf to see if I had gotten all of the factors correct in the cross sections. I'm pretty sure, but their (119) is written in a slightly different way because they have written the Δ's slightly differently from me.

$$Q_{scat} = \frac{4}{(k_1 a)^2} \sum_{l=1}^{\infty} (2l+1) \left| \frac{\Delta_1}{\Delta_0} \right|^2 \quad (6.43)$$

$$Q_{ext} = \frac{4}{(k_1 a)^2} \Re \sum_{l=1}^{\infty} (2l+1) \left(\frac{\Delta_1}{\Delta_0} \right) \quad (6.44)$$

The extinction efficiency, Q_{ext}, is the fraction of the incident energy that's removed due to scattering plus absorption for a single scatterer, but if the suspension is dilute enough, the total energy reduction is just proportional to this, that is, you multiply by the volume density of the scatterers or some such thing. The point for Mie is that you can plot Q_{ext} as a function of $k_1 a$ for a given set of sphere parameters (permeability, permittivity, and conductivity) and see how the different colors of light are muted. Since white light contains all colors, to get red stained glass you would need to scatter and absorb the other colors. As long as you don't have to do the computations by hand, it's pretty straightforward to use this model to "design" whatever colors you want.

I feel like after all those equations, I somehow owe you an anecdote about stained glass. If you're ever in Boston and you take a Duckboat tour or whatever, one of the stories you'll hear is about purple windowpanes on Beacon Hill. Across from the Boston Common, up the hill a bit from the *Cheers* bar, there's some rather expensive real estate. In a few windows, there are a few panes of accidentally pale purple glass.[6]

> Back in the 1800s, when the glass was shipped over from France, it arrived crystal clear. But soon after, the windows' exposure to sunlight resulted in a violet tint, thanks to an excess of manganese oxide in the glass. This, of course, was hugely upsetting to homeowners at the time.

But since the glass was in the homes of snooty upper-class Bostonians, all the wannabees wanted purple glass too. The problem was the glass company didn't know what they had done to make the purple glass and didn't yet know how to compute Mie scattering cross sections, so they went bankrupt trying whatever they could think of to purple up their glass. The remaining purple windowpanes are highly prized. In addition, Duckboats are wicked pissah!

6.2 Acoustic Scattering from a Sphere

Next, let's consider the simplest closed-body 3D scattering problem in acoustics, a plane wave scattering from a fluid sphere. It could be an air bubble in water or a liquid drop in air or an oil drop in water or whatever. The solution will be perfectly general, except that the incident wave is a plane wave and the scatterer is a sphere and, of course, we're not going to include viscosity at this stage.[7]

---

6 https://www.bostonmagazine.com/property/2019/02/12/where-to-find-purple-windows.

7 Victor Anderson was born to missionary parents in Shanghai, China, on 31 March 1922. He enrolled at the University of California, Los Angeles as a graduate student in physics in 1946. After a year in residence at UCLA, he joined the University of California's Marine Physical Laboratory (MPL) at Scripps. His research at MPL in the study of the deep scattering layer completed the requirements for a PhD, which he received in 1953. The following year, Anderson was granted a postdoctoral fellowship at the Acoustics Research Laboratory at Harvard University in Cambridge, Mass. While at Harvard, he designed and developed a digital time compression technique (DELTIC) for application to acoustic signal processing. Anderson returned to Scripps' MPL in 1955 and continued his research in the field of acoustical signal processing and ocean engineering. Topics of his research included the spatial and temporal distribution of acoustic ambient noise in the ocean and development of the remote underwater manipulator (RUM). In addition, he invented the digital multibeam steering system (DIMUS), a computationally

You may or may not have just read the preceding section on Mie scattering. This is the exactly equivalent problem in acoustics, and I've tried to include all the details so you won't have to refer back to the previous section if you don't want to. It's a slightly different version of things, so that might even turn out to be helpful if some of the previous sections didn't quite do it for you.

My friends who do electromagnetic scattering consider acoustics to be merely a scalar version of electromagnetics, and thereby think themselves somehow superior to acousticians [10–12]. My friends who do acoustic scattering snort derisively about the linear nature of Maxwell's equations as compared to the inherently non-linear Navier–Stokes equations. I generally keep my politics to myself so I can work with whoever happens to be in power, and if I were inclined to offer commentary and/or advice on the issues of the day I would do it via a carefully curated sock puppet [13, 14].

Electromagnetic waves are transverse waves, so they are vectorial; acoustic waves are longitudinal waves, so they are scalar. As we'll see, elastodynamic waves have both transverse and longitudinal modes, so they are mathematically more complicated than either electromagnetic or acoustic waves. That's why the elastodynamic theory of the æther was dropped as soon as Maxwell's unified theory came along.

At the origin of our coordinate system, we have a sphere of radius a which has different material properties from the surrounding medium, just as in Figure 6.1. Call the region $r > a$ medium 1 and $r < a$ medium 2. We start with linearized equations of motion, continuity, and state for an acoustic medium:

$$\rho_0 \partial_t \vec{v} + \nabla p = 0 \tag{6.45}$$

$$\partial_t \rho + \rho_0 \nabla \cdot \vec{v} = 0 \tag{6.46}$$

$$\nabla p - c^2 \nabla \rho = 0 \tag{6.47}$$

where ρ_0 is the constant density, $\rho = \rho(\vec{r}, t)$ is the perturbative density, p, $\vec{v}$ and c stand for pressure, velocity vector, and speed of sound, respectively.

These equations are valid both in medium 1 and in medium 2 and will allow us to write down the general form of the solution both inside and outside of the sphere. We'll then enforce the boundary conditions at $r = a$ to solve for unknown modal coefficients. What we're about to do is called a "forward" solution because we'll specify both the incident wave field and the details of the scatterer and then solve for the scattered field that develops due to the presence of the scatterer. The "inverse" problem is when both the incident and scattered fields are known (to some extent) and the goal is to solve for the details of the scatterer. In simple terms, for the forward problem we know what we send in and know what's there, and then we solve for what comes back out. For the inverse problem, we know what we send in and know what comes back out, and we then want to figure out what's in there. Forward scattering models are often quite helpful when trying to solve inverse problems, but it's not as simple as using the forward model predictions to match up to measured scattered fields and thus know what the scatterer is. I argue these days that the forward scattering solutions can be helpful in identifying signal features that can be formed into feature vectors in a machine learning paradigm for attacking inverse problems [15].

efficient approach for preferentially listening in many directions at once. It is used in sonar systems on U.S. Navy ships and submarines. Anderson served as deputy director of MPL from 1976 until his retirement in 1989. Anderson was the recipient of the National Security Industrial Association's 1986 Admiral Charles B. Martell Technical Excellence Award for his work in the development of the DIMUS system. He also received the Navy's 1976 Distinguished Public Service Award.

Since the incident field is known and the scattered field is unknown, let's be careful to note that the total acoustic field in the region $r > a$ is the incident plus the scattered field. Inside the scatterer $r < a$ is the transmitted field. Since the boundary conditions for acoustics are that the normal fluid velocity and the pressure must be continuous across the interface, we'll have the following two boundary condition equations at $r = a$:

$$v_r^{inc} + v_r^{scat} = v_r^{trans} \tag{6.48}$$

$$p^{inc} + p^{scat} = p^{trans} \tag{6.49}$$

Since the velocity is an irrotational vector field $\nabla \times \vec{v} = 0$, we can introduce a scalar function such that the velocity is proportional to the gradient of that function. We write

$$\vec{v} = -\frac{1}{k^2}\nabla\psi \tag{6.50}$$

where $k = \omega/c$ is the wave number and we've inserted the factor of $-1/k^2$ for later convenience. Actually, that's a bit of a fib. The real story goes something like: after a lot of painful algebra, we might eventually figure out that the final expressions would have turned out a lot simpler if we had thought to insert a factor of $-1/k^2$ in front of the velocity potential when we initially defined it.

We consider an incident plane wave of unit amplitude propagating in the z-direction, which gives

$$v_z^{inc} = e^{ikz} \qquad v_x^{inc} = v_y^{inc} = 0 \tag{6.51}$$

and in spherical coordinates, we write

$$\begin{aligned} v_r &= e^{ik_1 r\cos\theta}\cos\theta \\ v_\theta &= -e^{ik_1 r\cos\theta}\sin\theta \\ v_\phi &= 0 \end{aligned} \tag{6.52}$$

In terms of the vector potential, we thus have

$$\begin{aligned} v_r &= -\frac{1}{k^2}\left(\frac{1}{r}\frac{\partial}{\partial r} - \frac{1}{r^2}\right)(r\psi) \\ v_\theta &= -\frac{1}{(kr)^2}\frac{\partial}{\partial\theta}(r\psi) \\ v_\phi &= -\frac{1}{(kr)^2}\frac{1}{\sin\theta}\frac{\partial}{\partial\phi}(r\psi) \end{aligned} \tag{6.53}$$

Combining these and using Bauer's formula

$$e^{ix\cos\theta} = \sum_{l=0}^{\infty} i^l(2l+1)j_l(x)P_l(\cos\theta) \tag{6.54}$$

gives

$$r\psi^{inc} = \sum_{l=0}^{\infty} i^l(2l+1)k_1 r j_l(k_1 r)P_l(\cos\theta) \tag{6.55}$$

This is of the appropriate form because the general solution to

$$(\nabla^2 + k^2)\psi = 0$$

is a linear combination of the three types of functions

$$r\psi = \left\{\begin{array}{c} krj_l(kr) \\ kr\eta_l(kr) \end{array}\right\}\left\{\begin{array}{c} P_l^{(m)}(\cos\theta) \\ Q_l^{(m)}(\cos\theta) \end{array}\right\}\left\{\begin{array}{c} e^{im\phi} \\ e^{-im\phi} \end{array}\right\} \tag{6.56}$$

where the spherical Bessel and Neumann functions are

$$j_l(x) = \sqrt{\frac{\pi}{2x}} J_{l+1/2}(x) \qquad \eta_l(x) = \sqrt{\frac{\pi}{2x}} Y_{l+1/2}(x) \tag{6.57}$$

and $P_l^{(m)}(\cos\theta)$ and $Q_l^{(m)}(\cos\theta)$ are the associated Legendre functions of the first and second kind.

Exercise 6.1 Do the aforementioned algebra to derive $r\psi^{inc}$ as written earlier, starting from $v_z^{inc} = e^{ikz}$.

The scattered and transmitted fields need to be written in an appropriate form that satisfies the Helmholtz equation in spherical coordinates. There's a bit of an art to this, but it's not magic. I'll skip a bunch of the details and mathematical justifications, but don't be too freaked out by the "special functions" that we're using because they aren't really all that exotic. You may have noticed that many equations are named for people, which you may have thought to be either kind of cool or really pathetic. I have to assume it's a point of honor among mathematicians to have an equation or two named after them. I suppose it's like an X-gamer having a trick named after them, which also could be considered either cool or pathetic. Either way, it provides a shorthand for describing the equations and/or tricks you're trying to master.

Now think back to elementary school when you were into prealgebra and skateboarding. Very often, the method that you were taught for solving equations was "guess and check." Now flash-forward to college when you first learned about ordinary differential equations, and presumably had long since given up skateboarding. Your ODE professor probably didn't call it guess and check, but that's more or less the way to attack differential equations because otherwise you have to assume a series solution and grind out the answer that way. If you happen to grind out the solution to some important new equation, two things will happen. You'll get that equation named after you, and your series solution will be given its own symbol(s), which will henceforth be known as (insert your name here) functions. That's where special functions come from. So far, we've come across Bessel, Neumann, and Lengendre functions. There's lots of them.

Back to our problem at hand. Note that the incident field, when expanded in spherical harmonics, is a linear combination of Bessel and Legendre functions. That sum is only over l and is for $m = 0$, which explains why the $e^{\pm im\phi}$ terms aren't present. To guess at the forms for the scattered fields, let's write similar functional forms:

$$r\psi^{scat} = \sum_{l=0}^{\infty} C_1 k_1 r h_l^{(1)}(k_1 r) P_l(\cos\theta) \tag{6.58}$$

$$r\psi^{trans} = \sum_{l=0}^{\infty} C_2 k_1 r j_l(k_2 r) P_l(\cos\theta) \tag{6.59}$$

In these, I've included unknown modal coefficients C_1 and C_2, which we're hoping to solve for using the boundary conditions. For the angular part, I've used Legendre polynomials, $P_l(\cos\theta) = P_l^{(0)}(\cos\theta)$, which is what the associated Legendre function is called when $m = 0$. For the radial functions, I've used the spherical Hankel function of the first kind, $h_l^{(1)}(x) = j_l(x) + i\eta_l(x)$, since that function has the right mathematical property to make the scattered field behave like an expanding spherical wave when $r \to \infty$. I've also been careful to use k_1 since the subscript indicates that we're talking about fluid 1. For the transmitted field, I have used the same angular function, but only $j_l(x)$ for the radial function because $\eta_l(x)$ has the property that it's infinite for all values of l when $x \to 0$. Both of those choices for the radial functions are pretty easy to justify because we're trying to model actual physical phenomena. The incident wave is of unit amplitude, so it makes good sense that the field inside the scatterer will have to be finite, even if the scatterer does somehow focus the waves or some such thing. The scattered field obviously has to die off as it expands out in 3D space from a finite-sized scatterer, which is sometimes called the Sommerfeld radiation condition. Similarly to the spherical Bessel and Neumann functions, the spherical Hankel functions of the first and second kind are defined in terms of the corresponding half-order cylindrical functions:

$$h_l^{(1,2)}(x) = \sqrt{\frac{\pi}{2x}} H_{l+1/2}^{(1,2)}(x) = \sqrt{\frac{\pi}{2x}} \left[J_{l+1/2}(x) \pm iN_{l+1/2}(x)\right] \tag{6.60}$$

The lowest-order spherical functions can be expressed in terms of sines and cosines as:

$$j_0(x) = \frac{\sin x}{x}$$

$$j_1(x) = \frac{\sin x}{x^2} - \frac{\cos x}{x}$$

$$j_2(x) = \left(\frac{3}{x^3} - \frac{1}{x}\right)\sin x - \frac{3}{x^2}\cos x$$

$$j_3(x) = \left(\frac{15}{x^4} - \frac{6}{x^2}\right)\sin x - \left(\frac{15}{x^3} - \frac{1}{x}\right)\cos x$$

$$\eta_0(x) = -\frac{\cos x}{x}$$

$$\eta_1(x) = -\frac{\cos x}{x^2} - \frac{\sin x}{x}$$

$$\eta_2(x) = -\left(\frac{3}{x^3} - \frac{1}{x}\right)\cos x - \frac{3}{x^2}\sin x$$

$$\eta_3(x) = -\left(\frac{15}{x^4} - \frac{6}{x^2}\right)\cos x - \left(\frac{15}{x^3} - \frac{1}{x}\right)\sin x$$

which is more or less how your mathematical software deals with them. The point is that you can think of special functions like you do other functions which get their own buttons on your calculator.

For small argument, $\sin x \approx x$ and $\cos x \approx 1$, so we see that the Neumann functions are singular. For large arguments $h_l(kr) \sim e^{ikr}/kr$, which is an outgoing spherical wave. That's why we chose the radial functions for the scattered and transmitted fields the way we did.

The Legendre polynomial is the Associated Legendre function when $m = 0$ and if we use $x = \cos\theta$, the first few of them are

$$P_0(x) = 1 \qquad P_1(x) = x \qquad P_2(x) = \frac{1}{2}(3x^2 - 1)$$

These are orthogonal functions, which is why we have chosen $m = 0$ terms for the scattered and transmitted fields. We can calculate the rest of the Legendre polynomials from the recursion relation

$$(2l+1)xP_l(x) = (l+1)P_{l+1}(x) + lP_{l-1}(x)$$

and we can also write

$$P_l^{(m)}(x) = (1 - x^2)^{m/2} \frac{d^m P_l(x)}{dx^m}$$

to find the associated Legendre functions.

There are lots of excellent sources to look up the many and various properties of all the special functions, but I still recommend the "*Handbook of Mathematical Functions*" by Abramowitz and Stegun. Sometimes you can still find the old hardcover versions at used book sales and such, but it's also available in paperback from Dover. Or you can download it.

Now back to the scattering problem. We have velocity components in terms of ψ and we can also write

$$p = \frac{i\rho c^2}{\omega}\psi \tag{6.61}$$

where we've dropped the subscript on ρ_0 since we're going to want to have subscripts that indicate whether it's the density of medium 1 or 2. The boundary conditions then become

$$\begin{bmatrix} \frac{1}{k_1^2}\left(\frac{1}{r}\frac{\partial}{\partial r}-\frac{1}{r^2}\right) & -\frac{1}{k_2^2}\left(\frac{1}{r}\frac{\partial}{\partial r}-\frac{1}{r^2}\right) \\ \rho_1 c_1^2 & -\rho_2 c_2^2 \end{bmatrix} \begin{Bmatrix} r\psi^s \\ r\psi^t \end{Bmatrix} = -\begin{Bmatrix} \frac{1}{k_1^2}\left(\frac{1}{r}\frac{\partial}{\partial r}-\frac{1}{r^2}\right) \\ \rho_1 c_1^2 \end{Bmatrix} r\psi^{inc}$$

To solve this system of equations, plug in the expressions for the $r\psi$s as defined and take the derivatives indicated. Then plug in $r = a$ and solve the algebraic system by Cramer's rule. After doing that we find

$$C_l^{(1)} = -i^{l+1}(2l+1)\left(\frac{\Delta_1}{\Delta_0}\right) \tag{6.62}$$

$$C_l^{(2)} = -i^{l+1}(2l+1)\left(\frac{\Delta_2}{\Delta_0}\right) \tag{6.63}$$

where the three Cramer's rule determinants are given by

$$\begin{aligned} \Delta_0 &= \frac{k_2 a j_l'(k_2 a)}{j_l(k_2 a)} - \frac{\rho_2}{\rho_1}\frac{k_1 a h_l'(k_1 a)}{h_l(k_1 a)} \\ \Delta_1 &= \frac{j_l(k_1)}{h_l(k_1 a)}\left\{\frac{k_2 a j_l'(k_2 a)}{j_l(k_2 a)} - \frac{\rho_2}{\rho_1}\frac{k_1 a j_l'(k_1 a)}{j_l(k_1 a)}\right\} \\ \Delta_2 &= \frac{k_2}{k_1}\frac{j_l(k_1 a)}{j_l(k_1 a)}\left\{\frac{k_1 a h_l'(k_1 a)}{h_l(k_2 a)} - \frac{\rho_2}{\rho_1}\frac{k_1 a j_l'(k_1 a)}{j_l(k_1 a)}\right\} \end{aligned} \tag{6.64}$$

where the prime indicates differentiation with respect to argument

$$j_l'(x) = \frac{\partial}{\partial x} j_l(x) \qquad h_l'(x) = \frac{\partial}{\partial x} h_l(x)$$

and we note that for either $j_l(x)$ or $h_l(x)$

$$f_{l-1}(x) + f_{l+1}(x) = (2l+1)f_l(z)/x$$

$$l f_{l-1}(x) - (l+1)f_{l+1}(x) = (2l+1)f_l'(x)$$

$$(l+1)f_l(x)/x + f_l'(x) = f_{l-1}(x)$$

$$l f_l(x)/x - f_l'(x) = f_{l+1}(x)$$

Exercise 6.2 Check to see that these results match those of V. Anderson [16]. Note the acknowledgments where he thanks by name the two computers who made his plots.[8]

8 Computers usually labored in anonymity, rarely getting Wikipedia pages of their own. To wit, Creola Katherine Johnson (née Coleman; 26 August 1918–24 February 2020) was an American mathematician whose calculations of orbital mechanics as a NASA employee were critical to the success of the first and subsequent U.S. crewed spaceflights. During her 33-year career at NASA and its predecessor, she earned a reputation for mastering complex manual calculations and helped pioneer the use of computers to perform the tasks. The space agency noted her "historical role as one of the first African-American women to work as a NASA scientist." The Katherine G. Johnson Computational Research Facility at NASA LaRC was dedicated 22 September 2017 with a ribbon-cutting ceremony attended by family and friends of Johnson and her fellow "human computers," students from Black Girls Code and the 21st Century Community Learning Centers program, and special guests from across Virginia.

Now that we have solved the scattering problem, let's derive some physical quantities of interest. The energy flux vector is

$$\vec{\mathcal{F}} = p\vec{v} \tag{6.65}$$

and if we consider the acoustic energy flow through a large spherical surface surrounding the scatterer, we want the radial component of this

$$\mathcal{F}_r = pv_r \tag{6.66}$$

The intensity of the scattered wave is the time-averaged radial component of the energy flux vector for the scattered field

$$I = \langle \mathcal{F}_r \rangle = \frac{1}{2}\Re(pv_r^*) = \frac{1}{4}(pv_r^* + p^*v_r) \tag{6.67}$$

Let's consider this point for a moment. A harmonic time factor $e^{-i\omega t}$ is assumed in all fields, which suggests that

$$\begin{aligned}\vec{v}(\vec{r},t) &= \Re\left[\vec{v}_0(r)e^{-i\omega t}\right]\\ &= \frac{1}{2}\left[\vec{v}_0(r)e^{-i\omega t} + \vec{v}_0^*(r)e^{+i\omega t}\right]\end{aligned}$$

because $\Re z = \frac{1}{2}(z + z^*)$. Instantaneous values of any rapidly oscillating quantities can't be observed, but only their time average is taken over a time interval $-T' \le t \le T'$, which is large compared to the fundamental period $T = 2\pi/\omega$. Now, the time-averaged $\vec{\mathcal{F}}$ is given by

$$\langle \vec{\mathcal{F}} \rangle = \frac{1}{2T'}\int_{-T'}^{T'} p(\vec{r},t)\vec{v}(\vec{r},t)dt \tag{6.68}$$

in which we use

$$p(\vec{r},t) = \Re\left[p(\vec{r})e^{-i\omega t}\right] = \frac{1}{2}\left[p(\vec{r})e^{-i\omega t} + p^*(\vec{r})e^{i\omega t}\right]$$

$$\vec{v}(\vec{r},t) = \Re\left[\vec{v}(\vec{r})e^{-i\omega t}\right] = \frac{1}{2}\left[\vec{v}(\vec{r})e^{-i\omega t} + \vec{v}^*(\vec{r})e^{i\omega t}\right]$$

to get

$$\langle \vec{\mathcal{F}} \rangle = \frac{1}{2T'}\int_{-T'}^{T'} \frac{1}{4}\left[p\vec{v}e^{-2i\omega t} + p\vec{v}^* + p^*\vec{v} + p^*\vec{v}^*e^{2i\omega t}\right]dt \tag{6.69}$$

where the exponential terms drop out because

$$\frac{1}{2T'}\int_{-T'}^{T'} e^{\pm 2i\omega t}dt = \frac{1}{2T'}\frac{1}{2\omega}\sin 2\omega T' = \frac{1}{8\pi}\frac{T}{T'}\sin 2\omega T'$$

and we have assumed that $T/T' \ll 1$. Hence we have

$$\langle \vec{\mathcal{F}} \rangle = \frac{1}{4}(p\vec{v}^* + p * \vec{v}) = \frac{1}{2}\Re(p\vec{v}^*) \tag{6.70}$$

We also write

$$I = \langle \mathcal{F}_r \rangle = \frac{1}{2}\Re(pv_r^*) \tag{6.71}$$

Intensities represent the radiation per unit solid angle and give a measure of the directivity of the scattered acoustic energy. We'll almost always divide the scattered intensity by the incident intensity to write a differential scattering cross section

$$\frac{d\sigma_{scat}}{d\Omega} = \frac{1}{(k_1 r)^2}\left|\sum_{l=0}^{\infty}(2l+1)\left(\frac{\Delta_1}{\Delta_0}\right)P_l(\cos\theta)\right|^2 \tag{6.72}$$

In the backscatter direction this is

$$\sigma^b_{scat} = \left| \frac{1}{(k_1 r)} \sum_{l=0}^{\infty} (2l+1) \left(\frac{\Delta_1}{\Delta_0} \right) \right|^2 \tag{6.73}$$

and integrating the differential scattering cross section over a spherical surface of radius r gives the total scattering cross section as:

$$\sigma_{scat} = \frac{4\pi}{k_1^2} \sum_{l=0}^{\infty} (2l+1) \left| \frac{\Delta_1}{\Delta_0} \right|^2 \tag{6.74}$$

which we sometimes normalize $Q_{scat} = \sigma_{scat}/\pi a^2$ to write

$$Q_{scat} = \frac{4}{(k_1 a)^2} \sum_{l=0}^{\infty} (2l+1) \left| \frac{\Delta_1}{\Delta_0} \right|^2 \tag{6.75}$$

Note that we have used the orthogonality of the Legendre polynomial to write

$$\int_{-1}^{1} P_l(x) P_{l'}(x) dx = \frac{2}{2l+1} \delta_{ll'}$$

and that $P_l(-1) = (-1)^l$.

For small scatterers, we can use small argument approximations for the spherical Bessel and Neumann functions

$$j_l(x) = \frac{x^l}{(2l+1)!!} \left\{ 1 - \frac{x/2}{2l+3} + \frac{x^4/4}{2(2l+3)(2l+5)} \cdots \right\}$$

$$\eta_l(x) = -\frac{(2l-1)!!}{x^{l+1}} \left\{ 1 + \frac{x/2}{2l-1} + \frac{x^4/4}{2(2l-1)(2l-3)} \cdots \right\}$$

and we find

$$Q_{scat} \approx \frac{4}{9} (k_1 a)^4 \left\{ \left[\frac{\rho_1}{\rho_2} \left(\frac{k_2}{k_1} \right)^2 - 1 \right]^2 + 3 \left(1 - \frac{\rho_1}{\rho_2} \right)^2 \Big/ \left(2 + \frac{\rho_1}{\rho_2} \right)^2 \right] \tag{6.76}$$

which is the Rayleigh scattering result.

We can also consider special cases where we have: (i) an air bubble in water and (ii) a liquid drop in air. For (i), we have $\rho_2 \ll \rho_1$ and $c_2 \ll c_1$, while for (ii) the opposite holds.

For the air bubble in water, we find

$$\frac{\Delta_1}{\Delta_0} \approx \frac{j_l(k_1 a)}{h_l(k_1 a)} \tag{6.77}$$

and for the liquid drop in air, we find

$$\frac{\Delta_1}{\Delta_0} \approx \frac{k_1 a j_l'(k_1 a) + j_l(k_1 a)}{k_1 a h_l'(k_1 a) + h_l(k_1 a)} \tag{6.78}$$

Exercise 6.3 Show that the small-argument cross sections you get from these match with the corresponding small-argument cross sections mentioned earlier.

Exercise 6.4 Derive the scattering cross section from the differential scattering cross section, and the intensities from the scattered fields as given. Check my algebra for Δ_0, Δ_1, and Δ_2.

Exercise 6.5 Implement the aforementioned equations to reproduce Anderson's plots, as we've done in Figures 6.6 and 6.7, which were first done in 1949. Note that some of the curves are dashed lines over regions where the number of points calculated were "not sufficient to draw reliable" curves. Although you'll find Matlab does these calculations instantaneously, in 1949, it took two computers working essentially full time for two months to make these and a few other plots in

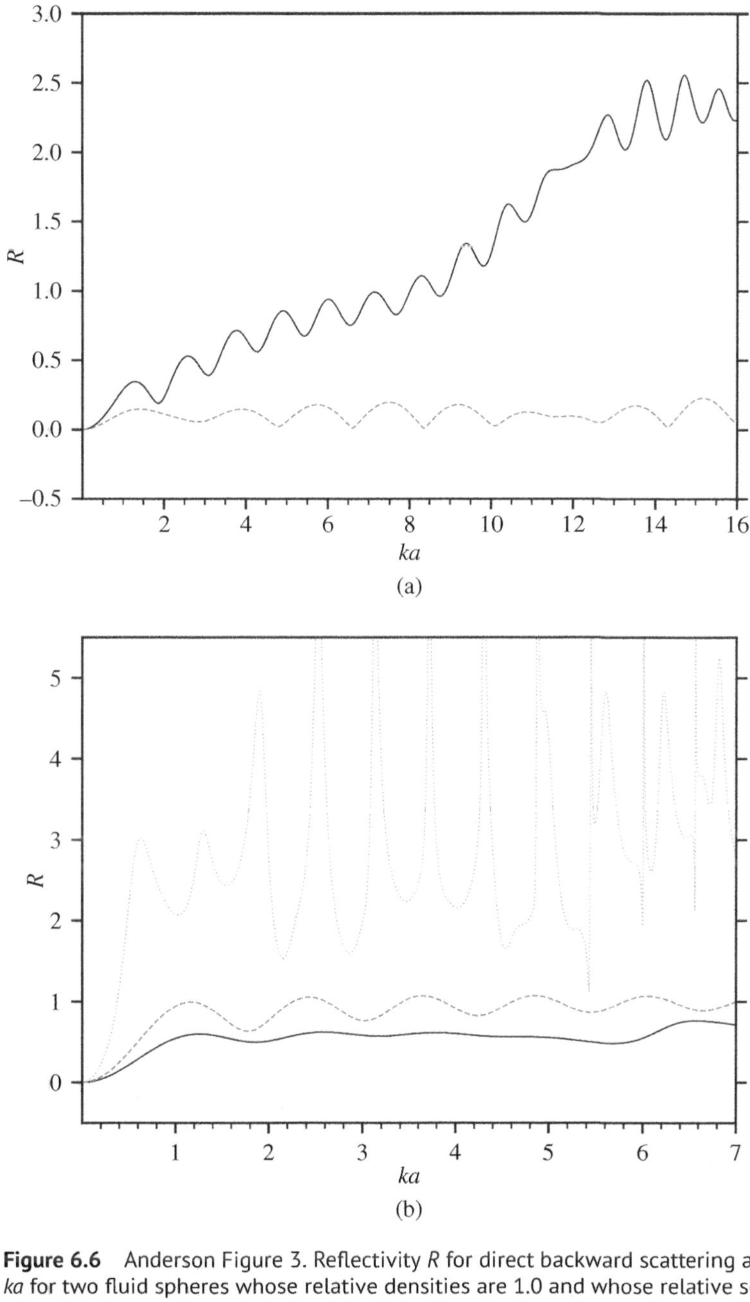

Figure 6.6 Anderson Figure 3. Reflectivity R for direct backward scattering as a function of acoustic radius ka for two fluid spheres whose relative densities are 1.0 and whose relative sound velocities are 1.2 and 0.8 (a) and Anderson Figure 4. Reflectivity R for direct backward scattering as a function of acoustic radius ka for various values of relative density g and relative velocity h (b).

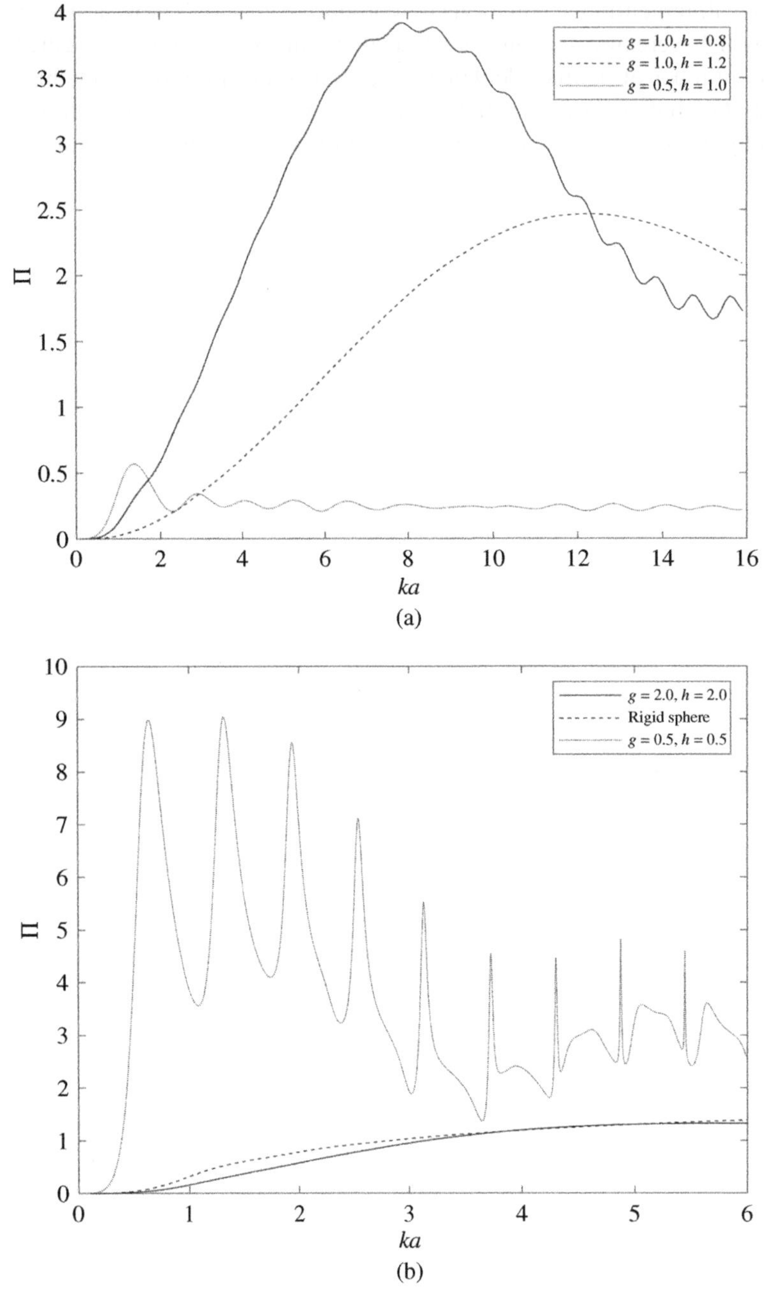

Figure 6.7 Anderson Figure 5. Total scattering Π as a function of acoustic radius *ka* for the two spheres considered in Figure 3 and also for a sphere with $g = 0.5, h = 1.0$ (a) and as a function of the acoustic radius *ka* for the three cases shown in Anderson Figure 4 (b).

Anderson's paper. Those two computers were named Misses Gwendolyn Roy and Gloria Slack. In 1949, *computer* was a job title, not a device, which I once pointed out to a student who was all smug that her Matlab code was instantaneous.[9]

Anderson's motivation was sonar, that is, scattering from the swim bladders of fish. More recently, we've used this formulation for acoustic scattering from air bubbles to improve the detection, counting, and sizing of gaseous emboli during heart bypass surgery. Some patients who have heart surgery end up with cognitive deficit. Sometimes it's temporary; sometimes it's permanent. What happens is that air gets into the heart-lung machine and isn't filtered out before the blood goes back into the body. If an embolism makes its way to the brain, it can lodge there and block blood flow to regions distal to the blockage. There is medication to help with this, but it's only used when needed. The EDACS system [17] uses backscattered ultrasound to detect air bubbles during surgery so appropriate action can be taken. Our challenge was to explore the scattering behavior to isolate simple rules of thumb that could be implemented on the existing FPGA in the FDA-approved device. Changing the hardware would have required going back for additional approvals, which is expensive and time consuming. After interacting with the clinicians to narrow down the acoustic parameters, we ran our models and gave the EDACS engineers the necessary simple rules to implement.

Although EDACS performed well as intended, they had difficulty convincing surgeons to adopt the technology. The near-universal response was, "My patients never have any problems. It's those other guys who you should talk to." A refined pitch was needed. We began to work on the scattering analysis that would enable enhanced removal of bubbles, using acoustic radiation force due to scattering to improve debubbling. The existing clinical devices all had a reservoir built into them where blood came in at the top and exited the bottom, with simple buoyancy removing some of the bubbles from the flow. By putting a large upward-pointing acoustic transducer at the bottom of the reservoir, the radiation force would give buoyancy an extra boost. The pitch for the clinician was that this would allow the debubbling reservoir to be smaller, and heart surgery would use less blood transfusion.

The primary technical challenge for us was that acoustic radiation force is a nonlinear scattering problem and viscosity seems to matter. The Anderson model that accurately described the backscattering from air bubbles needed to be augmented. As expected, it got complicated in a hurry. We implemented two models, one which accounted for viscosity and one which didn't. The blood was thinned during this surgery, so the clinicians felt that viscosity was an important parameter. The bottom line turned out to be that it didn't affect the radiation pressure enough to matter clinically. We were initially puzzled as to why the Russian mathematician who formulated the viscous model didn't include any plots in his papers. Once we got the model up and running, it became clear that he didn't have the computational power to make any plots.[10]

---

9 Dr. Alison Pouch is currently Assistant Professor of Radiology and Bioengineering at Penn, where she earned a PhD in Bioengineering and Biomedical Engineering. At W&M, she was a Physics-Anthropology double major with a minor in Mathematics and was president of the Juggling Club.

10 Implementation of the viscous model takes a significant amount of computational power. Numerical models were implemented with Matlab using part of the SciClone computing cluster at William and Mary with 72 dual-processor, dual-core Dell SC1435 servers running at 2.6 GHz. The viscous model required over 3000 CPU hours in order to calculate radiation force for one material combination. The computation was accomplished in less wall-clock time by breaking the code into pieces and running up to 65 jobs at once. Due to the time required by the viscous model, the infinite summation was computed only to $n = 30$. The inviscid model can be numerically implemented very quickly and takes only minutes to compute force for $n = 0 : 30$.

We were pleased that our modeling was able to help our former student in his commercialization efforts. Outside all of our control, though, was the collapse of the economy right about that time. The EDACS technology was about to be spun off to facilitate commercialization, but that got put on hold because the small company had just done an IPO and their stock tanked when everybody everywhere stopped buying the sorts of products they were selling. Cash flow from research grants doesn't help the quarterly sales numbers that affect your stock price. They may have also been a bit optimistic about the value of the EDACS IP in the IPO, and were insisting on valuations that prevented EDACS from spinning off.

Since we had included viscosity in both the embolism and the blood so as to be able to model lipid emboli [18], our radiation force model almost immediately found utility in a seemingly unrelated project: sorting micro algae with radiation force [19]. This was a huge, but short lived, project to use algae to both clean up the Chesapeake Bay and make biofuels.[11] We ran our models and published our results and waited for the biofuels superconsortium to get fully underway. The US$ 100 million proposal never quite got submitted because the company leading the project was cross that proposal preparation costs are not allowable charges at DoE. I heard afterward that DoE was crossed that our proposal didn't arrive as expected.

There are presumably other applications of this complicated acoustic radiation force analysis. Setting up standing waves will cause some particles to migrate to nodes and some to antinodes. Acoustic tweezers have been found useful for manipulating submillimeter particles for flow cytometry, cell separation, cell trapping, single-cell manipulation, and nanomaterial manipulation. Going way back, one of the promises for the International Space Station was containerless processing of exotic materials in microgravity. The idea was that acoustic levitation[12] could be used to hold blobs of unobtainium in place while ultrapure alloys, pharmaceuticals, etc. that couldn't be made on earth would be manufactured in orbit.

6.3 Elastic Wave Sphere Scattering

We next present the solution for plane longitudinal and transverse elastic wave scattering from an elastic sphere of radius $r = a$ in an infinite elastic medium. The geometry is the same as in the previous section. For an isotropic, homogeneous, and linearly elastic medium, the equation of motion is

$$\rho \partial_t^2 \vec{u} - \mu \nabla^2 \vec{u} - (\mu + \lambda) \nabla (\nabla \cdot \vec{u}) = 0 \tag{6.79}$$

where ρ is the constant medium density, μ and λ are elastic Lamé parameters and $\vec{u}$ is the displacement vector. After some manipulation and assuming harmonic time variation $e^{-i\omega t}$, the vector wave equation may be written as:

$$(\nabla^2 + K^2)\vec{u} - \left(1 - \frac{K^2}{k^2}\right) \nabla (\nabla \cdot \vec{u}) = 0 \tag{6.80}$$

11 Walter Adey received his BS in Geophysics from MIT, performed graduate studies at MIT and Harvard in Paleontology and Biology, and obtained his PhD in Marine Botany and Geology from the University of Michigan. Since 1977, he has been the Director of the Marine Systems Laboratory at the Museum of Natural History, Smithsonian Institution, now emeritus. US Patent 5,851,398 describes a novel algal energy scheme initially funded by Statoil, a Norwegian Oil Company: https://www.wm.edu/offices/economicdevelopment/_documents/100711vimschap.pdf.

12 Robert Apfel, Robert Higgin Professor of Mechanical Engineering, taught acoustics at Yale University for over 30 years, having trained at Harvard University with Frederic V. Hunt in a chain of advisors which can be loosely traced back to 1802. Although Bob's direct impact on the field of acoustics was prematurely cut short in 2002, his legacy and influence are still growing through the contributions of his students, and the students of his students.

Here $K = \omega/c_T$ and $k = \omega/c_L$ are the propagation constants for transverse and longitudinal elastic waves with $c_T = \sqrt{\mu/\rho}$ and $c_L = \sqrt{(\lambda + 2\mu)/\rho}$ defining the transverse and longitudinal wave speeds, respectively. Both the spherical scatterer and the surrounding medium will be considered to be elastic solids for which the above equations hold. The formulation of this problem dates to 1861, but it wasn't solved until about 1990 [20]. Clebsch was interested in trying to model the interaction of light with lenses using an elastodynamic theory of the aether. I was interested in modeling the interaction of ultrasonic waves in solids with flaws. Clebsch had to invent most of the math he needed. I had to find well-tested FORTRAN subroutines that evaluated the special functions. Both analyses start at exactly the same place.

In the case of an isolated spherical scatterer, it is natural to take a spherical coordinate system (r, θ, ϕ) with the scatterer at the origin of the coordinate system, as before, so that the sphere surface $r = a$ is a constant coordinate surface. To avoid dealing with the vector wave equation, we define three scalar generating functions $\pi_L, \pi_{SH}, \pi_{SV}$ in terms of the corresponding displacements

$$\begin{aligned} \vec{u}_L &= -\frac{1}{k^2}\nabla\pi_L & \left(\nabla^2 + k^2\right)\pi_L &= 0 \\ \vec{u}_{SH} &= \nabla\times(\vec{r}\pi_{SH}) & \left(\nabla^2 + K^2\right)\pi_{SH} &= 0 \\ \vec{u}_{SV} &= \frac{1}{K}\nabla\times\nabla\times(\vec{r}\pi_{SV}) & \left(\nabla^2 + K^2\right)\pi_{SV} &= 0 \end{aligned} \tag{6.81}$$

where the total displacement is written in three parts: $\vec{u} = \vec{u}_L + \vec{u}_{SH} + \vec{u}_{SV}$. Note that $\vec{u}_L$ is dilatational while $\vec{u}_{SH}$ and $\vec{u}_{SV}$ are solenoidal vectors, and that all of these scalar functions satisfy scalar wave equations. The stress components can be found from

$$\begin{aligned} \sigma_{rr} &= \lambda\nabla\cdot\vec{u} + 2\mu\partial_r u_r \\ \sigma_{r\theta} &= \mu\left(\partial_r u_\theta - \frac{u_\theta}{r} + \frac{1}{r}\partial_\theta u_r\right) \\ \sigma_{\theta\theta} &= \lambda\nabla\cdot\vec{u} + 2\mu\left(\frac{1}{r}\partial_\theta u_\theta + \frac{1}{r}u_r\right) \end{aligned} \tag{6.82}$$

where

$$\nabla\cdot\vec{u} = \frac{1}{r^2}\partial_r\left(r^2 u_r\right) + \frac{1}{r\sin\theta}\partial_\theta\left(\sin\theta u_\theta\right) + \frac{1}{r\sin\theta}\partial_\phi u_\phi$$

When a purely compressional plane wave of unit amplitude and with wave number k_1 is incident on the surface of the spherical scatterer, scattered as well as transmitted longitudinal and vertical shear modes will be generated. We set the following expressions for scalar potentials of incident, scattered, and transmitted waves

$$\begin{aligned} r\pi_L^i &= \sum_{l=0}^{\infty} i^{l+1}(2l+1)\psi_l(k_1 r)P_l(\cos\theta) \\ r\pi_L^s &= \sum_{l=0}^{\infty} C_l^{(1)}\zeta_l(k_1 r)P_l(\cos\theta) \\ r\pi_L^t &= \sum_{l=0}^{\infty} C_l^{(2)}\psi_l(k_2 r)P_l(\cos\theta) \\ r\pi_{SV}^s &= \frac{1}{K_1}\sum_{l=0}^{\infty} C_l^{(3)}\zeta_l(K_1 r)P_l^{(1)}(\cos\theta)\cos\phi \\ r\pi_{SV}^t &= \frac{1}{K_2}\sum_{l=0}^{\infty} C_l^{(4)}\psi_l(K_2 r)P_l^{(1)}(\cos\theta)\cos\phi \end{aligned} \tag{6.83}$$

where

$$\psi_l(x) = j_l(x) = \sqrt{x\pi/2}J_{l+1/2}(x) \qquad \zeta_l(x) = h_l^{(1)}(x) = \sqrt{x\pi/2}H_{l+1/2}^{(1)}(x)$$

with $J_{l+1/2}(x)$ and $H_{l+1/2}^{(1)}(x)$ the half-order cylindrical Bessel and Hankel functions. The horizontal shear components are not included because they are not excited by an incident compressional wave.

Using the boundary conditions of continuous displacement and normal traction at $r = a$, we write in matrix form:

$$\begin{pmatrix} \frac{-1}{2r}\left(\frac{K_1}{k_1}\right)^2 + \frac{[l(l+1)+2]}{k_1^2 r^3} - \frac{2\partial_r}{(k_1 r)^2} & \frac{\mu_2}{\mu_1}\left(\frac{1}{2r}\left(\frac{K_2}{k_2}\right)^2 - \frac{[l(l+1)+2]}{k_2^2 r^3} + \frac{2\partial_r}{(k_2 r)^2}\right) & -\frac{l(l+1)}{K_1 r^2}\left(\partial_r - \frac{2}{r}\right) & \frac{\mu_2}{\mu_1}\frac{l(l+1)}{K_2 r^2}\left(\partial_r - \frac{2}{r}\right) \\ -\frac{4}{k_1^2 r^3} + \frac{2\partial_r}{(k_1 r)^2} & \frac{\mu_2}{\mu_1}\left(\frac{4}{k_2^2 r^3} - \frac{2\partial_r}{(k_2 r)^2}\right) & -\frac{2l(l+1)}{K_1 r^3} + \frac{2\partial_r}{K_1 r^2} + \frac{K_1}{r} & \frac{\mu_2}{\mu_1}\left(\frac{2l(l+1)}{K_2 r^3} - \frac{2\partial_r}{K_2 r^2} - \frac{K_2}{r}\right) \\ -\frac{1}{(k_1 r)^2}\left(1 - r\partial_r\right) & \frac{1}{(k_2 r)^2}\left(1 - r\partial_r\right) & -\frac{l(l+1)}{K_1 r^2} & \frac{l(l+1)}{K_2 r^2} \\ \frac{1}{(k_1 r)^2} & -\frac{1}{(k_2 r)^2} & -\frac{\partial_r}{K_1 r} & \frac{\partial_r}{K_2 r} \end{pmatrix} \begin{bmatrix} (r\pi_L^s) \\ (r\pi_L^t) \\ (r\pi_{SV}^s) \\ (r\pi_{SV}^t) \end{bmatrix} = -\begin{bmatrix} \frac{1}{2r}\left(\frac{K_1}{k_1}\right)^2 - \frac{[l(l+1)+2]}{k_1^2 r^3} + \frac{2\partial_r}{(k_1 r)^2} \\ \frac{4}{k_1^2 r^3} - \frac{2\partial_r}{(k_1 r)^2} \\ \frac{1}{(k_1 r)^2}\left(1 - r\partial_r\right) \\ \frac{1}{(k_1 r)^2} \end{bmatrix} (r\pi_L^i) \tag{6.84}$$

In these equations, it is understood that after differentiation, r must be substituted by the boundary radius $r = a$. The identities

$$\frac{1}{\sin\theta}\frac{\partial}{\partial\theta}\left[\sin\theta\frac{\partial}{\partial\theta}(\cdot)\right]+\frac{1}{\sin^2\theta}\frac{\partial^2}{\partial\phi^2}(\cdot)=-l(l+1)(\cdot)$$
$$\frac{\partial^2}{\partial r^2}(\cdot)+\begin{Bmatrix}k^2\\K^2\end{Bmatrix}(\cdot)=\frac{l(l+1)}{r^2}(\cdot) \tag{6.85}$$

were used to simplify the expressions. Then, after necessary manipulations

$$C_l^{(1)}=i^{l+1}(2l+1)\left(\frac{\Delta_1^L}{\Delta_0}\right)\qquad C_l^{(3)}=i^{l+1}(2l+1)\left(\frac{\Delta_3^L}{\Delta_0}\right)\frac{P_l(\cos\theta)}{P_l^{(1)}(\cos\theta)\cos\phi}$$
$$C_l^{(2)}=i^{l+1}(2l+1)\left(\frac{\Delta_2^L}{\Delta_0}\right)\qquad C_l^{(4)}=i^{l+1}(2l+1)\left(\frac{\Delta_4^L}{\Delta_0}\right)\frac{P_l(\cos\theta)}{P_l^{(1)}(\cos\theta)\cos\phi}$$

where Δ_0, Δ_1^L–Δ_4^L are the expressions that follow. With these, we are able to define the scalar scattering potential functions by using the four modal coefficients $C_l^{(1)}$–$C_l^{(4)}$. The scattered and transmitted field components with known potentials $r\pi_L^s$, $r\pi_{SV}^s$ and $r\pi_L^t$, $r\pi_{SV}^t$ can easily be obtained from these equations. Note that the L and SV modes are coupled through and at the boundary, even though they propagate independently and it is clear that the boundary plays the role of a conversion mechanism. Upon the scattering of the incident longitudinal wave on the sphere, L and SV waves are generated in the scattered field. Therefore, scattered energy will be propagated in terms of L and SV waves. SH waves do not exist in the scattered field due to the fact that they can be excited only by an incident wave of the same type.

$$\begin{aligned}\Delta_0=&\left(\frac{\mu_2}{\mu_1}-1\right)^2[l(l+1)-2]\left[\frac{k_2aj_l'(k_2a)}{j_l(k_2a)}\frac{K_2a\psi_l'(K_2a)}{\psi_l(K_2a)}-l(l+1)\right]\left[\frac{k_1ah_l'(k_1a)}{h_l(k_1a)}\frac{K_1a\zeta_l'(K_1a)}{\zeta_l(K_1a)}-l(l+1)\right]\\&+\frac{1}{2}(K_1a)^2\left(\frac{\mu_2}{\mu_1}-1\right)\left\{\left[\frac{k_2aj_l'(k_2a)}{j_l(k_2a)}\frac{K_2a\psi_l'(K_2a)}{\psi_l(K_2a)}-l(l+1)\right]\left[\frac{K_1a\zeta_l'(K_1a)}{\zeta_l(K_1a)}+2\frac{k_1ah_l'(k_1a)}{h_l(k_1a)}-2l(l+1)\right]\right.\\&\left.-\frac{\rho_2}{\rho_1}\left[\frac{k_1ah_l'(k_1a)}{h_l(k_1a)}\frac{K_1a\zeta_l'(K_1a)}{\zeta_l(K_1a)}-l(l+1)\right]\left[\frac{K_2a\psi_l'(K_2a)}{\psi_l(K_2a)}+2\frac{k_2aj_l'(k_2a)}{j_l(k_2a)}-2l(l+1)\right]\right\}\\&+\frac{1}{4}(K_1a)^4\left\{l(l+1)\left[1-\frac{\rho_2}{\rho_1}\right]^2-\left[\frac{k_2aj_l'(k_2a)}{j_l(k_2u)}-\frac{\rho_2}{\rho_1}\frac{k_1ah_l'(k_1a)}{h_l(k_1u)}\right]\left[\frac{K_2a\psi_l'(K_2a)}{\psi_l(K_2a)}-\frac{\rho_2}{\rho_1}\frac{K_1a\zeta_l'(K_1a)}{\zeta_l(K_1a)}\right]\right\}\end{aligned}$$

$$\begin{aligned}\Delta_1^L=&\frac{j_l(k_1a)}{h_l(k_1a)}\left\{\left(\frac{\mu_2}{\mu_1}-1\right)^2[l(l+1)-2]\left[\frac{k_2aj_l'(k_2a)}{j_l(k_2a)}\frac{K_2a\psi_l'(K_2a)}{\psi_l(K_2a)}-l(l+1)\right]\left[\frac{k_1aj_l'(k_1a)}{j_l(k_1a)}\frac{K_1a\zeta_l'(K_1a)}{\zeta_l(K_1a)}-l(l+1)\right]\right.\\&+\frac{1}{2}(K_1a)^2\left(\frac{\mu_2}{\mu_1}-1\right)\left\{\left[\frac{k_2aj_l'(k_2a)}{j_l(k_2a)}\frac{K_2a\psi_l'(K_2a)}{\psi_l(K_2a)}-l(l+1)\right]\left[\frac{K_1a\zeta_l'(K_1a)}{\zeta_l(K_1a)}+2\frac{k_1aj_l'(k_1a)}{j_l(k_1a)}-2l(l+1)\right]\right.\\&\left.-\frac{\rho_2}{\rho_1}\left[\frac{k_1aj_l'(k_1a)}{j_l(k_1a)}\frac{K_1a\zeta_l'(K_1a)}{\zeta_l(K_1a)}-l(l+1)\right]\left[\frac{K_2a\psi_l'(K_2a)}{\psi_l(K_2a)}+2\frac{k_2aj_l'(k_2a)}{j_l(k_2a)}-2l(l+1)\right]\right\}\\&\left.+\frac{1}{4}(K_1a)^4\left\{l(l+1)\left[1-\frac{\rho_2}{\rho_1}\right]^2-\left[\frac{k_2aj_l'(k_2a)}{j_l(k_2a)}-\frac{\rho_2}{\rho_1}\frac{k_1aj_l'(k_1a)}{j_l(k_1a)}\right]\left[\frac{K_2a\psi_l'(K_2a)}{\psi_l(K_2a)}-\frac{\rho_2}{\rho_1}\frac{K_1a\zeta_l'(K_1a)}{\zeta_l(K_1a)}\right]\right\}\right\}\end{aligned}$$

$$\begin{aligned}\Delta_2^L=&\frac{j_l(k_1a)}{j_l(k_2a)}\left(\frac{k_1ah_l'(k_1a)}{h_l(k_1a)}-\frac{k_1aj_l'(k_1a)}{j_l(k_1a)}\right)\left(\frac{1}{4}(K_1a)^4\left[\frac{K_2a\psi_l'(K_2a)}{\psi_l(K_2a)}-\frac{\rho_2}{\rho_1}\frac{K_1a\zeta_l'(K_1a)}{\zeta_l(K_1a)}\right]\right.\\&\left.+\frac{1}{2}(K_1a)^2\left(\frac{\mu_2}{\mu_1}-1\right)\left\{\frac{K_1a\zeta_l'(K_1a)}{\zeta_l(K_1a)}\frac{K_2a\psi_l'(K_2a)}{\psi_l(K_2a)}-l(l+1)\left[\frac{K_1a\zeta_l'(K_1a)}{\zeta_l(K_1a)}+\frac{K_2a\psi_l'(K_2a)}{\psi_l(K_2a)}-2\right]\right\}\right)\end{aligned}$$

$$\Delta_3^L = \frac{K_1}{k_1}\frac{j_l(k_1a)}{h_l(K_1a)}\left(\frac{k_1aj_l'(k_1a)}{j_l(k_1a)} - \frac{k_1ah_l'(k_1a)}{h_l(k_1a)}\right)\left(\frac{1}{4}(K_1a)^4\left(\frac{\mu_2}{\mu_1}-1\right)\frac{\rho_2}{\rho_1}\left[1-\frac{\rho_2}{\rho_1}\right]\right.$$
$$+\frac{1}{2}(K_1a)^2\left(\frac{\mu_2}{\mu_1}-1\right)\left\{\left[\frac{k_2aj_l'(k_2a)}{j_l(k_2a)}\frac{K_2a\psi_l'(K_2a)}{\psi_l(K_2a)} - l(l+1)\right]\right.$$
$$\left.-\frac{\rho_2}{\rho_1}\left[\frac{K_2a\psi_l'(K_2a)}{\psi_l(K_2a)} + 2\frac{k_2aj_l'(k_2a)}{j_l(k_2a)} - 2l(l+1)\right]\right\}$$
$$\left.+\left(\frac{\mu_2}{\mu_1}-1\right)^2[l(l+1)-2]\left[\frac{k_2aj_l'(k_2a)}{j_l(k_2a)}\frac{K_2a\psi_l'(K_2a)}{\psi_l(K_2a)} - l(l+1)\right]\right)$$

$$\Delta_4^L = -\frac{K_2}{k_1}\frac{j_l(k_1a)}{j_l(K_2a)}\left(\frac{k_1aj_l'(k_1a)}{j_l(k_1a)} - \frac{k_1ah_l'(k_1a)}{h_l(k_1a)}\right)\left(\frac{1}{4}(K_1a)^4\left[1-\frac{\rho_2}{\rho_1}\right]\right.$$
$$\left.+\frac{1}{2}(K_1a)^2\left(\frac{\mu_2}{\mu_1}-1\right)\left\{l(l+1)-2+\left[\frac{K_2a\psi_l'(K_2a)}{\psi_l(K_2a)}-1\right]\left[\frac{K_1a\zeta_l'(K_1a)}{\zeta_l(K_1a)}-2\right]\right\}\right)$$

We also note that Δ_0 is the expression which describes the natural oscillations of an elastic sphere embedded in an infinitely extended elastic medium of different material properties. Indeed Δ_0 is the determinantal expression of the coefficient matrix of the unknown amplitudes $C_l^{(1)}$–$C_l^{(4)}$. In the absence of the incident wave, the right-hand column matrix disappears, and the left-hand side represents a homogeneous equation for scalar potentials. For all nonzero values of the amplitude parameters, the determinant must vanish, and a coupled equation for longitudinal and vertical shear types of oscillations can be obtained. In the case of the oscillation problem, this coupled equation indicates that energy can be converted from compressional elastic waves to the transverse waves and vice-versa. Oscillations are of purely dilatational type only when $l = 0$, which is expected since $l = 0$ is the only mode where no transverse wave-types exist. For $l \neq 0$ oscillations are coupled, and $\Delta_0 = 0$ is the general expression, which describes coupled oscillations of longitudinal and vertical shear wave modes.

If $\Delta_1^L = 0$ or $\Delta_3^L = 0$ while Δ_0 is nonzero, the amplitudes of the corresponding scattered L or SV waves will be zero, respectively. This indicates that for certain discrete sets of frequencies, either scattered L or scattered SV waves vanish. Hence there may also be certain discrete sets of frequencies for which the transmitted L or SV waves vanish. Moreover, whenever the frequency of the incident wave approaches a characteristic frequency which makes Δ_0 itself vanish, resonant phenomena will occur. However, the incident frequency is real and the characteristic frequencies are in general complex, so that in reality Δ_0 can be reduced to a minimum value but never quite to zero, so that the maximum amplitudes at resonance will be finite, not infinite.

The total flow of scattered elastic energy in the radial direction through a closed surface may be represented correctly by the radial component of the energy flux vector $\mathcal{F}_j = \sigma_{ij}\partial_t u_j$, which may be decomposed into longitudinal and transverse parts

$$\mathcal{F}_j = \mathcal{F}_j^L + \mathcal{F}_j^T = \sigma_{ij}^L\partial_t u_i^L + \sigma_{ij}^T\partial_t u_i^T \tag{6.86}$$

From the potentials π_L^s and π_{SV}^s the scattered field components can be obtained directly. Noting that L and SV modes are coupled through and at the boundary, they will propagate independently in the present linear field approximation. The boundary plays the role of conversion of energy, that is, the incident L mode is converted to the scattered SV mode and vice versa.

Significant quantities are intensities, differential, and total cross sections of the scattered field. Intensities for each mode represent the radiation-per-unit solid angle Ω. They are defined in terms of the time-averaged radial component of the far-field energy flux vector:

$$I_{scat}^{L} = \frac{\omega\mu_1}{2r}\left(\frac{K_1 r}{k_1 r}\right)^2 \frac{1}{k_1 r}\left|\sum_{l=0}^{\infty}(2l+1)\left(\frac{\Delta_1^L}{\Delta_0}\right)[P_l(\cos\theta)]\right|^2$$
$$I_{scat}^{SV} = \frac{\omega\mu_1}{2r}\frac{1}{K_1 r}\left|\sum_{l=0}^{\infty}(2l+1)\left(\frac{\Delta_3^L}{\Delta_0}\right)\left[\frac{\partial}{\partial\theta}P_l(\cos\theta)\right]\right|^2 \quad (6.87)$$
$$I_{inc}^{L} = \frac{\omega\mu_1}{2r}\frac{(K_1 r)^2}{k_1 r}$$

where I_{scat}^{L}, I_{scat}^{SV} and I_{inc}^{L} represent the intensities of the scattered L-type waves, SV-type waves, and the intensity of the incident L-wave which is propagating in the z-direction. Note that the scattered waves are independent of the polarization angle ϕ.

Dividing the intensity of the scattered wave by the intensity of the incident wave, we obtain the differential scattering cross sections of the longitudinal and transverse waves:

$$\frac{d\sigma^L}{d\Omega} = \frac{1}{(k_1 r)^2}\left|\sum_{l=0}^{\infty}(2l+1)\left(\frac{\Delta_1^L}{\Delta_0}\right)[P_l(\cos\theta)]\right|^2$$
$$\frac{d\sigma^{SV}}{d\Omega} = \frac{k_1 r}{(K_1 r)^3}\left|\sum_{l=1}^{\infty}(2l+1)\left(\frac{\Delta_3^L}{\Delta_0}\right)\left[\frac{\partial}{\partial\theta}P_l(\cos\theta)\right]\right|^2 \quad (6.88)$$

Integrating these expressions over a spherical surface with radius r, we obtain the total scattering cross sections

$$\sigma_{scat}^{L} = \frac{4\pi r^2}{(k_1 r)^2}\sum_{l=0}^{\infty}(2l+1)\left|\frac{\Delta_1^L}{\Delta_0}\right|^2$$
$$\sigma_{scat}^{SV} = \frac{4\pi r^2}{(K_1 r)^2}\frac{k_1}{K_1}\sum_{l=1}^{\infty}(2l+1)l(l+1)\left|\frac{\Delta_3^L}{\Delta_0}\right|^2 \quad (6.89)$$

Other important quantities are the extinction and absorption cross section. By a known procedure [5], they can be obtained

$$\sigma_{ext} = \frac{4\pi r^2}{(k_1 r)^2}\sum_{l=0}^{\infty}(2l+1)\Re\left(\frac{\Delta_1^L}{\Delta_0}\right) \quad (6.90)$$

and by definition $\sigma_{abs} = \sigma_{ext} - (\sigma_{scat}^{L} + \sigma_{scat}^{SV})$. Therefore, by knowing quantities $C_l^{(1)}$ and $C_l^{(3)}$, we are able to calculate the scattering quantities. Since the absorption cross section is identically zero for lossless scattering we then have $\sigma_{ext} = \sigma_{scat}$, which provides a detailed consistency check on our results.

If we set the shear modulus, μ_1, equal to zero in the medium surrounding the sphere, no shear waves will be present in the scattered field and our results reduce to those of the scattering of ordinary acoustic waves from an elastic sphere. In this case, $\Delta_3^L = 0$ and

$$\Delta_0 = -\frac{1}{2}(K_2a)^2\frac{\rho_1}{\rho_2}\left[\frac{k_2aj_l'(k_2a)}{j_l(k_2a)}\frac{K_2a\psi_l'(K_2a)}{\psi_l(K_2a)} - l(l+1) + \frac{1}{2}(K_2a)^2\frac{k_2aj_l'(k_2a)}{j_l(k_2a)}\right]$$
$$+\frac{k_1ah_l'(k_1a)}{h_l(k_1a)}\left\{\left[\frac{k_2aj_l'(k_2a)}{j_l(k_2a)}\frac{K_2a\psi_l'(K_2a)}{\psi_l(K_2a)} - l(l+1)\right][2-l(l+1)]\right.$$
$$\left.+\frac{1}{2}(K_2a)^2\left[\frac{K_2a\psi_l'(K_2a)}{\psi_l(K_2a)} + 2\frac{k_2aj_l'(k_2a)}{j_l(k_2a)} - 2l(l+1) + \frac{1}{2}(K_2a)^2\right]\right\}$$
$$\Delta_1^L = \left\{-\frac{1}{2}(K_2a)^2\frac{\rho_1}{\rho_2}\left[\frac{k_2aj_l'(k_2a)}{j_l(k_2a)}\frac{K_2a\psi_l'(K_2a)}{\psi_l(K_2a)} - l(l+1) + \frac{1}{2}(K_2a)^2\frac{k_2aj_l'(k_2a)}{j_l(k_2a)}\right]\right.$$
$$+\frac{k_1aj_l'(k_1a)}{j_l(k_1a)}\left\{\left[\frac{k_2aj_l'(k_2a)}{j_l(k_2a)}\frac{K_2a\psi_l'(K_2a)}{\psi_l(K_2a)} - l(l+1)\right][2-l(l+1)]\right.$$
$$\left.\left.+\frac{1}{2}(K_2a)^2\left[\frac{K_2a\psi_l'(K_2a)}{\psi_l(K_2a)} + 2\frac{k_2aj_l'(k_2a)}{j_l(k_2a)} - 2l(l+1) + \frac{1}{2}(K_2a)^2\right]\right\}\right\}\frac{j_l(k_1a)}{h_l(k_1a)}$$

with these substitutions, the expressions given earlier for longitudinal intensities and cross sections are valid and agree with those of J.J. Faran [21] who first derived the exact solution for acoustic wave scattering from an elastic sphere.

The consistency of the present formulation can be further checked by comparing the limit where the shear moduli of both media vanish with the known results of classical acoustic (hydrodynamic) theory. If the shear modulus, μ, is set to zero everywhere, the solutions become those of the scattering of acoustic waves from a fluid sphere. The problem is no longer an elastic problem, but is an acoustic (hydrodynamic) problem and the characteristic coefficients of the scattered wave now reduce to $\Delta_3^L = 0$ and

$$\Delta_0 = \frac{k_2aj_l'(k_2a)}{j_l(k_2a)} - \frac{\rho_2}{\rho_1}\frac{k_1ah_l'(k_1a)}{h_l(k_1a)} \quad \Delta_1^L = \frac{j_l(k_1a)}{h_l(k_1a)}\left\{\frac{k_2aj_l'(k_2a)}{j_l(k_2a)} - \frac{\rho_2}{\rho_1}\frac{k_1aj_l'(k_1a)}{j_l(k_1a)}\right\}$$

which are precisely the coefficients for the scattering problem of acoustic waves from a fluid bubble [16].

No transmitted shear waves are arrived at by the limit: $\mu_2 \to 0$, but $\mu_2K_2^2$ finite. Our results reduce to those of the scattering of ordinary acoustic waves from an elastic sphere.

$$\Delta_0 = -\frac{k_2aj_l'(k_2a)}{j_l(k_2a)}\left\{\left[\frac{k_1ah_l'(k_1a)}{h_l(k_1a)}\frac{K_1a\zeta_l'(K_1a)}{\zeta_l(K_1a)} - l(l+1)\right]\right.$$
$$\left.+\frac{1}{2}(K_1a)^2\left[\frac{K_1a\zeta_l'(K_1a)}{\zeta_l(K_1a)} + 2\frac{k_1ah_l'(k_1a)}{h_l(k_1a)} - 2l(l+1) + \frac{1}{2}(K_1a)^2\right]\right\}$$
$$+\frac{1}{2}(K_1a)^2\frac{\rho_2}{\rho_1}\left[\frac{k_1ah_l'(k_1a)}{h_l(k_1a)}\frac{K_1a\zeta_l'(K_1a)}{\zeta_l(K_1a)} - l(l+1) + \frac{1}{2}(K_1a)^2\frac{k_1ah_l'(k_1a)}{h_l(k_1a)}\right]$$

and so on.

If the scatterer is a cavity there will be no transmitted fields (Figure 6.8), so we take the limit where λ_2, μ_2, ρ_2 vanish. We find that

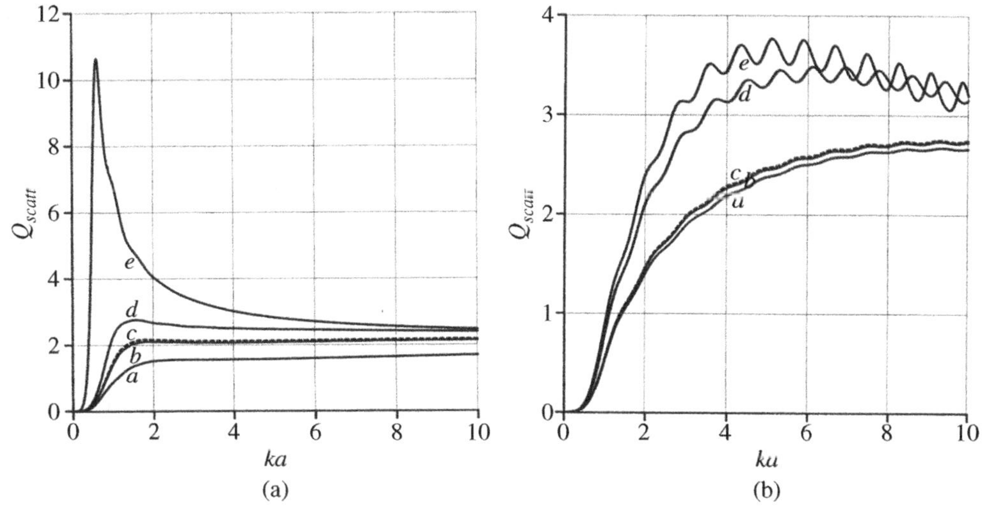

Figure 6.8 Scattering cross sections for spherical cavities in various media for incident L waves (a) and T waves (b) from Hinders (1990). These were drawn using an HP plotter, which uses actual ink pens to draw the figures, which were then literally cut and pasted into the document. I'm a little surprised that the paste lasted 30 years and that the ink hadn't faded or the paper yellowed too much.

$$\Delta_0 = [2 - l(l+1)]\left[\frac{k_1 a h_l'(k_1 a)}{h_l(k_1 a)} - \frac{K_1 a \zeta_l'(K_1 a)}{\zeta_l(K_1 a)}\right]$$
$$+\frac{1}{2}(K_1 a)^2\left[\frac{K_1 a \zeta_l'(K_1 a)}{\zeta_l(K_1 a)} + 2\frac{k_1 a h_l'(k_1 a)}{h_l(k_1 a)} - 2l(l+1) + \frac{1}{2}(K_1 a)^2\right]$$

and so on.

Finally, if the scatterer is perfectly rigid and infinitely dense, the boundary conditions will be zero displacements at $r = a$, which is technically nonphysical, but at least the algebra is simple. You can check for yourself that the answer is

$$\Delta_0 = \frac{k_1 a h_l'(k_1 a)}{h_l(k_1 a)} \frac{K_1 a \zeta_l'(K_1 a)}{\zeta_l(K_1 a)} - l(l+1)$$

Clebsch gave these results in 1861. I suppose technically the correct limit is $\lambda_2 \to \infty, \mu_2 \to \infty$, but with the density finite, to give a rigid but movable scatterer. We might as well just plug numbers for actual materials into the results for elastic scatterers.

6.4 Incident Transverse Wave

If a transverse wave with wave number $K_1 = \omega\sqrt{\rho_1/\mu_1}$ is incident upon the surface of the sphere, scattered and transmitted longitudinal, vertical shear, and horizontal shear waves will be excited. By using the boundary conditions, we write the following two systems of equations. The first is for coupled L and SV while the second is for SH. Note that Δ_0 is the same as for the incident longitudinal wave.

$$
\left(\begin{matrix}
\frac{-1}{2r}\left(\frac{K_1}{k_1}\right)^2 + \frac{[l(l+1)+2]}{k_1^2 r^3} - \frac{2\partial_r}{(k_1 r)^2} & \frac{\mu_2}{\mu_1}\left(\frac{1}{2r}\left(\frac{K_2}{k_2}\right)^2 - \frac{[l(l+1)+2]}{k_2^2 r^3} + \frac{2\partial_r}{(k_2 r)^2}\right) \\
-\frac{4}{k_1^2 r^3} + \frac{2\partial_r}{(k_1 r)^2} & \frac{\mu_2}{\mu_1}\left(\frac{4}{k_2^2 r^3} - \frac{2\partial_r}{(k_2 r)^2}\right) \\
-\frac{1}{(k_1 r)^2}\left(1 - r\partial_r\right) & \frac{1}{(k_2 r)^2}\left(1 - r\partial_r\right) \\
\frac{1}{(k_1 r)^2} & -\frac{1}{(k_2 r)^2}
\end{matrix}\right. \tag{6.91}
$$

$$
\left.\begin{matrix}
-\frac{l(l+1)}{K_1 r^2}\left(\partial_r - \frac{2}{r}\right) & \frac{\mu_2}{\mu_1}\frac{l(l+1)}{K_2 r^2}\left(\partial_r - \frac{2}{r}\right) \\
-\frac{2l(l+1)}{K_1 r^3} + \frac{2\partial_r}{K_1 r^2} + \frac{K_1}{r} & \frac{\mu_2}{\mu_1}\left(\frac{2l(l+1)}{K_2 r^3} - \frac{2\partial_r}{K_2 r^2} - \frac{K_2}{r}\right) \\
-\frac{l(l+1)}{K_1 r^2} & \frac{l(l+1)}{K_2 r^2} \\
-\frac{\partial_r}{K_1 r} & \frac{\partial_r}{K_2 r}
\end{matrix}\right)
\begin{bmatrix} (r\pi_L^s) \\ (r\pi_L^t) \\ (r\pi_{SV}^s) \\ (r\pi_{SV}^t) \end{bmatrix}
= -\begin{bmatrix}
\frac{l(l+1)}{K_1 r^2}\left(\partial_r - \frac{2}{r}\right) \\
\frac{2l(l+1)}{K_1 r^3} - \frac{2\partial_r}{K_1 r^2} - \frac{K_1}{r} \\
\frac{l(l+1)}{K_1 r^2} \\
\frac{\partial_r}{K_1 r}
\end{bmatrix}(r\pi_{SV}^i)
$$

$$
\begin{pmatrix}
-\left(\frac{\partial_r}{r} - \frac{2}{r^2}\right) & \frac{\mu_2}{\mu_1}\left(\frac{\partial_r}{r} - \frac{2}{r^2}\right) \\
-1 & 1
\end{pmatrix}
\begin{bmatrix} (r\pi_{SH}^s) \\ (r\pi_{SH}^t) \end{bmatrix}
= -\begin{bmatrix} \left(\frac{\partial_r}{r} - \frac{2}{r^2}\right) \\ 1 \end{bmatrix}(r\pi_{SH}^i)
$$

where

$$
\begin{aligned}
r\pi_{SV}^i &= \frac{1}{K_1}\sum_{l=1}^{\infty} i^{l-1}\left(\frac{2l+1}{l(l+1)}\right)\psi_l(K_1 r)P_l^{(1)}(\cos\theta)\cos\phi \\
r\pi_{SH}^i &= \frac{1}{K_1}\sum_{l=1}^{\infty} i^{l}\left(\frac{2l+1}{l(l+1)}\right)\psi_l(K_1 r)P_l^{(1)}(\cos\theta)\sin\phi \\
r\pi_L^s &= \sum_{l=0}^{\infty} B_l^{(1)}\zeta_l(k_1 r)P_l(\cos\theta) \\
r\pi_L^t &= \sum_{l=0}^{\infty} B_l^{(2)}\psi_l(k_2 r)P_l(\cos\theta) \\
r\pi_{SV}^s &= \frac{1}{K_1}\sum_{l=1}^{\infty} B_l^{(3)}\zeta_l(K_1 r)P_l^{(1)}(\cos\theta)\cos\phi \\
r\pi_{SV}^t &= \frac{1}{K_2}\sum_{l=1}^{\infty} B_l^{(4)}\psi_l(K_2 r)P_l^{(1)}(\cos\theta)\cos\phi \\
r\pi_{SH}^s &= \frac{1}{K_1}\sum_{l=1}^{\infty} A_l^{(1)}\zeta_l(K_1 r)P_l^{(1)}(\cos\theta)\sin\phi \\
r\pi_{SH}^t &= \frac{1}{K_2}\sum_{l=1}^{\infty} A_l^{(2)}\psi_l(K_2 r)P_l^{(1)}(\cos\theta)\sin\phi
\end{aligned} \tag{6.92}
$$

These potentials have been defined to give identical forms as those for the corresponding plasma problem [22], which is formally quite similar to the present problem since instead of L, SH, SV fields one must consider L, TE, TM fields in a plasma.[13] These equations are written for $r = a$, at the boundary surface separating the elastic sphere from the surrounding medium. As in the previous case, unknown amplitudes $A_l^{(1)}, A_l^{(2)}$ and also $B_l^{(1)}$–$B_l^{(4)}$ can be determined from these two separate sets of equations:

$$B_l^{(1)} = i^{l-1}\left(\frac{2l+1}{l(l+1)}\right)\left(\frac{\Delta_1^T}{\Delta_0}\right)\frac{P_l^{(1)}(\cos\theta)\cos\phi}{P_l(\cos\theta)} \qquad B_l^{(3)} = i^{l-1}\left(\frac{2l+1}{l(l+1)}\right)\left(\frac{\Delta_3^T}{\Delta_0}\right)$$

$$B_l^{(2)} = i^{l-1}\left(\frac{2l+1}{l(l+1)}\right)\left(\frac{\Delta_2^T}{\Delta_0}\right)\frac{P_l^{(1)}(\cos\theta)\cos\phi}{P_l(\cos\theta)} \qquad B_l^{(4)} = i^{l-1}\left(\frac{2l+1}{l(l+1)}\right)\left(\frac{\Delta_4^T}{\Delta_0}\right)$$

$$A_l^{(1)} = i^{l}\left(\frac{2l+1}{l(l+1)}\right)\Delta_5^T \qquad A_l^{(2)} = i^{l}\left(\frac{2l+1}{l(l+1)}\right)\Delta_6^T \tag{6.93}$$

Upon scattering of the incident transverse elastic waves on the surface of the spherical scatterer SV, SH, L waves will be generated in the scattered and transmitted regions. SV and L modes are coupled at the boundary and the generation of L waves is due to this boundary coupling. Since the incident energy is carried in both the SV- and SH-type waves, L is excited by the incident wave. In the case of incident L waves, SH waves were not excited, since there was no SH-wave mode participation in the incident wave modes, and since the boundary does not couple SH waves with either the L or the SV.

$$\begin{aligned}\Delta_1^T = \frac{k_1}{K_1}\frac{j_l(K_1a)}{h_l(k_1a)}\left[\frac{K_1a\psi_l'(K_1a)}{\psi_l(K_1a)} - \frac{K_1a\zeta_l'(K_1a)}{\zeta_l(K_1a)}\right]\Bigg(\left(\frac{\mu_2}{\mu_1}-1\right)^2[l(l+1)-2]\left[\frac{k_2aj_l'(k_2a)}{j_l(k_2a)}\frac{K_2a\psi_l'(K_2a)}{\psi_l(K_2a)} - l(l+1)\right]\\ +\frac{1}{2}(K_1a)^2\left(\frac{\mu_2}{\mu_1}-1\right)\Bigg\{\left[\frac{k_2aj_l'(k_2a)}{j_l(k_2a)}\frac{K_2a\psi_l'(K_2a)}{\psi_l(K_2a)} - l(l+1)\right]\\ -\frac{\rho_2}{\rho_1}\left[\frac{K_2a\psi_l'(K_2a)}{\psi_l(K_2a)} + 2\frac{k_2aj_l'(k_2a)}{j_l(k_2a)} - 2l(l+1)\right]\Bigg\} + \frac{1}{4}(K_1a)^4\left(\frac{\mu_2}{\mu_1}-1\right)\frac{\rho_2}{\rho_1}\left[1-\frac{\rho_2}{\rho_1}\right]\Bigg)\end{aligned}$$

$$\begin{aligned}\Delta_2^T = \frac{k_2}{K_1}\frac{j_l(K_1a)}{j_l(k_2a)}\left[\frac{K_1a\zeta_l'(K_1a)}{\zeta_l(K_1a)} - \frac{K_1a\psi_l'(K_1a)}{\psi_l(K_1a)}\right]\Bigg(\frac{1}{4}(K_1a)^4\left[1-\frac{\rho_2}{\rho_1}\right]\\ +\frac{1}{2}(K_1a)^2\left(\frac{\mu_2}{\mu_1}-1\right)\left\{\left[\frac{K_1a\zeta_l'(K_1a)}{\zeta_l(K_1a)} - 2\right]\left[\frac{k_2aj_l'(k_2a)}{j_l(k_2a)} - 1\right] + l(l+1) - 2\right\}\Bigg)\end{aligned}$$

$$\begin{aligned}\Delta_3^T = \frac{j_l(K_1a)}{h_l(K_1a)}\Bigg\{\left(\frac{\mu_2}{\mu_1}-1\right)^2[l(l+1)-2]\left[\frac{k_2aj_l'(k_2a)}{j_l(k_2a)}\frac{K_2a\psi_l'(K_2a)}{\psi_l(K_2a)} - l(l+1)\right]\left[\frac{k_1ah_l'(k_1a)}{h_l(k_1a)}\frac{K_1a\psi_l'(K_1a)}{\psi_l(K_1a)} - l(l+1)\right]\\ +\frac{1}{2}(K_1a)^2\left(\frac{\mu_2}{\mu_1}-1\right)\Bigg\{\left[\frac{k_2aj_l'(k_2a)}{j_l(k_2a)}\frac{K_2a\psi_l'(K_2a)}{\psi_l(K_2a)} - l(l+1)\right]\left[\frac{K_1a\psi_l'(K_1a)}{\psi_l(K_1a)} + 2\frac{k_1ah_l'(k_1a)}{h_l(k_1a)} - 2l(l+1)\right]\\ -\frac{\rho_2}{\rho_1}\left[\frac{k_1ah_l'(k_1a)}{h_l(k_1a)}\frac{K_1a\psi_l'(K_1a)}{\psi_l(K_1a)} - l(l+1)\right]\left[\frac{K_2a\psi_l'(K_2a)}{\psi_l(K_2a)} + 2\frac{k_2aj_l'(k_2a)}{j_l(k_2a)} - 2l(l+1)\right]\Bigg\}\\ +\frac{1}{4}(K_1a)^4\left\{l(l+1)\left[1-\frac{\rho_2}{\rho_1}\right]^2 - \left[\frac{k_2aj_l'(k_2a)}{j_l(k_2a)} - \frac{\rho_2}{\rho_1}\frac{k_1ah_l'(k_1a)}{h_l(k_1a)}\right]\left[\frac{K_2a\psi_l'(K_2a)}{\psi_l(K_2a)} - \frac{\rho_2}{\rho_1}\frac{K_1a\psi_l'(K_1a)}{\psi_l(K_1a)}\right]\right\}\Bigg\}\end{aligned}$$

---

13 Indeed, when I sat down to write up the results from my dissertation for publication, I followed the structure and mimicked the language of my then-deceased advisor and submitted the manuscript to the journal where he had published the plasma sphere scattering paper many years before. It was accepted without revision, which I now know to be rather unusual. What you'll find strange is that in 1990, I submitted the manuscript to that Italian journal via mail. Once the manuscript was accepted, the editor sent it out (to India) for typesetting where the text and equations were retyped using some proprietary system and the galley proofs were mailed to me to check for errors. I have a clear memory of sitting on my back steps, checking through everything carefully. Your first solo-author publication is a pretty big deal, BTW.

$$\Delta_4^T = \frac{K_2}{K_1}\frac{j_l(K_1a)}{j_l(K_2a)}\left[\frac{K_1a\zeta_l'(K_1a)}{\zeta_l(K_1a)} - \frac{K_1a\psi_l'(K_1a)}{\psi_l(K_1a)}\right]\left(\frac{1}{4}(K_1a)^4\left[\frac{k_2aj_l'(k_2a)}{j_l(k_2a)} - \frac{\rho_2}{\rho_1}\frac{k_1ah_l'(k_1a)}{h_l(k_1a)}\right]\right.$$
$$\left. + \frac{1}{2}(K_1a)^2\left(\frac{\mu_2}{\mu_1}-1\right)\left\{2\frac{k_1ah_l'(k_1a)}{h_l(k_1a)}\frac{k_2aj_l'(k_2a)}{j_l(k_2a)} - l(l+1)\left[\frac{k_1ah_l'(k_1a)}{h_l(k_1a)} + \frac{k_2aj_l'(k_2a)}{j_l(k_2a)} - 1\right]\right\}\right)$$

$$\Delta_5^T = \frac{j_l(K_1a)}{h_l(K_1a)}\left\{\frac{K_2a\psi_l'(K_2a)}{\psi_l(K_2a)} - 2 - \frac{\mu_2}{\mu_1}\left[\frac{K_1a\psi_l'(K_1a)}{\psi_l(K_1a)} - 2\right]\right\}\Big/\left\{\frac{K_1a\zeta_l'(K_1a)}{\zeta_l(K_1a)} - 2 - \frac{\mu_2}{\mu_1}\left[\frac{K_2a\psi_l'(K_2a)}{\psi_l(K_2a)} - 2\right]\right\}$$

$$\Delta_6^T = \frac{j_l(K_1a)}{j_l(K_2a)}\left\{\frac{K_1a\zeta_l'(K_1a)}{\zeta_l(K_1a)} - \frac{K_1a\psi_l'(K_1a)}{\psi_l(K_1a)}\right\}\Big/\left\{\frac{K_1a\zeta_l'(K_1a)}{\zeta_l(K_1a)} - 2 - \frac{\mu_2}{\mu_1}\left[\frac{K_2a\psi_l'(K_2a)}{\psi_l(K_2a)} - 2\right]\right\}$$

As in the previous calculations, we use the elastic energy flux vector to obtain the expressions for important scattering quantities. Intensities of the incident and scattered waves are written:

$$I_{inc}^T = \frac{\omega\mu_1}{2r}K_1r$$
$$I_{scat}^L = \frac{\omega\mu_1}{2r}\left(\frac{K_1}{k_1}\right)^2\frac{1}{k_1r}\left|\sum_{l=1}^{\infty}\left(\frac{2l+1}{l(l+1)}\right)\left(\frac{\Delta_1^T}{\Delta_0}\right)\left[P_l^{(1)}(\cos\theta)\right]\cos\phi\right|^2$$
$$I_{scat}^{SV} = \frac{\omega\mu_1}{2r}\frac{1}{K_1r}\left|\sum_{l=1}^{\infty}\left(\frac{2l+1}{l(l+1)}\right)\left(\frac{\Delta_3^T}{\Delta_0}\right)\times\left(\left[P_l^{(1)}(\cos\theta)\right]\sin\phi + \left[\frac{\partial}{\partial\theta}P_l^{(1)}(\cos\theta)\right]\cos\phi\right)\right|^2 \tag{6.94}$$
$$I_{scat}^{SH} = \frac{\omega\mu_1}{2r}\frac{1}{K_1r}\left|\sum_{l=1}^{\infty}\left(\frac{2l+1}{l(l+1)}\right)(\Delta_5^T)\times\left(\frac{\left[P_l^{(1)}(\cos\theta)\right]}{\sin\theta}\cos\phi + \left[\frac{\partial}{\partial\theta}P_l^{(1)}(\cos\theta)\right]\sin\phi\right)\right|^2$$

It can be seen from these expressions that scattered waves of all types include the polarization angle ϕ so that the scattered waves do not radiate azimuthally symmetric as they did for the case of incident longitudinal waves. We can now write the differential cross sections for the three types of waves for incident transverse waves as:

$$\frac{d\sigma}{d\Omega}^L = \frac{1}{(k_1r)^2}\frac{K_1r}{k_1r}\left|\sum_{l=1}^{\infty}\left(\frac{2l+1}{l(l+1)}\right)\left(\frac{\Delta_1^T}{\Delta_0}\right)\left[P_l^{(1)}(\cos\theta)\right]\cos\phi\right|^2$$
$$\frac{d\sigma}{d\Omega}^{SV} = \frac{1}{(K_1r)^2}\left|\sum_{l=1}^{\infty}\left(\frac{2l+1}{l(l+1)}\right)\left(\frac{\Delta_3^T}{\Delta_0}\right)\times\left(\frac{\left[P_l^{(1)}(\cos\theta)\right]}{\sin\theta}\sin\phi + \left[\frac{\partial}{\partial\theta}P_l^{(1)}(\cos\theta)\right]\cos\phi\right)\right|^2 \tag{6.95}$$
$$\frac{d\sigma}{d\Omega}^{SH} = \frac{1}{(K_1r)^2}\left|\sum_{l=1}^{\infty}\left(\frac{2l+1}{l(l+1)}\right)(\Delta_5^T)\times\left(\frac{\left[P_l^{(1)}(\cos\theta)\right]}{\sin\theta}\cos\phi + \left[\frac{\partial}{\partial\theta}P_l^{(1)}(\cos\theta)\right]\sin\phi\right)\right|^2$$

The scattering cross sections will be obtained by integrating these expressions over a surface of radius $r > a$:

$$\sigma_{scat}^{L} = \frac{2\pi r^2}{(k_1 r)^2}\frac{K_1}{k_1}\sum_{l=1}^{\infty}(2l+1)l(l+1)\left|\frac{\Delta_1^T}{\Delta_0}\right|^2$$

$$\sigma_{scat}^{SV} = \frac{2\pi r^2}{(K_1 r)^2}\sum_{l=1}^{\infty}(2l+1)\left|\frac{\Delta_3^T}{\Delta_0}\right|^2 \tag{6.96}$$

$$\sigma_{scat}^{SH} = \frac{2\pi r^2}{(K_1 r)^2}\sum_{l=1}^{\infty}(2l+1)\left|\Delta_5^T\right|^2$$

As one can see from the total scattering cross-sections expressions, there will be no radiation of scattered waves at zero mode. This means that transverse waves do not excite the zeroth mode of scattered waves as they did in the case of the incident L waves. Extinction and absorption cross sections are found as before, we write

$$\sigma_{ext} = \frac{2\pi r^2}{(K_1 r)^2}\sum_{l=1}^{\infty}(2l+1)\Re\left(\frac{\Delta_3^T}{\Delta_0} + \Delta_5^T\right) \tag{6.97}$$

which will be equal to the total scattering cross section for lossless scattering. Since $\sigma_{abs} = \sigma_{ext} - (\sigma_{scat}^{L} + \sigma_{scat}^{SV} + \sigma_{scat}^{SH})$, we can use the lossless case where $\sigma_{abs} = 0$ to check our results for consistency. I have a clear memory of those zeros scrolling down my screen on a sunny morning, which I took to mean both that I had done the algebra correctly and that my FORTRAN77 code was running properly.

The Rayleigh-range scattering cross sections are given in a simple form here. When the scatterer is small, the radial functions can be replaced by the corresponding small argument approximations and higher powers of the size parameters are neglected. Normalizing the scattering cross sections for incident longitudinal and transverse waves by the geometric cross section of the scatterer, πa^2, we find the following Rayleigh-limit expressions:

$$Q_{scat}^{L} = \frac{4}{9}(k_1 a)^4\left\{\left(\frac{(K_1/k_1)^2}{\left[(K_2/k_2)^2 - 4/3\right]\mu_2/\mu_1 + 4/3} - 1\right)^2 + \frac{1}{3}\left[1 + 2\left(\frac{K_1}{k_1}\right)^3\right]\left[1 - \frac{\rho_2}{\rho_1}\right]^2\right.$$
$$\left. + 40\left[2 + 3\left(\frac{K_1}{k_1}\right)^5\right]\left[\frac{\mu_2/\mu_1 - 1}{2\left[3(K_1/k_1)^2 + 2\right]\mu_2/\mu_1 + \left(9(K_1/k_1)^2 - 4\right)}\right]^2\right\}$$

$$Q_{scat}^{T} = \frac{8}{9}(K_1 a)^4\left\{\frac{1}{3}\left[1 + \frac{1}{2}\left(\frac{k_1}{K_1}\right)^3\right]\left[1 - \frac{\rho_2}{\rho_1}\right]^2\right.$$
$$\left. + 30\left[3 + 2\left(\frac{k_1}{K_1}\right)^5\right]\left[\frac{\mu_2/\mu_1 - 1}{2\left[3 + 2(k_1/K_1)^2\right]\mu_2/\mu_1 + \left(9 - 4(k_1/K_1)^2\right)}\right]^2\right\}$$

As one can see here, the scattering cross sections are proportional with the fourth power of the wave number, which agrees with the results of Rayleigh scattering. It is to be noted that in the small-sphere limit the contribution of the scattered SH-field to the transverse scattering cross section is negligible compared to that for the SV-field. Both of these results correspond exactly to the well-known results of Truell and coworkers [23, 24] and his students.[14]

One of the things that I'm forever harping on is to compare your new results with well-established ones in the literature. The reason for showing all the special cases above is that most of those results are in the literature, and consistency with them is a good check on more general results. But more importantly, when you code up some new results of your own, or even just the results of others' from the literature, you need to make sure that you've done that properly. The equations are complicated and there's no particular reason to think that you can look at your plots and tell whether they are right or not. Reproducing plots of others that have withstood some years of scrutiny is good practice (Figures 6.8 and 6.9). Here's a cautionary tale.

In 1972, McBride and Kraft coded up the equations of [24] for transverse elastic wave scattering from an elastic sphere and published a paper with a bunch of plots. That would have been rather tough to do in the early 1970s, so the paper was noteworthy for the number of plots they showed. Their goal was to do enough plots to begin to categorize scattering behavior(s). They concluded that, "The scattering behavior for an incident transverse wave was seen to be markedly different from that of an incident longitudinal wave." The problem is that they propagated a typo from [24] and so their results, and hence their conclusions, weren't meaningful. Woopsie. I just checked the online version[15] and there's no indication there that the plots and text are in error. Sooner or later, everybody who remembers this botch will be dead or retired and future researchers might assume that what's shown must be correct because it's been around for so long.

Special cases to the present problem may be obtained by letting the shear modulus, μ, equal zero both inside and outside of the sphere.

6.5 Scattering from Spherical Shells

The interaction of an elastic wave with a layered spherical inclusion has been the focus of many authors concerned with nondestructive testing and prediction of dynamical effects in composite materials. The difficulties encountered in measuring the properties of the interface between the particles and the matrix, make analyses of these types of problems useful in predicting the bulk behavior of the composite material. Many different analyses have been made, including those using finite element, integral equation, and eigenfunction approaches. The latter is the only method that avoids repeated numerical matrix inversions and provides exact and analytic, closed-form algebraic solutions. It is this eigenfunction method that is the focus of our analysis.

The longitudinal scattering cross section can be written as:

$$Q^L_{scatt} = \frac{4}{(k_1 b)^2} \sum_{l=0}^{\infty} (2l+1) \left\{ \left| \frac{\Delta^L_1}{\Delta_0} \right|^2 + l(l+1) \frac{K_1}{k_1} \left| \frac{\Delta^L_2}{\Delta_0} \right|^2 \right\} \tag{6.98}$$

14 C.F. Ying was the founder and pioneer of ultrasonics research in China. In 1951, Ying started to work for Prof. Rohn Truell's Metals Research Laboratory of Brown University and co-authored with Prof. Truell a series of research papers, including [23]. In 1956, Ying came back to China and entered into the Chinese Academy of Sciences (CAS), served as researcher at Institute of Applied Physics; then in 1957, director of research at the Institute of Electronics; and then in 1964, director of research at the newly founded Institute of Acoustics. He dedicated himself to promoting the research and applications of ultrasonics in China despite the turbulences of the political movements beginning to overwhelm the Chinese society since early 1960s. During the Cultural Revolution Ying was tortured, imprisoned and once attempted suicide, and his first spouse died of cancer in a camp in southern China. Nevertheless, Ying made numerous breakthroughs in research, education, and was granted multiple national awards.

15 https://pubs.aip.org/aip/jap/article/43/12/4853/168864/Scattering-of-a-transverse-elastic-wave-by-an.

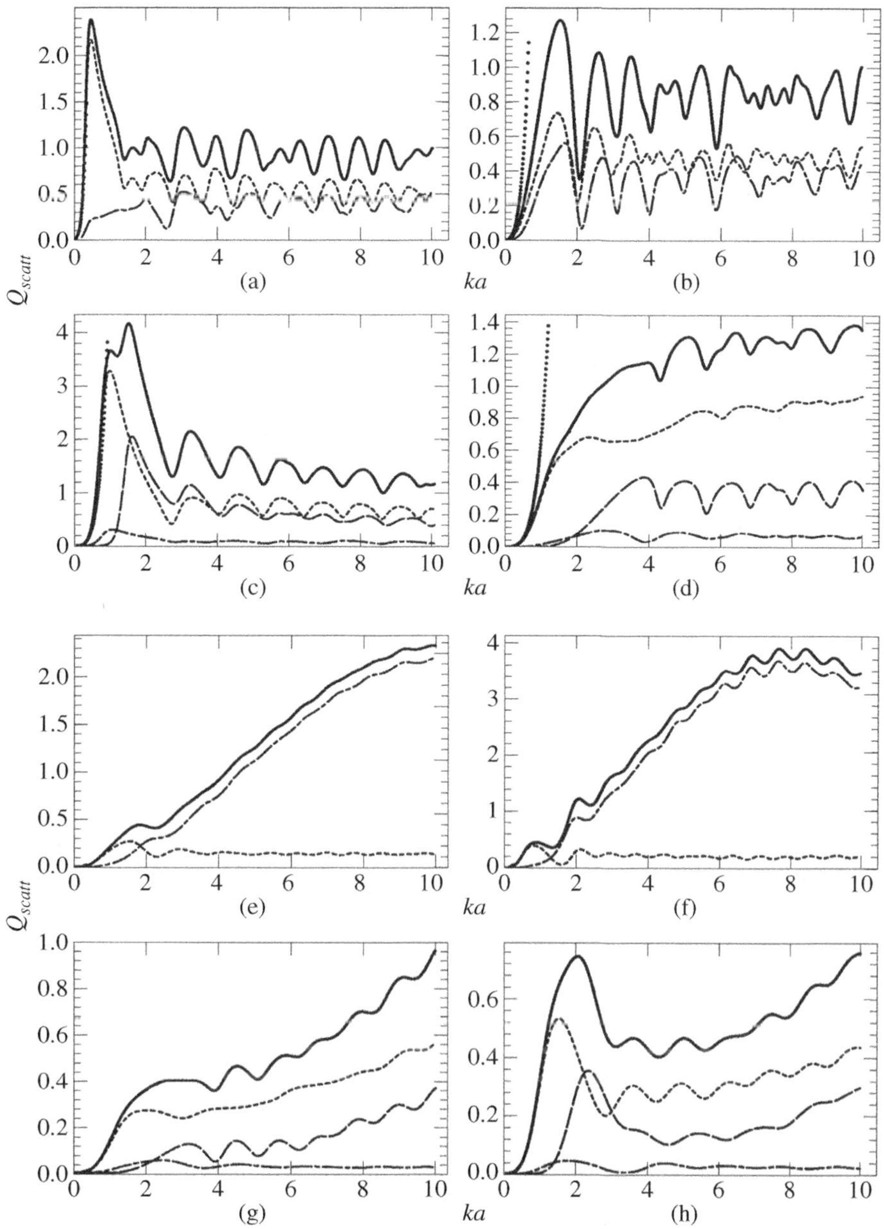

Figure 6.9 Scattering cross sections: short dashed line is the SV-wave component, long dashed line is the SH-wave component, and short-long dashed line is the L-wave component. (a) Incident longitudinal waves for sphere of aluminum in germanium; (b) incident longitudinal waves for sphere of germanium in aluminum; (c) incident transverse waves for a sphere of aluminum in germanium; (d) incident transverse waves for sphere of germanium in aluminum. The dotted lines are the Rayleigh-scattering approximations for small scatterers. (e) Incident longitudinal waves for sphere of stainless steel in magnesium; (f) incident longitudinal waves for sphere of magnesium in stainless steel; (g) incident transverse waves for a sphere of stainless steel in magnesium; (h) incident transverse waves for sphere of magnesium in stainless steel.

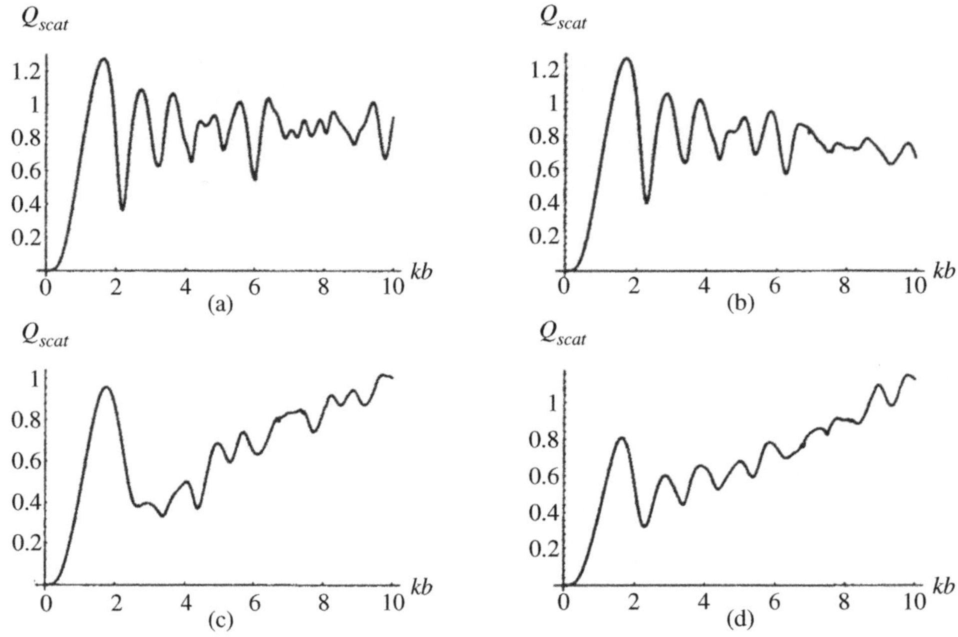

Figure 6.10 Elastic wave scattering from a two-layer spherical elastic inclusion in stainless steel. Plot (a) is for a zero-thickness shell surrounding a magnesium core. Plot (b) is for an aluminum shell with thickness 10% of the radius of the magnesium core. Plot (c) is for an aluminum shell with a thickness equal to that of the radius of the magnesium core. Plot (d) is for a zero-radius core so that the inclusion is aluminum.

This is essentially the same expression we arrived at for elastic wave scattering from an elastic sphere, except that the Δ's are a lot more complicated. I know because we did that algebra [25] and made the plots in Figure 6.10.

I realize that many of the expressions in this chapter are a bit extra. Some do go on for several pages, but I think it's important to point out that these are exact expressions. Getting plots out of them is a straightforward, if nontrivial, exercise in typing them into whichever computer language you prefer and summing things up to get the answer(s). As I was putting this book together, it was suggested that "A website with Matlab code for implementing the methods described in the book would be extremely helpful." My response was that while I agree that could be helpful for many readers, what I've found is that the languages change rather rapidly, that is, FORTRAN begets C which begets Matlab which begets Python. Part of my goal here is to help readers know how to find what they need when they need it, with an appreciation for how rapidly such things change. Once upon a time, you simply asked around the lab to see if anybody had the subroutine you needed. Not long ago, the answer used to be to have a 3.5 diskette or code listing in the back of the book. Then it was a CD and/or a webpage tied to the text/class. Now the answer seems to be GitHub, but we're all worried about that since Microsoft gobbled it up. I don't pretend to know what the right answer is going to be going forward, but my sneaking suspicion is that sometime soon some new LLM will be able to translate the equations off a page into code to make plots. If that turns out to be ridiculously easy someday, feel free to send me a postcard saying "neener-neener" and I'll put it on the bulletin board in my lab so my graduate students can glance over at it and smirk when I start going on about FORTRAN or whatever. I'm generally quite a good sport about such things.

If I included a comprehensive bibliography for scattering from spheres, it would be many, many pages long. Instead, I've included some of the key reference texts and important individual citations, but I do want to include three valuable, but obscure books you should be on the look out for. The first is something that I made myself a photocopy of more than 30 years ago and have had it on my bookshelf ever since [26]. The other two [27, 28] should be available on Internet Archive or discarded by libraries; they have lots of obscure details you might just need someday.

I feel like I should remind you that this is a book about *data engineering*. The plots that I've included for the various versions of scattering from spheres are mostly scattering cross sections. That's because those are the sorts of things that people have plotted, going all the way back to the days when computers were humans and it took months of effort to make them. Once you've gotten these various solutions coded up, you'll be able to make all kinds of plots, of course, but do start with cross sections so you can benchmark your output against the established literature. Then plot whatever you like because the whole point of the analysis is to be able to use the solutions to explore the scattering behavior. In particular, the idea is to use them to understand what features in the scattering behavior are most likely to be exploitable for machine learning. Computationally there's no longer any restrictions on fully exploring the scattering behavior, so exercise your code(s) in that way so as to gain insight into designing both measurement schemes and signal processing approaches for ML. That's what data engineering is.

References

1 Mie, G. (1908). Beiträge zur Optik trüber Medien. *Annalen der Physik* 25: 377.

2 Logan, N.A. (1965). Survey of some early studies of the scattering of plane waves by a sphere. *Proceedings of the IEEE* 53 (8): 773–785. https://doi.org/10.1109/PROC.1965.4055.

3 Clebsch, A. (1863). Ueber die Reflexion an einer Kugelfläche. *Journal für die reine und angewandte Mathematik* 61: 195–262. https://doi.org/10.1515/crll.1863.61.195.

4 Lorenz, L. (1890). On the reflection and refraction of light by a transparent sphere. *Videnskab. Skrifter.* 6: 1–62.

5 Born, M. and Wolf, E. (2013). *Principles of Optics: Electromagnetic Theory of Propagation, Interference and Diffraction of Light*. Elsevier.

6 Bohren, C.F. and Huffman, D.R. (2008). *Absorption and Scattering of Light by Small Particles*. Wiley.

7 Kerker, M. (1969). *The Scattering of Light and Other Electromagnetic Radiation, Physical Chemistry: A Series of Monographs*, vol. 16. New York: Academic Press.

8 van de Hulst, H.C. (1957). *Light Scattering by Small Particles*. Wiley. Also there's a version: Dover Books on Physics, 1981.

9 Hopke, P.K., Davis, E.J., and Wagner, P. (2016). Milton Kerker (1920–2016). *Aerosol Science and Technology* 50 (7): 758. https://doi.org/10.1080/02786826.2016.1190590.

10 Bowman, J.J., Senior, T.B.A., and Uslenghi, P.L.E. (1987). *Electromagnetic and Acoustic Scattering by Simple Shapes*. New York: Hemisphere Publishing.

11 Varadan, V.V., Lakhtakia, A., and Varadan, V.K. (eds.) (1991). *Field Representations and Introduction to Scattering*. New York: North-Holland.

12 Newton, R.G. (1966). *Scattering Theory of Waves and Particles*. Springer-Verlag.

13 Nowitall, I.B. (2020). *Scientific Adulting and BS Detection*. W&N Edutainment Press.

14 Nowitall, I.B. (2024). *Post-Pandemic Adulting*. W&N Edutainment Press.

15 Hinders, M.K. (2020). *Intelligent Feature Selection for Machine Learning Using the Dynamic Wavelet Fingerprint*. Springer Nature.

16 Anderson, V.C. (1950). Sound scattering from a fluid sphere. *The Journal of the Acoustical Society of America* 22 (4): 426–431.

17 Lynch, J.E., Pouch, A., Sanders, R. et al. (2007). Gaseous microemboli sizing in extracorporeal circuits using ultrasound backscatter. *Ultrasound in Medicine & Biology* 33 (10): 1661–1675. https://doi.org/10.1016/j.ultrasmedbio.2007.04.008.

18 Leckey, C.A.C. and Hinders, M.K. (2011). Newtonian viscous effects in ultrasonic emboli removal from blood. *Ultrasound in Medicine & Biology* 37 (8): 1340–1349. https://doi.org/10.1016/j.ultrasmedbio.2011.05.009.

19 Leckey, C.A.C. and Hinders, M.K. (2012). Viscous effects in the acoustic manipulation of algae for biofuel production. *Journal of Applied Phycology* 24: 145–156. https://doi.org/10.1007/s10811-011-9662-7.

20 Hinders, M.K. (1991). Plane-elastic-wave scattering from an elastic sphere. *Il Nuovo Cimento B* 106 (7): 799–818.

21 Faran, J.J. Jr. (1951). Sound scattering by solid cylinders and spheres. *The Journal of the Acoustical Society of America* 23 (4): 405–418.

22 Yildiz, A. (1963). Scattering of plane plasma waves from a plasma sphere. *Il Nuovo Cimento* 30: 1182–1207. https://doi.org/10.1007/BF02828781.

23 Ying, C.F. and Truell, R. (1956). Scattering of a plane longitudinal wave by a spherical obstacle in an isotropically elastic solid. *Journal of Applied Physics* 27 (9): 1086–1097.

24 Einspruch, N.G., Witterholt, E.J., and Truell, R. (1960). Scattering of a plane transverse wave by a spherical obstacle in an elastic medium. *Journal of Applied Physics* 31 (5): 806–818.

25 Bogan, S.D. and Hinders, M.K. (1994). *Interface Effects in Elastic Wave Scattering*. Springer.

26 Pao, Y.H. and Mow, C.C. (1973). *Diffraction of Elastic Waves and Dynamic Stress Concentrations*. New York: Krane Russak.

27 Hobson, E.W. (1931). *The Theory of Spherical and Ellipsoidal Harmonics*. Cambridge University Press.

28 MacRobert, T.M. (1967). *Spherical Harmonics: An Elementary Treatise on Harmonic Functions, with Applications*. Dover reprint of 1945 2nd Edition.

7

Scattering from Cylinders

Don't tell anybody, but a question I usually put on the PhD qualifying exam is: Why do the powerlines above I-95 in Quantico, VA have orange spheres on them? The first part of the question tests general awareness, that is, knowing that Quantico is a Marine Base. It's pretty hard not to know that if you've driven that stretch of I95 because the traffic always seems to choke right by the National Museum of the Marine Corps which you can't help noticing from the road. Slightly more awareness is tested by knowing that the Marines fly helicopters. That they fly them at night. That they fly them close to the ground. That powerlines and low-flying helicopters don't mix. I would expect a PhD student to conclude that the orange spheres have something to do with helicopters avoiding powerlines at night. During the day, the orange color is helpful because the powerlines themselves might be tough to see. At night, the orange color is not relevant. Hence, the second part of the question is related to backscattering of the radar that helicopters are using to avoid obstacles. So, the answer that I'm looking for is a discussion about electromagnetic scattering from spheres vs. cylinders. In particular, radar is transversely polarized and will scatter most strongly from long, thin wires only when the wires line up with that polarization, which could be a real problem. Spheres, on the other hand, have no orientation, so they will backscatter the same from all directions for all polarizations. Students get bonus points if they make a wild guess at the wavelengths of the radar marine helicopters might use and how that relates to the radii of the wires and spheres. Double bonus points if they speculate about how the magnetic fields of the transmission lines would introduce eddy currents in the metallic spheres which would tend to heat up just a bit because of the finite conductivity of the steel the hemispheres are stamped from and that slight temperature elevation would show up on infrared imagers that helicopter pilots probably also have....

7.1 Electromagnetic Wave Scattering

I once gave a talk with the title, "What ever became of the aether?" which must have been well received because I got the job that I've held for more than 30 years. The short version of the story is that before Maxwell's equations came into widespread use for modeling the propagation and scattering of light and assorted electromagnetic radiation [1–5], everybody had to use the elastic solid theory of the aether,[1] which was cumbersome enough that there was a collective, "Oh thank God. This is so much simpler!" when people figured out that they could start with Maxwell's equations instead. Part of the issue is that experiments had shown conclusively that light was a transverse wave, whereas elastic waves can be longitudinal and transverse with mode coupling between them

1 https://en.wikipedia.org/wiki/A_History_of_the_Theories_of_Aether_and_Electricity.

A Data Engineering Approach to Wave Scattering Analysis: with Applications in Radar, Sonar, Medical Diagnostics, Structural Flaw Detection and Intelligent Robotics, First Edition. Mark K. Hinders.

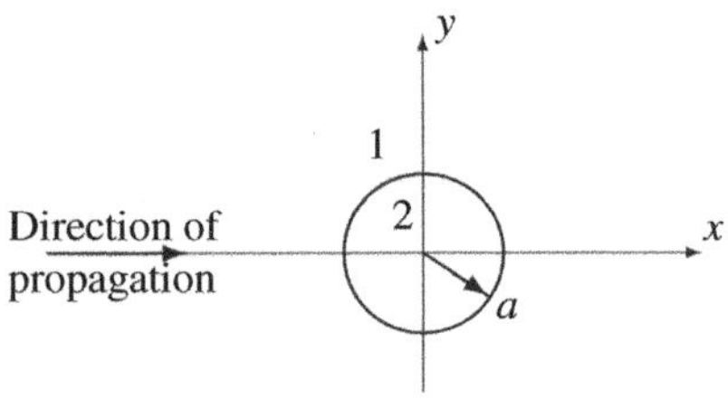

Figure 7.1 Problem geometry for scattering from a cylinder.

during reflection, refraction, and scattering. The answer to the question posed in the title of my 1993 job talk is that the mathematical machinery needed to model light propagating in the aether is now used to model things like ultrasonic nondestructive evaluation.

Let's start with electromagnetic wave scattering from cylinders, and then we'll do acoustic and elastic wave scattering from cylinders. As in Chapter 6, I'll go through the math and then show you some representative plots. I've made an attempt to dig up well-validated results from classic textbooks, and will then exhort you to get your own code running and reproduce the plots that I've mimicked here.

The problem geometry is shown Figure 7.1, with the infinite cylinder of radius $r = a$ oriented in the z-direction that we assume has different material properties from the surrounding medium. Call the region $r > a$ medium 1 and $r < a$ medium 2.

Since we're going to solve the problem in frequency domain, we will suppress the $e^{-i\omega t}$ time dependence, and write Maxwell's equations as:

$$\nabla \times \vec{H} = -i\omega\epsilon\vec{E} \qquad \nabla \times \vec{E} = i\omega\mu\vec{H} \tag{7.1}$$

along with $\nabla \cdot \vec{E} = 0$ and $\nabla \cdot \vec{H} = 0$, where the wave numbers are

$$K_1 = \omega\sqrt{\epsilon_1\mu_1} \qquad K_2 = \omega\sqrt{\left(\epsilon_2 + i\frac{\sigma_2}{\omega}\right)\mu_2} \tag{7.2}$$

and we note that the finite conductivity of the cylinder turns out to be really important because infinite conductivity means that the cylinder will be a perfect reflector. Since that's an excellent approximation for metals at radar frequencies, quite a lot of the literature makes this simplifying assumption. We won't do that, but will consider it as a limiting case to compare with standard results in the literature.

Now take the curl of the first two of Maxwell's equations and use my favorite vector identity

$$\nabla \times (\nabla \times \vec{A}) = \nabla(\nabla \cdot \vec{A}) - \nabla^2\vec{A} \tag{7.3}$$

to get the vector wave equation for both $\vec{E}$ and $\vec{H}$

$$\nabla^2\vec{E} + K^2\vec{E} = 0 \qquad \nabla^2\vec{H} + K^2\vec{H} = 0 \tag{7.4}$$

Since we're going to want a plane wave propagating in the x-direction, the incident wave solution is

$$E_z^{inc} = e^{iK_1x} \qquad E_x^{inc} = E_y^{inc} = 0 \tag{7.5}$$

or

$$E_y^{inc} = e^{iK_1x} \qquad E_x^{(i)} = E_z^{inc} = 0 \tag{7.6}$$

with the magnetic field components computed via Maxwell's equations. Note that we have recognized that we're talking about transversely polarized waves here, and have assumed that one case has the electric field polarized in the xz-plane, while the other has the electric field polarized in the

yz-plane. This is one way where scattering from a cylinder is a bit more complicated than scattering from a sphere. Another is that the incident wave can be oblique, but I'm not going to consider that (much), so we can focus on the scattering issues most of interest. In addition, if we want a different polarization, we can simply consider a linear combination of these two, so let's not worry about that right now either.

We'll then assume appropriate functional forms for the scattered electric and magnetic fields in region 1 ($r > a$) as well as in region 2 ($r < a$). Those solutions are going to be in cylindrical functions, of course. Then we'll apply the boundary conditions of continuity of tangential electric and magnetic fields at the boundary $r = a$ and so some algebra.

As in scattering by a sphere, we start with the scalar wave equation

$$\nabla^2\Psi + K^2\Psi = 0 \tag{7.7}$$

and in cylindrical coordinates, this becomes

$$\frac{1}{r}\frac{\partial}{\partial r}\left(r\frac{\partial\Psi}{\partial r}\right) + \frac{1}{r^2}\frac{\partial^2\Psi}{\partial\phi^2} + \frac{\partial^2\Psi}{\partial z^2} + K^2\Psi = 0 \tag{7.8}$$

In Chapter 2, there was an exercise for you to write the general solution to this via separation of variables, so here I'm simply going to assert that the general solution is a linear combination of

$$\Psi_n(r,\phi,z) = Z_n(\rho)e^{in\phi}e^{ihz} \tag{7.9}$$

for $n = 0, \pm1, \pm2, \dots$. Of course, Z_n is a solution of Bessel's equation

$$\rho\frac{d}{d\rho}\left(\rho\frac{d}{d\rho}Z_n\right) + (\rho^2 - n^2)Z_n = 0 \tag{7.10}$$

and $\rho = r\sqrt{K^2 - h^2}$, where h is a separation constant. Note that the variations in ϕ and z are both sinusoidal, which is kind of a relief for us because the only special functions involved so far are the plain old, integral-order Bessel functions. There is one key point to make right here, though. Electric and magnetic waves are vector fields, but we've just written the general solution to the scalar wave equation. We have to introduce vector cylindrical harmonics:

$$\vec{M}_n = \nabla\times(\hat{e}_z\Psi_n) \qquad \vec{N}_n = \nabla\times\left(\frac{\nabla\times\vec{M}_n}{K}\right) \tag{7.11}$$

I assume that you've long-since identified a convenient source for things like curl in cylindrical coordinates, but for your convenience here are the vector harmonics written out

$$\vec{M}_n = \sqrt{K^2 - h^2}\left(in\frac{Z_n(\rho)}{\rho}\hat{e}_r - Z_n'(\rho)\hat{e} + \phi\right)e^{i(n\phi+hz)}$$

$$\vec{N}_n = \frac{\sqrt{K^2 - h^2}}{K}\left(ihZ_n'(\rho)\hat{e}_r - hn\frac{Z_n(\rho)}{\rho}\hat{e}_\phi + \sqrt{K^2 - h^2}Z_n(\rho)\hat{e}_z\right)e^{i(n\phi+hz)}$$

Note that these vector harmonics are orthogonal, which is going to do something very important for us. When we apply the boundary conditions we'll invoke orthogonality to get rid of the pesky summation over all n on both sides of the equations, and then the boundary condition equations will be valid for each value of n. It's a subtle point that is usually glossed over until it comes back to bite you. Stay tuned for that.

Meanwhile, we have a general solution in cylindrical coordinates for $\Psi_n(\rho)$ and so we can write the general forms of the vector harmonics easily enough. The next step is to write something along the lines of

$$\vec{E} = \sum_{n=-\infty}^{\infty} \left[A_n \vec{M}_n + B_n \vec{N}_n \right] \tag{7.12}$$

for the incident, scattered, and transmitted fields. We'll want the incident wave to be a plane wave propagating in the positive x-direction and polarized in either the z- or y-direction, in both cases at normal incidence to the cylinder axis. That will simplify things quite a bit as we'll see in just a moment. For the scattered wave, we'll choose the appropriate Hankel function $H_n(\rho)$, so that the scattered field obeys the Sommerfeld radiation condition, that is, it decays like an expanding cylindrical wave at large distances from the origin. I hope that makes good sense to you. For the transmitted wave, which is the one that exists for $r < a$, we will note that the Neumann function $Y_n(\rho)$ is singular at the origin, so on physical grounds, we'll choose just the Bessel function $J_n(\rho)$ for that. This is all exactly analogous to what we did for spheres, except that it's cylindrical Bessel functions rather than spherical Bessel functions now. We can thus pencil in $J_n(\rho)$ or $H_n(\rho)$, where we've written $Z_n(\rho)$ in the expressions for $\vec{M}_n$ and $\vec{N}_n$ earlier.

As promised, we'll restrict things to normal incidence in just a moment, but most authors can't resist starting with the more general case of oblique incidence, so consider ζ to define the angle between the z-axis and the direction of propagation of the incident plane wave. To be just a bit pedantic about it, we then have

$$\vec{E}_{inc} = \vec{E}_0 e^{iK(\sin\zeta \hat{e}_x - \cos\zeta \hat{e}_z)} \tag{7.13}$$

where the polarization is determined by the vector $\vec{E}_0$ and that will be either parallel to the xz-plane or perpendicular to it. Of course, $\zeta = 90°$ is the normal-incidence case that I keep promising.

7.1.1 Incident E-Field Parallel to the *xz*-Plane

For the case of the incident electric field parallel to the xz-plane we write the incident electric field as:

$$\vec{E}_{inc} = E_0 \left(\sin\zeta \hat{e}_z - \cos\zeta \hat{e}_x \right) e^{-iK(r \sin\zeta \cos\phi + z\cos\zeta)} \tag{7.14}$$

and the separation constant that's been floating around must be $h = -K\cos\zeta$ so you can see why things are going to get quite a lot simpler when $\zeta = 90°$. To quote [2], "All that is required now is a good bit of patience to show that" the appropriate expansion of the incident electromagnetic field is

$$\vec{E}_{inc} = \frac{E_0}{K\sin\zeta} \sum_{n=-\infty}^{\infty} (-i)^n \vec{N}_n^{(1)} \qquad \vec{H}_{inc} = \frac{-iE_0}{\omega\mu\sin\zeta} \sum_{n=-\infty}^{\infty} (-i)^n \vec{M}_n^{(1)} \tag{7.15}$$

where E_0 is the amplitude of the incident plane wave which we'll usually take as unity because we'll ultimately plot things scaled by that anyway. The notation $\vec{M}_n^{(1)}, \vec{N}_n^{(1)}$ is intended to remind ourselves that we need to use the Bessel function J_n for the incident wave in the expressions for $\vec{M}_n$ and $\vec{N}_n$.

We can then write down the general forms for the scattered and transmitted fields as

$$\vec{E}_{scat} = -\frac{E_0}{K\sin\zeta} \sum_{n=-\infty}^{\infty} (-i)^n \left[b_n \vec{N}_n^{(3)} + i a_n \vec{M}_n^{(3)} \right] \tag{7.16}$$

$$\vec{H}_{scat} = -\frac{E_0}{\omega\mu\sin\zeta} \sum_{n=-\infty}^{\infty} (-i)^n \left[b_n \vec{M}_n^{(3)} + i a_n \vec{N}_n^{(3)} \right] \tag{7.17}$$

$$\vec{E}_{trans} = -\frac{E_0}{K'\sin\zeta} \sum_{n=-\infty}^{\infty} (-i)^n \left[d_n \vec{M}_n^{(1)} + c_n \vec{N}_n^{(1)} \right] \tag{7.18}$$

$$\vec{H}_{trans} = -\frac{E_0}{\omega\mu' \sin\zeta} \sum_{n=-\infty}^{\infty} (-i)^n \left[d_n \vec{M}_n^{(1)} + c_n \vec{N}_n^{(1)} \right] \tag{7.19}$$

The notation $\vec{M}_n^{(3)}, \vec{N}_n^{(3)}$ is intended to remind ourselves that we need to use the appropriate Hankel function H_n for the scattered wave in the expressions for $\vec{M}_n$ and $\vec{N}_n$. There's a bit of a trap here, because the Hankel function is defined as $H_n = J_n \pm iY_n$ and you have to pick the one that satisfies the radiation condition for the harmonic time convention you're using. Recall that we have suppressed factors of $e^{\pm i\omega t}$ throughout. Physicists tend to use "minus" while electrical engineers tend to use "plus" and also for some reason, use j as the imaginary unit. In addition, don't forget that the cylinder has different material properties than the surrounding medium, which we can indicate via a prime or a subscript or whatever is convenient.

We've got four unknown modal coefficients, a_n, b_n, c_n, and d_n, which we solve for by setting $\left[\vec{E}_{inc} + \vec{E}_{scat} - \vec{E}_{trans}\right] \times \hat{e}_r = 0$ and $\left[\vec{H}_{inc} + \vec{H}_{scat} - \vec{H}_{trans}\right] \times \hat{e}_r = 0$ for $r = a$. The math is a little tricky because of the cylindrical coordinates, of course, but it's just a statement that the ϕ- and z-components of the electric and magnetic fields are continuous at the surface of the scatterer. Note that the total field exterior to the scatterer is the incident plus the scattered, so those two together balance the transmitted field at $r = a$. Four equations allow you to solve for four unknowns, but typically what you most want to solve for are only the modal coefficients that determine the scattered field.

Exercise 7.1 Apply the boundary conditions of continuity of the tangential electric and magnetic fields to solve for a_n and b_n.

In doing that exercise, I hope that you wrote the system of four equations as a matrix equation, and then did Cramer's rule to solve for the coefficients. I'm not going to write that all out for you here because I am transcribing the treatment of [2] who write:

$$a_n = \frac{C_n V_n - B_n D_n}{W_n V_n + i(Dn)^2} \qquad b_n = \frac{W_n B_n + iD_n C_n}{W_n V_n + i(Dn)^2} \tag{7.20}$$

where we define $\xi = Ka \sin\zeta$, $\eta = Ka\sqrt{m^2 - \cos^2\zeta}$, and have assumed neither the cylinder nor the surrounding medium is magnetic, so $\mu = \mu'$. The index of refraction of the cylinder is m. The rest of the details are

$$D_n = n \cos\zeta \eta J_n(\eta) H_n^{(1)}(\xi) \left(\frac{\xi^2}{\eta^2} - 1\right)$$
$$B_n = \xi \left[m^2 \xi J_n'(\xi) J_n(\xi) - \eta J_n(\eta) J_n'(\xi)\right]$$
$$C_n = n \cos\zeta \eta J_n(\eta) J_n(\xi) \left(\frac{\xi^2}{\eta^2} - 1\right)$$
$$V_n = \xi \left[m^2 \xi J_n'(\eta) H_n^{(1)}(\xi) - \eta J_n(\eta) H_n'^{(1)}(\xi)\right]$$
$$W_n = i\xi \left[\eta J_n(\eta) H_n'^{(1)}(\xi) - \xi J_n'(\eta) H_n^{(1)}(\xi)\right]$$

Note carefully that prime indicates differentiation with respect to argument for the Bessel and Hankel functions, and we've indicated with the superscript (1) which of the two versions of the Hankel function we're using. In addition, notice that the denominators of a_n and b_n are identical, which should make good sense to you if you solved the system via Cramer's rule; the denominator is the determinant of the coefficient matrix, of course.

I realize that you've probably lost hope that I was ever going to invoke $\zeta = 90°$ for the normal-incidence case and simplify everything. Go ahead and simplify the above expressions and see if you find that $a_n = 0$ and with $x = Ka$

$$b_n = \frac{J_n(mx)J'_n(x) - mJ'_n(mx)J_n(x)}{J_n(mx)H'^{(1)}_n(x) - mJ'_n(mx)H^{(1)}_n(x)} \tag{7.21}$$

Numerical implementation of that is pretty straightforward these days. It used to matter a lot, numerically, to write the Bessel and Hankel functions in logarithmic-derivative forms, that is, $J'_n(x)/J_n(x)$ and so on. I don't know whether that will matter to you, but wanted to mention it just in case. Sometimes when you plot the scattering behavior, it looks for all the world like numerical garbage when in reality the many spikey resonances are real physical behavior. It could be numerical garbage, though.

7.1.2 Incident E-Field Perpendicular to the *xz*-Plane

That was the polarization where the electric field is parallel to the cylinder axis. It's the one where your radar signal will give the strongest backscatter from a powerline at normal incidence, which matters if you're flying in a helicopter at night, but this scattering behavior isn't very interesting. The good news is that we can go through this same procedure and find the answer for the other polarization. If you had solved that one using Cramer's rule, you'll see from that how similar things are.

For this polarization, we write the incident electric field as:

$$\begin{aligned} \vec{E}_{inc} &= E_0 \hat{e}_y e^{-ik(r \sin\zeta \cos\phi + z\cos\zeta)} \\ &= -i\frac{E_0}{K\sin\zeta} \sum_{n=-\infty}^{\infty} (-i)^n \vec{M}_n^{(1)} \end{aligned} \tag{7.22}$$

and the curl of this gives the incident magnetic field. We assume the same forms for the scattered and transmitted fields as before

$$\vec{E}_{scat} = -\frac{E_0}{K\sin\zeta} \sum_{n=-\infty}^{\infty} (-i)^n \left[b_n \vec{N}_n^{(3)} + ia_n \vec{M}_n^{(3)}\right] \tag{7.23}$$

$$\vec{H}_{scat} = -\frac{E_0}{\omega\mu\sin\zeta} \sum_{n=-\infty}^{\infty} (-i)^n \left[b_n \vec{M}_n^{(3)} + ia_n \vec{N}_n^{(3)}\right] \tag{7.24}$$

$$\vec{E}_{trans} = -\frac{E_0}{K'\sin\zeta} \sum_{n=-\infty}^{\infty} (-i)^n \left[d_n \vec{M}_n^{(1)} + c_n \vec{N}_n^{(1)}\right] \tag{7.25}$$

$$\vec{H}_{trans} = -\frac{E_0}{\omega\mu'\sin\zeta} \sum_{n=-\infty}^{\infty} (-i)^n \left[d_n \vec{M}_n^{(1)} + c_n \vec{N}_n^{(1)}\right] \tag{7.26}$$

and after some manipulations, we find that for this polarization

$$a_n = -\frac{A_n V_n - iC_n D_n}{W_n V_n + i(Dn)^2} \qquad b_n = -i\frac{W_n C_n + iD_n A_n}{W_n V_n + i(Dn)^2} \tag{7.27}$$

with $A_n = i\xi\left[\xi J'_n(\eta)J_n(\xi) - \eta J_n(\eta)J'_n(\xi)\right]$ and the others the same as before. At normal incidence b_n vanishes and we have

$$a_n = \frac{mJ_n'(x)J_n(mx) - J_n(x)J_n'(mx)}{mJ_n(mx)H_n'^{(1)}(x) - J_n'(mx)H_n^{(1)}(x)} \tag{7.28}$$

OK, so now we have exact expressions for the electric and magnetic fields scattered by an infinite circular cylinder. You should be able to code them up and make some plots to explore how the scattering behavior changes as the size parameter, Ka, and index of refraction, m, are varied. The problem you'll almost immediately face is whether or not your results are correct. Bessel and Hankel functions are a little unfamiliar, and there's always the chance that a typo or three has crept in along the way. It's important to have classic results to compare against. To wit, Figure 7.2 shows the special case where the cylinder is a perfect electrical conductor.

But, of course, the question is what exactly is being plotted? Backscattering means $\phi = 180°$, but the key thing to know is that the Hankel function has been replaced with its large-argument approximation:

$$H_n^{(1)}(Kr) \rightarrow \sqrt{\frac{2}{\pi Kr}} e^{iKr}(-i)^n e^{-i\pi/4} \tag{7.29}$$

and so people typically don't plot things as a function of Kr because this is the only place it shows up in the expression for the scattered field. Think about the 1D problems we considered, where it was the reflection and transmission coefficients that we cared about. Sometimes people would plot the absolute value of these and call them reflectance and transmission, but some authors plot the magnitude squared. The backscatter cross section in Figure 7.2 is probably the magnitude squared of what's left when you plug in $\phi = 180°$ and then invoke the large-argument approximation for the Hankel function to yank the r-dependence out from the infinite summation. There are probably also some normalization factors. Everybody seems to do this a bit differently.

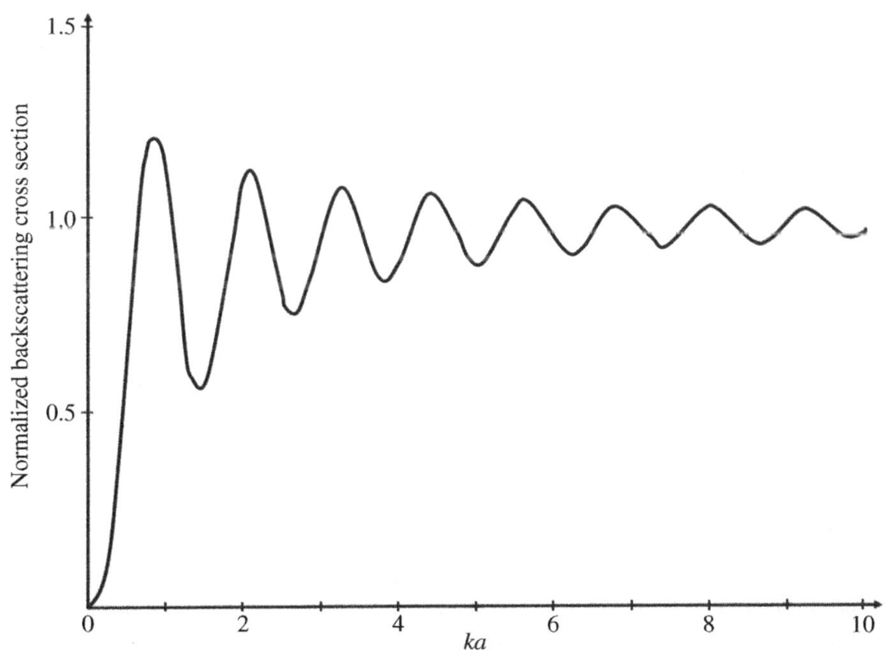

Figure 7.2 Backscattering from a perfectly electrically conducting (PEC) cylinder, with the electric field polarized perpendicular to the cylinder axis. The plot for parallel polarization isn't very interesting, so it's not shown here.

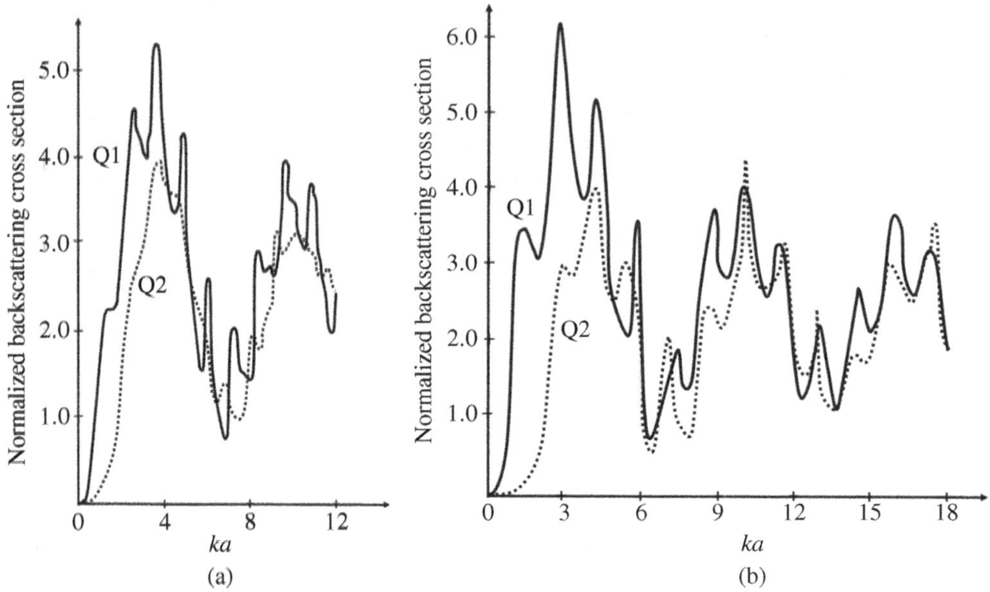

Figure 7.3 Normalized scattering cross sections for a dielectric cylinder with $m = 2.0$ (a) and $m = 2.5$ (b) for both polarizations of the normally-incident plane wave.

Exercise 7.2 Go back and repeat the aforementioned analysis but with the cylinder a perfect conductor, and then reproduce the plot in Figure 7.2. You may also then want to make the corresponding plot for the other polarization to confirm/rebut my assertion that it's boring.

In Figure 7.3, I've sketched from Kerker [3] the backscatter cross sections for dielectric cylinders with index of refraction $m = 2$ and $m = 2.5$. Both polarizations are shown, although I haven't indicated which is the parallel and which is perpendicular. I'm hoping that you'll be interested enough to reproduce the plots, figuring out by trial and error exactly what Kerker is plotting and what normalization factors he's using. I'm deliberately being a little coy about such things because it's useful to be able to figure that sort of thing out as you're getting your code running and then verifying it.

One more bit of esoterica. Our summation was over n from $-\infty$ to ∞, but many authors sum over n from zero to ∞, which is equivalent because they introduce the symbol ϵ_n that has the value 1 when $n = 0$ and 2 otherwise. They will then write $\sin n\phi$ or $\cos n\phi$ instead of $e^{in\phi}$.

Another common thing to plot is the scattering cross section. Bear with me for just a minute. The Poynting vector is $\vec{S} = \frac{1}{2}\Re\left\{\vec{E}\times\vec{H}^*\right\}$ and if we integrate the Poynting vector for the scattered field of a unit length over a cylindrical surface at a large distance from the cylinder and suitably normalize it, we get the scattering cross section

$$Q^{scat} = \frac{2}{(Ka)^2}\sum_{n=0}^{\infty}\epsilon_n\left(|a_n|^2 + |b_n|^2\right) \tag{7.30}$$

which can then be plotted as a function of Ka for various values of the index of refraction. If the index of refraction is complex, then typically it will be the extinction cross section which is plotted, in order to account for scattering plus absorption due to the lossy scatterer. The extinction cross sections will be

$$Q_z^{ext} = \frac{2}{(Ka)^2}\Re\sum_{n=0}^{\infty}\epsilon_n a_n \qquad Q_y^{ext} = \frac{2}{(Ka)^2}\Re\sum_{n=0}^{\infty}\epsilon_n b_n \tag{7.31}$$

When there is no absorption because the index of refraction is real, the corresponding scattering and extinction cross sections are equivalent. This often provides a convenient consistency check on your results. One last simplification and then you can go make some plots. Here are the scattering cross sections for normal incidence on a perfectly conducting cylinder, for example, a power line at radar frequencies. For the z-polarized incident electric field

$$\sigma = \frac{4}{K}\sum_{n=0}^{\infty}\epsilon_n\left|\frac{J_n(Ka)}{H_n^{(1)}(Ka)}\right|^2 \tag{7.32}$$

and for the y-polarized incident electric field

$$\sigma = \frac{4}{K}\sum_{n=0}^{\infty}\epsilon_n\left|\frac{J_n'(Ka)}{H'^{(1)}_n(Ka)}\right|^2 \tag{7.33}$$

I hope you noticed that for the scattering cross sections, the magnitude-squared happens inside the infinite summation. For the backscatter cross section, the magnitude squared computation happens outside the infinite summation. If you are having trouble matching your plots to the ones I included earlier, double check that isn't an issue. Once you do get your code up and running and validated by matching these sorts of plots from standard texts, go ahead and make a bunch more plots. I haven't included any for complex index of refraction, but you might go back to the sphere chapter and try some of those values to see whether scattering from cylinders is akin to scattering from spheres. It might be. It might not be.

7.2 Elastic Wave Scattering

Because solid materials which are opaque to light are transparent to ultrasonic (elastic) waves, ultrasound has been found to be a powerful tool for noninvasively probing the interior of structural materials. Ultrasound is routinely used to detect, locate, and qualitatively characterize defects such as voids, cracks, and inclusions. It is commonly employed for manufacturing quality control and on-line monitoring of manufacturing processes, as well as the safety and integrity of structural members. In advanced composite materials ultrasound is used to detect delaminations, irregularities in reinforcing fibers or particles, as well as unwanted inclusions or voids. Ultrasound has advantages over competing nondestructive evaluation technologies in that the required equipment is relatively portable, simple to operate and inexpensive, it is safe for the operator, and can be used for a variety of test specimen shapes.

A quantitative theory of ultrasonic nondestructive evaluation is necessarily more complex than corresponding electromagnetic or pure acoustic wave propagation and scattering theories. This is because in the propagation and scattering of elastic waves, a number of other effects must be considered in a complete and physically accurate theory. Mode-converted waves are generated during reflection and refraction at material interfaces as well as during diffraction from sharp discontinuities such as crack tips. From an analytical standpoint, difficulties in the rigorous solution of even the simplest of forward problems in ultrasonic NDE (given an input transducer signal and a specific defect shape, what is the output transducer signal) in general prevents the quantitative solution of the inverse problem (given the input and output transducer signals, what is the shape and location of the material defect), which is the real aim of NDE.

The widespread introduction of advanced composite materials created a need for further development of the theoretical models for describing bulk mechanical properties and prediction of material deterioration and failure. Advanced composite materials and structural members present unique difficulties for the prediction of both their mechanical properties and their probable modes of failure. Although phenomenological and experiment-based models abound, it is most desirable – and in the long run, most accurate and least expensive – to develop field-theoretical models which predict bulk properties and failure of advanced composites directly from the known material configuration (microstructural) parameters. The scattering of elastic waves from fibers embedded in an infinite elastic medium of different elastic material parameters can be deployed to study the ultrasonic wave interactions in fiber-reinforced organic-matrix composite materials. Unlike inviscid fluids, elastic solids have rigidity – they resist shear or distortional deformations as well as compressional deformations – and as a consequence, shear or transverse waves propagate in solid elastic media in addition to the familiar compressional sound waves. Neglecting shear waves in the ultrasonic nondestructive evaluation of solids, for example, leads to incorrect results, because when purely compressional ultrasonic waves encounter a material discontinuity where reflection, refraction, diffraction, etc. occur, the resulting *scattered* waves are in general always combinations of compressional and shear waves. This mode coupling is a direct consequence of the boundary conditions, which must be satisfied at the material discontinuity and cannot be avoided or ignored in a correct description of wave phenomena in solid media. For example, when ultrasound is used to detect flaws in homogeneous and isotropic materials, such as metals whose grain structure can be ignored for many frequencies, although the incident wave is purely compressional, the field scattered from the defect is part compressional and part shear. When there is no scattering, the two types of waves do propagate independently, with different wave velocities, and can be considered separately. However, any time there is scattering both types of waves and their mode coupling at the boundary must be considered.

In reinforced composite materials, coatings on the reinforcements and interface layers between the fibers and the binder are present. The coatings may be purposefully applied in order to enhance material lifetime, and curing processes generate an interface layer between the two media. This is important because the coating/interface alters the stresses surrounding the reinforcements, and if designed properly may be useful in minimizing material failure. More importantly, the interfacial properties change with environmental deterioration of fiber-reinforced organic matrix composites.

Recall that for an isotropic, homogeneous, and linearly elastic medium, the equation of motion is

$$\rho\partial_t^2\vec{u} - \mu\nabla^2\vec{u} - (\mu+\lambda)\nabla(\nabla\cdot\vec{u}) = 0 \tag{7.34}$$

where ρ is the constant medium density, μ and λ are elastic Lamé parameters and $\vec{u}$ is the displacement vector. After some manipulation and assuming harmonic time variation $e^{-i\omega t}$, the vector wave equation may be written as:

$$(\nabla^2 + K^2)\vec{u} - \left(1 - \frac{K^2}{k^2}\right)\nabla(\nabla\cdot\vec{u}) = 0 \tag{7.35}$$

Here $K = \omega/c_T$ and $k = \omega/c_L$ are the propagation constants for transverse and longitudinal elastic waves with $c_T = \sqrt{\mu/\rho}$ and $c_L = \sqrt{(\lambda+2\mu)/\rho}$ defining the transverse and longitudinal wave speeds, respectively.

Of course, we don't want to deal with the vector wave equation, so we invoke Clebsch's theorem, which says that any vector field can be written as the sum of a longitudinal and a transverse part $\vec{u} = \vec{u}_L + \vec{u}_T$. The longitudinal vector field has no curl and the transverse vector field has no divergence. This will allow us to write the longitudinal field as the gradient of a scalar generating function and the two polarizations of the transverse field as the curl and then curl-curl of other generating functions. This is exactly analogous to what we did for elastic wave scattering from spheres.

Both the reinforcing fibers (cylindrical scatterers) and the binder of the composite will be considered to be elastic solids for which the aforementioned equations hold. The binder will be referred to as medium 1 and the fibers (cylinders of radius $r = a$) will be medium 2. Later we'll briefly consider the region $a < r < b$ to be a distinct third elastic medium representing an interface layer between the fiber and the matrix.

We take a cylindrical coordinate system (r, θ, z) and consider an infinite elastic cylinder, whose axis coincides with the z-axis so that the cylindrical surface coincides with constant coordinate surface $r = a$.

For scattering from cylindrical inclusions, we define three scalar generating functions $\Pi_L, \Pi_{SH}, \Pi_{SV}$ in terms of the displacements as:

$$\begin{aligned}
\vec{u}_L &= \frac{1}{k}\nabla\Pi_L & \left(\nabla^2 + k^2\right)\Pi_L &= 0 \\
\vec{u}_{SV} &= \frac{1}{K}\nabla \times (\hat{z}\Pi_{SV}) & \left(\nabla^2 + K^2\right)\Pi_{SV} &= 0 \\
\vec{u}_{SH} &= \left(\frac{1}{K}\right)\nabla \times \nabla \times (\hat{z}\Pi_{SH}) & \left(\nabla^2 + K^2\right)\Pi_{SH} &= 0
\end{aligned} \tag{7.36}$$

where the displacement is a linear combination of longitudinal and transverse displacement modes $\vec{u} = \vec{u}_L + \vec{u}_{SH} + \vec{u}_{SV}$.

The displacement and stress components can be found from

$$\begin{aligned}
u_r &= \left(\frac{-1}{k}\right)\frac{\partial}{\partial r}(\Pi_L) + \frac{1}{r}\frac{\partial}{\partial\theta}(\Pi_{SV}) + \frac{1}{K}\frac{\partial^2}{\partial r\partial z}(\Pi_{SH}) \\
u_\theta &= \left(\frac{-1}{k}\right)\frac{1}{r}\frac{\partial}{\partial\theta}(\Pi_L) + \frac{\partial}{\partial r}(\Pi_{SV}) - \frac{1}{K}\frac{1}{r}\frac{\partial^2}{\partial z\partial\theta}(\Pi_{SH}) \\
u_z &= \frac{-1}{k}\frac{\partial}{\partial z}(\Pi_L) + 0 - \frac{1}{K}\left(\frac{1}{r}\frac{\partial}{\partial r} + \frac{\partial^2}{\partial r^2} + \frac{\partial^2}{\partial\theta^2}\right)(\Pi_{SH})
\end{aligned} \tag{7.37}$$

$$\begin{aligned}
\sigma_{rr} &= (\lambda + 2\mu)\frac{\partial u_r}{\partial r} + \lambda\left(\frac{1}{r}\frac{\partial u_\theta}{\partial\theta} + \frac{u_r}{r} + \frac{\partial u_z}{\partial z}\right) \\
\sigma_{r\theta} &= \mu\left(\frac{1}{r}\frac{\partial u_r}{\partial\theta} + \frac{\partial u_\theta}{\partial r} - \frac{u_\theta}{r}\right) \\
\sigma_{rz} &= \mu\left(\frac{\partial u_z}{\partial r} + \frac{\partial u_r}{\partial z}\right)
\end{aligned} \tag{7.38}$$

It's probably worth writing the stress components out in terms of the scalar generating functions because we're going to need them when we apply the boundary conditions.

$$\begin{aligned}
\sigma_{rr} &= -\lambda\frac{1}{k}\nabla^2\Pi_L + 2\mu\left[-\frac{1}{k}\partial_r^2\Pi_L + \partial_r\left(\frac{1}{4}\partial_\theta\Pi_{SV}\right) + \frac{1}{K}\partial_r^2\partial_z\Pi_{SH}\right] \\
\sigma_{r\theta} &= 2\mu\left[-\frac{1}{kr}\partial_r\partial_\theta\Pi_L + \frac{1}{kr^2}\partial_\theta\Pi_L + \frac{1}{Kr}\partial_r\partial_\theta\partial_z\Pi_{SH} - \frac{1}{Kr^2}\partial_\theta\partial_z\Pi_{SH}\right] \\
&\quad +\mu\left[\frac{1}{r^2}\partial_\theta^2\Pi_{SV} - r\partial_r\left(\frac{1}{r}\partial_r\Pi_{SV}\right)\right] \\
\sigma_{rz} &= \mu\left[-\frac{2}{k}\partial_r\partial_z\Pi_L + \frac{1}{r}\partial_\theta\partial_z\Pi_{SV}\right] + \frac{\mu}{K}\left[2\partial_r\partial_z^2\Pi_{SH} - \partial_r\nabla^2\Pi_{SH}\right]
\end{aligned} \tag{7.39}$$

where

$$\nabla^2\Pi = \frac{1}{r}\partial_r(r\partial_r\Pi) + \frac{1}{r^2}\partial_\theta^2\Pi + \partial_z^2\Pi$$

Exercise 7.3 Although I've tried to be careful in type-setting the above equations, you should certainly check them to make sure that I haven't made a botch along the way. You could cross-check with [6–8].

The incident plane wave will be taken to be one of three types. For incident L-waves, we write

$$\vec{u}_L^{inc} = (\hat{x}\sin\phi + \hat{z}\cos\phi)\, e^{ik_1(x\sin\phi + z\cos\phi)} \tag{7.40}$$

from which we see that the directions of propagation and displacement are the same. For incident SV-waves, we write

$$\vec{u}_{SV}^{inc} = -i\hat{y}\sin\phi^{iK_1(r\sin\phi + z\cos\phi)} \tag{7.41}$$

and for incident SH-waves, we write

$$\vec{u}_{SH}^{inc} = \left(-\hat{x}\cos\phi\sin\phi + \hat{z}\sin^2\phi\right) e^{iK_1(x\sin\phi + z\cos\phi)} \tag{7.42}$$

For the two transverse waves, the displacement is *transverse* to the direction of propagation. Note that for normal incidence ($\phi = 90°$), we get

$$\vec{u}_L^{inc} = \hat{x}e^{ik_1x} \qquad \vec{u}_{SV}^{inc} = -i\hat{y}e^{iK_1x} \qquad \vec{u}_{SH} = \hat{z}e^{iK_zx} \tag{7.43}$$

The exponential terms can be expanded in terms of cylindrical Bessel functions according to

$$e^{i\alpha\cos\theta} = \sum_{n=-\infty}^{\infty} i_n J_n(\alpha) e^{in\theta} \tag{7.44}$$

so we can write the three incident plane wave potentials in terms of cylinder functions as:

$$\begin{aligned}
\Pi_L^{inc} &= \sum_n i^{n+1} J_n(k_1 r\sin\phi) e^{in\theta} e^{ik_1 z\cos\phi} \\
\Pi_{SV}^{inc} &= \frac{1}{K_1}\sum_n i^n J_n(K_1 r\sin\phi) e^{in\theta} E^{iK_1 z\cos\phi} \\
\Pi_{SH}^{inc} &= \frac{1}{K_1}\sum_n i^n J_n(K_1 r\sin\phi) e^{in\theta} e^{iK_1 z\cos\phi}
\end{aligned} \tag{7.45}$$

The boundary conditions for "welded contact" require that the displacements and the normal surface tractions be continuous at $r = a$. Since the field exterior to the cylinder is the sum of the incident and scattered fields and the field inside the scatterer is only the transmitted field, we write

$$\sigma_{rr}^{inc} + \sigma_{rr}^{scat} = \sigma_{rr}^{trans} \qquad \sigma_{r\theta}^{inc} + \sigma_{r\theta}^{scat} = \sigma_{r\theta}^{trans} \tag{7.46}$$

and

$$u_r^{inc} + u_r^{scat} = u_r^{trans} \qquad u_\theta^{inc} + u_\theta^{scat} = u_\theta^{trans} \tag{7.47}$$

where $\sigma_{rz}^{inc} + \sigma_{rz}^{scat} = \sigma_{rz}^{trans}$ and $u_z^{inc} + u_z^{scat} = u_z^{trans}$ are redundant and will be dropped.

7.2.1 Scattering Due to an Incident L-Wave

When a purely compressional plane wave of unit amplitude and with wave number k_1 is incident on the surface of the cylindrical scatterer, scattered as well as transmitted longitudinal *and* vertical shear modes will be generated. We can write the following expressions for the scalar potentials of the incident, scattered and transmitted waves

$$\Pi_L^{inc} = \sum_n i^{n+1} J_n(k_1 r\sin\phi) e^{in\theta} e^{ik_1 z\cos\phi}$$

$$
\begin{aligned}
\Pi_L^{scat} &= \sum_n i^{n+1} a_n^L H_n^{(1)}(k_1 r \sin\phi) e^{in\theta} e^{ik_1 z\cos\phi} \\
\Pi_{SV}^{scat} &= \frac{1}{K_1}\sum_n i^n b_n^L H_n^{(1)}(K_1 r \sin\phi) e^{in\theta} e^{iK_1 z\cos\phi} \\
\Pi_L^{trans} &= \sum_n i^{n+1} c_n^L J_n(k_2 r \sin\phi) e^{in\theta} e^{ik_2 z\cos\phi} \\
\Pi_{SV}^{trans} &= \frac{1}{K_2}\sum_n i^n d_n^L J_n(K_2 r \sin\phi) e^{in\theta} e^{iK_2 z\cos\phi}
\end{aligned}
\tag{7.48}
$$

Here a_n^L–d_n^L are four unknown modal coefficients to be determined from the continuity of stress and displacement at $r = a$. After some manipulations, we can write these four equations as a matrix equation. After some more manipulations, we can solve the matrix equation via Cramer's rule to find the modal coefficients. Here you go:

$$
\begin{pmatrix}
\frac{1}{2}(K_1a)^2 - n^2 + \frac{k_1 a H_n^{(1)\prime}(k_1a)}{H_n^{(1)}(k_1a)} & n\left(1 - \frac{K_1 a H_n^{(1)\prime}(K_1a)}{H_n^{(1)}(K_1a)}\right) & \frac{\mu_2}{\mu_1}\left[\frac{1}{2}(K_2a)^2 - n^2 + \frac{k_2 a J_n'(k_2a)}{J_n(k_2a)}\right] & n\frac{\mu_2}{\mu_1}\left(1 - \frac{K_2 a J_n'(K_2a)}{J_n(K_2a)}\right) \\
n\left(1 - \frac{k_1 a H_n^{(1)\prime}(k_1a)}{H_n^{(1)}(k_1a)}\right) & n^2 - \frac{1}{2}(K_1a)^2 - \frac{K_1 a H_n^{(1)\prime}(K_1a)}{H_n^{(1)}(K_1a)} & n\frac{\mu_2}{\mu_1}\left(1 - \frac{k_2 a J_n'(k_2a)}{J_n(k_2a)}\right) & \frac{\mu_2}{\mu_1}\left[n^2 - \frac{1}{2}(K_2a)^2 - \frac{K_2 a J_n'(K_2a)}{J_n(K_2a)}\right] \\
\frac{k_1 a H_n^{(1)\prime}(k_1a)}{H_n^{(1)}(k_1a)} & -n & \frac{k_2 a J_n'(k_2a)}{J_n(k_2a)} & -n \\
-n & \frac{K_1 a H_n^{(1)\prime}(K_1a)}{H_n^{(1)}(K_1a)} & -n & \frac{K_2 a J_n'(K_2a)}{J_n(K_2a)}
\end{pmatrix}
\begin{bmatrix}
a_n^L H_n^{(1)}(k_1a) \\
b_n^L H_n^{(1)}(K_1a) \\
-c_n^L J_n(k_2a) \\
-d_n^L J_n(K_2a)
\end{bmatrix}
= i^{n+1} J_n(k_1a)
\begin{bmatrix}
\frac{1}{2}(K_1a)^2 - n^2 + \frac{k_1 a J_n'(k_1a)}{J_n(k_1a)} \\
n\left(1 - \frac{k_1 a J_n'(k_1a)}{J_n(k_1a)}\right) \\
\frac{k_1 a J_n'(k_1a)}{J_n(k_1a)} \\
-n
\end{bmatrix}
\tag{7.49}
$$

$$
\begin{aligned}
\Delta_0 = -n^2\frac{\mu_2}{\mu_1}&\left(\frac{k_2aJ_n'(k_2a)}{J_n(k_2a)} - \frac{k_1aH_n^{(1)\prime}(k_1a)}{H_n^{(1)}(k_1a)}\right)\left[\frac{1}{2}(K_1a)^2\left(\frac{K_2aJ_n'(K_2a)}{J_n(K_2a)} - 1\right)\right. \\
&\left.-\frac{1}{2}(K_2a)^2\left(\frac{K_1aH_n^{(1)\prime}(K_1a)}{H_n^{(1)}(K_1a)} - 1\right)\right] - n^2\frac{\mu_2}{\mu_1}\left(\frac{K_2aJ_n'(K_2a)}{J_n(K_2a)} - \frac{K_1aH_n^{(1)\prime}(K_1a)}{H_n^{(1)}(K_1a)}\right) \\
&\times\left[\frac{1}{2}(K_1a)^2\left(\frac{k_2aJ_n'(k_2a)}{J_n(k_2a)} - 1\right) - \frac{1}{2}(K_2a)^2\left(\frac{k_1aH_n^{(1)\prime}(k_1a)}{H_n^{(1)}(k_1a)} - 1\right)\right] \\
&-\left(\frac{k_2aJ_n'(k_2a)}{J_n(k_2a)}\frac{K_2aJ_n'(K_2a)}{J_n(K_2a)} - n^2\right)\left\{n^2\left(\frac{k_1aH_n^{(1)\prime}(k_1a)}{H_n^{(1)}(k_1a)} - 1\right)\left(\frac{K_1aH_n^{(1)\prime}(K_1a)}{H_n^{(1)}(K_1a)} - 1\right)\right. \\
&\left.-\left[\frac{k_1aH_n^{(1)\prime}(k_1a)}{H_n^{(1)}(k_1a)} + \frac{1}{2}(K_1a)^2 - n^2\right]\left[\frac{K_1aH_n^{(1)\prime}(K_1a)}{H_n^{(1)}(K_1a)} + \frac{1}{2}(K_1a)^2 - n^2\right]\right\}
\end{aligned}
$$

$$+\frac{\mu_2}{\mu_1}\left(\frac{k_2aJ_n'(k_2a)}{J_n(k_2a)}\frac{K_1aH_n^{(1)\prime}(K_1a)}{H_n^{(1)}(K_1a)}-n^2\right)\left\{n^2\left(\frac{k_1aH_n^{(1)\prime}(k_1a)}{H_n^{(1)}(k_1a)}-1\right)\left(\frac{K_2aJ_n'(K_2a)}{J_n(K_2a)}-1\right)\right.$$

$$\left.-\left[\frac{k_1aH_n^{(1)\prime}(k_1a)}{H_n^{(1)}(k_1a)}+\frac{1}{2}(K_1a)^2-n^2\right]\left[\frac{K_2aJ_n'(K_2a)}{J_n(K_2a)}+\frac{1}{2}(K_2a)^2-n^2\right]\right\}$$

$$+\frac{\mu_2}{\mu_1}\left(\frac{k_1aH_n^{(1)\prime}(k_1a)}{H_n^{(1)}(k_1a)}\frac{K_2aJ_n'(K_2a)}{J_n(K_2a)}-n^2\right)\left\{n^2\left(\frac{k_2aJ_n'(k_2a)}{J_n(k_2a)}-1\right)\left(\frac{K_1aH_n^{(1)\prime}(K_1a)}{H_n^{(1)}(K_1a)}-1\right)\right.$$

$$\left.-\left[\frac{k_2aJ_n'(k_2a)}{J_n(k_2a)}+\frac{1}{2}(K_2a)^2-n^2\right]\left[\frac{K_1aH_n^{(1)\prime}(K_1a)}{H_n^{(1)}(K_1a)}+\frac{1}{2}(K_1a)^2-n^2\right]\right\}$$

$$-\left(\frac{\mu_2}{\mu_1}\right)^2\left(\frac{k_1aH_n^{(1)\prime}(k_1a)}{H_n^{(1)}(k_1a)}\frac{K_1aH_n^{(1)\prime}(K_1a)}{H_n^{(1)}(K_1a)}-n^2\right)\left\{n^2\left(\frac{k_2aJ_n'(k_2a)}{J_n(k_2a)}-1\right)\right.$$

$$\left.\times\left(\frac{K_2aJ_n'(K_2a)}{J_n(K_2a)}-1\right)-\left[\frac{k_2aJ_n'(k_2a)}{J_n(k_2a)}+\frac{1}{2}(K_2a)^2-n^2\right]\left[\frac{K_2aJ_n'(K_2a)}{J_n(K_2a)}+\frac{1}{2}(K_2a)^2-n^2\right]\right\}$$

$$a_n^L=\left\{-n^2\frac{\mu_2}{\mu_1}\left(\frac{k_2aJ_n'(k_2a)}{J_n(k_2a)}-\frac{k_1aJ_n'(k_1a)}{J_n(k_1a)}\right)\left[\frac{1}{2}(K_1a)^2\left(\frac{K_2aJ_n'(K_2a)}{J_n(K_2a)}-1\right)\right.\right.$$

$$\left.-\frac{1}{2}(K_2a)^2\left(\frac{K_1aH_n^{(1)\prime}(K_1a)}{H_n^{(1)}(K_1a)}-1\right)\right]-n^2\frac{\mu_2}{\mu_1}\left(\frac{K_2aJ_n'(K_2a)}{J_n(K_2a)}-\frac{K_1aH_n^{(1)\prime}(K_1a)}{H_n^{(1)}(K_1a)}\right)$$

$$\times\left[\frac{1}{2}(K_1a)^2\left(\frac{k_2aJ_n'(k_2a)}{J_n(k_2a)}-1\right)-\frac{1}{2}(K_2a)^2\left(\frac{k_1aJ_n'(k_1a)}{J_n(k_1a)}-1\right)\right]$$

$$-\left(\frac{k_2aJ_n'(k_2a)}{J_n(k_2a)}\frac{K_2aJ_n'(K_2a)}{J_n(K_2a)}-n^2\right)\left\{n^2\left(\frac{k_1aJ_n'(k_1a)}{J_n(k_1a)}-1\right)\left(\frac{K_1aH_n^{(1)\prime}(K_1a)}{H_n^{(1)}(K_1a)}-1\right)\right.$$

$$\left.-\left[\frac{k_1aJ_n'(k_1a)}{J_n(k_1a)}+\frac{1}{2}(K_1a)^2-n^2\right]\left[\frac{K_1aH_n^{(1)\prime}(K_1a)}{H_n^{(1)}(K_1a)}+\frac{1}{2}(K_1a)^2-n^2\right]\right\}$$

$$+\frac{\mu_2}{\mu_1}\left(\frac{k_2aJ_n'(k_2a)}{J_n(k_2a)}\frac{K_1aH_n^{(1)\prime}(K_1a)}{H_n^{(1)}(K_1a)}-n^2\right)\left\{n^2\left(\frac{k_1aJ_n'(k_1a)}{J_n(k_1a)}-1\right)\left(\frac{K_2aJ_n'(K_2a)}{J_n(K_2a)}-1\right)\right.$$

$$\left.-\left[\frac{k_1aJ_n'(k_1a)}{J_n(k_1a)}+\frac{1}{2}(K_1a)^2-n^2\right]\left[\frac{K_2aJ_n'(K_2a)}{J_n(K_2a)}+\frac{1}{2}(K_2a)^2-n^2\right]\right\}$$

$$+\frac{\mu_2}{\mu_1}\left(\frac{k_1aJ_n'(k_1a)}{J_n(k_1a)}\frac{K_2aJ_n'(K_2a)}{J_n(K_2a)}-n^2\right)\left\{n^2\left(\frac{k_2aJ_n'(k_2a)}{J_n(k_2a)}-1\right)\left(\frac{K_1aH_n^{(1)\prime}(K_1a)}{H_n^{(1)}(K_1a)}-1\right)\right.$$

$$\left.-\left[\frac{k_2aJ_n'(k_2a)}{J_n(k_2a)}+\frac{1}{2}(K_2a)^2-n^2\right]\left[\frac{K_1aH_n^{(1)\prime}(K_1a)}{H_n^{(1)}(K_1a)}+\frac{1}{2}(K_1a)^2-n^2\right]\right\}$$

$$-\left(\frac{\mu_2}{\mu_1}\right)^2\left(\frac{k_1aJ_n'(k_1a)}{J_n(k_1a)}\frac{K_1aH_n^{(1)\prime}(K_1a)}{H_n^{(1)}(K_1a)}-n^2\right)\left\{n^2\left(\frac{k_2aJ_n'(k_2a)}{J_n(k_2a)}-1\right)\left(\frac{K_2aJ_n'(K_2a)}{J_n(K_2a)}-1\right)\right.$$

$$\left.\left.-\left[\frac{k_2aJ_n'(k_2a)}{J_n(k_2a)}+\frac{1}{2}(K_2a)^2-n^2\right]\left[\frac{K_2aJ_n'(K_2a)}{J_n(K_2a)}+\frac{1}{2}(K_2a)^2-n^2\right]\right\}\right\}\frac{J_n(k_1a)}{H_n^{(1)}(k_1a)}\frac{1}{\Delta_0}$$

$$b_n^L = n\left(\frac{K_1aH_n^{(1)\prime}(K_1a)}{H_n^{(1)}(K_1a)} - \frac{K_1aJ_n'(K_1a)}{J_n(K_1a)}\right)\left\{\left(\frac{\mu_2}{\mu_1}\right)^2\left[n^2\left(\frac{k_2aJ_n'(k_2a)}{J_n(k_2a)} - 1\right)\left(\frac{K_2aJ_n'(K_2a)}{J_n(K_2a)} - 1\right)\right.\right.$$

$$\left. - \left(\frac{k_2aJ_n'(k_2a)}{J_n(k_2a)} + \frac{1}{2}(K_2a)^2 - n^2\right)\left(\frac{K_2aJ_n'(K_2a)}{J_n(K_2a)} + \frac{1}{2}(K_2a)^2 - n^2\right)\right]$$

$$+ \frac{\mu_2}{\mu_1}\left[\left(\frac{k_2aJ_n'(k_2a)}{J_n(k_2a)} + \frac{1}{2}(K_2a)^2 - n^2\right)\left(n^2 - \frac{K_2aJ_n'(K_2a)}{J_n(K_2a)}\right)\right.$$

$$\left. - \left(\frac{K_2aJ_n'(K_2a)}{J_n(K_2a)} + \frac{1}{2}(K_2a)^2 - n^2\right)\left(n^2 - \frac{k_2aJ_n'(k_2a)}{J_n(k_2a)}\right)\right]$$

$$+ \frac{\mu_2}{\mu_1}\frac{1}{2}(K_1a)^2\left(\frac{1}{2}(K_2a)^2 + \frac{k_2aJ_n'(k_2a)}{J_n(k_2a)}\frac{K_2aJ_n'(K_2a)}{J_n(K_2a)} - n^2\right)$$

$$\left. - \left(\frac{k_2aJ_n'(k_2a)}{J_n(k_2a)}\frac{K_2aJ_n'(K_2a)}{J_n(K_2a)} - n^2\right)\left(\frac{1}{2}(K_1a)^2 + 1 - n^2\right)\right\}\frac{J_n(K_1a)}{H_n^{(1)}(k_1a)}\frac{1}{\Delta_0}$$

I haven't included the transmitted wave amplitudes, c_n^L and d_n^L but they can be found here [8]. In addition, note that there are no SH modes present because they are not coupled to the L and SV waves during scattering.

Recall that for the electromagnetic scattering case, we calculated the Poynting vector. Here, the total flow of scattered elastic energy in the radial direction through a closed surface of unit length in z can be represented by the radial component of the energy flux vector $\mathcal{F}_j = \sigma_{ij}\partial_t u_j$, which may be decomposed into longitudinal and transverse parts.

The differential scattering cross sections are then defined as the time-averaged radial component of the far-field energy flux vector, normalized by the intensity of the incident wave. These are often plotted vs. θ to show the angular variation of the scattering. The backscatter cross section is the differential scattering cross section for $\theta = 180°$. There is no mode-coupled term for the backscatter cross section.

Integrating over a cylindrical surface with radius $r \gg a$ and dividing by the geometric cross section (πa^2), we obtain the normalized scattering cross section for a unit length of the cylinder

$$q_{scat}^L = \frac{4}{(k_1a)^2}\sum_n\left(\left|\frac{\Delta_1^L}{\Delta_0}\right|^2 + \left(\frac{k_1}{K_1}\right)^3\left|\frac{\Delta_3^L}{\Delta_0}\right|^2\right) \tag{7.50}$$

The extinction cross section also accounts for energy removed via absorption in the scatterer

$$q_{ext}^L = \frac{4}{(k_1a)^2}\sum_n\Re\left(\frac{\Delta_1^L}{\Delta_0}\right) \tag{7.51}$$

and in the absence of absorption $q_{scat} = q_{ext}$, which can provide a helpful consistency check given the complex equations for the Δ's involved.

You would be within your rights to ask a question along the lines of: How many terms in the infinite summation are needed? There are rules-of-thumb worked out in the days when computers where slow and precision was limited. Something along the lines of $n = x = 4x^{1/3} + 2$, where x is the size parameter of the cylinder, might work pretty well. The series will converge, so run things a bit and see what works best. It's not like you have to pedal the computer.

I assume you noticed that at some point during the aforementioned derivations I seem to have stopped considering oblique incidence, that is, began assuming that $\phi = 90°$. I could have argued that things are complicated enough without that, or that most of the interesting physical behavior is in the 2D problem. Both would be true, but then you might wonder why I didn't just start out with the $\phi = 90°$ case and try to improve the clarity of the exposition. Good point. There is some literature on this subject, of course, but most authors only show results for the 2D case. It might be wishful thinking to assume that not much happens for angles close to normal incidence because we remember a few chapters ago when our assumptions about reflection from an interface broke down beyond some critical angle when surface waves were generated. Here, we might expect leaky pseudo-Stoneley waves that travel on the matrix-cylinder interface to show up in some cases.

Exercise 7.4 Go back through the analysis and see if you can figure out if there's some sort of critical angle ϕ_{crit} at which the way we've formulated the scattering problem breaks down.

There are presumably many applications where oblique-incidence scattering from a cylindrical inclusion would matter. Figure 7.4 has various plots of backscattering from cylinders done as part of work to model Integrated Backscatter (IB) ultrasound measurements of trans-laminar reinforced (TLR) composites [9] for aerospace applications. In order to make composite structures more damage tolerant, that is, tougher, through-thickness stitches or pins are used to help prevent delaminations after impact or other in-service wear and tear. The IB method, adapted from cardiac applications of medical ultrasonography, integrates the backscattered signal over a range of angles in order to produce a "parameter image" with the IB as an area is scanned. The clinical machines would have a button that the sonographer could push to show either the standard B-mode image or the IB parameter image and/or other parameter images. In our paradigm, the IB is a rudimentary form of data engineering and it seems a little silly to be toggling back and forth between different images individually formed from the collection of RF waveforms, but that's just because we've gotten used to the idea of ML systems being able to deal with lots and lots of data as inputs. Clinicians need to have the amount of information presented to them strictly limited to the capacity of the human visual system. Don't get too snooty though. Your fancy data-engineered, machine-learning expert system isn't licensed to practice medicine and therefore cannot, by law, make any diagnoses. The first AI system for reading mammograms got slapped down hard when they made claims about computer-aided diagnosis, CAD_X. They had to back track and only claim to suggest to the clinician spots that might be suspicious and where they could maybe take a second look with their licensed human eyeballs.

In order to compare models with experiments (for nonsentient composites) a series of hockey-puck-sized samples were produced with various fibers, wires, etc. molded into them. The backscatter was measured as the samples were rotated, and we found that there wasn't all that much interesting behavior once we integrated, which is why I'm just showing you the backscatter vs. *ka* for normal incidence later.

7.2.2 Scattering of Acoustic Waves from an Elastic Cylinder

If we set the shear modulus μ_1 equal to zero in the medium surrounding the cylinder, no shear waves will be present in the scattered field and our results reduce to those for an ordinary acoustic

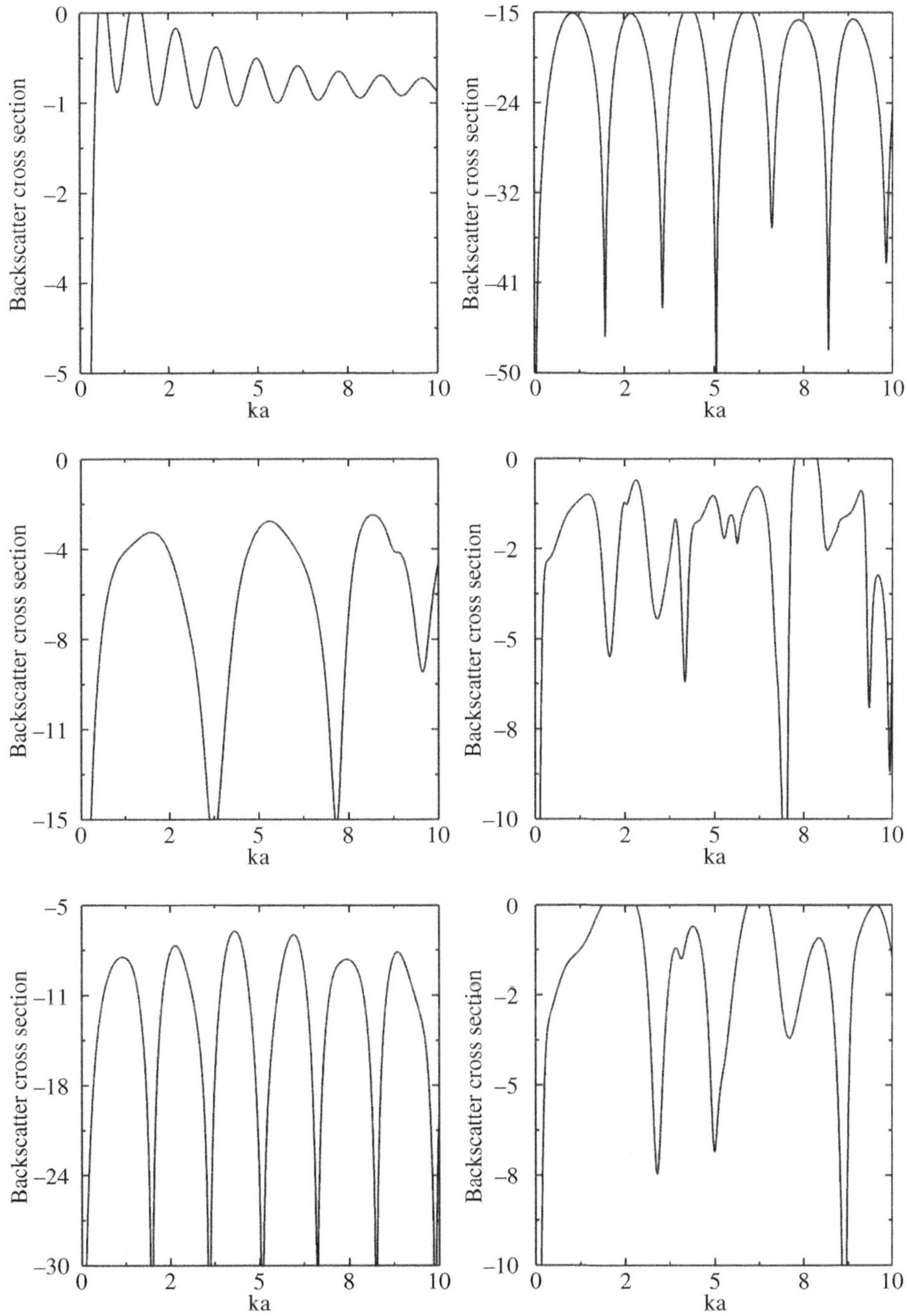

Figure 7.4 Backscattering from elastic cylinders. (a) is for an air-filled hole in plexiglas. (b) is for a glass fiber in acrylic. (c) is for a carbon fiber in acrylic. (d) is for a Kevlar fiber in lucite. (e) is for a copper wire in lucite. (f) is for a chrome wire in lucite.

wave scattering from an elastic cylinder. In this case $b_n^L = 0$ and

$$\Delta_0 = \frac{k_1 a H_n^{(1)\prime}(k_1 a)}{H_n^{(1)}(k_1 a)} \left\{ n^2 \left(\frac{k_2 a J_n'(k_2 a)}{J_n(k_2 a)} - 1 \right) \left(\frac{K_2 a J_n'(K_2 a)}{J_n(K_2 a)} - 1 \right) \right.$$
$$\left. - \left[\frac{k_2 a J_n'(k_2 a)}{J_n(k_2 a)} + \frac{1}{2}(K_2 a)^2 - n^2 \right] \left[\frac{K_2 a J_n'(K_2 a)}{J_n(K_2 a)} + \frac{1}{2}(K_2 a)^2 - n^2 \right] \right\}$$
$$-\frac{1}{2}(K_2 a)^2 \frac{\rho_1}{\rho_2} \left\{ \frac{k_2 a J_n'(k_2 a)}{J_n(k_2 a)} \left[\frac{K_2 a J_n'(K_2 a)}{J_n(K_2 a)} + \frac{1}{2}(K_2 a)^2 - n^2 \right] - n^2 \left(\frac{k_2 a J_n'(k_2 a)}{J_n(k_2 a)} - 1 \right) \right\}$$

$$a_n^L = \frac{J_n(k_1 a)}{H_n^{(1)}(k_1 a)} \left\{ \frac{k_1 a J_n'(k_1 a)}{J_n(k_1 a)} \left\{ n^2 \left(\frac{k_2 a J_n'(k_2 a)}{J_n(k_2 a)} - 1 \right) \left(\frac{K_2 a J_n'(K_2 a)}{J_n(K_2 a)} - 1 \right) \right. \right.$$
$$\left. - \left[\frac{k_2 a J_n'(k_2 a)}{J_n(k_2 a)} + \frac{1}{2}(K_2 a)^2 - n^2 \right] \left[\frac{K_2 a J_n'(K_2 a)}{J_n(K_2 a)} + \frac{1}{2}(K_2 a)^2 - n^2 \right] \right\}$$
$$\left. -\frac{1}{2}(K_2 a)^2 \frac{\rho_1}{\rho_2} \left\{ \frac{k_2 a J_n'(k_2 a)}{J_n(k_2 a)} \left[\frac{K_2 a J_n'(K_2 a)}{J_n(K_2 a)} + \frac{1}{2}(K_2 a)^2 - n^2 \right] - n^2 \left(\frac{k_2 a J_n'(k_2 a)}{J_n(k_2 a)} - 1 \right) \right\} \right\} \frac{1}{\Delta_0}$$

This agrees with the results of James Faran[2] who first derived the exact solution for acoustic wave scattering from an elastic cylinder [10]. Faran's scattering patterns for various metal cylinders in water are reproduced in Figures 7.5 and 7.6. We can be fairly confident because this paper has been cited more than 1200 times since 1951. Figure 7.7 illustrates an application to autonomous robotics.

The consistency of the results can be further checked by considering the further limit where the shear moduli of both media vanish, which returns the result for an acoustic wave scattering from an acoustic cylinder. We get $b_n^L = 0$ and

$$a_n^L = \frac{\rho_1 k_2 J_n(k_1 a) J_n'(k_2 a) - \rho_2 k_1 J_n(k_1 a) J_n'(k_2 a)}{\rho_1 k_2 H_n^{(1)}(k_1 a) J_n'(k_2 a) - \rho_2 k_1 H_n^{(1)}(k_1 a) J_n'(k_2 a)} \tag{7.52}$$

At this point, you're probably thinking that this special case of a fluid cylinder in another fluid has no physical application. While spherical bubbles in air or liquid drops in water or even immiscible liquids which don't mix, for example, oil and water, all make good sense, this one doesn't. Except that in medical ultrasound it's common to treat soft tissues acoustically, meaning that their solid nature doesn't affect things very much. Hence, a cylindrical blood vessel in the body could be treated as a cylinder of fluid in another fluid. There are innumerable clinical applications of this, of course, but let me mention just one. Angiogenesis is the development of new blood vessels. Cancerous tumors are especially crafty at causing angiogenesis in order to feed their rapacious growth. Mapping out these blood vessels can be important because the number and tortuosity of them can indicate malignancy. Remember, we're all about data engineering here. Using scattering analysis to help clinicians extract diagnostic information from medical ultrasound is exactly that.

---

2 Originally from Glendale, Ohio, James Faran graduated from Washington & Jefferson College and the Harvard School of Engineering. During World War II he worked in the Harvard Underwater Sound Laboratory, after which he worked as an electronics engineer for the General Radio Company of West Concord, MA for 36 years designing computer test equipment. Jim Faran played a variety of brass instruments including trombone, baritone horn, and tuba. He marched with the Concord Band, and was the author of an article published in Horizon magazine called "How to Buy a Tuba." He died at age 87 in 2008.

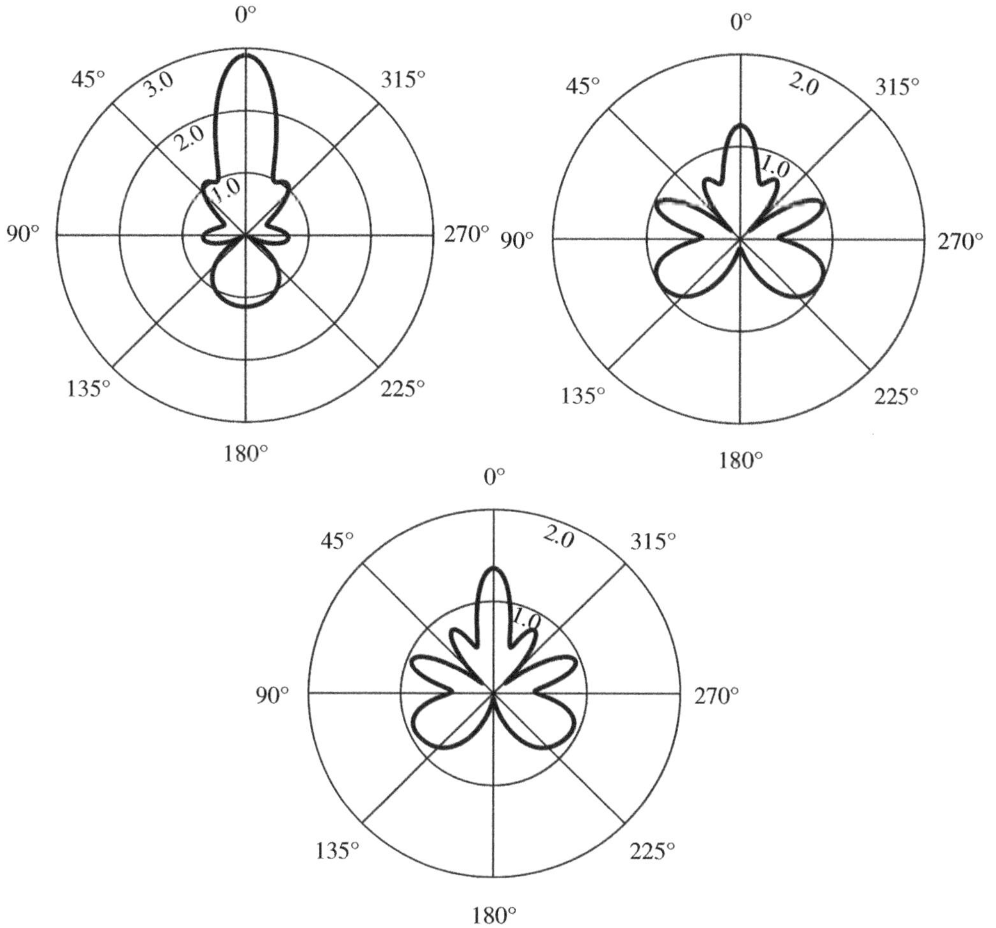

Figure 7.5 Scattering patterns for 0.093 in diameter brass (a) 0.09375 in diameter steel (b) and 0.0925 diameter aluminum (c) cylinders, with $f = 1$ MHz.

7.2.3 Scattering Due to an Incident T-Wave

If a transverse wave with wave number $K_1 = \omega\sqrt{\rho_1/\mu_1}$ is incident upon the surface of the cylinder, scattered and transmitted longitudinal, vertical shear, and horizontal shear waves will be excited. We can follow the same solution procedure as before, and it won't really even be all that much more complicated because only the L and SV will be coupled this time too. The SH waves will be all by themselves and we can do that algebra separately.

The incident, scattered, and transmitted waves are represented by

$$\Pi_{SV}^{inc} = \frac{1}{K_1}\sum_n i^n J_n(K_1 r \sin\phi)e^{in\theta}e^{iK_1 z\cos\phi}$$

$$\Pi_L^{scat} = \sum_n i^{n+1} a_n^T H_n^{(1)}(k_1 r \sin\phi)e^{in\theta}e^{ik_1 z\cos\phi}$$

$$\Pi_{SV}^{scat} = \frac{1}{K_1}\sum_n i^n b_n^T H_n^{(1)}(K_1 r \sin\phi)e^{in\theta}e^{iK_1 z\cos\phi}$$

$$\Pi_L^{trans} = \sum_n i^{n+1} c_n^T J_n(k_2 r \sin\phi)e^{in\theta}e^{ik_2 z\cos\phi}$$

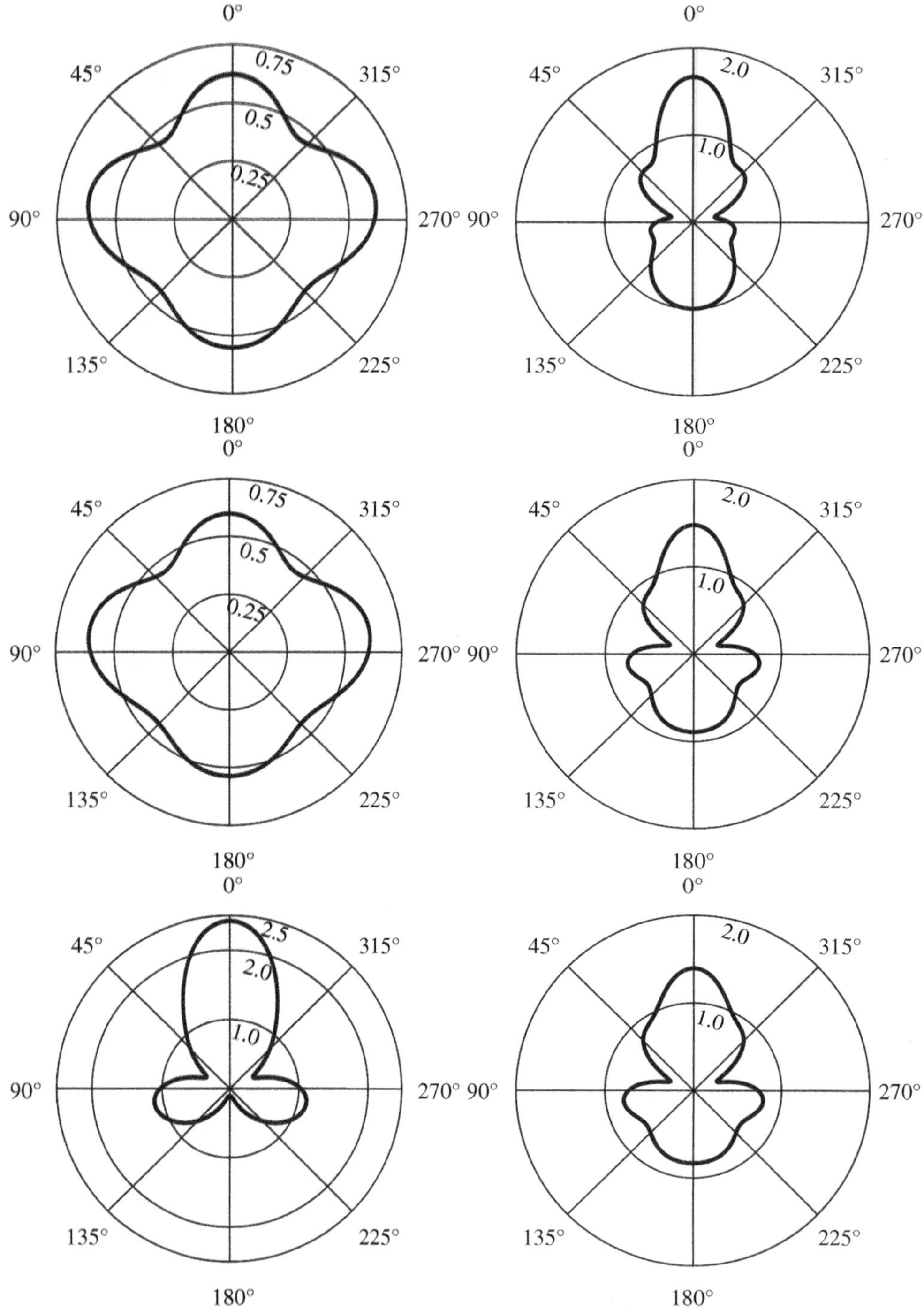

Figure 7.6 Scattering patterns for 0.032 in diameter brass (a) 0.032 in diameter steel (b) and 0.0625 diameter brass (c) cylinders. Scattering patterns for 0.0625 in diameter copper (d) 0.0625 in diameter steel (e) and 0.0625 diameter rigid (f) cylinders. $f = 1$ MHz.

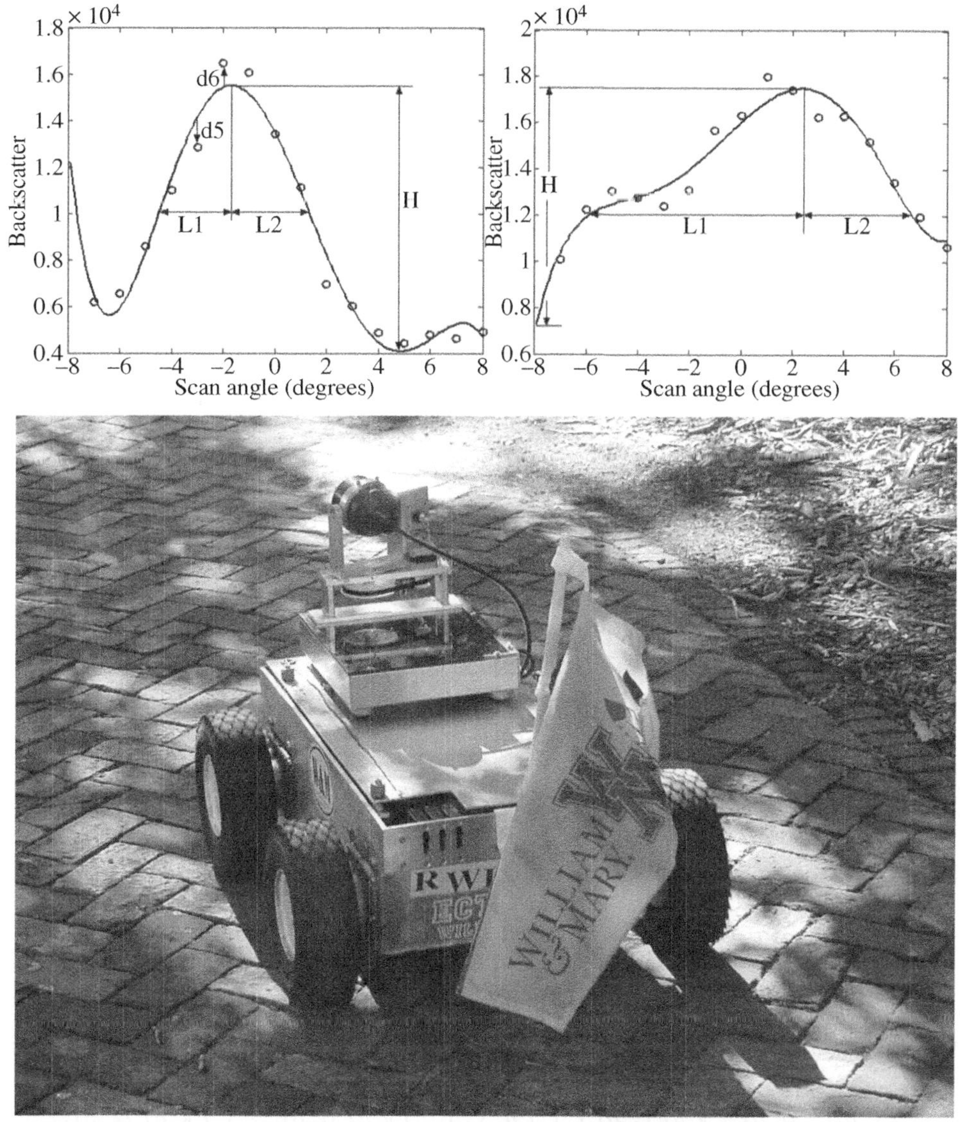

Figure 7.7 A mobile robot which we named rWilliam has a forward-looking 50 kHz pulse-echo scanner mounted in front. As the ultrasound beam sweeps across an upright cylindrical scatterer, the backscatter vs. angle will be symmetric near the peak for a round metal light pole, whereas the bark on a tree will introduce asymmetry into the backscattering because the roughness of the bark is similar in size to the wavelength of sound in air at 50 kHz. Such features can also be used to distinguish big trees from small trees based on backscattering [11] because even so-called "narrow beam" transducers have beams that are too divergent to image in the manner that is done for medical ultrasound.

$$
\begin{aligned}
\Pi_{SV}^{trans} &= \frac{1}{K_2}\sum_n i^n d_n^T J_n(K_2 r\sin\phi)e^{in\theta}e^{iK_2 z\cos\phi} \\
\Pi_{SH}^{inc} &= \frac{1}{K_1}\sum_n i^n J_n(K_1 r\sin\phi)e^{in\theta}e^{iK_1 z\cos\phi} \\
\Pi_{SH}^{scat} &= \frac{1}{K_1}\sum_n i^n e_n^T H_n^{(1)}(K_1 r\sin\phi)e^{in\theta}e^{iK_1 z\cos\phi} \\
\Pi_{SH}^{trans} &= \frac{1}{K_2}\sum_n i^n f_n^T J_n(K_2 r\sin\phi)e^{in\theta}e^{iK_2 z\cos\phi}
\end{aligned}
\tag{7.53}
$$

where a_n^T–f_n^T are six more modal coefficients to be found from the boundary conditions. The SH case will give a 2×2 system, which is straightforward to solve. The coupled L–SV system is 4×4 as before and somewhat more tedious.

$$
\begin{pmatrix}
\frac{1}{2}(K_1a)^2 - n^2 - \frac{k_1aH_n^{(1)\prime}(k_1a)}{H_n^{(1)}(k_1a)} & n\left(1-\frac{K_1aH_n^{(1)\prime}(K_1a)}{H_n^{(1)}(K_1a)}\right) & \frac{\mu_2}{\mu_1}\left[\frac{1}{2}(K_2a)^2 - n^2 + \frac{k_2aJ_n'(k_2a)}{J_n(k_2a)}\right] & n\frac{\mu_2}{\mu_1}\left(1-\frac{K_2aJ_n'(K_2a)}{J_n(K_2a)}\right) \\
n\left(1-\frac{k_1aH_n^{(1)\prime}(k_1a)}{H_n^{(1)}(k_1a)}\right) & n^2-\frac{1}{2}(K_1a)^2-\frac{K_1aH_n^{(1)\prime}(K_1a)}{H_n^{(1)}(K_1a)} & n\frac{\mu_2}{\mu_1}\left(1-\frac{k_2aJ_n'(k_2a)}{J_n(k_2a)}\right) & \frac{\mu_2}{\mu_1}\left[n^2-\frac{1}{2}(K_2a)^2-\frac{K_2aJ_n'(K_2a)}{J_n(K_2a)}\right] \\
\frac{k_1aH_n^{(1)\prime}(k_1a)}{H_n^{(1)}(k_1a)} & -n & \frac{k_2aJ_n'(k_2a)}{J_n(k_2a)} & \frac{\mu_2}{\mu_1}\left[n^2-\frac{1}{2}(K_2a)^2-\frac{K_2aJ_n'(K_2a)}{J_n(K_2a)}\right] \\
-n & \frac{K_1aH_n^{(1)\prime}(K_1a)}{H_n^{(1)}(K_1a)} & -n & \frac{K_2aJ_n'(K_2a)}{J_n(K_2a)}
\end{pmatrix}
\begin{bmatrix}
a_n^T H_n^{(1)}(k_1a) \\
b_n^T H_n^{(1)}(K_1a) \\
-c_n^T J_n(k_2a) \\
-d_n^T J_n(K_2a)
\end{bmatrix}
= i^n J_n(K_1a)
\begin{bmatrix}
n\left(1-\frac{K_1aJ_n'(K_1a)}{J_n(K_1a)}\right) \\
n^2-\frac{1}{2}(K_1a)^2-\frac{K_1aJ_n'(K_1a)}{J_n(K_1a)} \\
-n \\
\frac{K_1aJ_n'(K_1a)}{J_n(K_1a)}
\end{bmatrix}
$$

You should note in the matrix system above that the coefficient matrix is as we had for the L-wave incidence case, so that means that Δ_0 will be the same here too. The other parts of the coefficients are as follows.

$$
\begin{aligned}
a_n^T = n&\left(\frac{k_1aH_n^{(1)\prime}(k_1a)}{H_n^{(1)}(k_1a)} - \frac{k_1aJ_n'(k_1a)}{J_n(k_1a)}\right)\left\{\left(\frac{\mu_2}{\mu_1}\right)^2\left[n^2\left(\frac{k_2aJ_n'(k_2a)}{J_n(k_2a)}-1\right)\left(\frac{K_2aJ_n'(K_2a)}{J_n(K_2a)}-1\right)\right.\right. \\
&\left.-\left(\frac{k_2aJ_n'(k_2a)}{J_n(k_2a)}+\frac{1}{2}(K_2a)^2-n^2\right)\left(\frac{K_2aJ_n'(K_2a)}{J_n(K_2a)}+\frac{1}{2}(K_2a)^2-n^2\right)\right] \\
&+\frac{\mu_2}{\mu_1}\left[\left(\frac{k_2aJ_n'(k_2a)}{J_n(k_2a)}+\frac{1}{2}(K_2a)^2-n^2\right)\left(n^2-\frac{K_2aJ_n'(K_2a)}{J_n(K_2a)}\right)\right. \\
&\left.-\left(\frac{K_2aJ_n'(K_2a)}{J_n(K_2a)}+\frac{1}{2}(K_2a)^2-n^2\right)\left(n^2-\frac{k_2aJ_n'(k_2a)}{J_n(k_2a)}\right)\right] \\
&+\frac{\mu_2}{\mu_1}\frac{1}{2}(K_1a)^2\left(\frac{1}{2}(K_2a)^2+\frac{k_2aJ_n'(k_2a)}{J_n(k_2a)}\frac{K_2aJ_n'(K_2a)}{J_n(K_2a)}-n^2\right)
\end{aligned}
$$

$$+\left(\frac{k_2aJ_n'(k_2a)}{J_n(k_2a)}\frac{K_2aJ_n'(K_2a)}{J_n(K_2a)}-n^2\right)\left(\frac{1}{2}(K_1a)^2+1-n^2\right)\Bigg\}\frac{J_n(k_1a)}{H_n^{(1)}(K_1a)}\frac{1}{\Delta_0}$$

$$\begin{aligned}
b_n^T = &\left\{-n^2\frac{\mu_2}{\mu_1}\left(\frac{k_2aJ_n'(k_2a)}{J_n(k_2a)}-\frac{k_1aH_n^{(1)\prime}(k_1a)}{H_n^{(1)}(k_1a)}\right)\left[\frac{1}{2}(K_1a)^2\left(\frac{K_2aJ_n'(K_2a)}{J_n(K_2a)}-1\right)\right.\right.\\
&\left.-\frac{1}{2}(K_2a)^2\left(\frac{K_1aJ_n'(K_1a)}{J_n(K_1a)}-1\right)\right]-n^2\frac{\mu_2}{\mu_1}\left(\frac{K_2aJ_n'(K_2a)}{J_n(K_2a)}-\frac{K_1aJ_n'(K_1a)}{J_n(K_1a)}\right)\\
&\times\left[\frac{1}{2}(K_1a)^2\left(\frac{k_2aJ_n'(k_2a)}{J_n(k_2a)}-1\right)-\frac{1}{2}(K_2a)^2\left(\frac{k_1aH_n^{(1)\prime}(k_1a)}{H_n^{(1)}(k_1a)}-1\right)\right]\\
&-\left(\frac{k_2aJ_n'(k_2a)}{J_n(k_2a)}\frac{K_2aJ_n'(K_2a)}{J_n(K_2a)}-n^2\right)\left\{n^2\left(\frac{k_1aH_n^{(1)\prime}(k_1a)}{H_n^{(1)}(k_1a)}-1\right)\left(\frac{K_1aJ_n'(K_1a)}{J_n(K_1a)}-1\right)\right.\\
&\left.-\left[\frac{k_1aH_n^{(1)\prime}(k_1a)}{H_n^{(1)}(k_1a)}+\frac{1}{2}(K_1a)^2-n^2\right]\left[\frac{K_1aJ_n'(K_1a)}{J_n(K_1a)}+\frac{1}{2}(K_1a)^2-n^2\right]\right\}\\
&+\frac{\mu_2}{\mu_1}\left(\frac{k_2aJ_n'(k_2a)}{J_n(k_2a)}\frac{K_1aJ_n'(K_1a)}{J_n(K_1a)}-n^2\right)\left\{n^2\left(\frac{k_1aH_n^{(1)\prime}(k_1a)}{H_n^{(1)}(k_1a)}-1\right)\left(\frac{K_2aJ_n'(K_2a)}{J_n(K_2a)}-1\right)\right.\\
&\left.-\left[\frac{k_1aH_n^{(1)\prime}(k_1a)}{H_n^{(1)}(k_1a)}+\frac{1}{2}(K_1a)^2-n^2\right]\left[\frac{K_2aJ_n'(K_2a)}{J_n(K_2a)}+\frac{1}{2}(K_2a)^2-n^2\right]\right\}\\
&+\frac{\mu_2}{\mu_1}\left(\frac{k_1aH_n^{(1)\prime}(k_1a)}{H_n^{(1)}(k_1a)}\frac{K_2aJ_n'(K_2a)}{J_n(K_2a)}-n^2\right)\left\{n^2\left(\frac{k_2aJ_n'(k_2a)}{J_n(k_2a)}-1\right)\left(\frac{K_1aJ_n'(K_1a)}{J_n(K_1a)}-1\right)\right.\\
&\left.-\left[\frac{k_2aJ_n'(k_2a)}{J_n(k_2a)}+\frac{1}{2}(K_2a)^2-n^2\right]\left[\frac{K_1aJ_n'(K_1a)}{J_n(K_1a)}+\frac{1}{2}(K_1a)^2-n^2\right]\right\}\\
&-\left(\frac{\mu_2}{\mu_1}\right)^2\left(\frac{k_1aH_n^{(1)\prime}(k_1a)}{H_n^{(1)}(k_1a)}\frac{K_1aJ_n'(K_1a)}{J_n(K_1a)}-n^2\right)\left\{n^2\left(\frac{k_2aJ_n'(k_2a)}{J_n(k_2a)}-1\right)\left(\frac{K_2aJ_n'(K_2a)}{J_n(K_2a)}-1\right)\right.\\
&\left.\left.-\left[\frac{k_2aJ_n'(k_2a)}{J_n(k_2a)}+\frac{1}{2}(K_2a)^2-n^2\right]\left[\frac{K_2aJ_n'(K_2a)}{J_n(K_2a)}+\frac{1}{2}(K_2a)^2-n^2\right]\right\}\right\}\frac{J_n(K_1a)}{H_n^{(1)}(K_1a)}\frac{1}{\Delta_0}
\end{aligned}$$

7.2.3.1 Scattering from an Acoustic Cylinder

In the limiting case where the shear rigidity of the cylinder vanishes $\mu_2 \to 0$ we can write the coefficients for shear wave scattering from a fluid-filled cylinder. I've even included the modal coefficient for the SH-wave incidence case.

$$\begin{aligned}
\Delta_0 = &-\frac{k_2aJ_n'(k_2a)}{J_n(k_2a)}\left\{n^2\left(\frac{k_1aH_n^{(1)\prime}(k_1a)}{H_n^{(1)}(k_1a)}-1\right)\left(\frac{K_1aH_n^{(1)\prime}(K_1a)}{H_n^{(1)}(K_1a)}-1\right)\right.\\
&\left.+\left[\frac{k_1aH_n^{(1)\prime}(k_1a)}{H_n^{(1)}(k_1a)}+\frac{1}{2}(K_1a)^2-n^2\right]\left[\frac{K_1aH_n^{(1)\prime}(K_1a)}{H_n^{(1)}(K_1a)}+\frac{1}{2}(K_1a)^2-n^2\right]\right\}\\
&+\frac{1}{2}(K_1a)^2\frac{\rho_2}{\rho_1}\left\{\frac{K_1aH_n^{(1)\prime}(K_1a)}{H_n^{(1)}(K_1a)}\left[\frac{K_1aH_n^{(1)\prime}(K_1a)}{H_n^{(1)}(K_1a)}+\frac{1}{2}(K_1a)^2-n^2\right]-n^2\right\}
\end{aligned}$$

$$a_n^T = n\frac{J_n(K_1a)}{H_n^{(1)}(k_1a)}\left(\frac{K_1aJ_n'(K_1a)}{J_n(K_1a)} - \frac{K_1aH_n^{(1)\prime}(K_1a)}{H_n^{(1)}(K_1a)}\right)\left[1 - n^2 + \frac{1}{2}(K_1a)^2\right]\frac{1}{\Delta_0}$$

$$\begin{aligned} b_n^T = &-\frac{J_n(K_1a)}{H_n^{(1)}(K_1a)}\left\{\frac{k_2aJ_n'(k_2a)}{J_n(k_2a)}\left\{n^2\left(\frac{k_1aH_n^{(1)\prime}(k_1a)}{H_n^{(1)}(k_1a)} - 1\right)\left(\frac{K_1aJ_n'(K_1a)}{J_n(K_1a)} - 1\right)\right.\right. \\ &+ \left.\left[\frac{k_1aH_n^{(1)\prime}(k_1a)}{H_n^{(1)}(k_1a)} + \frac{1}{2}(K_1a)^2 - n^2\right]\left[\frac{K_1aJ_n'(K_1a)}{J_n(K_1a)} + \frac{1}{2}(K_1a)^2 - n^2\right]\right\} \\ &+ \left.\frac{1}{2}(K_1a)^2\frac{\rho_2}{\rho_1}\left\{\frac{K_1aH_n^{(1)\prime}(K_1a)}{H_n^{(1)}(K_1a)}\left[\frac{K_1aJ_n'(K_1a)}{J_n(K_1a)} + \frac{1}{2}(K_1a)^2 - n^2\right] - n^2\right\}\right\}\frac{1}{\Delta_0} \end{aligned}$$

$$e_n^T = -\frac{J_n'(K_1a)}{H_n^{(1)\prime}(K_1a)} \tag{7.54}$$

Exercise 7.5 Find the expression for e_n^T for SH wave scattering from an elastic cylinder in an elastic medium, and then take the limit where $\mu_2/\mu_1 \to 0$ to verify the aforementioned expression for e_n^T.

The total flow of scattered elastic energy in the radial direction through a closed surface can be represented by the radial component of the energy flux vector as in the longitudinal case, which may be decomposed into longitudinal and transverse parts. Although the L and SV modes are coupled at the boundary, they will propagate independently in the present linear field approximation. As before, intensities are the time-averaged radial component of the far-field energy flux vector, and if we divide the scattered intensities by the intensity of the incident plane wave and integrate over θ we can write the scattering cross section, per unit length, as:

$$q_n^T = \frac{4}{(k_1a)^2}\sum_n\left(\left(\frac{K_1}{k_1}\right)^3|a_n^T|^2 + |b_n^T|^2 + |e_n^T|^2\right) \tag{7.55}$$

In Figures 7.8 and 7.9, we show the angular scattering patterns for boron cylinders in aluminum, for L and SV plane wave incidence, respectively, and for two different size parameters. Note that there are both L and SV scattered waves for either L or SV wave incidence. You might wonder whether there's some regularity to these scattering patterns such that we could develop some rules of thumb for various size parameters and/or materials combinations. Nope, sorry. You just have to plot them and see what you get. These plots are from a book that's about 30 years old [7], which has lots of the detailed derivations for scattering from cylinders (and spheres) and also includes FORTRAN code. That was a selling point back in the day, although the book was quite expensive and not all that widely known.

I've sketched the polar plots from the book so you can have something to compare against once you get your code working, but don't get too concerned about details of the shapes of the lobes. I may not have accurately reproduced them, and when making polar plots of sharp lobes, they can get a little distorted if too few angles were used.

You're probably wondering why boron cylinders in aluminum. The answer is probably that in those days there was quite a lot of research on advanced composite materials, particularly for high-speed aeronautics applications. NASA had a program to develop a Mach 2.4 civilian airliner, which had an emphasis on developing new materials that would be strong enough and light enough so that such aircraft could carry both passengers and fuel. Boron filaments in aluminum had been used for military applications, but cost would be a significant constraint for civilian uses.

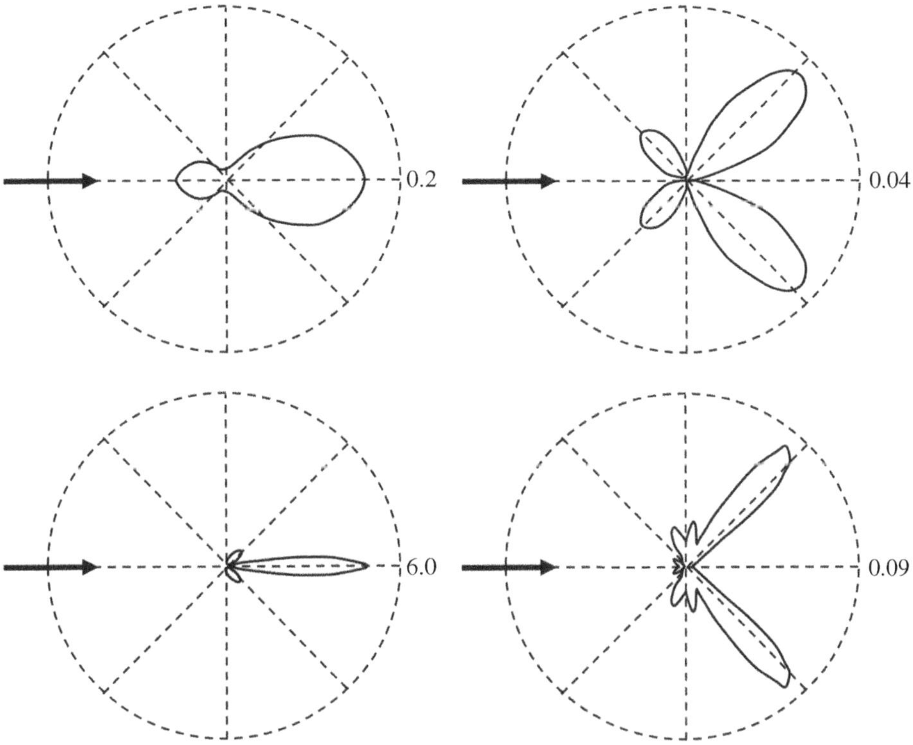

Figure 7.8 Differential scattering cross sections for boron cylinders in aluminum, with incident L plane wave indicated by the arrow. (a) and (b) are for $ka = 1$, while (c) and (d) are for $ka = 10$. (a) and (c) show the angular distribution of the scattered L wave. (b) and (d) show the corresponding mode-converted shear wave. Note the different scales indicated by the number.

In addition, such materials would need to be inspected, both when new and as they aged, and ultrasound could be employed for that.

Exercise 7.6 Plot the SH-incidence differential scattering cross sections for boron cylinders in aluminum.

7.2.4 Limiting Cases

We can consider two limiting cases where the cylinder is a void and rigid, respectively. For scattering from a cylindrical hole, the boundary conditions are that the normal surface tractions, σ_{rr} and $\sigma_{r\theta}$ are zero at $r = a$, which gives two equations to solve for the scattered L and SV wave amplitudes. SH waves are uncoupled, as always. We find that:

$$\Delta_0 = n^2 \left(\frac{k_1 a H_n^{(1)\prime}(k_1 a)}{H_n^{(1)}(k_1 a)} - 1 \right) \left(\frac{K_1 a H_n^{(1)\prime}(K_1 a)}{H_n^{(1)}(K_1 a)} - 1 \right)$$
$$- \left[\frac{k_1 a H_n^{(1)\prime}(k_1 a)}{H_n^{(1)}(k_1 a)} + \frac{1}{2}(K_1 a)^2 - n^2 \right] \left[\frac{K_1 a H_n^{(1)\prime}(K_1 a)}{H_n^{(1)}(K_1 a)} + \frac{1}{2}(K_1 a)^2 - n^2 \right]$$

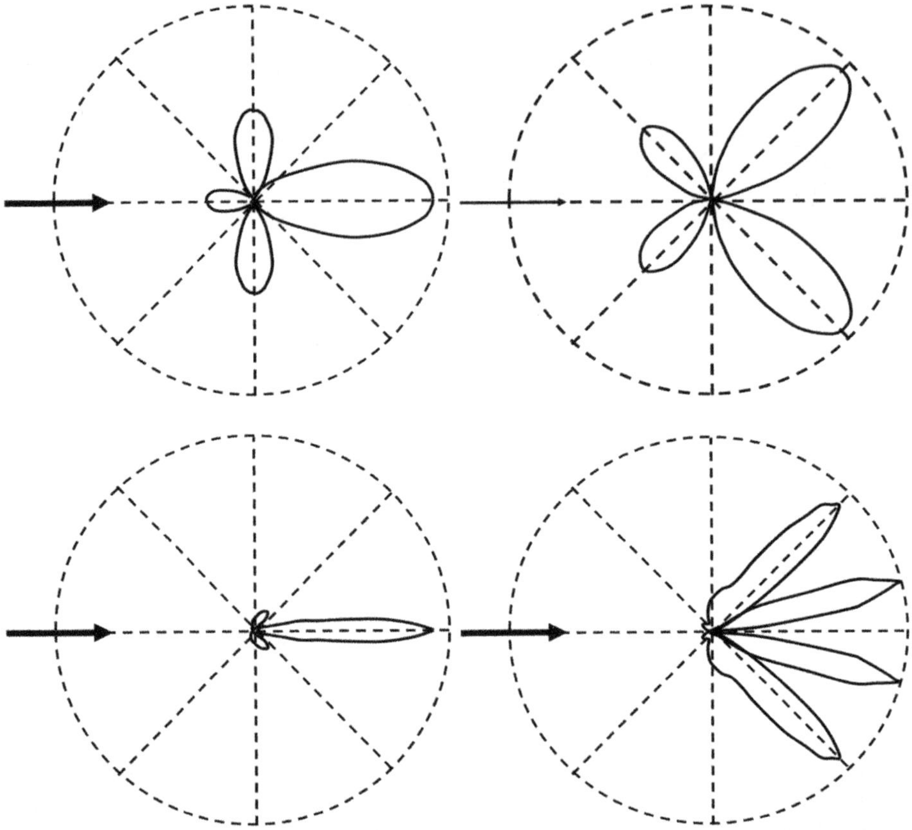

Figure 7.9 Differential scattering cross sections for boron cylinders in aluminum, with incident SV plane wave indicated by the arrow. (a) and (b) are for $Ka = 1$, while (c) and (d) are for $Ka = 10$. (a) and (c) show the angular distribution of the scattered SV wave. (b) and (d) show the corresponding mode-converted L wave. The plots are normalized to highlight the differences.

$$a_n^L = \frac{J_n(K_1a)}{H_n^{(1)}(K_1a)}\left\{n^2\left(\frac{k_1aJ_n'(k_1a)}{J_n(k_1a)}-1\right)\left(\frac{K_1aH_n^{(1)\prime}(K_1a)}{H_n^{(1)}(K_1a)}-1\right)\right.$$
$$\left.-\left[\frac{k_1aJ_n'(k_1a)}{J_n(k_1a)}+\frac{1}{2}(K_1a)^2-n^2\right]\left[\frac{K_1aH_n^{(1)\prime}(K_1a)}{H_n^{(1)}(K_1a)}+\frac{1}{2}(K_1a)^2-n^2\right]\right\}\frac{1}{\Delta_0}$$

$$b_n^L = \frac{J_n(K_1a)}{H_n^{(1)}(K_1a)}n\left[1-n^2+\frac{1}{2}(K_1a)^2\right]\left(\frac{k_1aH_n^{(1)\prime}(k_1a)}{H_n^{(1)}(k_1a)}-\frac{k_1aJ_n'(k_1a)}{J_n(k_1a)}\right)\frac{1}{\Delta_0}$$

$$a_n^T = \frac{J_n(K_1a)}{H_n^{(1)}(K_1a)}n\left[1-n^2+\frac{1}{2}(K_1a)^2\right]\left(\frac{k_1aJ_n'(k_1a)}{J_n(k_1a)}-\frac{k_1aH_n^{(1)\prime}(k_1a)}{H_n^{(1)}(k_1a)}\right)\frac{1}{\Delta_0}$$

$$b_n^T = \frac{J_n(K_1a)}{H_n^{(1)}(K_1a)}\left\{n^2\left(\frac{k_1aH_n^{(1)\prime}(k_1a)}{H_n^{(1)}(k_1a)}-1\right)\left(\frac{K_1aJ_n'(K_1a)}{J_n(K_1a)}-1\right)\right.$$
$$\left.-\left[\frac{k_1aH_n^{(1)\prime}(k_1a)}{H_n^{(1)}(k_1a)}+\frac{1}{2}(K_1a)^2-n^2\right]\left[\frac{K_1aJ_n'(K_1a)}{J_n(K_1a)}+\frac{1}{2}(K_1a)^2-n^2\right]\right\}\frac{1}{\Delta_0}$$

$$e_n^T = -\frac{J_n(K_1 a)}{H_n^{(1)}(K_1 a)}$$

For a rigid scatterer, the boundary conditions at $r = a$ are that the displacements, u_r and u_θ, vanish. Although the algebra is fairly simple, this is a little nonphysical because it also assumes that the scatterer is infinitely dense. Nevertheless, the scattering amplitudes come out as:

$$\Delta_0 = \frac{k_1 a H_n^{(1)\prime}(k_1 a)}{H_n^{(1)}(k_1 a)} \frac{K_1 a H_n^{(1)\prime}(K_1 a)}{H_n^{(1)}(K_1 a)} - n^2$$

$$a_n^L = \frac{J_n(K_1 a)}{H_n^{(1)}(k_1 a)} \left\{ \frac{k_1 a J_n'(k_1 a)}{J_n(k_1 a)} \frac{K_1 a H_n^{(1)\prime}(K_1 a)}{H_n^{(1)}(K_1 a)} - n^2 \right\} \frac{1}{\Delta_0}$$

$$b_n^L = \frac{J_n(K_1 a)}{H_n^{(1)}(k_1 a)} n \left\{ \frac{k_1 a H_n^{(1)\prime}(k_1 a)}{H_n^{(1)}(k_1 a)} - \frac{k_1 a J_n'(k_1 a)}{J_n(k_1 a)} \right\} \frac{1}{\Delta_0}$$

$$a_n^T = \frac{J_n(K_1 a)}{H_n^{(1)}(k_1 a)} n \left\{ \frac{K_1 a H_n^{(1)\prime}(K_1 a)}{H_n^{(1)}(K_1 a)} - \frac{K_1 a J_n'(K_1 a)}{J_n(K_1 a)} \right\} \frac{1}{\Delta_0}$$

$$b_n^T = \frac{J_n(K_1 a)}{H_n^{(1)}(k_1 a)} \left\{ \frac{k_1 a H_n^{(1)\prime}(k_1 a)}{H_n^{(1)}(k_1 a)} \frac{K_1 a J_n'(K_1 a)}{J_n(K_1 a)} - n^2 \right\} \frac{1}{\Delta_0}$$

$$e_n^T = 1 \tag{7.56}$$

Figure 7.10 illustrates the angular scattering patterns for a cylindrical hole in a metal, which I've reproduced from [12]. Figure 7.11 shows three plots scanned from that 1958 paper where computations were compared to measurements, done as a part of a doctoral dissertation at Harvard.[3] Both the experiments and the computations would have been exceedingly difficult. My doctoral advisor was doing a PhD at Yale in those days, and he once mentioned traveling to Boston "to

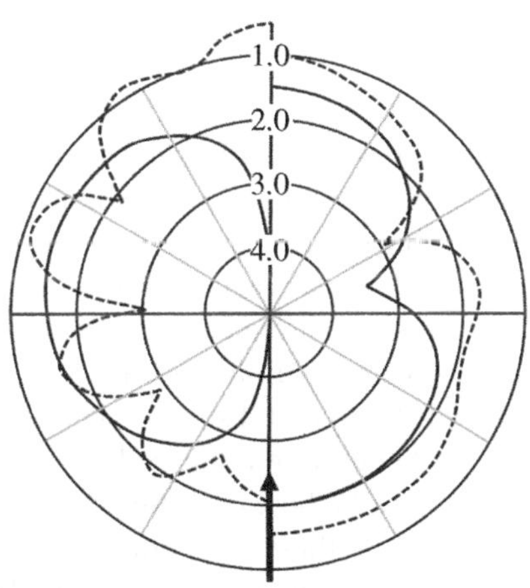

Figure 7.10 White's Figure 4 reproduced from [12] showing normalized angular scattering of elastic waves from a cylindrical hole in an elastic solid with Poisson's ratio of 1/3. According to the text of [12] "The solid curve on the left ... is the angular distribution for normally polarized shear waves incident and compressional waves scattered; the dashed curve on the left is for normally polarized shear waves incident and scattered. The solid curve on the right ... is for compressional waves incident and scattered, and the dashed curve on the right is for axially polarized shear waves incident and scattered." The size parameters are $ka = 2.0$ and $Ka = 4.0$ and the incident wave is vertical. Source: Adapted from [12].

3 Richard M. White was a professor (emeritus) in the UC Berkeley Department of Electrical Engineering and Computer Sciences and a Co-Founding Director of the Berkeley Sensor & Actuator Center. His research interests included wireless microsensors, energy scavenging devices for use in electric power systems, and portable particulate matter monitors for measuring concentrations of airborne aerosols and diesel exhaust particulates. Prior to joining the faculty at UC Berkeley in 1963, Professor White conducted microwave device research at General Electric. In addition to the 2003 Rayleigh Award of the IEEE for seminal contributions to surface acoustic wave technology, Professor White was a member of the National Academy of Engineering, a Fellow of the IEEE and the American Association for the Advancement of Science, and was the recipient of many academic awards. Professor White earned his AB and AM degrees in Engineering Science and Applied Physics from Harvard University, and his

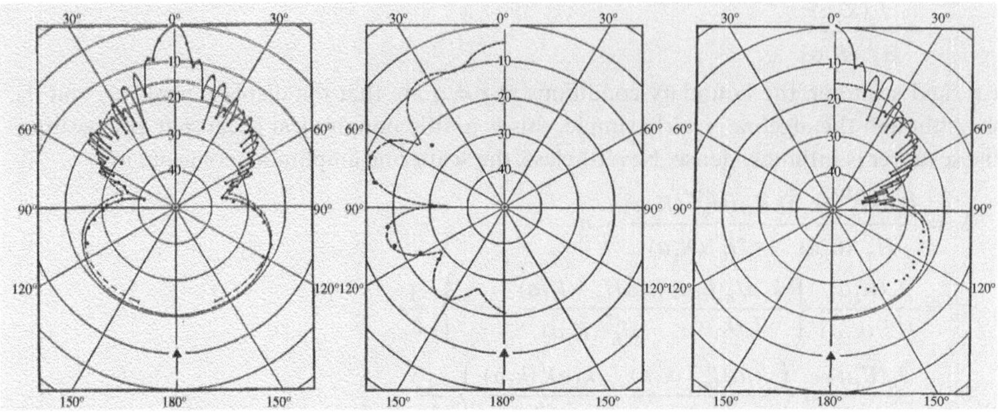

Figure 7.11 Figures 5, 6, and 7 from [12] showing angular scattering of elastic waves from an elastic cylinder. I've included scans of White's figures because I'm so impressed with how difficult it was in the 1950s to do both the experiments and the simulations in order to compare them. To quote Jesse Pinkman from Breaking Bad a bit out of context, "Yeah Mr. White! Yeah science!" In addition, drugs are bad, don't do drugs. And don't bury all seven barrels of money in the same hole, geez. (a) is measured and computed "compression-to-compression" scattering at an empty bore, measured at a frequency of 3.73 MHz. (b) compares the measured and computed normally polarized "shear-to-shear" scattering at an empty bore, also at a frequency of 3.73 MHz. Note that the computed scattering patterns here and in Figure 7.10 don't reproduce all the lobes in the forward-scattering quadrant, which isn't surprising given the computational limits in those days. White speculates that what's actually going on is "interference between the scattered and incident waves." He doesn't show many measured data points in (b), because presumably similar interference exists in the forward direction for the shear-to-shear scattering; "it was for this reason that the tedious shear pickup measurements were not made in that region." (c) shows the measured "compression-to-compression" scattering at a mercury-filled bore, again at a frequency of 3.73 MHz. The comparison is to computations for an empty bore, although White had formulated the equations for the case of a nonviscous fluid scatterer. Presumably Professor F.V. Hunt told him at some point that he already had enough to graduate and not to try to include the solid and fluid scatterer computations in his dissertation. Source: White [12]/with permission of AIP Publishing LLC.

use the computer." Although [12] sets up the problem for elastic wave scattering from a cylindrical elastic inclusion, his computations were only for the limiting case of a cylindrical void. Most of the measurements were also for a cylindrical void, although the third plot in Figure 7.11 is for a fluid-filled cylindrical void. The fluid is mercury. Yikes. I assume that soon-to-be-Dr. White wanted to get some measurements for elastic wave scattering from an elastic cylinder, but the exterior material was a large aluminum cylinder so there's no obvious way to get a solid elastic cylinder firmly embedded into that. I suppose he could have poured molten lead into the empty bore and let that harden. Other alloys that melt at a low-enough temperature could be done as well, presumably drilling out the bore each time before refilling it. There are all sorts of plastics and whatnot that come to mind these days, but we're talking the 1950s here, so that was in the future. In the movie "The Graduate" which is based on a 1963 book, there's a famous scene where Mr. McGuire takes Dustin Hoffman's character, Benjamin, off to the side and says "Plastics there's a great future in plastics." In addition, there's the obvious point that dissertation research can't go on forever. At some point, the student has accomplished enough and it's time to graduate.

Figure 7.12 shows several differential scattering cross sections for cylindrical holes in aluminum, but as a function of size parameter (ka) rather than angle. These are reproduced from a ca. 1991

PhD degree in Applied Physics in 1956, also from Harvard. Dick White was still active in his field when he died on 17 August 2020.

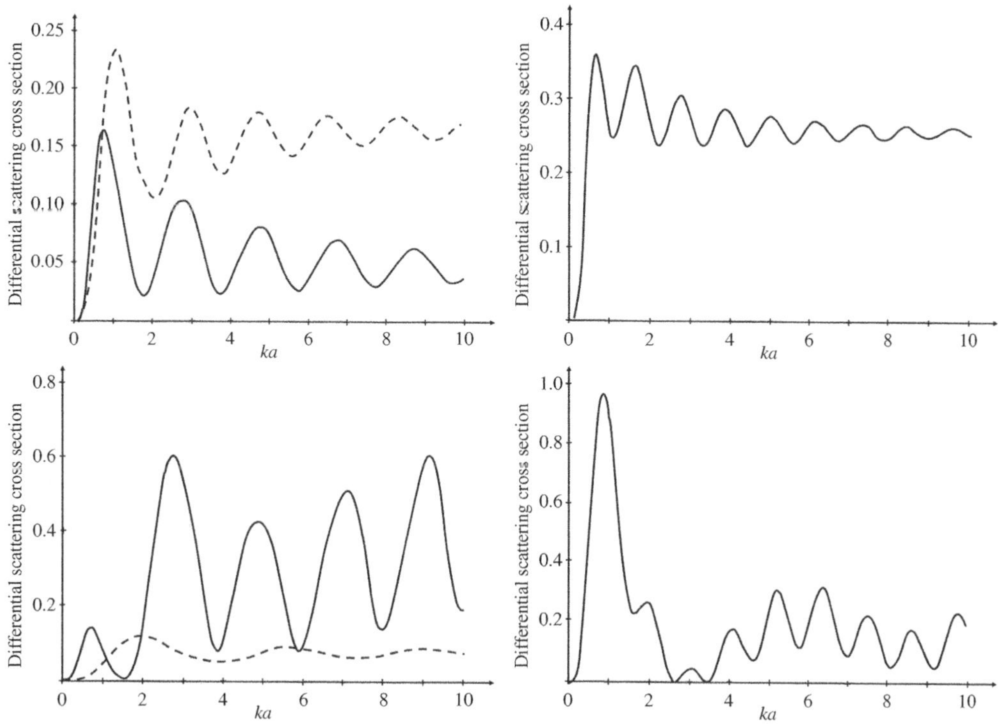

Figure 7.12 Differential scattering cross sections for cylindrical holes in aluminum. (a) and (b) are for incident L wave. (c) and (d) are for incident SV waves. (a) and (c) are scattering at 90°, where dashed line is the mode-converted waves. (b) and (d) is for backscattering, where there is no mode-converted wave.

book [7], which included FORTRAN code listings. Computationally these weren't much of a challenge to do in those days. It should be even easier for you because the formulas that you'll be plotting are written out explicitly earlier.

Exercise 7.7 Plot elastic wave scattering from cylindrical holes in Al.

The aforementioned analysis allows us to calculate dynamical properties of interest. Recall that we had suppressed $e^{-i\omega t}$ time variation, that is, we're doing the analysis in frequency domain. That means that the analysis also applies to consideration of a reinforcing fiber in a matrix, where the composite material is experiencing a vibration. Of interest in the development of new advanced composites is whether vibrations might lead to stress concentrations (Figures 7.13 and 7.14) at the interfaces.[4]

7.3 Plate Wave Scattering

Ultrasonic guided waves, Lamb waves, are useful for evaluating the integrity of plate and shell structures common in many applications. Tomographic reconstruction with Lamb waves allows

4 Subhendu Datta is Professor Emeritus of Mechanical Engineering at UC Boulder. His research deals with analytical and numerical modeling of wave propagation and scattering in solid media, including ultrasonic techniques for material characterization and nondestructive evaluation, elastic guided waves in layered, composite, and anisotropic plates and cylinders, and waves in thin layers and coatings.

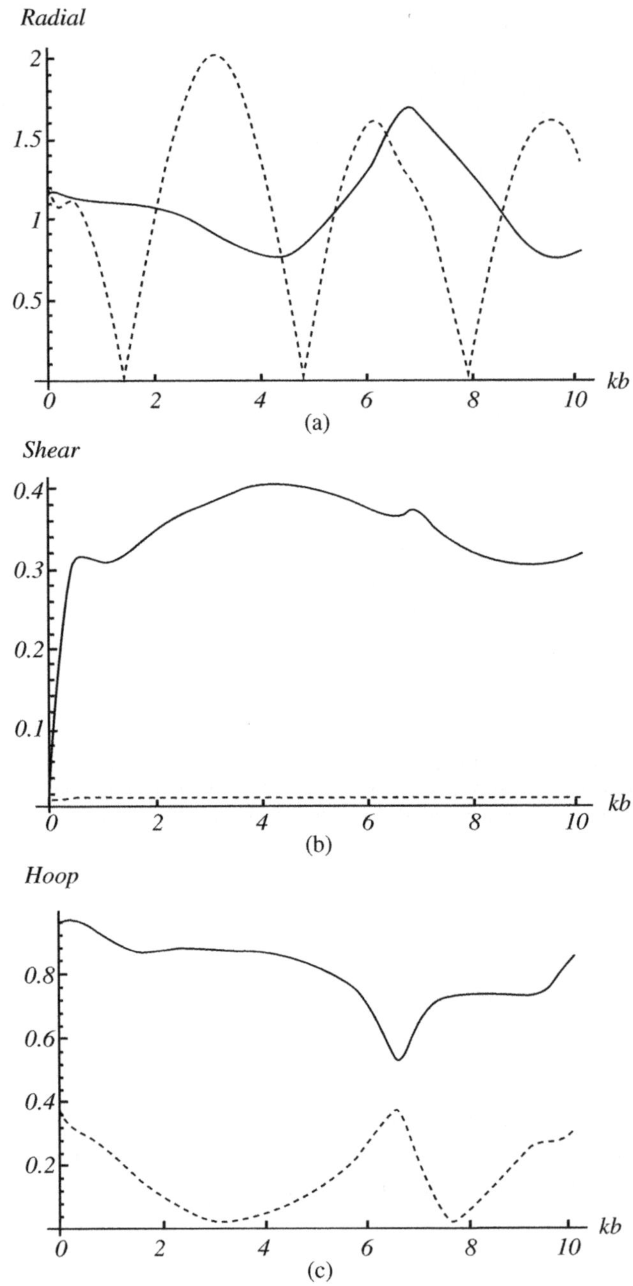

Figure 7.13 Dynamic stress concentrations for a boron fiber in epoxy. The three plots are for the radial (a), shear (b), and hoop (c) stresses at the interface. Solid lines are for lateral and backscattering directions, respectively.

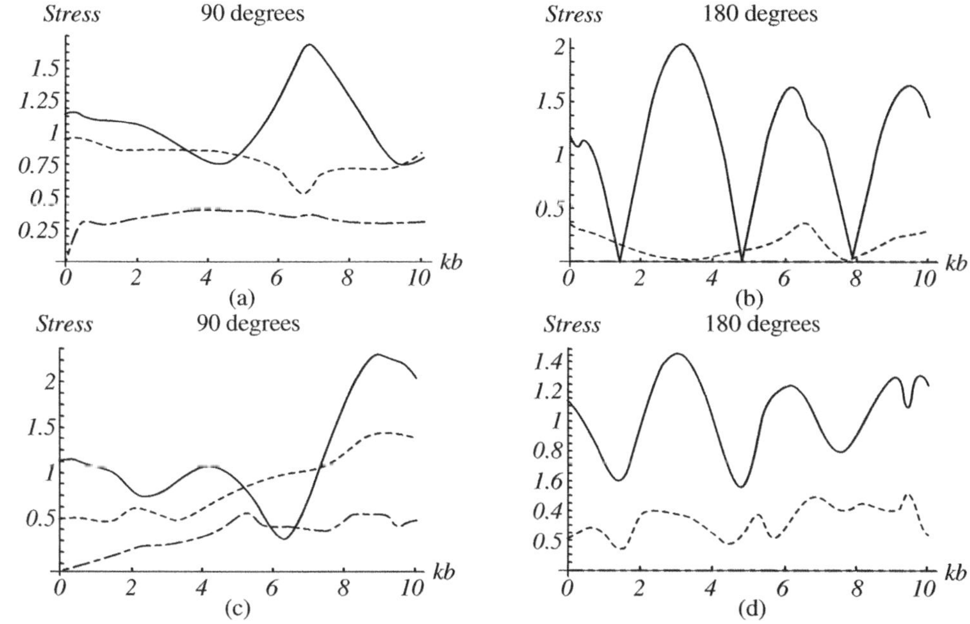

Figure 7.14 Solving the similar but more algebraically involved problem of scattering from a two-layer cylindrical shell allows us to calculate dynamic stress concentrations. In these plots, the radial (solid line) shear (dash-dot line), and hoop (short-dash line) stress concentrations are shown. Plots (a) and (b) are for a boron shell deposited on a tungsten core embedded in a matrix of epoxy. Plots (c) and (d) are for a silicon carbide fiber with a carbon core embedded in an aluminum matrix. See [13, 14] for details, or if you prefer you might instead look at Dr. Bogan's doctoral dissertation if that turns out to be more convenient for you. His Boston University doctoral dissertation has the same title as [14] and also includes FORTRAN code if you're interested in that sort of thing. Some university libraries have scanned all their dissertations and made them freely available. They're usually much better quality than the reduced-size, printed-from-microfilm versions we used to be satisfied with.

for the accurate reconstruction of the variation of quantities of interest throughout the investigated region, and it presents the data as a quantitative map. The location, shape, and extent of flaws can then be easily extracted from this tomographic image. The scattering of Lamb waves from severe flaws introduces artifacts in such reconstructions, however. When Lamb waves scatter from inhomogeneities, some of the energy is converted into other modes, which have different dispersion properties and propagate at different speeds. Moreover, the through-thickness displacement profiles are different for the different modes, which complicates rather severely the application of the boundary conditions. Fortunately, higher-order plate theory can be used to derive analytical solutions for the scattering of the lowest-order symmetric Lamb wave from a circular inclusion, such as an isolated through hole or in-plane disk (rivet) in plates.

We can use the Mindlin[5] higher-order plate theory [15–19] to model the scattering of the lowest Lamb waves. The simplest plate theories only describe the dispersionless region of the $S0$ curve, but in our measurements, we make use of the dispersion of the $S0$ wave, so we need a theory that more accurately models the behavior of the $S0$ Lamb wave. Using methods similar to the previous sections, analytical expressions for the scattering of $S0$ Lamb waves from a cylindrical inclusion in

5 See https://history.aip.org/phn/11607001.html or https://www.nytimes.com/1987/11/24/obituaries/raymond-d-mindlin-engineering-professor.html.

a plate are derived for an incident plane wave. The expressions are explicitly evaluated and plots shown for the cases of a hole in a plate and inserted disks. But first, flexural waves.

7.3.1 Flexural Wave Scattering from Cylinders

For the lowest-order antisymmetric Lamb wave mode, we can model the scattering from rivets by the following procedure. Other modes, especially the lowest order symmetric modes can be treated also as scattering problems, but the analysis differs since the plate equation is not appropriate for dilatational waves.

The displacement of a plate in bending is given by

$$D\nabla^2\nabla^2 w + 2b\rho\partial_t^2 w = q$$

where w is the transverse displacement. Here $2b$ is the thickness of the plate, ρ is its density, and D is given by

$$D = \frac{2\mu I}{1-\nu} = \frac{2Eb^3}{3(1-\mu^2)} \qquad I = 2b^3/3$$

where E, ν, μ are Young's modulus, Poisson's ratio, and rigidity of the material. Note the cylindrical scatterer of radius $r = a$ is at the origin. For $q = 0$ and harmonic wave motion solutions to the equation of motion are

$$w(x, y, t) = \left[W_1(x, y) + W_2(x, y)\right] e^{-i\omega t}$$

where $W_{1,2}$ satisfy

$$(\nabla^2 + \gamma^2)W_1 = 0 \qquad (\nabla^2 - \gamma^2)W_2 = 0$$

with $\gamma = \omega\sqrt{2b\rho/D}$. Note that W_1 represents the part of the flexural wave that travels with the speed $c_f = \gamma\sqrt{D/2b\rho}$ and W_2 represents the part attenuating as it progresses.

If the scatterer is at a large distance from the source, the attenuation will reduce W_2 to a negligible amount and only W_1 needs to be considered as the incident flexural wave. We write

$$w^{inc} = w_0 e^{i(\gamma_1 x - \omega t)} = w_0 \sum_{n=0}^{\infty} \epsilon_n i^n J_n(\gamma_1 r)\cos n\theta e^{-i\omega t}$$

In polar coordinates, the solutions for $W_{1,2}$ are of the form $H_n^{(1),(2)}(\gamma r)e^{\pm in\theta}$ and $H_n^{(1),(2)}(i\gamma r)e^{\pm in\theta}$, respectively, so we can write the scattered flexural wave

$$w^{scat} = w_0 \sum_{n=0}^{\infty} \epsilon_n i^n \left[A_n H_n^{(1)}(\gamma_1 r) + B_n H_n^{(1)}(i\gamma_1 r)\right] \cos n\theta e^{-i\omega t}$$

Inside the rivet, the refracted waves can be represented as:

$$w^{refr} = w_0 \sum_{n=0}^{\infty} \epsilon_n i^n \left[C_n J_n(\gamma_2 r) + D_n J_n(i\gamma_2 r)\right] \cos n\theta e^{-i\omega t}$$

where for the rivet, the physical properties are different, so the material parameters have the subscript 2. We will denote the region $r > a$ as 1 and $r < a$ as 2. In polar coordinates, the boundary conditions at $r = a$ are continuity of transverse displacements and shear forces as well as slope and bending moment. The needed bending moments and transverse shear forces are given by

$$M_{rr} = -D\left[\nu\nabla^2 w + (1-\nu)\partial_r^2 w\right]$$

and

$$V_r = -D\partial_r(\nabla^2 w) - \frac{1}{r}\partial_\theta\left\{D(1-\nu)\partial_r\left(\frac{1}{r}\partial_\theta w\right)\right\}$$

For a rivet the following four conditions must be satisfied at the boundary $r = a$:

$$w^{inc} + w^{scat} = w^{refr} \qquad \partial_r w^{inc} + \partial_r w^{scat} = \partial_r w^{refr}$$

$$M_{rr}^{inc} + M_{rr}^{scat} = M_{rr}^{refr} \qquad V_r^{inc} + V_r^{scat} = V_r^{refr}$$

These give rise to four simultaneous equations for the unknown coefficients A_n, B_n, C_n, D_n. Solving this system algebraically gives for $A_n = -\frac{J_n(\gamma_1 a)}{H_n^{(1)}(\gamma_1 a)}\frac{\Delta_1}{\Delta_0}$:

$$\begin{aligned}
\Delta_0 = &\left(\frac{\gamma_2 a J_n'(\gamma_2 a)}{J_n(\gamma_2 a)} + \frac{i\gamma_2 a J_n'(i\gamma_2 a)}{J_n(i\gamma_2 a)}\right)\left(\frac{\gamma_1 a H_n^{(1)'}(\gamma_1 a)}{H_n^{(1)}(\gamma_1 a)} + \frac{i\gamma_1 a H_n^{(1)'}(i\gamma_1 a)}{H_n^{(1)}(i\gamma_1 a)}\right)\\
&\times\left\{-2 + \left(\frac{\frac{\gamma_1 a H_n^{(1)'}(\gamma_1 a)}{H_n^{(1)}(\gamma_1 a)} - \frac{i\gamma_1 a H_n^{(1)'}(i\gamma_1 a)}{H_n^{(1)}(i\gamma_1 a)}}{\frac{\gamma_1 a H_n^{(1)'}(\gamma_1 a)}{H_n^{(1)}(\gamma_1 a)} + \frac{i\gamma_1 a H_n^{(1)'}(i\gamma_1 a)}{H_n^{(1)}(i\gamma_1 a)}}\right)\left(\frac{i\gamma_2 a J_n'(i\gamma_2 a)}{J_n(i\gamma_2 a)} - n^2\frac{p-1}{x_2^2}\right)\frac{p-1}{x_1^2}\right.\\
&\left.+\left(\frac{\frac{\gamma_2 a J_n'(\gamma_2 a)}{J_n(\gamma_2 a)} - \frac{i\gamma_2 a J_n'(i\gamma_2 a)}{J_n(i\gamma_2 a)}}{\frac{\gamma_2 a J_n'(\gamma_2 a)}{J_n(\gamma_2 a)} + \frac{i\gamma_2 a J_n'(i\gamma_2 a)}{J_n(i\gamma_2 a)}}\right)\frac{i\gamma_1 a H_n^{(1)'}(i\gamma_1 a)}{H_n^{(1)}(i\gamma_1 a)}\frac{p-1}{x_2^2}\right\}\\
&+\left(\frac{x_1}{x_2}\right)^2\left(\frac{\gamma_1 a H_n^{(1)'}(\gamma_1 a)}{H_n^{(1)}(\gamma_1 a)} - \frac{i\gamma_1 a H_n^{(1)'}(i\gamma_1 a)}{H_n^{(1)}(i\gamma_1 a)}\right)\left\{2n^2\frac{p-1}{x_2^2}\left[1 + \frac{i\gamma_2 a J_n'(i\gamma_2 a)}{J_n(i\gamma_2 a)}\frac{p-1}{x_2^2}\right]\right.\\
&\left.+\left(\frac{\gamma_2 a J_n'(\gamma_2 a)}{J_n(\gamma_2 a)} - \frac{i\gamma_2 a J_n'(i\gamma_2 a)}{J_n(i\gamma_2 a)}\right)\left[2 + \frac{i\gamma_1 a H_n^{(1)'}(i\gamma_1 a)}{H_n^{(1)}(i\gamma_1 a)}\frac{p-1}{x_1^2} + \frac{i\gamma_2 a J_n'(i\gamma_2 a)}{J_n(i\gamma_2 a)}\frac{p-1}{x_2^2}\right]\right\}
\end{aligned}$$

and

$$\begin{aligned}
\Delta_1 = &\left(\frac{\gamma_2 a J_n'(\gamma_2 a)}{J_n(\gamma_2 a)} + \frac{i\gamma_2 a J_n'(i\gamma_2 a)}{J_n(i\gamma_2 a)}\right)\left(\frac{\gamma_1 a J_n'(\gamma_1 a)}{J_n(\gamma_1 a)} + \frac{i\gamma_1 a H_n^{(1)'}(i\gamma_1 a)}{H_n^{(1)}(i\gamma_1 a)}\right)\\
&\times\left\{-2 + \left(\frac{\frac{\gamma_1 a J_n'(\gamma_1 a)}{J_n(\gamma_1 a)} - \frac{i\gamma_1 a H_n^{(1)'}(i\gamma_1 a)}{H_n^{(1)}(i\gamma_1 a)}}{\frac{\gamma_1 a J_n'(\gamma_1 a)}{J_n(\gamma_1 a)} + \frac{i\gamma_1 a H_n^{(1)'}(i\gamma_1 a)}{J_n^{(1)}(i\gamma_1 a)}}\right)\left(\frac{i\gamma_2 a J_n'(i\gamma_2 a)}{J_n(i\gamma_2 a)} - n^2\frac{p-1}{x_2^2}\right)\frac{p-1}{x_1^2}\right.\\
&\left.+\left(\frac{\frac{\gamma_2 a J_n'(\gamma_2 a)}{J_n(\gamma_2 a)} - \frac{i\gamma_2 a J_n'(i\gamma_2 a)}{J_n(i\gamma_2 a)}}{\frac{\gamma_2 a J_n'(\gamma_2 a)}{J_n(\gamma_2 a)} + \frac{i\gamma_2 a J_n'(i\gamma_2 a)}{J_n(i\gamma_2 a)}}\right)\frac{i\gamma_1 a H_n^{(1)'}(i\gamma_1 a)}{H_n^{(1)}(i\gamma_1 a)}\frac{p-1}{x_2^2}\right\}\\
&+\left(\frac{x_1}{x_2}\right)^2\left(\frac{\gamma_1 a J_n'(\gamma_1 a)}{J_n(\gamma_1 a)} - \frac{i\gamma_1 a H_n^{(1)'}(i\gamma_1 a)}{H_n^{(1)}(i\gamma_1 a)}\right)\left\{2n^2\frac{p-1}{x_2^2}\left[1 + \frac{i\gamma_2 a J_n'(i\gamma_2 a)}{J_n(i\gamma_2 a)}\frac{p-1}{x_2^2}\right]\right.\\
&\left.+\left(\frac{\gamma_2 a J_n'(\gamma_2 a)}{J_n(\gamma_2 a)} - \frac{i\gamma_2 a J_n'(i\gamma_2 a)}{J_n(i\gamma_2 a)}\right)\left[2 + \frac{i\gamma_1 a H_n^{(1)'}(i\gamma_1 a)}{H_n^{(1)}(i\gamma_1 a)}\frac{p-1}{x_1^2} + \frac{i\gamma_2 a J_n'(i\gamma_2 a)}{J_n(i\gamma_2 a)}\frac{p-1}{x_2^2}\right]\right\}
\end{aligned}$$

where we have defined $p = D_2/D_1$ and $x^2 = (\gamma a)^2/(\nu - 1)$. Similar expressions can be derived for $B_n - D_n$ in order to specify analytically the scattered and refracted fields that result from interaction of the incident flexural wave with the rivet.

In the far-field of the rivet, the scattered field will consist of only the propagating part, so B_n is not needed. In addition, for large $\gamma_1 r$ we have

$$H_n(\gamma_1 r) \to \sqrt{\frac{2}{i\pi\gamma_1 r}} e^{i(\gamma_1 r - n\pi/2)}$$

so we can write

$$w^{scatt} \to \sqrt{\frac{2}{i\pi\gamma_1 r}} e^{i(\gamma_1 r)} \sum_{n=0}^{\infty} \epsilon_n f_n(\theta) \tag{7.57}$$

where

$$f_n(\theta) = w_0 e^{-i\omega t} A_n \cos n\theta \tag{7.58}$$

7.3.2 Dilatational Wave Scattering

In order to model the scattering of the lowest-order symmetric Lamb wave modes from rivets, it is necessary to go to a slightly more complicated theory than was used for the antisymmetric modes. This is because the simplest plate theories give only that portion of the $S0$ curve which is dispersionless. Since we are interested in modeling measurements made on the shoulder of that curve, we need to account for the dispersive effects. Without resorting to a full three-dimensional elasticity treatment, Kane and Mindlin [15] provide us with an appropriate theory for dilatational plate wave scattering from rivets.

As in the previous section, we consider an infinite plate that is homogeneous, isotropic, and linearly elastic. We assume that the plate is bounded by the planes $z = \pm h$ and that there is a disk of radius $r = a$ at the origin of a cylindrical coordinate system. The model rivet (disk) is a similar material, but with arbitrarily different material parameters. We describe the plate by density ρ and Lamé parameters λ, μ. The rivet is described by ρ' and λ', μ'. The wave speeds for bulk longitudinal and transverse waves are

$$c_L = \sqrt{\frac{\lambda + 2\mu}{\rho}} \qquad c_T = \sqrt{\frac{\mu}{\rho}}$$

and the limiting value of the wave speed for plate waves is

$$c_P = \sqrt{\frac{4\mu(\lambda + \mu)}{\rho(\lambda + 2\mu)}}$$

For thin plates, we assume that the components of displacement in cylindrical coordinates are approximated sufficiently well by

$$u_r(r,\theta,t) = v_r(r,\theta,t) \qquad u_\theta(r,\theta,t) = v_\theta(r,\theta,t) \qquad u_z(r,\theta,t) = \frac{z}{h} v_z(r,\theta,t)$$

and then we introduce three independent displacement potentials $\phi_1(r,\theta)$, $\phi_2(r,\theta)$, $\psi(r,\theta)$ defined by

$$v_r = \left(\frac{\partial \phi_1}{\partial r} + \frac{\partial \phi_2}{\partial r} + \frac{1}{r}\frac{\partial \psi}{\partial \theta} \right) e^{-i\omega t} \tag{7.59}$$

$$v_\theta = \left(\frac{1}{r}\frac{\partial \phi_1}{\partial \theta} + \frac{1}{r}\frac{\partial \phi_2}{\partial \theta} - \frac{\partial \psi}{\partial r} \right) e^{-i\omega t}$$

$$v_z = \left(\sigma_1 \phi_1 + \sigma_2 \phi_2 \right) e^{-i\omega t}$$

We must then consider solutions of the scalar Helmholtz equation:

$$\left(\nabla^2 + k_1^2\right)\phi_1 = 0 \qquad \left(\nabla^2 + k_2^2\right)\phi_1 = 0 \qquad \left(\nabla^2 + K^2\right)\psi_1 = 0$$

In these equations, we have used the terms

$$\sigma_{1,2} = \frac{h(\lambda+\mu)}{\pi\lambda/\sqrt{12}}\left(k_{1,2}^2 - \frac{\omega^2}{c_L^2}\right) \qquad K^2 = \frac{\omega^2}{c_T^2}$$

$$k_{1,2}^2 = \frac{3}{2}\left(\frac{\pi^2}{12h}\right)^2\left\{\left(\frac{c_L^2}{c_T^2}+1\right)\frac{\omega^2}{\omega_0^2} - \frac{c_P^2}{c_T^2} \pm \sqrt{\left[\left(\frac{c_L^2}{c_T^2}+1\right)\frac{\omega^2}{\omega_0^2} - \frac{c_P^2}{c_T^2}\right]^2 + 4\frac{c_L^2}{c_T^2}\frac{\omega^2}{\omega_0^2}\left(1-\frac{\omega^2}{\omega_0^2}\right)}\right\}$$

where $\omega_0^2 = \pi\sqrt{c_L}/4h^2$. The stress components needed are then given by, suppressing $e^{-i\omega t}$ time variation,

$$\sigma_{rr} = (\lambda+2\mu)\frac{\partial v_r}{\partial r} + \lambda\left(\frac{1}{r}\frac{\partial v_\theta}{\partial\theta} + \frac{v_r}{r} + \frac{v_z}{h}\right) \tag{7.60}$$

$$\sigma_{r\theta} = \mu\left(\frac{1}{r}\frac{\partial v_r}{\partial\theta} + \frac{\partial v_\theta}{\partial r} - \frac{v_\theta}{r}\right)$$

$$\sigma_{rz} = \mu\left(\frac{z}{h}\frac{\partial v_z}{\partial r}\right)$$

Now consider an incident plane wave described by

$$\phi_1^{inc} = e^{ik_1x} \qquad \phi_2^{inc} = \psi^{inc} = 0 \tag{7.61}$$

and expand scattered and transmitted waves in terms of the general solutions of the scalar wave equation:

$$\begin{aligned}
\phi_1^{scat} &= \sum_{n=-\infty}^{\infty} A_n i^n H_n(k_1 r)e^{in\theta} \\
\phi_2^{scat} &= \sum_{n=-\infty}^{\infty} B_n i^n H_n(k_2 r)e^{in\theta} \\
\psi^{scat} &= \sum_{n=-\infty}^{\infty} C_n i^n H_n(Kr)e^{in\theta} \\
\phi_1^{trans} &= \sum_{n=-\infty}^{\infty} A'_n i^n J_n(k'_1 r)e^{in\theta} \\
\phi_2^{trans} &= \sum_{n=-\infty}^{\infty} B'_n i^n J_n(k'_2 r)e^{in\theta} \\
\psi^{trans} &= \sum_{n=-\infty}^{\infty} C'_n i^n J_n(K' r)e^{in\theta}
\end{aligned} \tag{7.62}$$

where $A_n, \dots, C'_n$ are unknown modal coefficients to be determined from the boundary conditions. Note that the incident plane wave potential can also be expanded in cylindrical functions:

$$\phi_1^{inc} = e^{ik_1 r\cos\theta} = \sum_{n=-\infty}^{\infty} i^n J_n(k_1 r)e^{in\theta} \tag{7.63}$$

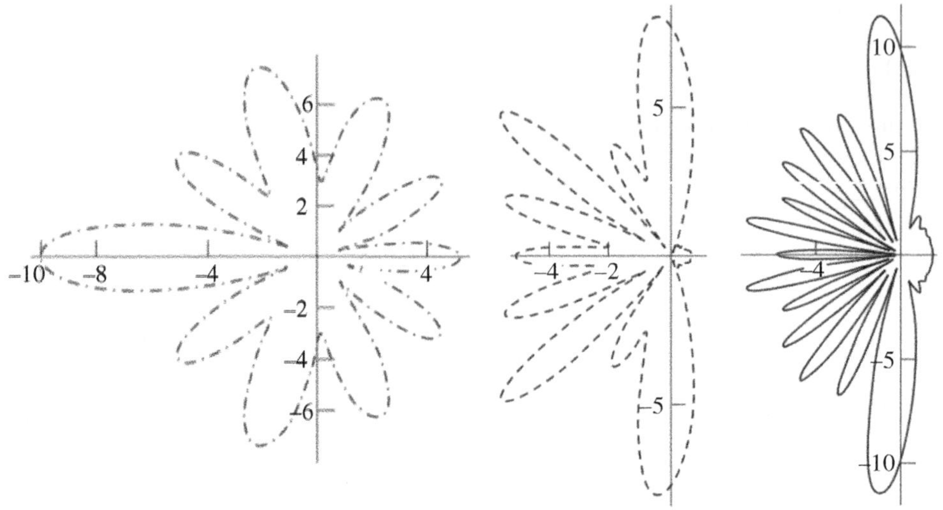

Figure 7.15 Polar plots of the magnitude of the scattered amplitude of a plane wave incident from the left on isolated holes with three different radii. Source: McKeon and Hinders [20]/with permission of Elsevier.

Applying boundary conditions, we find the coefficients to be $A_n = \frac{\Delta_1}{\Delta_0}$ and $B_n = \frac{\Delta_2}{\Delta_0}$, where

$$\beta_{1,2} = \frac{1}{2}\frac{c_L^2}{c_T^2}\left(\frac{\sigma_{1,2}}{h} - (k_{1,2}a)^2\right) + n^2 - \frac{\sigma_{1,2}}{h} - 1$$

and

$$\beta'_{1,2} = \frac{1}{2}\frac{c_L^2}{c_T^2}\left(\frac{\sigma'_{1,2}}{h} - (k'_{1,2}a)^2\right) + n^2 - \frac{\sigma'_{1,2}}{h} - 1$$

and the Cramer's rule determinants can be written in closed form (on the following page) as an exact solution to this approximate formulation of plate wave scattering from cylindrical inclusions. It's then a simple matter to put the pieces together and make plots. Figure 7.15 shows the angular distribution of scattering for three different-sized holes in a plate. Figure 7.16 is for a steel disc in an aluminum plate, as an approximation to Lamb wave scattering from rivets; Figure 7.17 is for a titanium disk. They show the scattering for varying values of ω/ω_0 and a/h.

$$\begin{aligned}\Delta_0 = &\left(\frac{k'_1 a J'_n(k'_1 a)}{J_n(k'_1 a)} - \frac{k'_2 a J'_n(k'_2 a)}{J_n(k'_2 a)}\right)\left[\left(\frac{k_2 a H'_n(k_2 a)}{H_n(k_2 a)} - 1\right)\beta_1 - \left(\frac{k_1 a H'_n(k_1 a)}{H_n(k_1 a)} - 1\right)\beta_2\right]\\ &+\frac{\mu'}{\mu}\left(\frac{k_1 a H'_n(k_1 a)}{H_n(k_1 a)} - \frac{k_2 a H'_n(k_2 a)}{H_n(k_2 a)}\right)\left[\left(\frac{k'_1 a J'_n(k'_1 a)}{J_n(k'_1 a)} - 1\right)\beta'_2 - \left(\frac{k'_2 a J'_n(k'_2 a)}{J_n(k'_2 a)} - 1\right)\beta'_1\right]\end{aligned} \tag{7.64}$$

$$\begin{aligned}\Delta_1 = \frac{J_n(k_1 a)}{H_n(k_1 a)}&\left\{\left(\frac{k'_1 a J'_n(k'_1 a)}{J_n(k'_1 a)} - \frac{k'_2 a J'_n(k'_2 a)}{J_n(k'_2 a)}\right)\left[\left(\frac{k_2 a H'_n(k_2 a)}{H_n(k_2 a)} - 1\right)\beta_1 - \left(\frac{k_1 a J'_n(k_1 a)}{J_n(k_1 a)} - 1\right)\beta_2\right]\right.\\ &\left.+\frac{\mu'}{\mu}\left(\frac{k_1 a J'_n(k_1 a)}{J_n(k_1 a)} - \frac{k_2 a H'_n(k_2 a)}{H_n(k_2 a)}\right)\left[\left(\frac{k'_1 a J'_n(k'_1 a)}{J_n(k'_1 a)} - 1\right)\beta'_2 - \left(\frac{k'_2 a J'_n(k'_2 a)}{J_n(k'_2 a)} - 1\right)\beta'_1\right]\right\}\end{aligned}$$

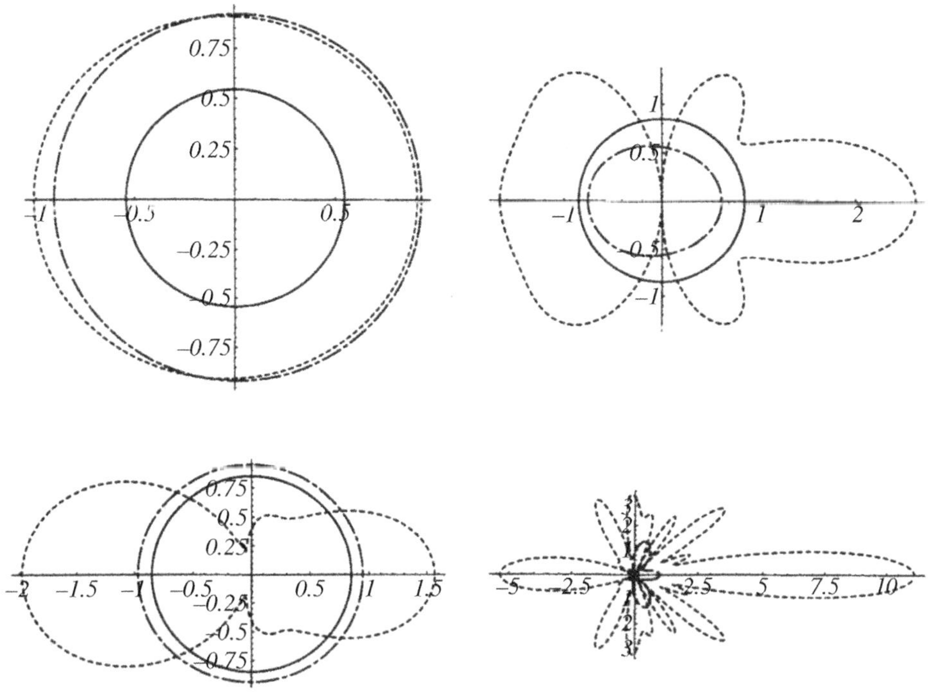

Figure 7.16 Angular far-field scattering for in-plane steel disk in aluminum plate. In each, the solid line is for $\omega/\omega_0 = 0.1$, the long-short dashed line is for $\omega/\omega_0 = 0.5$, and the dashed line is for $\omega/\omega_0 = 0.9$. The four plots are for (a) $a/h = 0.1$, (c) $a/h = 0.5$, (c) $a/h = 1.0$, and (d) $a/h = 10$.

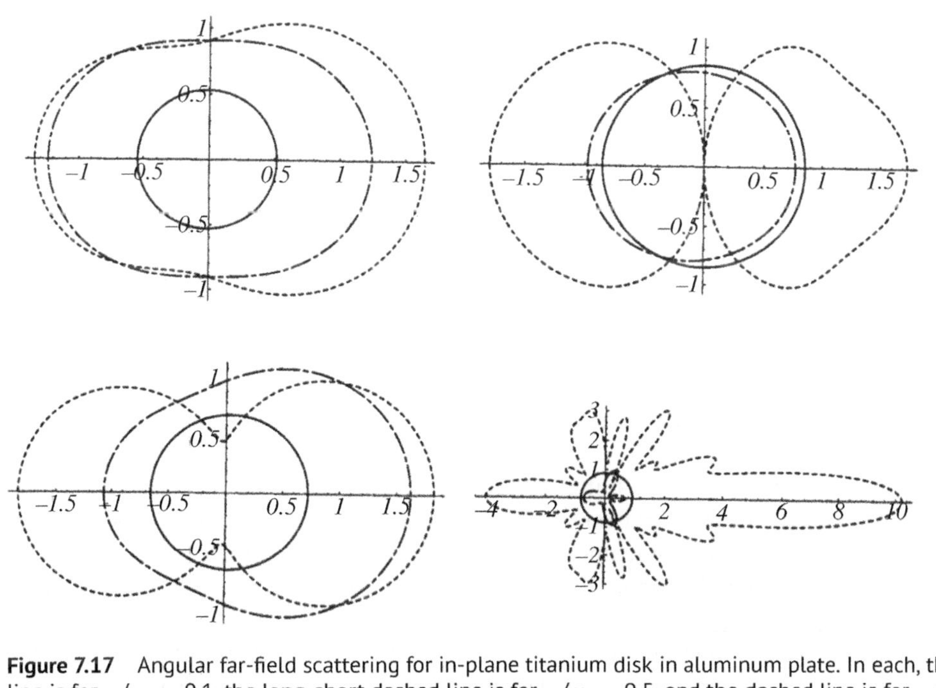

Figure 7.17 Angular far-field scattering for in-plane titanium disk in aluminum plate. In each, the solid line is for $\omega/\omega_0 = 0.1$, the long-short dashed line is for $\omega/\omega_0 = 0.5$, and the dashed line is for $\omega/\omega_0 = 0.9$. The four plots are for (a) $a/h = 0.1$, (b) $a/h = 0.5$, (c) $a/h = 1.0$, and (d) $a/h = 10$.

$$\Delta_2 = \frac{J_n(k_1a)}{H_n(k_2a)}\left(\frac{k_1'aJ_n'(k_1'a)}{J_n(k_1'a)} - \frac{k_2'aJ_n'(k_2'a)}{J_n(k_2'a)}\right)\left(\frac{k_1aJ_n'(k_1a)}{J_n(k_1a)} - \frac{k_1aH_n'(k_1a)}{H_n(k_1a)}\right)$$
$$\times\left\{\beta_1 - \frac{\mu'}{\mu}\left(\frac{k_1aH_n'(k_1a)}{H_n(k_1a)} - \frac{k_2aH_n'(k_2a)}{H_n(k_2a)}\right)\left[\left(\frac{k_1'aJ_n'(k_1'a)}{J_n(k_1'a)} - 1\right)\beta_2' - \left(\frac{k_2'aJ_n'(k_2'a)}{J_n(k_2'a)} - 1\right)\beta_1'\right]\right\}$$

7.4 Thermal "Wave" Scattering

If you've ever farted with a lapdog snoozing on your lap, you may have been surprised how fast the dog leaped off your lap and ran to the other side of the room. You may have thought that your dog was startled by the "plip" sound, but that's not what just happened. Think about the last time you walked that dog, perhaps expecting to get some exercise yourself along the way. Mostly your dog was stopping to sniff the smells along the way and every few feet or so contribute to the neighborhood smell story. Especially if your dog has just spent the afternoon watching out the window while you were at work, it's pretty important to fill in the rest of the story about who walked past earlier. I let my dog lead the way when we go out for a walk because it all smells the same to me anyway. Not to her, though. Her sense of smell is much more sensitive than mine, and she is able to pick up quite a lot of information from smells and how they dissipate over time.

When you farted, even if it wasn't loud at all, you overwhelmed the sensitive smeller of your lapdog. Think of it this way. If you're temporarily blinded by a flash of lightning nearby, there's nothing you can do but cover your ears to protect from the imminent clap of thunder. There will be a slight delay, which depends on distance because the speed of light is much faster than the speed of sound. The clap of your butt cheeks when you farted just now announced the imminent nose-blinding flash of your fart. You really should eat better, you know. Your dog jumped down from the sofa and ran across the room because she didn't want her nose to be overwhelmed. She wasn't scared. She's trying to retain the ability to track subtle variations in smells, especially their dissipations, which is how she tells time. I assume you've noticed that she's waiting at the window for you to come home and provide walks, food, and scritches. She wasn't pining for you all afternoon. She knew when you were coming. Even if she was snoozing when you pulled into the driveway, she heard your car and opened at least one eye to see you come up the front walk. You may even have heard her see you because she's a yippie little lapdog, after all.

The point of all that is that light and sound are wave phenomena of the sort we've been discussing, and wave propagation is fundamentally different from diffusion phenomena. But we often do our scattering analysis in frequency domain, so mathematically, the distinction between waves and diffusion comes down to whether the wave number is real or complex. If we're in the habit of making the wave number complex in order to introduce attenuation ad hoc or finite conductivity formally, it's a very small leap to apply our mathematical machinery to diffusion phenomena like the conduction of heat in solids.

Infrared thermography is a bread-and-butter NDE method, although it has undergone profound changes in recent years as the cost of IR cameras dropped dramatically. These cameras used to cost as much as my house; now they are attachments for smartphones that cost about the same as those earbuds you keep leaving behind in your Ubers. The essence of thermography is applying pulses of heat with flashlamps, or step functions of heat with quartz lamps, or time-harmonic heat flow with chopped lasers, etc. and then monitoring the heat flow into and back out of (or through) a solid structure with an IR camera. Back in the days when the cost of the camera was a significant barrier to entry, there weren't all that many people doing thermography and academic feuds tended to develop. I used to have to caution my graduate students to not get sucked into arguments like whether it was proper to use the term "thermal waves" or not.

One application of interest was using IR thermography to evaluate exotic new fiber-reinforced composite materials for high-speed (supersonic) aircraft structures, and chopped lasers were being used to apply heat in a controlled manner.[6] The graduate student who was doing the experiments was unsure how to model his results, so we went to the chalkboard and I sketched the problem for him and sent him off to do the algebra and make some plots.

The problem geometry assumes a heat source at some distance r_0 from a fiber of radius $r = a$ at the origin of a cylindrical coordinate system. The matrix is material 1 and the fiber embedded in the matrix is material 2. Since the excitation is via a chopped laser, we suppress a harmonic time variation $e^{i\omega t}$ and the heat equation becomes

$$\left(\nabla^2 - \frac{i\omega}{\alpha}\right) v = 0 \tag{7.65}$$

where $k = \sqrt{\omega/i\alpha}$ and α is the thermal diffusivity and v is the scalar temperature field. This allows us to deal with the Helmholtz equation as usual.

$$\nabla^2 v + k^2 v = 0 \tag{7.66}$$

which has the general solution in cylindrical coordinates of a linear combination of Bessel functions and sinusoids as before. The total field in material 1 will be incident plus scattered, with a transmitted field in the fiber. Boundary conditions are simply that the temperature and normal derivative of the temperature (heat flux) must be continuous at the surface, $r = a$. We can write the incident, scattered, and transmitted fields as:

$$v_{inc} = \sum_{n=0}^{\infty} \epsilon_n A_n J_n(k_1 r) \cos n\theta$$

$$v_{scat} = \sum_{n=0}^{\infty} \epsilon_n B_n H_n^{(2)}(k_1 r) \begin{pmatrix} \cos n\theta \\ \sin n\theta \end{pmatrix}$$

$$v_{scat} = \sum_{n=0}^{\infty} \epsilon_n C_n J_n(k_2 r) \begin{pmatrix} \cos n\theta \\ \sin n\theta \end{pmatrix}$$

where $A_n = (-1)^n H_n^{(1)}(-k_1 r_0)/4\alpha_1$ and B_n, C_n are unknown modal coefficients determined from the boundary conditions enforced at $r = a$

$$v_{inc} + v_{scat} = v_{trans} \qquad K_1\left(\frac{\partial v_{inc}}{\partial r} + \frac{\partial v_{scat}}{\partial r}\right) = K_2 \frac{\partial v_{trans}}{\partial r} \tag{7.67}$$

where K_1 and K_2 are the thermal conductivities of the matrix and fiber, respectively. The algebra is straightforward enough that I've just done it myself, because it simplifies things to write things in logarithmic-derivative form and I found the answers that Pierre derived a little confusing. Also, orthogonality means that we'll need to choose $\cos n\theta$ as the angular form for both the scattered

6 See Emeric and Winfree [21]. The last time I chatted at length with Dr. Emeric was in the Fall of 2019. He was just about to start a new job as Global Head of Research & Development, Monitoring and Analytics for Philips at their new Innovation Hub in Cambridge, MA and we were discussing several research projects we might-could do once he got settled in and knew what his budget for sponsoring university research was going to be. What neither of us knew was that COVID-19 was about to hit, and Philips was going to catch hell for not having gotten the stockpile of ventilators refreshed in time for the pandemic that nobody saw coming. In the summer of 2020, Dr. Emeric moved to Ortho Clinical Diagnostics as Global Head Of Platform Research & Development. We take COVID test kits for granted now, but think back to what it was like in the summer of 2020....

and transmitted waves. Feel free to check my algebra to verify that the scattered and transmitted amplitude modal coefficients are

$$B_n = A_n \frac{J_n(k_1 a)}{H_n^{(2)}(k_1 a)} \left(\frac{\frac{k_1 a J_n'(k_1 a)}{J_n(k_1 a)} - \frac{K_2}{K_1} \frac{k_2 a J_n'(k_2 a)}{J_n(k_2 a)}}{\frac{k_1 a H_n'(k_1 a)}{H_n(k_1 a)} + \frac{K_2}{K_1} \frac{k_2 a J_n'(k_2 a)}{J_n(k_2 a)}} \right) \tag{7.68}$$

$$C_n = A_n \frac{J_n(k_1 a)}{J_n(k_2 a)} \left(\frac{\frac{k_1 a H_n'(k_1 a)}{H_n(k_1 a)} - \frac{k_1 a J_n'(k_1 a)}{J_n(k_1 a)}}{\frac{k_1 a H_n'(k_1 a)}{H_n(k_1 a)} + \frac{K_2}{K_1} \frac{k_2 a J_n'(k_2 a)}{J_n(k_2 a)}} \right) \tag{7.69}$$

Exercise 7.8 After you check my algebra, and perhaps notice that this looks pretty similar to the analogous acoustic case, make some plots of the scattered temperature distribution for a fiber with $a = 75\,\mu\text{m}$ and excitation frequencies in the range 1–20 Hz. You'll have to look up some typical properties for thermal conductivity and diffusivity.

Scattering plots aren't going to be all that interesting because heat conduction is diffusive, of course. What you'll find, though, is that if the thermal "wave propagation" for the fiber is slower than the matrix then the heat will tend to build up in front of the fiber, whereas if the fiber conducts heat better than the matrix the heat will build up behind the fiber. Make some plots, but take care that you're implementing the Bessel functions properly because their both $k_1 a$ and $k_2 a$ are complex. After you've made some plots, you can then form an opinion about whether it's appropriate to use the term "thermal wave" scattering for this. I'm pretty sure that Skip Favro would still say no. Steve Shepard (https://www.thermalwave.com) would probably be fine with that.

7.5 Scattering from a Semicircular Gap in a Ground Plane

There's an issue that we haven't had to worry about quite yet that is about to matter quite a lot: *orthogonality*. For both the spherical and cylindrical scatterers, the eigenfunctions that we've used are orthogonal functions, and it turned out that the summations of terms didn't really make things any more complicated. At some point, we just solved the boundary value problems as if those summations didn't matter. Because they didn't. Rather than explaining all about orthogonality, I think it's better for me to show you a situation where the orthogonality is incomplete and we can't just solve the boundary-value problem term-by-term.

The issue came up in the late 1980s as we were working to understand more and more subtle features of radar scattering, either to make one's own air vehicles stealthier or to try to identify scattering behaviors that would allow one to identify others' already stealthy air vehicles. Any vehicle is going to have various doors and hatches and whatnot that open and close, and there's always going to be some gap around the edges when they are closed. If your vehicle gets stealthy enough, the scattering contribution from these sorts of cracks and gaps might be the thing that gives you away. In some cases, you can fill the gaps with goop or tape them over in order to mask them electromagnetically, but that's not really a practical solution. In addition, before going to that sort of trouble, it seems sensible to see if a scattering model could be used to identify potential vulnerabilities.

The "crack" problem has been studied by many people including Lord Rayleigh, who considered the problem of plane wave scattering from a half-cylindrical indentation in a ground plane in 1907 [22]. Of course, he wasn't concerned about counter-low observables problems. He was interested in optical diffraction. He also couldn't quite figure out how to attack the cylindrical indentation problem, so he solved a similar problem but then didn't feel the need to go back and update his title to make it appropriately descriptive. We'll give him a pass on that because he is Lord Rayleigh after all, and I find that whenever I read one of his papers, I come away with both technical insights and new vocabulary words.

The problem that Lord Rayleigh actually solved was electromagnetic scattering from a half-cylindrical protuberance on a perfectly conducting ground plane. The vocab word is: *excrescence*. It means protuberance, which is also an excellent word but not quite as esoteric. Lord Rayleigh's technical insight was to set up the problem with both an incident plane wave and another plane wave that mirrors it such that the boundary condition for a perfect conductor is automatically satisfied at the ground plane. All that's needed is to solve the relatively straightforward problem of scattering from a perfectly conducting cylinder, which we did earlier in this chapter, and the cylindrical excrescence solution pops right out. The thing that makes this formulation work is that the scatterer is impenetrable, so the "extra" fields for $r < a$ and $y < 0$ don't matter. Incomplete orthogonality doesn't matter either.

What we're going to do is rather similar, but we'll have to extend Rayleigh's method via a dual-series eigenfunction approach to solve the scattering of electromagnetic plane waves from a half-cylindrical (circular) indentation in a ground plane. As Lord Rayleigh did, we will use to our advantage the coincidence of the problem geometry and a constant coordinate surface in circular coordinates. The main difference between this method and typical eigenfunction problems lies in the incomplete orthogonality of the sinusoids over a half-space, and the use of two separate regions of the problem because we will have to consider fields in the region $r < a$. We consider two polarizations: incident transverse magnetic (TM) and incident transverse electric (TE). The analysis for each is done similarly, but they are presented separately to try to make things easier to follow.

We consider a TM plane wave which is polarized in the z-direction and makes an angle φ^{inc} with the positive x-semi-axis. It is incident upon and scatters from a half-cylindrical indentation, or channel, in the ground plane, as shown in Figure 7.18.

The channel is of radius $r = a$ and of infinite extent in the z-direction. All angles are measured positive in the counter-clockwise direction starting with $\varphi = 0$ along the positive x-semiaxis. Angles are in the range $0 < \varphi < 2\pi$. The channel is described by $\pi < \varphi < 2\pi$ and $r = a$. The surface $0 < \varphi < \pi$ and $r = a$ is referred to as the aperture. It is the region complementary to the channel. The problem will be formulated so that the boundary conditions along the ground plane will be automatically satisfied, and then three boundary conditions will be applied at the surface $r = a$: one on the channel and two across the aperture. The region $r < a$ is the interior region and is denoted

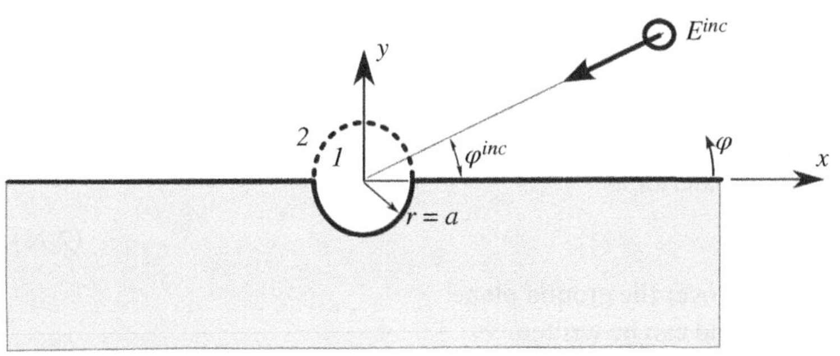

Figure 7.18 An incident TM plane wave at angle φ^{inc} to a semicircular channel of radius $r = a$ in a perfectly conducting ground plane.

by the superscript 1. The region $r > a$ and $0 < \varphi < \pi$ is the exterior region and is denoted by the superscript 2.

The incident field is a plane wave as described earlier. We write

$$E_z^{inc} = e^{ikr\cos(\varphi-\varphi^{inc})} \tag{7.70}$$

where unit amplitude has been assumed and $e^{i\omega t}$ time dependence suppressed. This plane wave has the following well-known expansion in terms of cylindrical Bessel functions

$$E_z^{inc} = \sum_{n=-\infty}^{\infty} i_n J_n(kr) e^{in(\varphi-\varphi^{inc})} \tag{7.71}$$

In the exterior region, the scattered field will be considered to be made up of two parts. The first is the reflected wave that would be present if there were no channel, and the second is the deviation from that caused by diffraction at the channel. Thus, in the exterior region, we refer to the incident, reflected, and diffracted waves. We are, of course, most interested in the diffracted wave.

The reflected wave is also a plane wave and is well known. We have $\varphi^{ref} = 2\pi - \varphi^{inc}$ so that the reflected wave can be written as:

$$E_z^{ref} = -e^{ikr\cos(\varphi+\varphi^{inc})} \tag{7.72}$$

We now note that $\left(E_z^{inc} + E_z^{ref}\right)_{\varphi=0,\pi} = 0$, which means that they together satisfy the appropriate boundary conditions on the ground plane. Don't take my word for that, though. For $\varphi = 0$, we have

$$E_z^{inc} + E_z^{ref} = e^{ikr\cos(-\varphi^{inc})} - e^{ikr\cos(\varphi^{inc})} = 0 \tag{7.73}$$

since $\cos(-\alpha) = \cos(\alpha)$. For $\varphi = \pi$, we have

$$E_z^{inc} + E_z^{ref} = e^{ikr\cos(\pi-\varphi^{inc})} - e^{ikr\cos(\pi+\varphi^{inc})} = 0 \tag{7.74}$$

and since $\cos(a \pm b) = \cos a \cos b \mp \sin a \sin b$ we have $\cos(\pi \pm \varphi^{inc}) = -\cos\varphi^{inc}$ and thus

$$E_z^{inc} + E_z^{ref} = e^{-ikr\cos(\varphi^{inc})} - e^{-ikr\cos(\varphi^{inc})} = 0 \tag{7.75}$$

Writing the reflected field in terms of the cylindrical Bessel functions, we have

$$E_z^{ref} = -\sum_{n=-\infty}^{\infty} i^n J_n(kr) e^{in(\varphi+\varphi^{inc})} \tag{7.76}$$

Finally, for the diffracted exterior field, we write

$$E_z^{dif} = \sum_{n=0}^{\infty} A_n H_n^{(2)}(kr) \sin n\varphi \tag{7.77}$$

where A_n is an unknown modal coefficient and we have expanded the outgoing wave in terms of Hankel functions (of the second kind) and also $\sin(n\phi)$. Noting that because $\sin 0 = \sin(n\pi) = 0$, the diffracted wave vanishes on the ground plane and so satisfies the appropriate boundary condition there. If we write the total field in the exterior as:

$$E_z^2 = E_z^{inc} + E_z^{ref} + E_z^{dif} \tag{7.78}$$

we see that $E_z^2 = 0$ for $\varphi = 0, \pi$, that is, over the ground plane.

In the interior region, the total E_z-field can be written

$$E_z^1 = \sum_{n=-\infty}^{\infty} D_n J_n(kr) e^{in\varphi} \tag{7.79}$$

where D_n is an unknown modal coefficient. However, it will be more convenient for our analysis if we write this as:

$$E_z^1 = \sum_{n=-\infty}^{\infty} J_n(kr)\left[B_n \cos(n\varphi) + C_n \sin(n\varphi)\right] \tag{7.80}$$

where B_n, C_n are the modal coefficients that will be determined from the boundary conditions. Sorry for introducing D_n and then immediately replacing it with B_n and C_n. Some authors write the angular part in terms of $e^{in\varphi}$ and some do it in terms of $\sin(n\varphi)$ and $\cos(n\varphi)$, which can sometimes throw one a little. We've got three unknown modal coefficients and so we're going to need three boundary condition equations. First a little more trigonometry, though.

We now need to write the incident and reflected E_z-fields as series over $n = 0$ to ∞ rather than over all n. We note that $J_{-n}(kr) = (-1)^n J_n(kr)$ and $i^{-n} = (-1)^n i^n$ so that

$$E_z^{inc} = -J_0(kr) + 2\sum_{n=0}^{\infty} i^n J_n(kr) \cos n(\varphi - \varphi^{inc}) \tag{7.81}$$

and similarly for the reflected wave

$$E_z^{ref} = J_0(kr) + 2\sum_{n=0}^{\infty} i^n J_n(kr) \cos n(\varphi + \varphi^{inc}) \tag{7.82}$$

Hence

$$E_z^{inc} + E_z^{ref} = 2\sum_{n=0}^{\infty} i^n J_n(kr)\left[\cos n(\varphi - \varphi^{inc}) - \cos n(\varphi + \varphi^{inc})\right] \tag{7.83}$$

and if we again use the trigonometric identity $\cos(a \pm b) = \cos a \cos b \mp \sin a \sin b$ we find

$$E_z^{inc} + E_z^{ref} = 4\sum_{n=0}^{\infty} i^n J_n(kr) \sin n\varphi \tag{7.84}$$

Recalling the expression for the diffracted field, we write the total exterior field

$$E_z^2 = \sum_{n=0}^{\infty} \left(4i^n J_n(kr) \sin n\varphi^{inc} + A_n H_n^{(2)}(kr)\right) \sin n\varphi \tag{7.85}$$

In this expression, A_n is the modal coefficient that we want to solve for.

The φ-component of the magnetic field is also of interest in this problem. It is related to E_z by

$$H_\varphi = \frac{-i}{\omega\mu} \partial_r E_z \tag{7.86}$$

so that the exterior and interior H_φ-fields are

$$H_\varphi^2 = \frac{-ik}{\omega\mu} \sum_{n=0}^{\infty} \left(4i^n J_n'(kr) \sin n\varphi^{inc} + A_n H_n^{'(2)}(kr)\right) \sin n\varphi \tag{7.87}$$

$$H_\varphi^1 = \frac{-ik}{\omega\mu} \sum_{n=0}^{\infty} J_n'(kr)\left(B_n \cos n\varphi + C_n \sin n\varphi\right) \tag{7.88}$$

where $J_n'(z) = \partial_z J_n(z)$ is differentiation with respect to the argument of the Bessel function.

The required boundary conditions for solving this TM problem are

$$E_z^2 = 0 \text{ for } r = a \text{ and } \pi < \varphi < 2\pi \tag{7.89}$$

$$E_z^2 = E_z^1 \text{ for } r = a \text{ and } 0 < \varphi < \pi \tag{7.90}$$

$$H_\varphi^2 = H_\varphi^1 \text{ for } r = a \text{ and } 0 < \varphi < \pi \tag{7.91}$$

The first condition enforces zero tangential electric field on the surface of the perfectly reflecting channel and the last two ensure continuous tangential electric and magnetic field across the aperture, or (imaginary) surface that is complementary to the channel.

The boundary conditions give the following three equations:

$$\sum_{n=0}^{\infty} J_n(ka)B_n \cos n\varphi + \sum_{n=0}^{\infty} J_n(ka)C_n \sin n\varphi = 0 \tag{7.92}$$

$$\sum_{n=0}^{\infty} J_n(ka)B_n \cos n\varphi + \sum_{n=0}^{\infty} J_n(ka)C_n \sin n\varphi = \sum_{n=0}^{\infty}\left(4i^n J_n(ka)\sin n\varphi^{inc} + A_n H_n^{(2)}(ka)\right)\sin n\varphi^{inc} \tag{7.93}$$

$$\sum_{n=0}^{\infty} J'n(ka)B_n \cos n\varphi + \sum_{n=0}^{\infty} J_n'(ka)C_n \sin n\varphi = \sum_{n=0}^{\infty}\left(4i^n J_n'(ka)\sin n\varphi^{inc} + A_n H_n'^{(2)}(ka)\right)\sin n\varphi^{inc} \tag{7.94}$$

where we must keep in mind that the first is valid for $\pi < \varphi < 2\pi$ and the second two are valid for $0 < \varphi < \pi$.

So now I'm going to have to ask you to bear with me for a couple of pages. First, I'm going to define some functions that merely serve the purpose of containing the functional variation on things besides φ and then I'll do some trigonometry that will allow us to write all three of these boundary condition equations in the same form. Then we'll be able to isolate the key behavior of interest with regards to incomplete orthogonality. As I said, bear with me just a bit.

We define the following for convenience:

$$\mathcal{F}_n = 4i^n J_n(ka)\sin n\varphi^{inc} + A_n H_n^{(2)}(ka) - C_n J_n(ka)$$
$$\mathcal{G}_n = B_n J_n(ka)$$
$$\mathcal{R}_n = -C_n J_n(ka)$$

as well as $\mathcal{F}_n' = \frac{\partial}{\partial(ka)}\mathcal{F}_n$ and so on. The boundary conditions are then

$$\sum_{n=0}^{\infty} \mathcal{G}_n \cos n\varphi = \sum_{n=0}^{\infty} \mathcal{R}_n \sin n\varphi \qquad (\pi < \varphi < 2\pi)$$
$$\sum_{n=0}^{\infty} \mathcal{G}_n \cos n\varphi = \sum_{n=0}^{\infty} \mathcal{F}_n \sin n\varphi \qquad (0 < \varphi < \pi)$$
$$\sum_{n=0}^{\infty} \mathcal{G}_n' \cos n\varphi = \sum_{n=0}^{\infty} \mathcal{F}_n' \sin n\varphi \qquad (0 < \varphi < \pi)$$

In the first equation, we make the change of variables $\phi = \varphi - \pi$ so that

$$\sum_{n=0}^{\infty} \mathcal{G}_n \cos n(\phi + \pi) = \sum_{n=0}^{\infty} \mathcal{F}_n \sin n(\phi + \pi) \qquad (0 < \phi < \pi)$$

Noting that $\cos n(\phi + \pi) = (-1)^n \cos n\phi$ and $\sin n(\phi + \pi) = (-1)^n \sin n\phi$, we have

$$\sum_{n=0}^{\infty} (-1)^n \mathcal{G}_n \cos n\phi = \sum_{n=0}^{\infty} (-1)^n \mathcal{F}_n \sin n\phi \qquad (0 < \phi < \pi)$$

and assuming that you're still bearing with me as I include so very many of the small steps, now define $\overline{\mathcal{G}_n} = (-1)^n \mathcal{G}_n$ and such to write

$$\sum_{n=0}^{\infty} \overline{\mathcal{G}_n} \cos n\phi = \sum_{n=0}^{\infty} \overline{\mathcal{R}} \sin n\phi \qquad (0 < \phi < \pi)$$

which is exactly the same form in the angle variable as the other two boundary condition equations. Please do flip back a couple of pages and check that, since I went through the trouble of including mathematical details so you wouldn't even have to go look up simple trigonometric identities.

We now make use of the following orthogonality relations among sinusoids, which I'm typing here so you don't have to go look them up.

$$\int_0^{\pi} \sin n\varphi \sin m\varphi d\varphi = \frac{\pi}{2}\delta_{nm}$$

$$\int_0^{\pi} \sin n\varphi \cos m\varphi d\varphi = \begin{cases} 0 & \text{if } n-m \text{ is even} \\ \frac{2n}{n^2-m^2} & \text{if } n-m \text{ is odd} \end{cases}$$

So now take our "generic" boundary condition equation, multiply it by $\sin m\varphi$ and integrate from 0 to π.

$$\int_0^{\pi} \sum n = 0^{\infty} \mathcal{G}_n \cos n\phi d\varphi = \int_0^{\infty} \sum_{n=0}^{\infty} \mathcal{R} \sin n\phi d\varphi$$

Exchanging the sums and integrals and using the orthogonality relation gives

$$2\sum_n^{*} \left(\frac{m\mathcal{G}_n}{m^2 - n^2} \right) = \frac{\pi}{2} \mathcal{F}_m \tag{7.95}$$

where $\sum_n^*$ is used to indicate a sum over n from zero to infinity where $(n-m)$ is odd. Yes, I know that's strange. It won't be difficult to program, though, so let's just keep going. Ok, fine. Here's this strange even–odd summation written out explicitly

$$\mathcal{F}_{m_e} = \sum_{n=1,3,5}^{\infty} \frac{m\mathcal{G}_n}{m^2 - n^2}$$

$$\mathcal{F}_{m_o} = \sum_{n=0,2,4}^{\infty} \frac{m\mathcal{G}_n}{m^2 - n^2}$$

Referring to our previous definitions of the variables we defined for convenience, the three boundary conditions give us

$$4i^m J_n(ka) \sin m\varphi^{inc} + A_m H_m^{(2)}(ka) - C_m J_m(ka) = \frac{4m}{\pi} \sum_n^{*} \frac{B_n J_n(ka)}{m^2 - n^2}$$

$$4i^m Jl_n(ka) \sin m\varphi^{inc} + A_m H_m^{\prime(2)}(ka) - C_m J_m'(ka) = \frac{4m}{\pi} \sum_n^{*} \frac{B_n J_n'(ka)}{m^2 - n^2}$$

$$C_m J_m(ka) = \frac{4m}{\pi} \sum_n^{*} \frac{B_n J_n(ka)}{m^2 - n^2}$$

where we've used the fact that $n - m$ is odd in the last of these, which took care of a minus sign for us. These three equations will be used to solve for the three modal coefficients A_n, B_n, C_n. Of course, all we really want is A_n since that is the one for the diffracted field, but we can't just solve for it. Go ahead and stare at the three equations. I'll wait.

I don't know what algebra you just tried to do, but what I did was to solve the first two equations for A_n and then equated them. I then used the Wronskian relationship $\Delta\left(J_m(ka), H_m^{(2)}(ka)\right) = \frac{-2i}{\pi ka}$ and substituted in for C_m to end up writing an expression that only contains B_n

$$\sum_n^* \frac{B_n}{m^2 - n^2}\left\{J_n(ka) + \frac{i\pi ka}{2}\left(J_n(ka)H_m'^{(2)}(ka) - J_n'(ka)H_m^{(2)}(ka)\right)J_m(ka)\right\} = \frac{\pi i^m}{m} J_m(ka)\sin m\varphi^{inc} \tag{7.96}$$

I may have skipped some steps, but I'm fully confident you can fill in the missing ones. One more bit of algebra and then we're pretty much home free. Recall the two equations

$$4i^m J_n(ka)\sin m\varphi^{inc} + A_m H_m^{(2)}(ka) - C_m J_m(ka) = \frac{4m}{\pi}\sum_n^* \frac{B_n J_n(ka)}{m^2 - n^2}$$

$$C_m J_m(ka) = \frac{4m}{\pi}\sum_n^* \frac{B_n J_n(ka)}{m^2 - n^2}$$

Plug the second in to the first and do just a tiny bit more algebra to write

$$A_m = \frac{1}{H_m^{(2)}(ka)}\left\{\frac{8m}{\pi}\sum_n^* \frac{B_n J_n(ka)}{m^2 - n^2} - 4i^m J_m(ka)\sin m\varphi^{inc}\right\} \tag{7.97}$$

Thus, once B_n is solved for from the seemingly simple equation above, we can use this expression to solve for the modal coefficients, A_m, that determine the diffracted fields in the exterior:

$$E_z^{dif} = \sum_{m=1}^{\infty} A_m H_m^{(2)}(kr)\sin m\varphi \tag{7.98}$$

I said seemingly simple because we can't just solve for B_n in the way we want to. Because the orthogonality was incomplete. I also said that the strange even–odd summation wouldn't be difficult to program. I found it pretty simple to do using FORTRAN77 back in the 1980s. That goes double if you're using something like Matlab that's naturally suited to solve matrix equations, which is what we're dealing with. I still have a hardcopy of that F77 code and it's only a couple dozen lines, with the rest of it just a matrix inversion subroutine that somebody gave me. Yes, I'm so old that I remember the days when you asked people you worked with if they had any good matrix inversion routines.

I'm also so old that my instinct about what to do next was to consider the special case of a small-diameter channel where the special functions could be replaced with their small-argument approximations and with ka small you'd only need a few terms in the infinite summations. Don't judge. The VAX 11/750 mainframe I was using was only as fast as the 286 PC that I had on my desk. That PC didn't have a math coprocessor. The good news is that I'm just young enough that I never had to use punch cards. My mother is in her mid-80s and she tells me about having to drive an hour to use the computer to analyze the data for her dissertation. She ultimately paid someone to key the punchcards for her so her program would actually run. Kids these days walk around with semidisposable supercomputers in their pockets....

When $ka \to 0$, the cylinder functions have the simple forms

$$J_n(ka) \to \frac{(ka/2)^n}{n!}$$

$$H_n^{(2)}(ka) \to \frac{i}{n\pi}\frac{n!}{(ka/2)^n}$$

The three equations that we must solve also become much simpler, and we can separate the *ka*-dependence from the necessary numerical matrix inversion. Performing that necessary matrix inversion, we find a leading factor of −0.2 which appears to make good sense because the well-established result for diffraction from a narrow slit gives −0.25 for that factor. Here are the two answers together to reinforce that point

$$E_z^{dif} \approx 0.2\sqrt{2\pi}(ka)^2 e^{i3\pi/4}\frac{e^{-ikr}}{\sqrt{kr}}\sin\varphi\sin\varphi^{inc} \qquad \text{(channel)} \tag{7.99}$$

$$E_z^{dif} \approx 0.25\sqrt{2\pi}(ka)^2 e^{i3\pi/4}\frac{e^{-ikr}}{\sqrt{kr}}\sin\varphi\sin\varphi^{inc} \qquad \text{(slit)} \tag{7.100}$$

The transverse electric polarization for this problem is done in a very similar way, except that the incident *H*-wave is polarized in the *z*-direction. In the small-argument limit, the answer that we're ultimately after turns out to be

$$H_z^{dif} \approx \sqrt{\frac{\pi}{2}}(ka)^2 e^{i3\pi/4}\frac{e^{-ikr}}{\sqrt{kr}}\sin\varphi\sin\varphi^{inc}\left(-1+\cos\varphi\cos\varphi^{inc}\right) \tag{7.101}$$

Here we have neglected $\frac{1}{2}(ka)^2\ln(ka)$ with respect to unity. In the TM case, we were able to compare our result to those for TM wave scattering from a thin slit and we obtained excellent agreement because of the following similarity in both cases: the electric field vector was oriented along the channel/slit and the "open circuit" of the slit does not manifest itself. For the TE polarization, however, the electric field vector is oriented *normal* to the discontinuity, and here the currents excited will be greatly affected by the difference between the channel and the slit. Simply put, the currents can travel up to, down into, out of, and away from the channel. Certainly, there will be scattering from the slope changes of the surface, but that will be less than that caused by the slit, where the currents are stopped completely by the "open circuit."

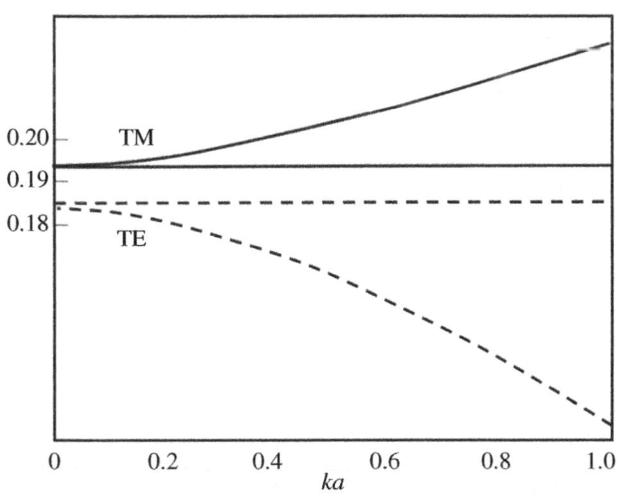

Figure 7.19 Amplitude parameter $\pm Q/\pi(Ka^2)$ plotted vs. *Ka* for TM and TE cases. Note that as *Ka* gets smaller, both cases approach their small-argument approximations, shown by the horizontal lines.

In [23], we plot the exact dual-series eigenfunction solutions for the TM and TE cases and show that they reduce to the low-frequency asymptotic solutions for small channels.[7] In Figure 7.19, I plot just the amplitude parameters to show that they approach the low-frequency approximation derived earlier.

References

1 Born, M. and Wolf, E. (2013). *Principles of Optics: Electromagnetic Theory of Propagation, Interference and Diffraction of Light*. Elsevier.

2 Bohren, C.F. and Huffman, D.R. (2008). *Absorption and Scattering of Light by Small Particles*. Wiley.

3 Kerker, M. (1969). *The Scattering of Light and Other Electromagnetic Radiation, Physical Chemistry: A Series of Monographs*, vol. 16. New York: Academic Press.

4 van de Hulst, H.C. (1957). *Light Scattering by Small Particles*. Wiley. Also there's a version: Dover Books on Physics, 1981.

5 Bowman, J.J., Senior, T.B.A., and Uslenghi, P.L.E. (1987). *Electromagnetic and Acoustic Scattering by Simple Shapes*. New York: Hemisphere Publishing.

6 Pao, Y.H. and Mow, C.C. (1973). *Diffraction of Elastic Waves and Dynamic Stress Concentrations*. New York: Krane Russak.

7 Varadan, V.V., Lakhtakia, A., and Varadan, V.K. (eds.) (1991). *Field Representations and Introduction to Scattering*. New York: North-Holland.

8 Hinders, M.K. (1993). Plane elastic wave scattering from an elastic cylinder. *Il Nuovo Cimento* 108B (3): 285.

9 Dickinson, L.C., Farley, G.L., and Hinders, M.K. (1999). Translaminar reinforced composites: a review. *Composites Technology and Research* 21 (1): 3–15.

10 Faran, J.J. Jr. (1951). Sound scattering by solid cylinders and spheres. *The Journal of the Acoustical Society of America* 23 (4): 405–418.

11 Gao, W. and Hinders, M.K. (2005). Mobile robot sonar interpretation algorithm for distinguishing trees from poles. *Robotics and Autonomous Systems* 53 (2): 89–98.

12 White, R.M. (1958). Elastic wave scattering at a cylindrical discontinuity in a solid. *The Journal of the Acoustical Society of America* 30 (8): 772–785.

13 Bogan, S.D. and Hinders, M.K. (1993). Dynamic stress concentrations in fiber-reinforced composites with interface layers. *Journal of Composite Materials* 27 (11): 1272.

14 Bogan, S.D. and Hinders, M.K. (1994). *Interface Effects in Elastic Wave Scattering, Lecture Notes in Physics: Monographs*, vol. M19. Springer-Verlag.

15 Kane, T.R. and Mindlin, R.D. (1956). High-frequency extensional vibrations of plates. *Journal of Applied Mechanics* 78: 277.

16 Kane, T.R. (1957). Reflection of dilatational waves at the edge of a plate. *Journal of Applied Mechanics* 79: 219.

17 Mindlin, R.D. and Medick, M.A. (1959). Extensional vibrations of elastic plates. *Journal of Applied Mechanics* 80: 561.

---

7 My co-author, Dr. Yaghjian, didn't actually use computers, so he had Margaret Woodworth make the plots for all of his papers. I deferred to his judgment back in those days, but it seems wrong to me now that Margaret wasn't a coauthor on this paper. I googled her just now and see that she was the first author on [24] and has several other papers that come up in Google Scholar. I should also point out that whereas Dr. Yaghjian was an Air Force civil servant, Ms. Woodward was an Air Force contractor; it's a two-tier system and historically women were more likely to be in that second tier, which is not OK.

18 Vemula, C. and Norris, A.N. (1997). Flexural wave propagation and scattering on thin plates using Mindlin theory. *Wave Motion* 26: 1–12.

19 Mindlin, R.D. and Yang, J. (2006). *An Introduction to the Mathematical Theory of Vibrations of Elastic Plates*. World Scientific.

20 McKeon, J.C.P. and Hinders, M.K. (1999). Lamb wave scattering from a through hole. *Journal of Sound and Vibration* 224 (5): 843–862.

21 Emeric, P.R. and Winfree, W.P. (1996). Characterization of composite materials from temporal thermal response. In: *Review of Progress in Quantitative Nondestructive Evaluation*, vol. 15 (ed. D.O. Thompson and D.E. Chimenti), 1275–1282. Boston, MA: Springer. https://doi.org/10.1007/978-1-4613-0383-1_166.

22 Rayleigh, L. (1907). On the light dispersed from fine lines ruled upon reflecting surfaces or transmitted by very narrow slits. *The London, Edinburgh, and Dublin Philosophical Magazine and Journal of Science* 14 (81): 350–359. https://doi.org/10.1080/14786440709463690.

23 Hinders, M.K. and Yaghjian, A.D. (1991). Dual-series solution to scattering from a semicircular channel in a ground plane. *IEEE Microwave and Guided Wave Letters* 1 (9): 239–242.

24 Woodworth, M.B. and Yaghjian, A.D. (1994). Multiwavelength three-dimensional scattering with dual-surface integral equations. *Journal of the Optical Society of America A* 11: 1399–1413.

8

Scattering from Spheroids and Elliptic Cylinders

In for a penny, in for a pound. If you've made it this far, I'm hoping that you'll be willing to go just a bit farther. Maybe I should say further rather than farther because the equations are going to get deep. Not deep, as in deeply meaningful, but deep as in you may feel a bit like you're drowning in the notation. As a way of tossing you a life preserver, be (re)assured that we're going to be following exactly the same solution procedure as before. We'll start by doing separation of variables for the Helmholtz equation to get three ODEs instead of one PDE. Those ODEs will often have names associated with them as will the special functions that are their solutions. We then write plane waves in terms of these orthogonal functions, assume appropriate forms for scattered and transmitted fields in terms of these new-to-us functions, and apply the boundary conditions at the constant coordinate surface which corresponds to the scatterer. The rest is just algebra, and in some cases that algebra is quite simple. In some cases, it's not, of course. Getting numbers out of the functions we're just about to start getting comfortable with used to be virtually impossible, but these days, this part is not particularly difficult. Most of the time. The mathematical functions we get when dealing with the wave equation in spheroidal and elliptic cylinder coordinates sometimes don't have good manners, numerically speaking. But let's not worry about that just yet.

Please keep in mind that the perfect sort of problem is one that's so difficult nobody else can solve it, while simultaneously it's so important that money is no object. Easy problems are no challenge, and lots of people can be found to solve those cheaply. What we're often looking for is some sort of barrier to competition. That's often mathematics. To paraphrase Einstein, most people hold a secret grudge against arithmetic and vegetables. I added the part about vegetables, which all the diets and weight loss plans agree are important to eat. Like vegetables, most people turn up their noses at even simple math. It makes them uncomfortable. I'm not even talking trigonometry here. I'm talking calculating the tip after a dinner out. I'm a little surprised that as a society we haven't all agreed to adjust servers' wages so that an easy-to-guess-at 10% tip would become the norm. I think some people tip 20% because they can't do 15% in their heads. I always just let my wife pay because she likes to travel and gets hotel points on her credit card, and remembers her time as a cocktail waitress in college so probably tips better than me. It all comes out of the same joint account anyway.

It's quite a small fraction of the population that has gotten comfortable with the math we used in the first four or five chapters of this book. Once we introduced special functions, we lost just about everybody else. Some people come across Bessel functions briefly as undergrads or master's students, but generally speaking special functions are the sorts of things that only show up in PhD-level work. Congratulations, you're not just a one-percenter, mathematically speaking, you're a tenth-percenter. What I mean by that is it's about a tenth of a percent of the cohort you went to

A Data Engineering Approach to Wave Scattering Analysis: with Applications in Radar, Sonar, Medical Diagnostics, Structural Flaw Detection and Intelligent Robotics, First Edition. Mark K. Hinders.

high school with who will have come across the math we've been doing. My high school was fairly large, so my graduating class was about 500 kids. Hence, my math works out that I'm the only one who half knows about special functions.

But if the only special functions you know are the ones you get in spherical and cylindrical coordinates, ya' basic. What we're about to do is rare. I was going to say esoteric, but that's kind of a big word, so let me just again make the point that only a handful of people on the planet are comfortable with the special functions we'll get when we do separation of variables in spheroidal and elliptic cylindrical coordinates. That creates a natural barrier to competition for you. The inexorable march of time means that most of the people who know how to use this mathematics are retired or otherwise aging out of the system. I'm always a little afraid to google people who were big names back in the day[1] because obituaries are often the top search results. (You may or may not have noticed that a large percentage of the old-timers I've mentioned in footnotes seemed to have died during the COVID-19 pandemic.) My point is merely that it's going to be worth getting comfortable with the math below because when you need it, you need it and if you find an important application that needs it, you aren't going to have much competition. You may have to be careful about writing these equations while on an airplane because your seatmates might push the call button and get you put on the terror watch list.[2]

8.1 Scalar Wave Equation in Elliptic Cylinder Coordinates

I think I've been a little too dismissive of Wikipedia. It's not that I fondly remember the days of looking things up in an outdated set of encyclopedias when I needed to do a book report for social studies. It's that I expect treatments of highly technical topics to be superficial. Somehow I feel better when I have on hand a definitive monograph on a highly specialized subject of study. Something that a mathematician or scientist spent a career learning about, and then eventually got around to typing up and publishing so you would have the esoteric bits you need to solve your problem at hand. I may have just looked up Mathieu functions on Wikipedia and the harshest criticism I could levy is that I prefer different symbols for the elliptic cylinder coordinates. I prefer to write the relations between the Cartesian (x, y, z) and elliptic cylinder (ξ, η, z) coordinates as:

$$\begin{aligned} x &= q \cosh \xi \cos \eta \\ y &= q \sinh \xi \sin \eta \\ z &= z \end{aligned} \tag{8.1}$$

which are illustrated in Figure 8.1. Ok, so here we go. Buckle up.

---

1 I googled Cavour W. Yeh because I was reading his 1963 paper on scattering from a penetrable ribbon and found that he got "what federal prosecutors said was the stiffest punishment ever for research grant fraud nationwide," namely two years in prison plus US$ 1.75 million in fines and restitution. Yeh, a UCLA professor since 1967, also resigned his tenured post. He apparently put two of his sisters and a brother on the UCLA payroll, paid by research grants, and they each got three years in prison because they kicked back some of the salary to their brother, which is stupid. The judge who sentenced him, repeatedly referred to him as a genius who produced good scientific results on the grants. Yeh said that he only hired his family members because nobody else could do the research properly. As if facility with elliptic cylinder wave functions is somehow genetic. After Yeh got out of prison his wife divorced him and a few years later, his son died in a camping accident.

2 https://www.theguardian.com/us-news/2016/may/07/professor-flight-delay-terrorism-equation-american-airlines.

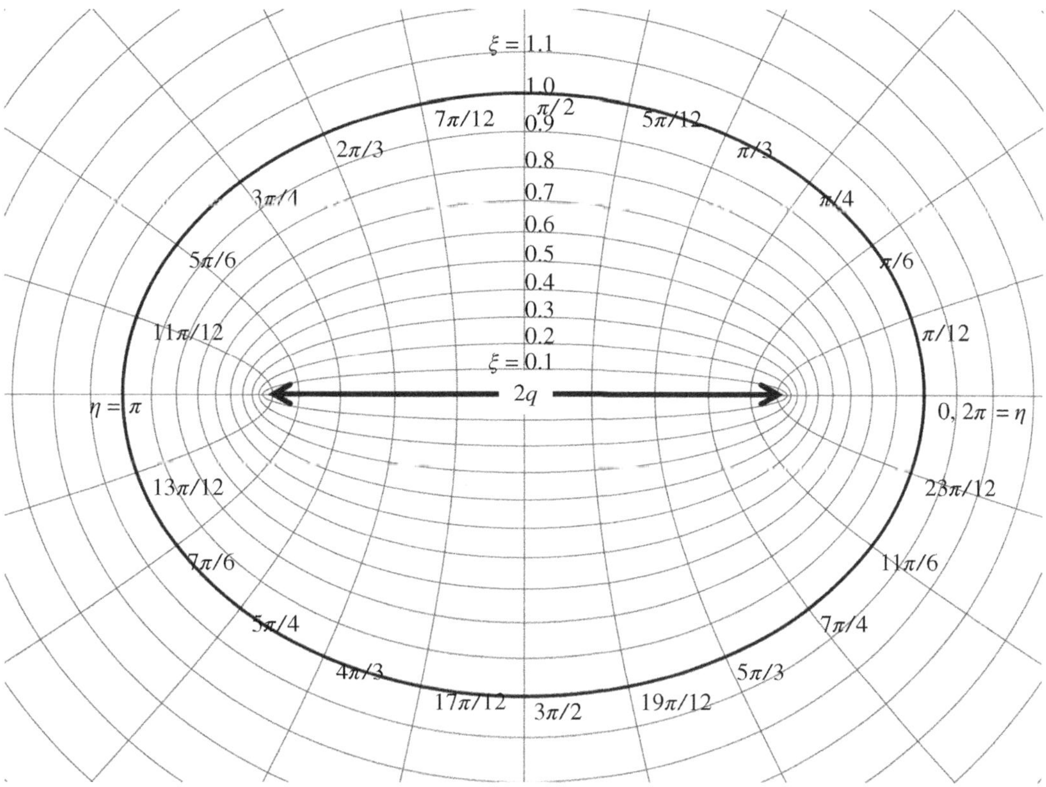

Figure 8.1 Elliptical coordinates (ξ, η) with the radial lines (ellipses) shown from 0 to 1.1 and the angular coordinate lines (parabolas) shown from 0.0 to 2π. Note that the distance between the two foci is $2q$.

The scalar wave equation can be written as:

$$\nabla^2 \phi - \frac{1}{c^2} \partial_t^2 \phi = 0 \tag{8.2}$$

or, taking the Fourier transform in time with kernel $e^{i\omega t}$ and writing $k = \omega/c$

$$\left(\nabla^2 + k^2\right) \phi = 0 \tag{8.3}$$

where ϕ is a scalar function, k is the wave number and in Cartesian coordinates the Laplacian operator is $\nabla^2 = \partial_x^2 + \partial_y^2 + \partial_z^2$. Of course, we're dealing with elliptic cylinder coordinates, so we must transform the Laplacian operator from Cartesian to confocal coordinates. The two systems are related by

$$x + iy = q \cosh(\xi + i\eta) \;\; z = z$$

Now, since $\cosh(a + b) = \cosh a \cosh b + \sinh a \sinh b$ and also $\cosh ix = \cos x$ and $\sinh ix = i \sin x$ we write

$$\begin{aligned} x + iy &= q \cosh \xi \cosh i\eta + q \sinh \xi \sinh i\eta \\ &= q \cosh \xi \cos \eta + iq \sinh \xi \sin \eta \end{aligned}$$

so that we have the relations between Cartesian (x, y, z) and elliptic cylinder (ξ, η, z) given above.

If we write $w = x + iy$ and $\zeta = \xi + i\eta$ we have: $\overline{w} = x - iy$; $\overline{\zeta} = \xi - i\eta$; $w\overline{w} = x^2 + y^2$; $\zeta\overline{\zeta} = \xi^2 + \eta^2$ as well as $w = q\cosh\zeta$ and $\overline{w} = q\cosh\overline{\zeta}$. Also note

$$\begin{aligned} w\overline{w} = x^2 + y^2 &= q^2\{\cosh(\xi + i\eta)\cosh(\xi - i\eta)\} \\ &= q^2\{\cosh\xi\cos\eta + i\sinh\xi\sin\eta\}\{\cosh\xi\cos\eta - i\sinh\xi\sin\eta\} \\ &= q^2\{[\cosh^2\xi\cos^2\eta - \sinh^2\xi\sin^2\eta] + i[\sinh\xi\cosh\xi\sin\eta\cos\eta \\ &\quad - \cosh\xi\cos\eta\sinh\xi\sin\eta]\} \end{aligned}$$

Hence

$$w\overline{w} = q^2\left\{\cosh^2\xi\cos^2\eta - \sinh^2\xi\sin^2\eta\right\} \tag{8.4}$$

Furthermore

$$r\frac{\partial^2 w\overline{w}}{\partial w\partial\overline{w}} = \left(\partial_x^2 + \partial_y^2\right) z\overline{z}$$

$$r\frac{\partial^2 \zeta\overline{\zeta}}{\partial\zeta\partial\overline{\zeta}} = \left(\partial_\xi^2 + \partial_\eta^2\right)\zeta\overline{\zeta}$$

since $\partial_x^2(x^2 + y^2) = 2$, $\partial_y^2(x^2 + y^2) = 2$ and

$$\frac{\partial^2}{\partial w\partial_{\overline{x}}}(w\overline{x}) = \partial_w(w\partial_{\overline{w}}\overline{w}) = \partial_w(w) = 1$$

with similar relations for $\zeta\overline{\zeta}$. We can, of course, write

$$\partial_\zeta w = q\sinh\zeta \qquad \partial_{\overline{\zeta}}\overline{w} = q\sinh\overline{\zeta}$$

so that

$$\partial_w\zeta = \frac{1}{q\sinh\zeta} \qquad \partial_{\overline{w}}\overline{\zeta} = \frac{1}{q\sinh\overline{\zeta}}$$

and by the chain rule

$$\partial_w = \frac{1}{q\sinh\zeta}\partial_\zeta \qquad \partial_{\overline{w}} = \frac{1}{h\sinh\overline{\zeta}}\partial_{\overline{\zeta}}$$

Hence

$$\partial_w\partial_{\overline{w}} = \frac{1}{q^2\sinh\zeta\sinh\overline{\zeta}}\partial_\zeta\partial\overline{\zeta}$$

and substituting, we get

$$\partial_x^2 + \partial_y^2 = \frac{1}{q^2\sinh\zeta\sinh\overline{\zeta}}\partial_\zeta\partial_{\overline{\zeta}} \tag{8.5}$$

which allows us to write

$$\partial_x^2 + \partial_y^2 = \frac{1}{q^2\sinh\zeta\sinh\overline{\zeta}}\left(\partial_\xi^2 + \partial_\eta^2\right)$$

or, rearranging and adding back in the z-variation

$$\left(\partial_\xi^2 + \partial_\eta^2 + \partial_z^2\right)\phi + k^2h^2\sinh\zeta\sinh\overline{\zeta}\phi = 0 \tag{8.6}$$

A half of a page or so of trig allows us to write

$$\sinh\zeta\sinh\overline{\zeta} = \frac{1}{2}[\cosh(2\xi) - \cos(2\eta)]$$

so that the scalar wave equation in elliptic cylinder coordinates becomes, with $\kappa = 2kh$

$$\left(\partial_\xi^2 + \partial_\eta^2 + \partial_z^2\right)\phi + 2\kappa^2\left[\cosh 2\xi - \cos 2\eta\right]\phi = 0 \tag{8.7}$$

which is the desired result. Whew!

8.1.1 Separation of Variables

If it's OK with you, I'm going to do the separation of variables in 2D. We all know that the z-variation is going to give us sinusoids, so let's put a pin in that part for now and do the hard part. The approach is going to be familiar, but the special functions will not be familiar at all. They have names though: Mathieu functions. Ready?

Consider the two-dimensional wave equation in elliptic coordinates

$$\left(\partial_\xi^2 + \partial_\eta^2\right)\phi + 2k^2\left[\cosh 2\xi - \cos 2\eta\right]\phi = 0 \tag{8.8}$$

To solve this, we assume a solution of the form $\phi(\xi,\eta) = F(\xi)G(\eta)$, where F is a function of ξ alone and G is a function of η alone. Substituting, we get

$$GF'' + F\ddot{G} + 2k^2\left[\cosh 2\xi - \cos 2\eta\right]FG = 0$$

or

$$\frac{F''}{F} + \frac{\ddot{G}}{G} + 2k^2\left[\cosh 2\xi - \cos 2\eta\right]$$

Grouping together the terms dependent on ξ and also grouping terms dependent on η we write

$$\frac{F''}{F} + 2k^2\cosh 2\xi = -\frac{\ddot{G}}{G} + 2k^2\cos 2\eta$$

Since the LHS is independent of η and the RHS is independent of ξ, each side must be a constant, say a. Accordingly, we obtain the two ordinary differential equations

$$F'' + (a - 2k^2\cosh 2\xi)F = 0$$

$$\ddot{G} - (a - 2k^2\cos 2\eta)G = 0$$

where a is separation constant. Note that if we write $\eta = \pm i\xi$ in the second equation it is equivalent to the first. Similarly, if we write $\xi = \pm i\eta$ in the first it is equivalent to the second. These two equations are connected with the name of Mathieu.[3] You'll be unsurprised to know that the solutions to Mathieu's equations are Mathieu functions. According to Wikipedia, these are not to be confused with Massieu functions. Apparently, it is perfectly OK to confuse Massieu function with

3 If it were asked what tyranny in this world has the least foundation in reason and is at the same time most overbearing and capricious, none could be found to answer better to this description than fashion; that fashion which makes us admire to-day what but yesterday would have excited astonishment, and which may provoke ridicule to-morrow. We all know that this sovereign whose iron rule is so much more keenly felt on account of its injustice governs the thousand and one details of every-day life; that it is supreme in literature and in the arts. But those who have not watched closely the life of the scientific world may perhaps be surprised to hear that even there if you would please you must bend the knee to fashion. What? might exclaim the stranger to the world of science, can it be true that the mathematician knows other laws than the inflexible rules of logic? Does he care to obey other orders than the invariable commands of reason? – Well, yes. Of course, to have a mathematical production accepted as correct, it is sufficient that it conforms to the precepts of logic, but to have it admired as beautiful, as interesting, as of importance, to gain honor and success by it, more is required: it must then satisfy the manifold and varying exactions imposed by the prevailing taste of the day, by the preferences of prominent men, by the preoccupations of the public (from Duhem [1]).

Gibbs function or even to sometimes share credit by using the term Massieu–Gibbs functions.[4] I had never heard of Massieu before, but I thought it would be amusing to cite a footnote in a footnote. It's important to stop and laugh just a bit along the way through a complicated derivation. Back to work.

A solution for the wave equation thus comprises the product of any two functions, which are solutions of the two aforementioned equations, respectively, for the same values of a and $q = k^2$. We consider solutions given by ordinary and modified Mathieu functions.

The functions $ce_n(\eta)$ and $se_n(\eta)$ are cosine-elliptic (even) and sine-elliptic (odd) Mathieu functions. The functions $Ce_n(\xi)$ and $Se_n(\xi)$ are cosine-elliptic and sine-elliptic modified Mathieu functions. The general solution of the wave equation can thus be written as:

$$\phi = \sum_{n=0}^{\infty} \{A_n Ce_n(\xi) ce_n(\eta) + B_n Se_n(\xi) se_n(\eta)\} \tag{8.9}$$

where A_n and B_n are modal coefficients. It's not obvious at this point, but it will be convenient to write this in general form as:

$$\begin{aligned}\phi = \sum_{n=0}^{\infty} \{&A_{2n} Ce_{2n}(\xi) ce_{2n}(\eta) + B_{2n+2} Se_{2n+2}(\xi) se_{2n+2}(\eta) \\ &+ A_{2n+1} Ce_{2n+1}(\xi) ce_{2n+1}(\eta) + B_{2n+1} Se_{2n+1}(\xi) se_{2n+1}(\eta)\}\end{aligned} \tag{8.10}$$

where A_{2n}, A_{2n+1}, B_{2n+1} and B_{2n+2} are modal coefficients.

8.2 Scattering from a Perfectly-Conducting Elliptic Cylinder

There's a lot more we could say about Mathieu functions, and I started typing out a bunch of it, but I think that it would be best to simply proceed with the solution to a few scattering problems in elliptic cylinder coordinates. I'm going to pay some homage to Cavour W. Yeh because he did some really nice work just before I was born and I don't want his later fall from grace to erase that. Also, he follows the notation of McLachlan, who followed the notation of Inge and McLachlan's book is available as a downloadable PDF, which you probably should get a copy of if you're going to deal with Mathieu functions. You can buy a paperback version on Amazon if you prefer. I'm hoping I'll come across a hardcopy in a used bookstore someday.

The simplest problem we can start with turns out to be pretty simple, at least on paper. Consider an incident plane transverse-magnetic wave incident at an angle θ on a perfectly conducting elliptic cylinder whose surface corresponds to the ellipse $\xi = \xi_0$. McLachlan helpfully tells us to write

$$\begin{aligned}E_z^{inc} &= E_0 e^{x\cos\theta + y\sin\theta} \\ &= E_0 e^{ik_0 q(\cosh\xi\cos\eta\cos\theta + \sinh\xi\sin\eta\sin\theta)} \\ &= 2E_0 \sum_{n=0}^{\infty} \left[\frac{1}{p_{2n}} Ce_{2n}(\xi) ce_{2n}(\eta) ce_{2n}(\theta) + \frac{1}{s_{2n+2}} Se_{2n+2}(\xi) se_{2n+2}(\eta) se_{2n+2}(\theta)\right. \\ &\left. + \frac{i}{p_{2n+1}} Ce_{2n+1}(\xi) ce_{2n+1}(\eta) ce_{2n+1}(\theta) + \frac{i}{s_{2n+1}} Se_{2n+1}(\xi) se_{2n+1}(\eta) se_{2n+1}(\theta)\right]\end{aligned} \tag{8.11}$$

4 Massieu, as discussed in the first footnote to the abstract of Gibbs' *Equilibrium*, "appears to have been the first to solve the problem of representing all the properties of a body of invariable composition which are concerned in reversible processes by means of a single function."

As usual, the scattered field is a linear combination of the appropriate elliptical functions, and as before, we have to choose the particular functions that will satisfy the Sommerfeld radiation condition at $\xi \to \infty$ because the scattered field spreads out more and more as it gets farther and farther from the scatterer. Recall that for the spheres and circular cylinders, that meant that we had to choose an appropriate Hankel function for the radial part of the solution. That's what the functions $Me^{(1)}(\xi)$ and $Ne^{(1)}(\xi)$ are in the following expression for the z-component of the scattered electric field:

$$E_z^{scat} = 2E_0 \sum_{n=0}^{\infty} \left[\frac{A_{2n}}{p_{2n}} Me_{2n}^{(1)}(\xi) ce_{2n}(\eta) ce_{2n}(\theta) + \frac{B_{2n+2}}{s_{2n+2}} Ne_{2n+2}^{(1)}(\xi) se_{2n+2}(\eta) se_{2n+2}(\theta) \right.$$
$$\left. + \frac{iA_{2n+1}}{p_{2n+1}} Me_{2n+1}^{(1)}(\xi) ce_{2n+1}(\eta) ce_{2n+1}(\theta) + \frac{iB_{2n+1}}{s_{2n+1}} Ne_{2n+1}^{(1)}(\xi) se_{2n+1}(\eta) se_{2n+1}(\theta) \right] \tag{8.12}$$

In these, A's and B's represent the modal coefficients that we want to solve for by application of the boundary condition that $E_z^{inc} + E_z^{scat} = 0$ for $\xi = \xi_0$ and all values of η from 0 to 2π. The algebra turns out to be quite simple because the Mathieu functions are orthogonal. What that means is that we multiply the boundary condition equation by $ce_m(\eta)$ or $se_m(\eta)$ and integrate over η from 0 to 2π. The only terms in the summation over n that survive will be when $n = m$. Hence we have

$$A_{2m} = \frac{Ce_{2m}(\xi_0)}{Me_{2m}^{(1)}(\xi_0)} \tag{8.13}$$

and similarly for the other modal coefficients.

It's then straightforward, albeit nontrivial, to go back and compute E_z^{scat}. One of the things that makes this tricky is the numerics of implementing the Mathieu functions. When I first looked at this class of problems in the late 1980s, I came up empty asking around if anybody I knew had any F77 Mathieu functions subroutines they could share. Even today, it can be tricky. Two years ago, we noticed errors in the *PyHub* routines for some of the functions, which nobody had cared about for at least six years before that. My point is merely that I think we should pause here a moment and look at the behavior of the special functions we're about to use to make plots of TM scattering from a PEC elliptic cylinder.

A very mature thing for you to do right about now would be to reproduce McLachlan's plots of the Mathieu functions from that book you downloaded a few pages ago. It's the simplest way to make sure that you're using these new-to-you functions properly. The notation can be quite confusing, so it would be easy to mix up some of them or some such thing.

Exercise 8.1 Reproduce McLachlan's plots of the Mathieu functions. I shouldn't have had to tell you twice.

In 1964, Twersky[5] solved the Mathieu function series solution for the scattering of a plane wave by a perfectly conducting elliptic cylinder [2]. They obtained closed-form approximations for the

5 Victor Twersky was one of the world's leading authorities on the scattering of radiation of all kinds, from light, radar, and sonar to seismic waves in the Earth. His work is used to analyze the clutter produced by the scattering of radar waves from chaff, to describe the effect of dust on the propagation of light in the atmosphere, to analyze the scattering of light by the red cells in the blood and by biological tissue, etc. Twersky earned a BS from City College of New York in 1943, an MA from Columbia University in 1948, and a PhD from New York University in 1950, all in physics. From 1950 to 1966, he was an Engineering Specialist and Senior Scientist in the Electronic Defense Laboratories of Sylvania-GTE in Mountain View, California and was also a Visiting Lecturer at Stanford during part of that time. Twersky joined the faculty of the U. Illinois Chicago Department of Mathematics, Statistics, and Computer Science in 1966 as a full professor and retired 1990.

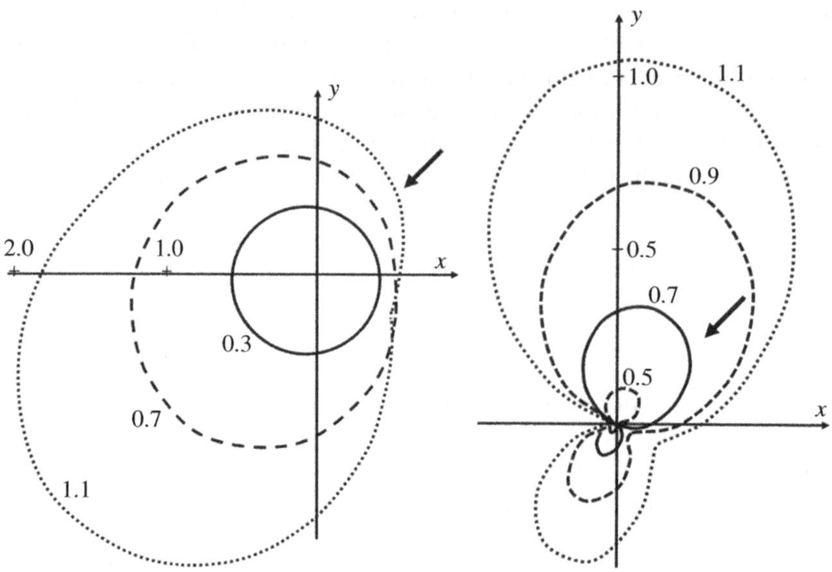

Figure 8.2 Polar plots of normalized bistatic scattering cross section for a PEC elliptic cylinder, sketched from [3]. Arrows indicate direction of incident beam. (a) is for E^{inc} parallel to the cylinder axis. (b) is for H^{inc} parallel to the cylinder axis. The different curves are for various values of $c\xi_1$. The elliptic cylinder has an axes ratio $\xi_1\sqrt{\xi_1^2 - 1} = 2$. Source: [3]/NASA/Public domain.

scattering coefficients of the elliptic modes and for the far-field scattering amplitudes. The corresponding series for the scattered intensity and total scattering cross section are given for E parallel to a generator and for E perpendicular. Analogous results are also obtained for the semielliptic protuberance on a ground plane. For this case, the series for the intensities and cross sections for both polarizations. Numerical results obtained from series approximations, from closed forms, and from tables of functions, are given.

We can have good confidence in this result because [3] cite this paper when they include plots for bistatic scattering cross sections for an elliptic cylinder, which are sketched in Figure 8.2.

You should be able to reproduce these plots from [2] easily enough, but please have some respect for people who went through agony to make plots back when I was a small six-week-early premie and computers were room-sized behemoths. I assume that in late 1963 you'd have to code up your own Mathieu function subroutines, kind of like how Dr. Frost said to my mother, "You might as well take him home. You can do as much for him there as we can here in the hospital. Be sure to wake him up and feed him every two hours."

We can go back to the prehistory of computations when computer was a job description. The Mathieu function agony is minimized if we consider the PEC elliptic cylinder to tend to the limit of a flat ribbon. In the 1940s, McLachlan would have had to do the computations for Figures 8.3 and 8.4 by hand, unless he had computers working for him and they did the computations by hand. I looked carefully just now in McLachlan's book, and didn't see any acknowledgments to computers who made the figures. Although there were electronic computers in those days, they were entirely taken up by the war effort, that is, doing ballistics-trajectory calculations and code-breaking.

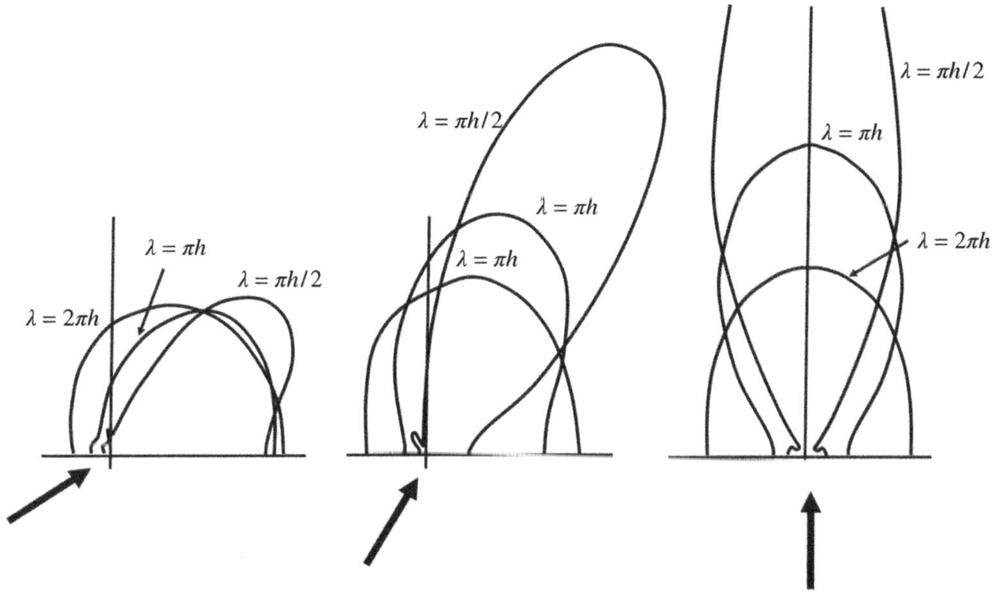

Figure 8.3 Polar plots of EM scattering from a long PEC ribbon of width *h*. Arrows indicate direction of incident beam.

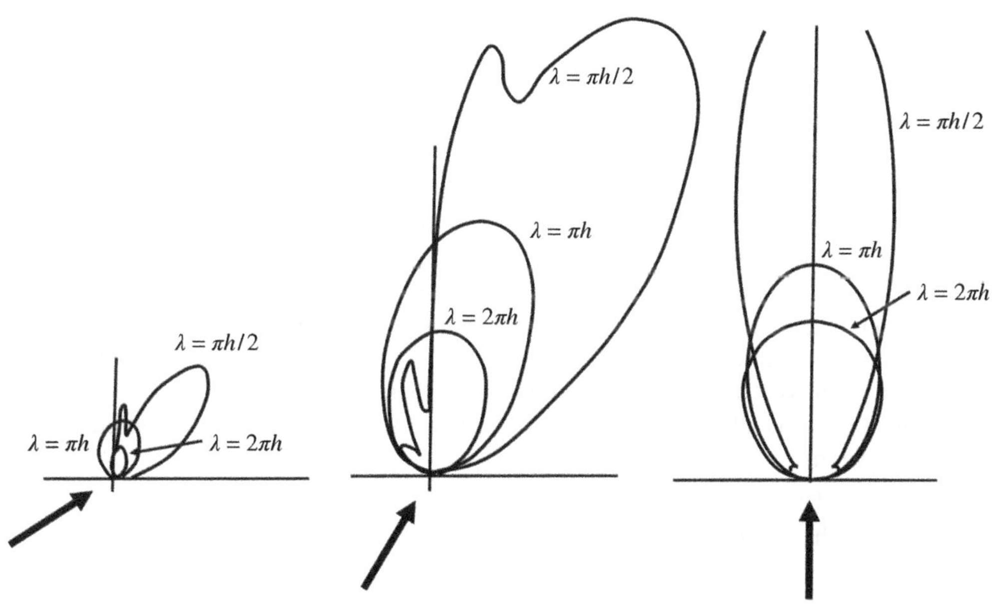

Figure 8.4 Polar plots of sound scattering from a long rigid ribbon of width *h*. Arrows indicate direction of incident beam.

8.3 Scattering from a Dielectric Elliptic Cylinder

The next obvious step is to make the elliptic scatterer penetrable, that is, a dielectric, so that we'll need to account for incident, scattered, and transmitted waves according to our usual procedure. The dielectric will have a different permeability and permittivity than the region surrounding the scatterer, and our boundary conditions will be continuity of tangential electric and magnetic fields at $\xi = \xi_0$. Let's start as before with the TM case, so that the incident and scattered fields are written as:

$$\begin{aligned} E_z^{inc} &= E_0 e^{x\cos\theta + y\sin\theta} \\ &= E_0 e^{ik_0 q(\cosh\xi\cos\eta\cos\theta + \sinh\xi\sin\eta\sin\theta)} \\ &= 2E_0 \sum_{n=0}^{\infty} \left[\frac{1}{p_{2n}} Ce_{2n}(\xi) ce_{2n}(\eta) ce_{2n}(\theta) + \frac{1}{s_{2n+2}} Se_{2n+2}(\xi) se_{2n+2}(\eta) se_{2n+2}(\theta) \right. \\ &\left. + \frac{i}{p_{2n+1}} Ce_{2n+1}(\xi) ce_{2n+1}(\eta) ce_{2n+1}(\theta) + \frac{i}{s_{2n+1}} Se_{2n+1}(\xi) se_{2n+1}(\eta) se_{2n+1}(\theta) \right] \end{aligned} \tag{8.14}$$

and

$$\begin{aligned} E_z^{scat} = 2E_0 \sum_{n=0}^{\infty} &\left[\frac{A_{2n}}{p_{2n}} Me_{2n}^{(1)}(\xi) ce_{2n}(\eta) ce_{2n}(\theta) + \frac{B_{2n+2}}{s_{2n+2}} Ne_{2n+2}^{(1)}(\xi) se_{2n+2}(\eta) se_{2n+2}(\theta) \right. \\ &\left. + \frac{iA_{2n+1}}{p_{2n+1}} Me_{2n+1}^{(1)}(\xi) ce_{2n+1}(\eta) ce_{2n+1}(\theta) + \frac{iB_{2n+1}}{s_{2n+1}} Ne_{2n+1}^{(1)}(\xi) se_{2n+1}(\eta) se_{2n+1}(\theta) \right] \end{aligned} \tag{8.15}$$

When we assume a functional form for the transmitted fields, we'll need to make sure the radial functions are finite at $\xi = 0$ as usual, so we write

$$\begin{aligned} E_z^{trans} = 2E_0 \sum_{n=0}^{\infty} &\left[\frac{C_{2n}}{p_{2n}} Ce_{2n}^{*}(\xi) ce_{2n}^{*}(\eta) ce_{2n}(\theta) + \frac{D_{2n+2}}{s_{2n+2}} Se_{2n+2}^{*}(\xi) se_{2n+2}^{*}(\eta) se_{2n+2}(\theta) \right. \\ &\left. + \frac{iC_{2n+1}}{p_{2n+1}} Ce_{2n+1}^{*}(\xi) ce_{2n+1}^{*}(\eta) ce_{2n+1}(\theta) + \frac{iD_{2n+1}}{s_{2n+1}} Se_{2n+1}^{*}(\xi) se_{2n+1}^{*}(\eta) se_{2n+1}(\theta) \right] \end{aligned} \tag{8.16}$$

where C and D are two more modal coefficients, and I've rather begrudgingly followed Yeh's notation using an asterisk to indicate where the Mathieu functions include a dependence on the material properties of the scatterer. We obviously elided this issue when we were considering just the PEC scatterer. Here we'll have both $k = \omega\sqrt{\mu_0\epsilon_0}$ for free space and $k_1 = \omega\sqrt{\mu_1\epsilon_1}$ for the dielectric scatterer. One of our boundary condition equations will be that $E_z^{inc} + E_z^{scat} = E_z^{trans}$ for $\xi = \xi_0$ and all values of η from 0 to 2π.

The tangential magnetic field is also continuous at the boundary, so we need to write $H_\eta^{inc} + H_\eta^{scat} = H_\eta^{trans}$ for $\xi = \xi_0$ and all values of η from 0 to 2π. Since Maxwell's equations in elliptic cylinder coordinates gives us

$$H_\eta = \frac{-i\omega\epsilon}{k^2 q\sqrt{\sinh^2\xi + \sin^2\eta}} \partial_\xi E_z$$

we can write

$$\begin{aligned} H_\eta^{inc} = 2E_0 \frac{-i\omega\epsilon_0}{k^2 q\sqrt{\sinh^2\xi + \sin^2\eta}} \sum_{n=0}^{\infty} &\left[\frac{1}{p_{2n}} Ce_{2n}'(\xi) ce_{2n}(\eta) ce_{2n}(\theta) \right. \\ &+ \frac{1}{s_{2n+2}} Se_{2n+2}(\xi) se_{2n+2}'(\eta) se_{2n+2}(\theta) + \frac{i}{p_{2n+1}} Ce_{2n+1}(\xi) ce_{2n+1}'(\eta) ce_{2n+1}(\theta) \\ &\left. + \frac{i}{s_{2n+1}} Se_{2n+1}(\xi) se_{2n+1}'(\eta) se_{2n+1}(\theta) \right] \end{aligned} \tag{8.17}$$

and

$$H_\eta^{scat} = 2E_0 \frac{-i\omega\epsilon_0}{k^2 q\sqrt{\sinh^2\xi + \sin^2\eta}} \sum_{n=0}^{\infty} \left[\frac{A_{2n}}{p_{2n}} Me_{2n}^{'(1)}(\xi) ce_{2n}(\eta) ce_{2n}(\theta) \right.$$
$$+ \frac{B_{2n+2}}{s_{2n+2}} Ne_{2n+2}^{'(1)}(\xi) se_{2n+2}(\eta) se_{2n+2}(\theta) + \frac{iA_{2n+1}}{p_{2n+1}} Me_{2n+1}^{'(1)}(\xi) ce_{2n+1}(\eta) ce_{2n+1}(\theta)$$
$$\left. + \frac{iB_{2n+1}}{s_{2n+1}} Ne_{2n+1}^{'(1)}(\xi) se_{2n+1}(\eta) se_{2n+1}(\theta) \right] \tag{8.18}$$

and also

$$H_\eta^{trans} = 2E_0 \frac{-i\omega\epsilon_1}{k_1^2 q\sqrt{\sinh^2\xi + \sin^2\eta}} \sum_{n=0}^{\infty} \left[\frac{C_{2n}}{p_{2n}} Ce_{2n}^{*}(\xi) ce_{2n}^{*}(\eta) ce_{2n}(\theta) \right.$$
$$+ \frac{D_{2n+2}}{s_{2n+2}} Se_{2n+2}^{*\prime}(\xi) se_{2n+2}^{*}(\eta) se_{2n+2}(\theta) + \frac{iC_{2n+1}}{p_{2n+1}} Ce_{2n+1}^{*\prime}(\xi) ce_{2n+1}^{*}(\eta) ce_{2n+1}(\theta)$$
$$\left. + \frac{iD_{2n+1}}{s_{2n+1}} Se_{2n+1}^{*\prime}(\xi) se_{2n+1}^{*}(\eta) se_{2n+1}(\theta) \right] \tag{8.19}$$

where prime indicates differentiation with respect to ξ.

At this point, it looks like we can simply proceed as before, using orthogonality to get rid of the pesky summation from 0 to ∞ and then canceling common terms and doing a bit of algebra. It's just two equations, after all. Not so fast there, skippy. This issue we're about to face is somewhat obscured by the notation we're using. Don't be too hard on Prof. Yeh, though. He published his 1963 paper [4] in a journal with a two-column format, so he had to use a notation that could fit on those pages. Recall that we're using an asterisk to indicate where the Mathieu functions include a dependence on the material properties of the scatterer. What that means is when we multiply the boundary condition equations by $ce_m(\eta)$ or $se_m(\eta)$ and integrate over η from 0 to 2π it won't necessarily be the case that the only terms in the summation over n that survive will be when $n = m$. Put another way, $ce_m(\eta)$ and $se_m(\eta)$ aren't orthogonal with $ce_m^*(\eta)$ and $se_m^*(\eta)$, respectively. Feel free to curse under your breath about that. Then remind yourself about the whole barrier to competition and get ready with your PDF or paperback copy of McLachlan because the mathematical gymnastics are going to be in there.

8.3.1 Important Tea About Orthogonality

As I may have mentioned, I learned differential equations in a movie theater. It wasn't some sort of on-line course because this was the early 1980s. My urban university had worked a deal where a small multiplex was used for classes during the day and showed art films at night. I saw many movies there, and noticed that at night the chalkboards were covered by curtains. During the day I noticed that the movie screens were covered by curtains and the snack bar was closed. I also noticed that there were two basic solutions to ODEs. The first was to guess an exponential solution and see if that worked. The second was to try a series solution. Special functions like those attributed to Mathieu started out as series solutions but eventually got their own symbols. So, without further ado, here they are from page 21 in McClachlan.

$$ce_{2n}(\eta, k) = \sum_{r=0}^{\infty} A_{2r}^{(2n)} \cos 2r\eta \tag{8.20}$$

$$ce_{2n+1}(\eta, k) = \sum_{r=0}^{\infty} A_{2r+1}^{(2n+1)} \cos(2r+1)\eta \tag{8.21}$$

$$se_{2n+1}(\eta, k) = \sum_{r=0}^{\infty} B_{2r+1}^{(2n+1)} \sin(2r+1)\eta \tag{8.22}$$

$$se_{2n+2}(\eta, k) = \sum_{r=0}^{\infty} B_{2r+2}^{(2n+2)} \sin(2r+2)\eta \tag{8.23}$$

In these series, the A's and B's *are* functions of wave number k and hence material. Note especially that the sinusoids *aren't*! This will allow us to now write

$$\begin{aligned}
\int_0^{2\pi} ce_{2n}(\eta, k_0) ce_{2m}(\eta, k_1) d\eta &= \sum_{r=0}^{\infty} A_{2r}^{(2n)} \sum_{r'=0}^{\infty} A_{2r'}^{*(2m)} \int_0^{2\pi} \cos 2r\eta \cos 2r'\eta d\eta \\
&= 2\pi A_0^{(2n)} A_0^{*(2m)} + \pi \sum_{r=1}^{\infty} A_{2r}^{(2n)} \sum_{r'=1}^{\infty} A_{2r'}^{*(2m)} \delta_{rr'} \\
&= 2\pi A_0^{(2n)} A_0^{*(2m)} + \pi \sum_{r=1}^{\infty} A_{2r}^{(2n)} A_{2r}^{*(2m)}
\end{aligned} \tag{8.24}$$

where we've separated out the $r = 0$ term in the summation because $\cos 0 = 1$. But what are the A's you might ask? You might even wonder about the B's it looks like we're going to need it in just a minute. Chapter 3 in McClachlan [5] has the title "Calculation of Characteristic Coefficients and Numbers" which is kind of an ominous sign for us, especially given that there are a bunch of figures illustrating the variation in certain parameters in stable and unstable regions. I vote we try not to panic just yet and continue on to get down on paper an answer to the problem at hand. Keep tight to the knowledge that Prof. Yeh made plots with computers from these equations in the early 1960s, so it shouldn't be too tough these days. I do feel the need to mention that one of Prof. Yeh's students seemed to have found a typo in his code 25 years later and they together published a few corrected plots in an erratum to the early paper [4]. But let's keep going for now.

$$\begin{aligned}
\int_0^{2\pi} ce_{2n+1}(\eta, k_0) ce_{2m+1}(\eta, k_1) d\eta &= \sum_{r=0}^{\infty} A_{2r+1}^{(2n+1)} \sum_{r'=0}^{\infty} A_{2r'+1}^{*(2m+1)} \int_0^{2\pi} \cos(2r+1)\eta \cos(2r'+1)\eta d\eta \\
&= \pi \sum_{r=0}^{\infty} A_{2r+1}^{(2n+1)} \sum_{r'=0}^{\infty} A_{2r'+1}^{*(2m+1)} \delta_{rr'} \\
&= \pi \sum_{r=0}^{\infty} A_{2r+1}^{(2n+1)} A_{2r+1}^{*(2m+1)}
\end{aligned} \tag{8.25}$$

$$\begin{aligned}
\int_0^{2\pi} se_{2n+1}(\eta, k_0) se_{2m+1}(\eta, k_1) d\eta &= \sum_{r=0}^{\infty} B_{2r+1}^{(2n+1)} \sum_{r'=0}^{\infty} B_{2r'+1}^{*(2m+1)} \int_0^{2\pi} \sin(2r+1)\eta \sin(2r'+1)\eta d\eta \\
&= \pi \sum_{r=0}^{\infty} B_{2r+1}^{(2n+1)} \sum_{r'=0}^{\infty} A_{2r'+1}^{*(2m+1)} \delta_{rr'} \\
&= \pi \sum_{r=0}^{\infty} B_{2r+1}^{(2n+1)} B_{2r+1}^{*(2m+1)}
\end{aligned} \tag{8.26}$$

$$\begin{aligned}
\int_0^{2\pi} se_{2n+2}(\eta, k_0) se_{2m+2}(\eta, k_1) d\eta &= \sum_{r=0}^{\infty} B_{2r+2}^{(2n+2)} \sum_{r'=0}^{\infty} B_{2r'+2}^{*(2m+2)} \int_0^{2\pi} \sin(2r+2)\eta \sin(2r'+2)\eta d\eta \\
&= \pi \sum_{r=0}^{\infty} B_{2r+2}^{(2n+2)} \sum_{r'=0}^{\infty} B_{2r'+2}^{*(2m+2)} \delta_{rr'} \\
&= \pi \sum_{r=0}^{\infty} B_{2r+2}^{(2n+2)} B_{2r+2}^{*(2m+2)}
\end{aligned} \tag{8.27}$$

We'll also need these, which are derived similarly

$$\int_0^{2\pi} ce_{2n}^2(\eta,k)d\eta = 2\pi\left(A_0^{(2m)}\right)^2 + \pi\sum_{r=1}^{\infty}\left(A_{2r}^{(2n)}\right)^2 \tag{8.28}$$

$$\int_0^{2\pi} ce_{2n+1}^2(\eta,k)d\eta = \pi\sum_{r=0}^{\infty}\left(A_{2r+1}^{(2n+1)}\right)^2 \tag{8.29}$$

$$\int_0^{2\pi} se_{2n+1}^2(\eta,k)d\eta = \pi\sum_{r=0}^{\infty}\left(B_{2r+1}^{(2n+1)}\right)^2 \tag{8.30}$$

$$\int_0^{2\pi} se_{2n+2}^2(\eta,k)d\eta = \pi\sum_{r=0}^{\infty}\left(B_{2r+2}^{(2n+2)}\right)^2 \tag{8.31}$$

OK, so here we go. We first multiply the boundary condition equation for continuity of tangential electric field by $ce_{2m}(\eta)$ and integrate over η

$$\int_0^{2\pi} ce_{2m}(\eta)\left(E_z^{inc} + E_z^{scat} = E_z^{trans}\right)_{\xi=\xi_0} d\eta \tag{8.32}$$

which gives

$$\frac{Ce_{2m}(\xi_0)}{p_{2m}}ce_{2m}(\theta)\left[2(A^{(2m)})^2 + \sum_{r=1}^{\infty}\left(A_{2r}^{(2m)}\right)^2\right] + A_{2m}\frac{Me_{2m}(\xi_0)}{p_{2m}}ce_{2m}(\theta)\left[2(A^{(2m)})^2 + \sum_{r=1}^{\infty}\left(A_{2r}^{(2m)}\right)^2\right]$$
$$= \sum_{n=0}^{\infty} C_{2n}\frac{Ce_{2m}^*(\xi_0)}{p_{2m}}ce_{2m}(\theta)\left[2A_0^{(2n)}A_0^{(2m)} + \sum_{r=1}^{\infty}A_{2r}^{(2n)}A_{2r}^{*(2m)}\right] \tag{8.33}$$

This can be rewritten as:

$$Ce_{2m}(\xi_0) + A_{2m}Me_{2m}(\xi_0) = \sum_{n=0}^{\infty} C_{2n}Ce_{2n}^*(\xi_0)\frac{p_{2m}}{p_{2n}}\frac{ce_{2n}(\theta)}{ce_{2m}(\theta)} \times \left[\frac{2A_0^{(2n)}A_0^{*(2m)} + \sum_{r=1}^{\infty}A_{2r}^{(2n)}A_{2r}^{*(2m)}}{2\left(A_0^{(2m)}\right)^2 + \sum_{r=1}^{\infty}\left(A_{2r}^{(2m)}\right)^2}\right] \tag{8.34}$$

The corresponding boundary condition equation for continuity of tangential electric field becomes

$$Ce_{2m}'(\xi_0) + A_{2m}Me_{2m}'(\xi_0) = \frac{\mu_0}{\mu_1}\sum_{n=0}^{\infty} C_{2n}Ce_{2n}^{*\prime}(\xi_0)\frac{p_{2m}}{p_{2n}}\frac{ce_{2n}(\theta)}{ce_{2m}(\theta)}\left[\frac{2A_0^{(2n)}A_0^{*(2m)} + \sum_{r=1}^{\infty}A_{2r}^{(2n)}A_{2r}^{*(2m)}}{2\left(A_0^{(2m)}\right)^2 + \sum_{r=1}^{\infty}\left(A_{2r}^{(2m)}\right)^2}\right] \tag{8.35}$$

Don't worry. We don't have to sum to infinity, so we're going to be able to write this as a matrix equation that can be inverted to solve for C_{2n}. First, though, we can divide by $Me_{2m}(\xi_0)$ and $Me_{2m}'(\xi_0)$, respectively, in these two equations and subtract the second from the first to eliminate A_{2m} and get an equation that's the one we're going to write as a matrix equation and invert.

$$C_{2n} = \mathcal{R}_{2m,2n}^{-1}\left(\frac{Ce_{2m}(\xi_0)}{Me_{2m}(\xi_0)} - \frac{Ce_{2m}'(\xi_0)}{Me_{2m}'(\xi_0)}\right) \tag{8.36}$$

where

$$\mathcal{R}_{2m,2n} = \left(\frac{Ce^*_{2m}(\xi_0)}{Me_{2m}(\xi_0)} - \frac{\mu_0}{\mu_1}\frac{Ce^{*\prime}_{2m}(\xi_0)}{Me'_{2m}(\xi_0)}\right)\frac{p_{2m}}{p_{2n}}\frac{ce_{2n}(\theta)}{ce_{2m}(\theta)}\left[\frac{2A_0^{(2n)}A_0^{*(2m)} + \sum_{r=1}^{\infty} A_{2r}^{(2n)}A_{2r}^{*(2m)}}{2\left(A_0^{(2m)}\right)^2 + \sum_{r=1}^{\infty}\left(A_{2r}^{(2m)}\right)^2}\right]$$

and then the modal coefficient of the scattered field, which as you may or may not remember is the thing we've been trying to find, can be written

$$A_{2m} = -\frac{1}{M_{2m}(\xi_0)}\left\{Ce_{2m}(\xi_0) + \sum_{n=0}^{\infty} C_{2n}Ce^*_{2n}(\xi_0)\frac{p_{2m}}{p_{2n}}\frac{ce_{2n}(\theta)}{ce_{2m}(\theta)}\left[\frac{2A_0^{(2n)}A_0^{*(2m)} + \sum_{r=1}^{\infty} A_{2r}^{(2n)}A_{2r}^{*(2m)}}{2\left(A_0^{(2m)}\right)^2 + \sum_{r=1}^{\infty}\left(A_{2r}^{(2m)}\right)^2}\right]\right\} \tag{8.37}$$

So now we're left with calculating the A's and doing a matrix inversion, except that I haven't yet defined the joining factors p_2n along with the other three that we're going to need after we've multiplied our boundary condition equations by ce_{2n+2}, se_{2n+1}, se_{2n+2}, respectively, and integrated over η to get expressions for B_{2m+2}, A_{2m+1} and B_{2m+1}, and then doing the same sort of algebra we just did to find A_{2m}. Fortunately, McLachlan defines these "joining factors" for us in his Appendix I as:

$$\begin{aligned}
p_{2n} &= ce_{2n}(0,q)ce_{2n}\left(\frac{\pi}{2},q\right)/A_0^{(2n)} \\
p_{2n+1} &= -ce_{2n+1}(0,q)ce'_{2n+1}\left(\frac{\pi}{2},q\right)/\sqrt{q}A_1^{(2n+1)} \\
s_{2n+1} &= se'_{2n+1}(0,q)se_{2n+1}\left(\frac{\pi}{2},q\right)/\sqrt{q}B_1^{(2n+1)} \\
s_{2n+2} &= se'_{2n+2}(0,q)se'_{2n+2}\left(\frac{\pi}{2},q\right)/qB_2^{(2n+2)}
\end{aligned}$$

Let's put aside for the moment the numerical implementation of these, and get an expression for B_{2m+2} by multiplying the boundary condition equations by $se_{2m+2}(\eta)$ and integrating over η from 0 to 2π with $\xi = \xi_0$, pretty much as we did before. For the tangential electric field, we get

$$Se_{2m+2}(\xi_0) + B_{2m+2}Ne^{(1)}_{2m+2}(\xi_0) = \sum_{n=0}^{\infty} D_{2n+2}Se^*_{2n+2}\frac{s_{2m+2}}{s_{2n+2}}\frac{se^*_{2n+2}(\theta)}{se_{2m+2}(\theta)}\left[\frac{\sum_{r=0}^{\infty} B_{2r+2}^{(2n+2)}B_{2r+2}^{*(2m+2)}}{\sum_{r=0}^{\infty}\left(B_{2r+2}^{(2m+2)}\right)^2}\right] \tag{8.38}$$

and for the tangential magnetic field, we get

$$Se'_{2m+2}(\xi_0) + B_{2m+2}Ne^{(1)\prime}_{2m+2}(\xi_0) = \frac{\mu_0}{\mu_1}\sum_{n=0}^{\infty} D_{2n+2}Se^{*\prime}_{2n+2}\frac{s_{2m+2}}{s_{2n+2}}\frac{se^*_{2n+2}(\theta)}{se_{2m+2}(\theta)}\left[\frac{\sum_{r=0}^{\infty} B_{2r+2}^{(2n+2)}B_{2r+2}^{*(2m+2)}}{\sum_{r=0}^{\infty}\left(B_{2r+2}^{(2m+2)}\right)^2}\right] \tag{8.39}$$

Divide these by $Ne^{(1)}_{2m+2}(\xi_0)$ and $Ne^{(1)\prime}_{2m+2}(\xi_0)$, respectively, and then subtract to get

$$D_{2m+2} = \mathcal{Q}^{-1}_{2m,2n}\left(\frac{Se_{2m+2}(\xi_0)}{Se^*_{2m+2}(\xi_0)} - \frac{Se'_{2m+2}(\xi_0)}{N^{(1)\prime}_{2m+2}(\xi_0)}\right) \tag{8.40}$$

where the matrix we need to invert is

$$\mathcal{Q}_{2m,2n} = \sum_{n=0}^{\infty} D_{2n+2} \frac{s_{2m+2}}{s_{2n+2}} \frac{se^*_{2n+2}(\theta)}{se_{2m+2}(\theta)} \left[\frac{\sum_{r=0}^{\infty} B^{(2n+2)}_{2r+2} B^{*(2m+2)}_{2r+2}}{\sum_{r=0}^{\infty} \left(B^{(2m+2)}_{2r+2} \right)^2} \right] \left(\frac{Se^*_{2m+2}(\xi_0)}{Se_{2m+2}(\xi_0)} - \frac{\mu_0}{\mu_1} \frac{Se^{*\prime}_{2m+2}(\xi_0)}{N^{(1)*\prime}_{2m+2}(\xi_0)} \right) \tag{8.41}$$

after which we can find the second of our four modal coefficients for the scattered field

$$B_{2m+2} = -\frac{1}{Ne^{(1)}_{2m+2}(\xi_0)} \left\{ Se_{2m+2}(\xi_0) + \sum_{n=0}^{\infty} D_{2n+2} Se^*_{2n+2} \frac{s_{2m+2}}{s_{2n+2}} \frac{se^*_{2n+2}(\theta)}{se_{2m+2}(\theta)} \left[\frac{\sum_{r=0}^{\infty} B^{(2n+2)}_{2r+2} B^{*(2m+2)}_{2r+2}}{\sum_{r=0}^{\infty} \left(B^{(2m+2)}_{2r+2} \right)^2} \right] \right\} \tag{8.42}$$

There are two more of these, A_{2m+1} and B_{2m+1}, but I feel confident that you can work out what those want to be. It's not just that I'm a little too lazy to type them out for you, but I think it's important for you to work those out for yourself and then while you're at it *check my algebra*. The Mathieu functions are unfamiliar and the notation is rather confusing. It's good practice to assume that there might could be a typo. I've tried to be careful and get things right, but in science, skeptical disbelief is good manners. Politicians often get offended when people don't automatically believe everything they say, but scientists expect other scientists to question their assertions and help correct any typos. I suppose we could adopt a famous Cold War expression that President Reagan used *vis-a-vis* the Soviet Union: trust, but verify.

So now, assuming that you've checked my work and worked out the last two expressions for the A's and B's, it's time to figure out how to calculate them and then do the necessary matrix inversions and make a few plots. As I mentioned, Chapter 3 in McLachlan discusses numerical issues, but remember that this book was written at the start of the Cold War when computers were still mostly humans, so my advice is to ask around for somebody who has made available code to calculate them. Presumably, you'll find it together with code you hunted up for the Mathieu functions themselves.

It's always good practice to start by reproducing plots of earlier authors that have stood the test of time. Coming soon are some plots from the Erratum to Yeh's early work on scattering from a dielectric ribbon. After a quarter century, one of Yeh's students was apparently trying to reproduce his early plots and couldn't. They identified a typo in Yeh's original code, rechecked things carefully (we hope) and then published some new plots [4].

Exercise 8.2 Reproduce the corrected figures from the Erratum to Yeh's foundational paper. Although that 1998 paper is available online [4], the copy you'll find there isn't great. Note that the plots I've included in Figure 8.5–8.8 are actually sketches, drawn from that low-quality PDF. I could have walked across campus to the physics library and taken the bound volume back across campus to scan the pages and then returned the bound volume. It's kind of a nice day today, but there are two buildings under construction between here and there and the particular volume I want is probably at off-site storage these days and it's technically a holiday here today so the librarians are who-knows-where. Hence, some of the smaller backscattering/lateral lobes aren't drawn. When you make actual plots, pay special attention to the details that I haven't reproduced. You'll come across this issue often because classic results from the predigital days are often scanned at poor resolution and when you're trying to benchmark your new code against them, it can be frustrating.

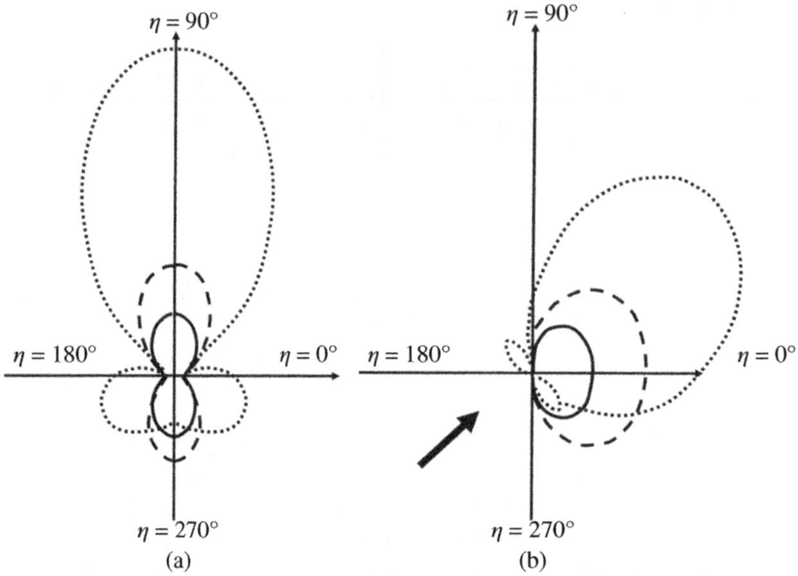

Figure 8.5 Polar plots of scattering of $|H_\eta|$ from a dielectric ribbon with $(k_0q)^2 = 10$. The incident electric vector is polarized in the axial direction. Solid line is for $\xi_0 = 0.1$; dashed line is for $\xi_0 = 0.2$; dotted line is for $\xi = 0.5$. Incident wave is vertical in (a) and at 45° as indicated by the arrow in (b). Note that in (b), I'm confident of the forward-scattering lobes, but not the smaller back-scattered lateral lobes. You should check those.

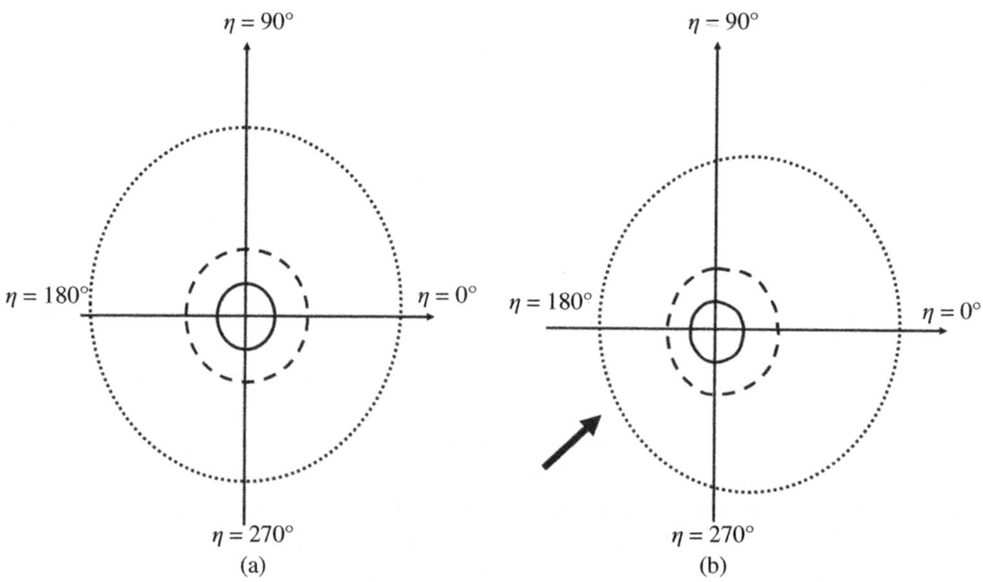

Figure 8.6 Polar plots of scattering of $|H_\eta|$ from a dielectric ribbon with $(k_0q)^2 = 1.0$. The incident electric vector is polarized in the axial direction. Solid line is for $\xi_0 = 0.1$; dashed line is for $\xi_0 = 0.2$; dotted line is for $\xi = 0.5$. Incident wave is vertical in (a) and at 45° as indicated by the arrow in (b).

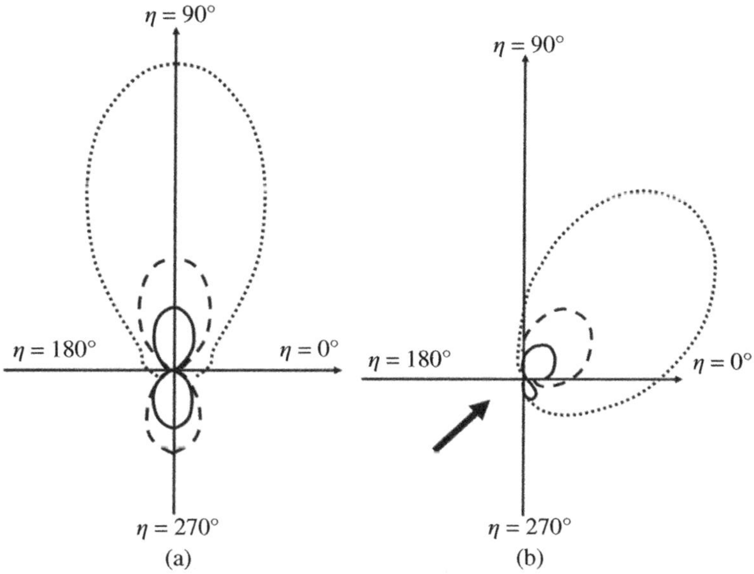

Figure 8.7 Polar plots of scattering of $|E_\eta|$ from a dielectric ribbon with $(k_0 q)^2 = 10$. The incident magnetic vector is polarized in the axial direction. Solid line is for $\xi_0 = 0.1$; dashed line is for $\xi_0 = 0.2$; dotted line is for $\xi = 0.5$. Incident wave is vertical in (a) and at 45° in (b).

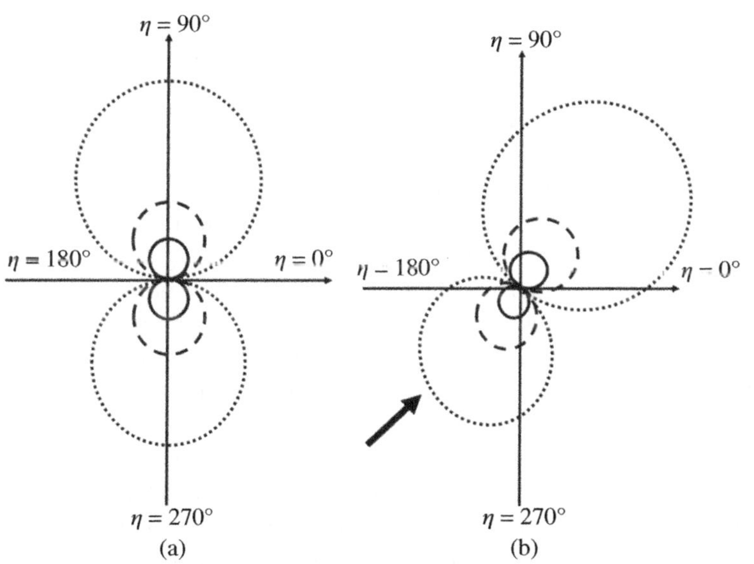

Figure 8.8 Polar plots of scattering of $|E_\eta|$ from a dielectric ribbon with $(k_0 q)^2 = 1.0$. The incident magnetic vector is polarized in the axial direction. Solid line is for $\xi_0 = 0.1$; dashed line is for $\xi_0 = 0.2$; dotted line is for $\xi = 0.5$. Incident wave is vertical in (a) and at 45° in (b).

8.3.2 Numerical Implementation of Mathieu Functions

One of my graduate students just rolled his eyes at me. He's about the same age as my daughter, so I'm used to it. I had just pulled from a bookshelf in the lab a 1996 book *Computation of Special Functions* (https://www.amazon.com/Computation-Special-Functions-Shanjie-Zhang/dp/0471119636) and showed him that there was a 3.5-in diskette in a pocket in the back with FORTRAN code. Also, my wife presumably rolls her eyes at me when I talk excitedly about finding a newly published review paper that gives a historical perspective on the computation and application of Mathieu functions [6]. My wife and I have been together for more than 40 years, so she knows to turn her head slightly when she rolls her eyes. I make no comment about her daily Talbots delivery.

I'm feeling much better today about numerical issues surrounding Mathieu functions. The FORTRAN from that 1996 book is now readily available via a direct translation, performed using f2matlab, of the original FORTRAN-77 implementations from the floppy disk at the back of the book, although there are complaints about ease of use and potential bugs. The particular things that we'll need are the expansion coefficients for the Mathieu functions (the A's and B's) and then the Mathieu functions and modified Mathieu functions and derivatives. Before starting to make plots, it's probably good practice to poke around a bit and see if there's something more recent. General Mathieu functions with arbitrary parameters V1.0[6] or Mathieu Functions Toolbox v.1.0[7] which is a computational toolbox that includes routines for the characteristic values, the expansion coefficients, and the four categories of angular and radial Mathieu functions together with their derivatives: even–even, even–odd, odd–even, and odd–odd. My preliminary conclusion is that making plots of the scattering from a dielectric elliptic cylinder is straightforward numerically. It's not trivial, though.

I feel like I should also include some discussion from the 2020 review paper [6] that states in the introduction that they "believe that there is still no fully satisfactory code available" although they do seem to be fans of the Digital Library of Mathematical Functions (https://dlmf.nist.gov/28.1). They further opine that the use of Mathieu functions and modified Mathieu functions in the solution of physical problems is attractive because the functions are analogous to harmonic functions, and expansions in terms of them can be efficient in comparison with direct numerical solution of the PDE model, but in practice the available software might be restricted to real arguments or use a normalization different from the one desired (apparently there are at least three normalizations in common use) or the software may fail to be accurate for "difficult" values of the problem parameters. I get it. Mathieu functions are tricky. I think I'll stick to my positive spin that this trickiness provides a barrier to competition. My wife understands this, which is why she turns her head before she rolls her eyes.

A bit more from [6] seems appropriate. It was Sir Edmund Taylor Whittaker who bestowed the name Mathieu equation and Mathieu functions. "Attributing a person's name to an equation or a function is a significant event in mathematics because people (even mathematicians) are social animals, and we simply pay more attention when a person's name is involved. Such namings often get it wrong, of course: 'Stigler's Law of Eponymy' states that no scientific discovery is named after its original discoverer." In this case, I agree that Mathieu deserves the credit. I also agree that "A

6 https://www.mathworks.com/matlabcentral/fileexchange/27101-general-mathieu-functions-with-arbitrary-parameters-v1-0, MATLAB Central File Exchange.
7 https://www.mathworks.com/matlabcentral/fileexchange/22081-mathieu-functions-toolbox-v-1-0, MATLAB Central File Exchange.

Course of Modern Analysis" by Whittaker and Watson is an excellent place to start any serious study of Mathieu functions.

Also, since the book by McLachlan that I've mentioned repeatedly owes much to the work by Edward Lindsay Ince, a research student of Whittaker who died of leukemia at age 49, I think it's fitting to include this quote included in [6] from the obituary that Whittaker wrote about Inge's sensibility with regard to computation:

> 'Ince held that an important part of a pure mathematician's duty is to provide tables for the use of physicists and astronomers, and he was well aware that the possibility of constructing such tables without a colossal expenditure of time and energy depends on the progress of theoretical analysis.' Whittaker remarks that Ince's 1932 tables of the Mathieu functions with their zeros and turning points was 'A splendid piece of work, performed single-handed save for some help by an Egyptian assistant.' Ince gave the name of his helper, when he acknowledged Mansy Shehata, who was then an Assistant in Pure Mathematics at the Egyptian University in Cairo. Ince also acknowledged grant support in purchasing calculating machines, which seemed to be of significant use.

8.4 Scattering of Elastic Waves by an Elliptic Cylindrical Inclusion

The 1971 Rand Corporation report by Pao and Mow [7] is now available as a downloadable PDF, so you should get yourself a copy. I still have a version that I photocopied more than 30 years ago and then recently ran back through a high-end scanner to get a PDF. Although the copy is clean and it scanned quite well, it's a little hard to read because it was typewritten. On a typewriter. Probably one of those IBM Ball Selectric typewriters with the interchangeable balls to do the Greek characters in equations. I remember as a graduate student helping my advisor put together drafts of proposals for the woman, named Anne, who used such a typewriter to turn a monstrosity that we had assembled by literally cutting and pasting from other documents and then annotated by hand and inserted handwritten text and equation and whatnot. Anne would turn it into a polished document ready to take to the copy center. Then I foolishly showed what I could do with Tex, the precursor to LaTex, and inadvertently volunteered myself to type all subsequent proposals. Oops.

Pao and Mow include a chapter on Elliptic Cylinder Problems, where they discuss elastic wave scattering from elastic cylinders of elliptical cross section. Their opening paragraph is worth transcribing. "Basically, the diffraction of waves by an elliptic cylinder is not much different from the diffraction caused by a circular cylinder, especially when the eccentricity of the elliptical cross section is small. But in mathematical analysis, because of the geometry of the scatterer, an entirely different wave function is used, involving products of Mathieu functions; the angular Mathieu function contains the wave number in its argument. Thus at the boundary of an ellipse, the wave functions which are the sum of the P and S wave parts do not form an orthogonal set, and the boundary conditions cannot be satisfied exactly by using the wave-function expansion method. This imposes a serious difficulty in calculating stresses and displacements near the scatterer." That sounds rather ominous, of course, but in the second paragraph, they reassure us. "We shall, however, carry out the analysis as far as we can in a manner analogous to that in the previous chapter, and indicate how to get around this particular difficulty." They also state the perhaps obvious point. "The literature concerning the diffraction of elastic waves is much less abundant for elliptic cylinders than for circular cylinders or spheres." I'd also like to add that this paragraph looks nicer when properly typeset.

Since we're going to follow the same solution procedure as for elastic wave scattering from cylinders and spheres, we know that we're going to be needing certain mathematical quantities in this new coordinate system. Pao and Mow helpfully include them, so I'm simply going to typeset them here for when we need them. The first quantity is a factor that shows up so often it wants to have its own symbol J, where

$$J^2 = \cosh^2\xi - \cos^2\eta = \frac{1}{2}(\cosh 2\xi - \cos 2\eta)$$

This will make expressions for grad, div, curl, and all that a bit simpler

$$\nabla f = \frac{1}{qJ}\left(\hat{e}_\xi \partial_\xi f + \hat{e}_\eta \partial_\eta f\right) + \hat{e}_z \partial_z f \tag{8.43}$$

$$\nabla \cdot \vec{F} = \frac{1}{q^2J^2}\partial_\xi\left(qJF_\xi\right) + \frac{1}{q^2J^2}\partial_\eta\left(qJF_\eta\right) + \partial_z F_z \tag{8.44}$$

$$\nabla \times \vec{F} = \hat{e}_\xi\left(\frac{1}{qJ}\partial_\eta F_z - \partial_z F_\eta\right) + \hat{e}_\eta\left(\partial_z F_\xi - \frac{1}{qJ}\partial_\xi F_z\right) + \hat{e}_z\frac{1}{q^2J^2}\left[\partial_\xi(qJF_\eta) - \partial_\eta(qJF_\xi)\right]$$

$$\nabla^2 f = \frac{1}{q^2J^2}\left(\partial_\xi^2 f + \partial_\eta^2 f\right) + \partial_z^2 f \tag{8.45}$$

As is usual, we do a Helmholtz decomposition of the displacement field into longitudinal and transverse components (φ, ψ, χ), where the longitudinal part has no curl, so it can be written as the gradient of a scalar function and the transverse part has no divergence so its two polarizations can be written as the curl and curl–curl of two other scalar functions. Each of these scalar functions satisfy the scalar wave equation, which looks like this in elliptic cylinder coordinates

$$\left[\frac{1}{q^2J^2}\left(\partial_\xi^2 + \partial_\eta^2\right) + \partial_z^2\right](\varphi, \psi, \chi) = k^2(\varphi, \psi, \chi) \tag{8.46}$$

where k is the wave number, which is different for the longitudinal and transverse waves. Since the boundary conditions are going to be continuity of displacements and stresses (tractions) at the scatterer surface $\xi = \xi_0$ we're going to need the components of the displacement in terms of the potential functions and then the stress–displacement relations. Following the lead of Pao and Mow, we list the pertinent equations without further explanation

$$\begin{aligned} u_\xi &= \frac{1}{qJ}\left(\partial_\xi\varphi + \partial_\eta\psi + l\partial_\xi\partial_z\chi\right) \\ u_\eta &= \frac{1}{qJ}\left(\partial_\eta\varphi - \partial_\xi\psi + l\partial_\xi\partial_z\chi\right) \end{aligned} \tag{8.47}$$

and

$$\begin{aligned} \sigma_{\xi\xi} = \frac{\mu}{q^2J^2}\Bigg\{&\left(\frac{\lambda}{\mu}k^2 + 2\partial_\xi^2 + \frac{\sinh 2\xi}{J^2}\partial_x + \sin 2\eta\partial_\eta\right)\varphi + \left(2\partial_\xi\partial_\eta - \frac{\sin 2\xi}{J^2}\partial_\xi + \frac{\sin 2\eta}{J^2}\partial_\eta\right)\psi \\ &+ l\left(2\partial_\xi^2 - \frac{\sin 2\xi}{J^2}\partial_\xi + \frac{\sin 2\eta}{J^2}\partial_\eta\right)\xi\Bigg\} \end{aligned} \tag{8.48}$$

$$\begin{aligned} \sigma_{\xi\eta} = \frac{\mu}{q^2J^2}\Bigg\{&\left(2\partial_\xi\partial_\eta - \frac{\sin 2\eta}{J^2}\partial_\xi - \frac{\sinh 2\xi}{J^2}\partial_\eta\right)\psi + \left(\partial_\eta^2 - \partial_\xi^2 + \frac{\sinh 2\xi}{J^2}\partial_\xi - \frac{\sin 2\eta}{J^2}\partial_\xi\right)\psi \\ &+ l\partial_z\left(2\partial_\xi\partial_\eta - \frac{\sin 2\eta}{J^2}\partial_\xi - \frac{\sinh 2\xi}{J^2}\partial_\eta\right)\chi\Bigg\} \end{aligned} \tag{8.49}$$

We won't need the χ terms, though, since the *SH* modes are decoupled from the *L* and *SV* waves. The latter two are coupled at the boundary, so for either incident *L* or *SV* waves, we have to account for both *L* and *SV* modes in both the scattered and the transmitted fields. You should probably steel yourself at this point for some straightforward, but highly nontrivial algebraic manipulations. In addition, I think it's important to prepare emotionally for having to give up because the algebraic manipulations may not be worth it in the end. With that in mind, here are the *L* and *SV* potential functions for the incident, scattered, and transmitted fields. Let's consider first the case where the incident wave is a longitudinal wave

$$\varphi^{inc} = 2\varphi_0 \sum_{n=0}^{\infty} \left[Ce_n(\xi)ce_n(\eta)ce_n(\theta_0) + Se_n(\xi)se_n(\eta)se_n(\theta_0)\right] \tag{8.50}$$

$$\varphi^{scat} = \sum_{n=0}^{\infty} \left[B_n Me_n(\xi)ce_n(\eta)ce_n(\theta) + C_n Ne_n(\xi)se_n(\eta)se_n(\theta)\right] \tag{8.51}$$

$$\psi^{scat} = \sum_{n=0}^{\infty} \left[D_n Me_n(\xi)ce_n(\eta)ce_n(\theta) + E_n Ne_n(\xi)se_n(\eta)se_n(\theta)\right] \tag{8.52}$$

and noting with the $*$ that the scatterer has different material properties

$$\varphi^{trans} = \sum_{n=0}^{\infty} \left[F_n Ce_n^*(\xi)ce_n^*(\eta)ce_n(\theta) + G_n Se_n^*(\xi)se_n^*(\eta)se_n(\theta)\right] \tag{8.53}$$

$$\psi^{trans} = \sum_{n=0}^{\infty} \left[H_n Ce_n^*(\xi)ce_n^*(\eta)ce_n(\theta) + I_n Se_n^*(\xi)se_n^*(\eta)se_n(\theta)\right] \tag{8.54}$$

In these B_n-I_n are the modal coefficients.

We can plug these into the expressions for the stress and displacement components and write the four boundary condition equations at $\xi = \xi_0$, and then sequentially multiply each of them by $ce_n(\eta), se_n(\eta)$ and integrate over η to see how much orthogonality does for us once we've expanded $ce_n(\eta)$ and $se_n(\eta)$ in Fourier series as before. Here's the part that we're going to need to have prepared emotionally for. We're used to the idea that η depends on material properties, because that issue came up when we considered electromagnetic scattering from a dielectric elliptic cylinder. What's not obvious from the notation that we're using is that orthogonality for the scattered mode-converted wave is going to have the same issue because the wave numbers are different for *L* and *SV* waves. I hope you're not too disappointed about that. It's quite a serious complication. Stay strong, though, because it may not be a deal breaker.

As much as I'm tempted to plow ahead at full speed, I think it makes good sense to redirect just a bit and consider the simpler case of elastic wave scattering from an elliptic cylindrical cavity. This problem will be plenty complicated, but also has innumerable real-world applications. There won't be any transmitted waves, of course, so that allows us to focus on the issue of how the presence of mode converted *SV* waves affects the orthogonality. Moreover, we'll get a bit of practice dealing with the shocking amount of complexity that's added when boundary conditions involve stresses. I promise that we'll come back to the problem of longitudinal wave scattering from an elastic elliptic cylinder. Actually, promise is too strong of a word to use here. What I actually mean is that I'm pretty sure that it's going to be too unwieldy to include in this chapter, but in principle, it probably could be done and so what I'll probably include in this chapter is acoustic scattering from a solid elliptic cylinder in a fluid. Sorry, not sorry.

Here's what the boundary condition equations look like for the elliptic cylinder cavity. The first is the normal stresses $\sigma_{\xi\xi}^{inc} + \sigma_{\xi\xi}^{scat} = 0$ for $\xi = \xi_0$

$$\frac{2\varphi_0\mu}{q^2J^2}\sum_{n=0}^{\infty} i^n\left\{\frac{\lambda k^2}{\mu}Ce_n(\xi_0)ce_n(\eta)ce_n(\theta)+2Ce''(\xi_0)ce_n(\eta)ce_n(\theta)+\frac{\lambda k^2}{\mu}Se_n(\xi_0)se_n(\eta)se_n(\theta)\right.$$
$$+2Se''(\xi_0)se_n(\eta)se_n(\theta)\frac{\sinh^2 2\xi_0}{J^2}Ce_n'(\xi_0)ce_n(\eta)ce_n(\theta)+\sin 2\eta Ce_n(\xi_0)ce_n'(\eta)ce_n(\theta)$$
$$\left.+\frac{\sinh^2 2\xi_0}{J^2}Se_n'(\xi_0)se_n(\eta)se_n(\theta)+\sin 2\eta Se_n(\xi_0)se_n'(\eta)se_n(\theta)\right\}$$
$$+\frac{\mu}{q^2J^2}\sum_{n=0}^{\infty}\left\{A_n\left[\frac{\lambda k^2}{\mu}Me_n(\xi_0)ce_n(\eta)ce_n(\theta)+2Me''(\xi_0)ce_n(\eta)ce_n(\theta)\right]\right.$$
$$+B_n\left[\frac{\lambda k^2}{\mu}Ne_n(\xi_0)se_n(\eta)se_n(\theta)+2Ne''(\xi_0)se_n(\eta)se_n(\theta)\right]$$
$$+A_n\left[\frac{\sinh^2 2\xi_0}{J^2}Me_n'(\xi_0)ce_n(\eta)ce_n(\theta)+\sin 2\eta Me_n(\xi_0)ce_n'(\eta)ce_n(\theta)\right]$$
$$+B_n\left[\frac{\sinh^2 2\xi_0}{J^2}Ne_n'(\xi_0)se_n(\eta)se_n(\theta)+\sin 2\eta Ne_n(\xi_0)se_n'(\eta)se_n(\theta)\right]$$
$$+C_n\left[2Me_n'(\xi_0)ce_n'(\eta_T)ce_n(\theta)-\frac{\sin 2\eta_T}{J^2}Me_n'(\xi_0)ce_n(\eta_T)ce_n(\theta)\right.$$
$$\left.+\frac{\sin 2\eta_T}{J^2}Me_n(\xi_0)ce_n'(\eta_T)ce_n(\theta)\right]$$
$$+D_n\left[2Ne_n'(\xi_0)se_n'(\eta_T)se_n(\theta)-\frac{\sin 2\eta_T}{J^2}Ne_n'(\xi_0)se_n(\eta_T)se_n(\theta)\right.$$
$$\left.\left.+\frac{\sin 2\eta_T}{J^2}Ne_n(\xi_0)se_n'(\eta_T)se_n(\theta)\right]\right\}$$

The second boundary condition equation is for the shear stresses $\sigma_{\xi\eta}^{inc}+\sigma_{\xi\eta}^{scat}=0$ for $\xi=\xi_0$

$$\frac{2\varphi_0\mu}{q^2J^2}\sum_{n=0}^{\infty} i^n\left\{2Ce_n'(\xi_0)ce_n'(\eta)ce_n(\theta)-\frac{\sin 2\eta}{J^2}Ce_n'(\xi_0)ce_n(\eta ce_n(\theta)-\frac{\sinh 2\xi_0}{J^2}Ce_n(\xi_0)ce_n'(\xi)ce_n(\theta)\right.$$
$$\left.+2Se_n'(\xi_0)se_n'(\eta)se_n(\theta)-\frac{\sin 2\eta}{J^2}Se_n'(\xi_0)se_n(\eta se_n(\theta)-\frac{\sinh 2\xi_0}{J^2}Se_n(\xi_0)se_n'(\xi)se_n(\theta)\right\}$$
$$+\frac{\mu}{q^2J^2}\sum_{n=0}^{\infty}\left\{B_n\left[2Me_n'(\xi_0)ce_n'(\eta)ce_n(\theta)-\frac{\sin 2\eta}{J^2}Me_n'(\xi_0)ce_n(\eta ce_n(\theta)\right.\right.$$
$$\left.-\frac{\sinh 2\xi_0}{J^2}Me_n(\xi_0)ce_n'(\xi)ce_n(\theta)\right]+C_n\left[2Ne_n'(\xi_0)se_n'(\eta)se_n(\theta)\right.$$
$$\left.-\frac{\sin 2\eta}{J^2}Ne_n'(\xi_0)se_n(\eta se_n(\theta)-\frac{\sinh 2\xi_0}{J^2}Ne_n(\xi_0)se_n'(\xi)se_n(\theta)\right]$$
$$+D_n\left[Me_n(\xi_0)ce_n''(\eta_T)ce_n(\theta)-Me_n''(\xi_0)ce_n(\eta_T)ce_n(\theta)\right.$$
$$\left.+\frac{\sinh 2\xi_0}{J^2}Me_n'(\xi_0)ce_n(\eta_T)ce_n(\theta)-\frac{\sinh 2\eta_T}{J^2}Me_n(\xi_0)ce_n'(\eta_T)ce_n(\theta)\right]$$
$$+E_n\left[Ne_n(\xi_0)se_n''(\eta_T)se_n(\theta)-Ne_n''(\xi_0)se_n(\eta_T)se_n(\theta)\right.$$
$$\left.\left.+\frac{\sinh 2\xi_0}{J^2}Ne_n'(\xi_0)se_n(\eta_T)se_n(\theta)-\frac{\sinh 2\eta_T}{J^2}Ne_n(\xi_0)se_n'(\eta_T)se_n(\theta)\right]\right\}$$

OK, so I'm breaking my promise from a couple of pages ago. You may or may not have even made it this far and were planning to look at some plots and call it good. I didn't include any plots, though. What I'm hoping is that your take-away message is: if I ever find an urgent need to solve these elliptic-cylinder scattering problems involving elastic waves, I had better budget some significant time for the analysis and computations and that it's probably doable, but really tricky.

8.5 Scattering from Spheroids

If the Mathieu functions didn't scare you off, then you're probably going to be OK with the eigenfunction solutions for scattering from spheroids. The analysis methods will be analogous to what we did for Mie scattering from spheres, of course, but the eccentricity of the scatterer and orientation relative to the incident field is going to matter. Spheroids come in two extremes: needle-like and disk-like. Not quite so extreme spheroids: cigar and pancake. Of course, cigar-shaped objects might be excellent first approximations to use for radar scattering from missiles and sonar scattering from submarines, so it's not surprising that there is a huge literature. Penny-shaped cracks and delaminations can be a big deal in nondestructive evaluation, so there's a sizable literature for that as well. We'll be able to write the equations to set up these various problems, but we may find that even today, there are numerical difficulties. We should probably steel ourselves that we might only be able to get the results we're after for special cases.

The prolate spheroidal coordinate system is defined with the relations

$$
\begin{aligned}
x &= \frac{1}{2}a \sinh\mu \sin\theta \cos\phi \\
y &= \frac{1}{2}a \sinh\mu \sin\theta \sin\phi \\
z &= \frac{1}{2}a \cosh\mu \cos\theta
\end{aligned}
\tag{8.55}
$$

and with the oblate coordinates obtained from these by formally changing a into $-ia$ and simultaneously changing $\cosh\mu$ into $i\sinh\mu$. Consequently, when solutions for the prolate case have been obtained, the oblate solutions invoke a fairly simple extension to imaginary values and coordinates. As I was typing that sentence just now, it seemed a little dismissive of the difficulties we may yet encounter, but I could be wrong so stay tuned.

The Helmholtz equation $\nabla^2\psi + k^2\psi = 0$ has the following representation in prolate spheroidal coordinates:

$$
\begin{aligned}
&\frac{1}{\sinh\mu}\frac{\partial}{\partial\mu}\left(\sinh\mu\frac{\partial\psi}{\partial\mu}\right) + \frac{1}{\sin\theta}\frac{\partial}{\partial\theta}\left(\sin\theta\frac{\partial\psi}{\partial\theta}\right) \\
&+\frac{\sinh^2\mu+\sin^2\theta}{\sinh^2\mu\sin^2\theta}\frac{\partial^2\psi}{\partial\phi^2} + \frac{1}{4}\left(\cosh^2\mu - \cos^2\theta\right)(ka)^2\psi = 0
\end{aligned}
\tag{8.56}
$$

Of course the next step is to do separation of variables. In the axial coordinate ϕ, the solutions are $\cos(m\phi)$ and $\sin(m\phi)$ for $m = 0, 1, 2, \ldots$ and $0 \le \phi \le 2\pi$. For the other two coordinates, we define $\xi = \cosh\mu$ and $\eta = \cos\theta$ and have the following ODEs:

$$
\frac{d}{d\xi}\left[\left(\xi^2-1\right)\frac{dJ}{d\xi}\right] - \left[A - h^2\xi^2 + \frac{m^2}{\xi^2-1}\right]J = 0 \tag{8.57}
$$

$$
\frac{d}{d\eta}\left[\left(1-\eta^2\right)\frac{dS}{d\eta}\right] + \left[A - h^2\eta^2 - \frac{m^2}{1-\eta^2}\right]S = 0 \tag{8.58}
$$

where $h = \frac{1}{2}ka$ and A is the separation constant. Hence, we can write the general solution as:

$$\psi(\phi,\eta,\xi) = \frac{\cos(m\phi)}{\sin(m\phi)} S(\eta)J(\xi) \tag{8.59}$$

Note that the two aforementioned ODEs are really the same. One is for $S(\eta)$ involving the behavior of the solutions between the singular points -1 and $+1$ and the other is for $J(\xi)$ for the range $+1$ to ∞. We can therefore consider the solutions of one equation for z, where η comprises the range -1 and $+1$ and ξ over the range from $+1$ to ∞.

The plane wave expansion in prolate spheroidal coordinates is

$$e^{i\vec{k}\cdot\vec{r}} = 2\sum_{m,l} \frac{\epsilon_m i^l}{\Lambda_{ml}(h)} S_{ml}(h,\cos\theta_0)\cos[m(\phi-\phi_0)]S_{ml}(h,\cos\theta)je_{ml}(h,\cosh\mu) \tag{8.60}$$

where ϵ_m is the Neumann factor, $\Lambda_{ml}(h)$ are the normalization constants, and

$$\begin{aligned} i\vec{k}\cdot\vec{r} &= k\left[z\cos\theta_0 + x\sin\theta_0\cos\phi_0 + y\sin\theta_0\sin\phi_0\right] \\ &= h\left[\cosh\mu\cos\theta\cos\theta_0 + \sinh\mu\sin\theta\sin\theta_0\cos(\phi-\phi_0)\right] \end{aligned}$$

In these expressions, $je_{ml}(h,\cosh\mu)$ are the spheroidal radial functions of the first kind and $S_{ml}(h,\cos\theta)$ are the prolate spheroidal angular functions. Note that for $h \to 0$ these will reduce to the usual spherical Bessel functions and Legendre polynomials.

Let's first consider the simplest possible case of a plane acoustic wave scattering from a rigid spheroid. Without loss of generality, we can set $\phi_0 = 0$ and write the incident plane wave of unit amplitude as:

$$\psi^{inc} = 2\sum_{m,l} \frac{\epsilon_m i^l}{\Lambda_{ml}(h)} S_{ml}(h,\cos\theta_0)S_{ml}(h,\cos\theta)je_{ml}(h,\cosh\mu)\cos[m\phi] \tag{8.61}$$

and we then assume a scattered field of the form

$$\psi^{scat} = 2\sum_{m,l} \frac{\epsilon_m i^l}{\Lambda_{ml}(h)} A_{ml}S_{ml}(h,\cos\theta)he_{ml}(h,\cosh\mu)\cos[m\phi] \tag{8.62}$$

where $he_{ml}(h,\cosh\mu)$ is the radial function of the third kind that I like to call the "spheroidal Hankel function" even though I'm pretty sure that would piss off some mathematician. Speaking of which, in 2011, a Cal State Northridge mathematics professor was accused of urinating on a colleague's door.[8]

Back to serious business. For a rigid spheroid, the boundary conditions are

$$\frac{\partial\psi^{inc}}{\partial\xi} = \frac{\partial\psi^{scat}}{\partial\xi} \tag{8.63}$$

as the surface $\xi = \xi_0$ and remembering that $\xi = \cosh\mu$, we can let prime indicate differentiation with respect to ξ and do the simple algebra to find

8 Tihomir Petrov, 43, was allegedly caught on surveillance camera urinating on his colleague's office door. It's a literal pee tape. "My client is a man of impeccable character," said his lawyer with a straight face. CBS News headlined their story, "Pissing Match between two Professors." The camera was hidden by an engineering colleague when officials suspected the puddles next to a certain math whiz's office were of urine. I wonder what the source of the argument between the two math profs was?

$$A_{ml} = S_{ml}(h, \cos\theta_0)\frac{je'_{lm}(h, \xi_0)}{he'_{lm}(h, \xi_0)} \tag{8.64}$$

which then fully determines the expression for the scattered field. What we're going to want, though, is the far-field scattering pattern, which for $r \to \infty$ should look like

$$\psi^{scat} \to f(\theta, \phi)e^{ikr}/r \tag{8.65}$$

For large arguments, we have

$$he_{ml}(h, \xi) = -i(-i)^{m+l}e^{ih\xi}/h\xi \tag{8.66}$$

where $h = ka/2$. We can thus write

$$|f(\theta, \phi)| = \left|2\sum_{lm}(-i)^{l+m}\frac{\epsilon_m}{\Lambda_{lm}(h)}S_{ml}(h, \eta)S_{lm}(h, \eta_0)\frac{je'_{lm}(h, \xi_0)}{he'_{lm}(h, \xi_0)}\cos m\phi\right| \tag{8.67}$$

In Figure 8.9 are shown two scattering patterns, adapted from [8] who did her[9] computations in 1951.[10] They note that the various wave functions needed had not yet been tabulated, "and therefore, it was necessary to compute tables of functions of sufficient extent and accuracy to enable us to complete the computation" which I find really impressive. The paper includes four sets of six plots, each with two scattering patterns on them. Figure 8.9 also shows scattering patterns for four different angles relative to the long axis of the spheroid. The work was funded by the Office of Naval Research, so the obvious application was sonar scattering from submarines. A rigid prolate spheroid is a pretty good first approximation for that.

Exercise 8.3 Since radar people say that acoustics is just a scalar version of electromagnetics, does this solution for acoustic scattering from a rigid spheroid apply to the corresponding EM scattering from a perfectly conducting spheroid?

The answer to this exercise seems to be a hard nope [9]. Sorry. Kerker [10] references a 1950 U. Michigan Engineering Research Institute report[11] by F.W. Schultz, "Scattering by a

9 I've been wondering about Sara Granger, who was presumably the student of Prof. R.D. (Bob) Spence and did these calculations. So far, no luck, but I did find the following about Helen Spence, "Helen Spence came to MSU as a student in 1936 and worked as a lab assistant in the physics department. After she graduated in early 1939, she taught math, physics and chemistry in a public school in Portland, Michigan. In 1942 she married Bob Spence and they moved out East where her husband taught radar to Army and Navy personnel while she worked in the theoretical division of the Massachusetts Institute of Technology (MIT) Radiation Lab on state-of-the-art computers (Monroes and Marchants mainly electrically-powered calculating machines that carried 10 digits and could add, subtract, multiply and divide but not figure square roots). Spence and her husband came to MSU in 1947 where he taught in the Physics Department and she was an instructor in the Computer Science Department, teaching Assembly languages. She helped program the first major computer on campus and eventually taught FORTRAN in the 1960s. Mrs. Spence retired in 1987." Her obituary tells a bit more about her life https://www.legacy.com/us/obituaries/lsj/name/helen-spence-obituary?id=18217236. That leads me to wonder whether the plots in [8] were done by hand or on an early computer. If it was an electronic computer rather than a human computer, that gadget would have been essentially hand built. Apparently the MSU machine shop has the following quote on a banner, which I find apropos, "If you're not building original research equipment, you're not doing original research." – Dr. Robert D. Spence.

10 My mother wasn't allowed to take math in high school in the 1950s, and then she needed inferential statistics to evaluate the data for her dissertation. She was being raised by her father's big sister, because my grandmother had died quite young, who insisted that my mother learn practical skills to be able to support herself when needed.

11 https://deepblue.lib.umich.edu/handle/2027.42/7482.

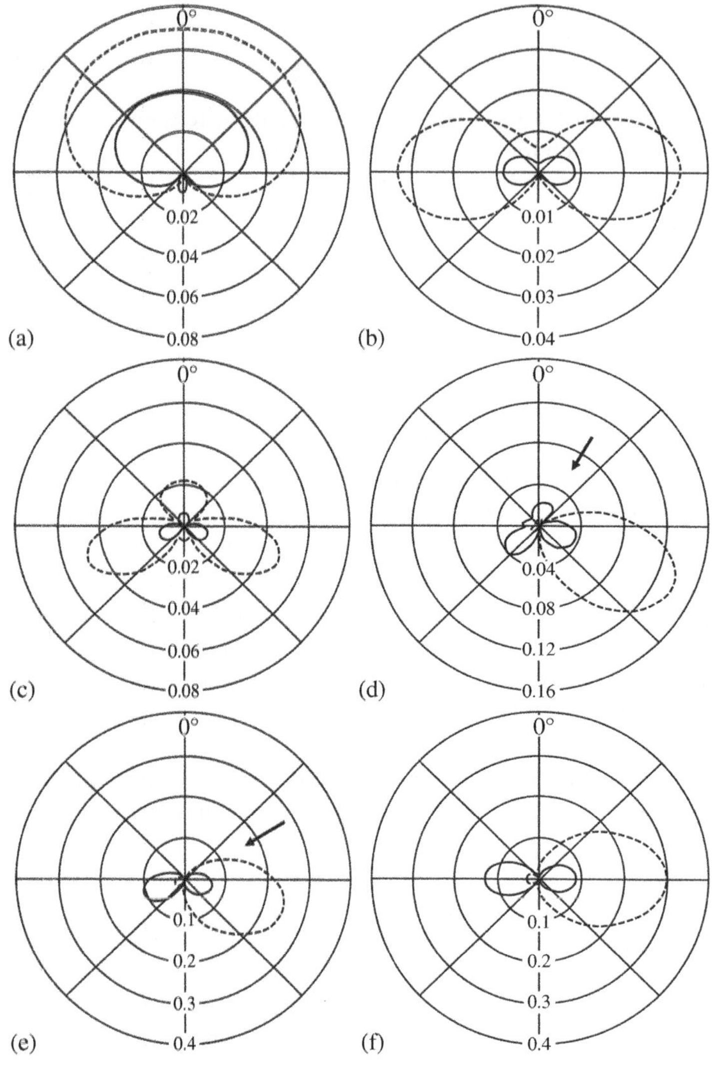

Figure 8.9 Scattering for acoustic wave incident along the axis of a rigid spheroid (a,b). (a) is for $\theta = 0$ and $h = 1$ with solid line for $\xi = 1.044$ and dashed line for $\xi = 1.077$. (b) is for $\theta = 0$ and $h = 2$ with solid line for $\xi = 1.005$ and dashed line for $\xi = 1.020$. The other four plots are for a rigid spheroid with $h = 3$ for different angles of incidence. In each, the solid line is for $\xi = 1.005$ and the dashed line for $\xi = 1.020$. (c) is for the incident along the axis of the spheroid. (d) is for 30° angle of incidence, as indicated by the arrow. (e) is for 60° angle of incidence, as indicated by the arrow. (f) is for 90° angle of incidence.

prolate spheroid" with the comment that "an explicit solution for the scattering coefficients has not been obtained except for the special case of the backscattering by a perfectly conducting prolate spheroid at nose-on incidence." I looked up that report and put the hyperlink in the footnote for you. It's amazing to read, particularly because the text is typed and the innumerable equations are handwritten so very beautifully. Here's the expression for the scattered field:

$$^{S}\underline{E} = \left\{ \underline{i}_{\eta} \left[\sum_{l=0}^{\infty} j^{l+2} \alpha'_{0l} S_{0l}^{(1)}(\eta) \sin\phi \right] \right.$$

$$\left. + \underline{i}_{\phi} \left[\sum_{l=0}^{\infty} j^{l} \alpha'_{0l} \eta S_{0l}^{(1)}(\eta) \cos\phi + j^{l+3} \beta'_{1l} (1-\eta^2)^{1/2} S_{1l}^{(1)}(\eta) \cos\phi \right] \right\}$$

$$\times ({}^{I}E_0 a_0) \times \frac{1}{r} e^{-j\frac{2\pi}{\lambda} r} . \qquad (112)$$

Here $\alpha'_{0l} = \frac{1}{{}^{I}E_0 a_0} \alpha_{0l}$ and $\beta'_{1l} = \frac{1}{{}^{I}E_0 a_0} \beta_{1l}$. We notice at once the absence of an $\underline{i}_{\xi}$ component, which is the longitudinal component for large values of ξ . This, of course, is a necessary condition that the wave be a purely transverse wave at a large distance from the spheroid, as it must be.

except that in these, the modal coefficients α_{0l} and β_{1l} have to be determined by numerical inversion of the infinite systems of equations:

$$\begin{array}{l}
\alpha_{00}C_{00} + 0 + 0 + 0 + \cdots + 0 + \beta_{11}D_{01} + 0 + \beta_{13}D_{03} + \cdots = ({}^{I}E_0 a_0)(B_{00} + 0 + 0 + 0 + \cdots) \\
0 + \alpha_{01}C_{11} + 0 + 0 + \cdots + \beta_{10}D_{10} + 0 + \beta_{12}D_{12} + 0 + \cdots = ({}^{I}E_0 a_0)(0 + B_{11} + 0 + 0 + \cdots) \\
0 + 0 + \alpha_{02}C_{22} + 0 + \cdots + 0 + \beta_{11}D_{21} + 0 + \beta_{13}D_{23} + \cdots = ({}^{I}E_0 a_0)(0 + 0 + B_{22} + 0 + \cdots) \\
0 + 0 + 0 + \alpha_{03}C_{33} + \cdots + \beta_{10}D_{30} + 0 + \beta_{12}D_{32} + 0 + \cdots = ({}^{I}E_0 a_0)(0 + 0 + 0 + B_{33} + \cdots) \\
\vdots
\end{array} \qquad (110)$$

$$\begin{array}{l}
0 + \alpha_{01}V_{01} + 0 + \alpha_{03}V_{03} + \cdots + \beta_{10}W_{00} + 0 + \beta_{12}W_{02} + 0 + \cdots = ({}^{I}E_0 a_0)(0 + U_{01} + 0 + U_{03} + \cdots) \\
\alpha_{00}V_{10} + 0 + \alpha_{02}V_{12} + 0 + \cdots + 0 + \beta_{11}W_{11} + 0 + \beta_{13}W_{13} + \cdots = ({}^{I}E_0 a_0)(U_{10} + 0 + U_{12} + 0 + \cdots) \\
0 + \alpha_{01}V_{21} + 0 + \alpha_{03}V_{23} + \cdots + \beta_{10}W_{20} + 0 + \beta_{12}W_{22} + 0 + \cdots = ({}^{I}E_0 a_0)(0 + U_{21} + 0 + U_{23} + \cdots) \\
\alpha_{00}V_{30} + 0 + \alpha_{02}V_{32} + 0 + \cdots + 0 + \beta_{11}W_{31} + 0 + \beta_{13}W_{33} + \cdots = ({}^{I}E_0 a_0)(U_{30} + 0 + U_{32} + 0 + \cdots) \\
\vdots
\end{array} \qquad (111)$$

Alas. Schultz does give the exact expression for the corresponding scalar case, but doesn't show any plots, so I guess in this rivalry, Michigan State beat U. Michigan. Again according to Kerker, T.B.A. Senior[12] carried out numerical studies in 1967, so advantage Blue.

12 Thomas B. A. Senior was known for his fundamental contributions to electromagnetic and acoustic scattering, for his remarkable legacy of service and leadership to the department and professional community, and for his excellence as an educator. Born 26 June 1928 in Yorkshire, England, Professor Senior received his MSc and PhD degrees in Applied Mathematics from Manchester and Cambridge Universities in 1950 and 1954, respectively. He joined the University of Michigan in 1957 as a researcher at the famed Willow Run Laboratories. He was specifically

Let's console ourselves by writing down the answer for sound scattering from an acoustically soft prolate spheroid.

$$|f(\theta,\phi)| = \left|2\sum_{lm}(-i)^{l+m}\frac{\epsilon_m}{\Lambda_{lm}(h)}S_{ml}(h,\eta)S_{lm}(h,\eta_0)\frac{je_{lm}(h,\xi_0)}{he_{lm}(h,\xi_0)}\cos m\phi\right| \tag{8.68}$$

which is pretty similar to what we had for the acoustically hard scatterer. You might have to stare at it for a moment to notice that the only difference is that there's no derivatives for the radial functions in the amplitude term. An example of this situation might be a spheroidal gas bubble in a liquid. There are a fair number of sonar and ultrasound situations where this could be relevant. Go ahead and make some plots, but do reproduce some or all of the hard-spheroid plots to make sure your code is running properly before doing these. One of the issues you should find a little frustrating is that the notation for these functions seems to be all over the map. Abromowitz and Stegun have a helpful table, which lays out the notation of various authors in the early days. Little things like, WTF is the "normalization" term $\Lambda_{lm}(h)$ that some other authors call $N_{m,l}$ and describe as "the norm of the angular wave function." I'm not deliberately torturing you by leaving things like that a little ambiguous; it's just that I want you to get into the habit of doing a little background reading to orient yourself on new-to-you special functions before coding up equations and making a bunch of plots.

Recall that I casually mentioned that the oblate coordinates are obtained from the prolate ones by formally changing a into $-ia$ and simultaneously changing $\cosh\mu$ into $i\sinh\mu$. I then went on to assert that when solutions for the prolate case have been obtained, the oblate solutions invoke a fairly simple extension to imaginary values and coordinates. I think I'm going to have to call bullshit on myself because [3] has separate chapters for each. It's true that the analysis is pretty similar for the two kinds of spheroids, but I think it makes good sense to go through them both. I'm not going to include the oblate case here, or suggest an exercise for you to do that analysis, but if you should feel excited to do it, I hope that you'll then show that the two cases match each other when they tend to the common limit of spheres. Do that limiting case analytically, and also show that for the acoustically soft scatterers they both match up with Anderson's results in the spherical bubble limits.

At this point, I had in mind discussing at length the important problem of penny-shaped cracks in nondestructive evaluation. On 19 July 1989, a DC-10 (United Airlines flight 232) crash-landed at the airport in Sioux City, Iowa, after suffering a catastrophic failure of its tail-mounted engine due to an unnoticed manufacturing defect in the engine's fan disk that resulted in the loss of many flight controls. Of the 296 passengers and crew on board only 112 died, which counts as a win. A hard-alpha inclusion had been present in the titanium billet, which was forged into the turbine rotor hub (fan disk) of the tail-mounted engine and that resulted in a penny-shaped crack that the inspectors subsequently missed. When that engine disintegrated explosively, the uncontained failure took out the triple-redundant hydraulic lines making the plane almost impossible to control. Miraculously, Dennis E. "Denny" Fitch, 46, a training-check airman was aboard Flight 232 as a passenger. He had a total flight time of around 23,000 hours under his belt, including 2987 hours of

recruited by Kip Siegel for his experience detecting V2 missiles, first launched in WWII. Professor Senior's research turned to the detection of stealth aircraft in the 1960s. Working in the days before computers, he created many of the analytical tools needed to predict how radar cross-section reduction could be accomplished using shaping and radar-absorbing materials. His research directly impacted the design of stealth aircraft in the U.S. "Electromagnetic and Acoustic Scattering by Simple Shapes" was a foundational work that emerged from Willow Run research on the radar detection of aircraft and missiles. Prof. Senior directed the RADLAB at a time when military research, especially classified research, on campus was a target for student activists. He passed away peacefully 24 November 2017 at the age of 89. See: http://ece.umich.edu/bicentennial/documents/radlab-history.pdf.

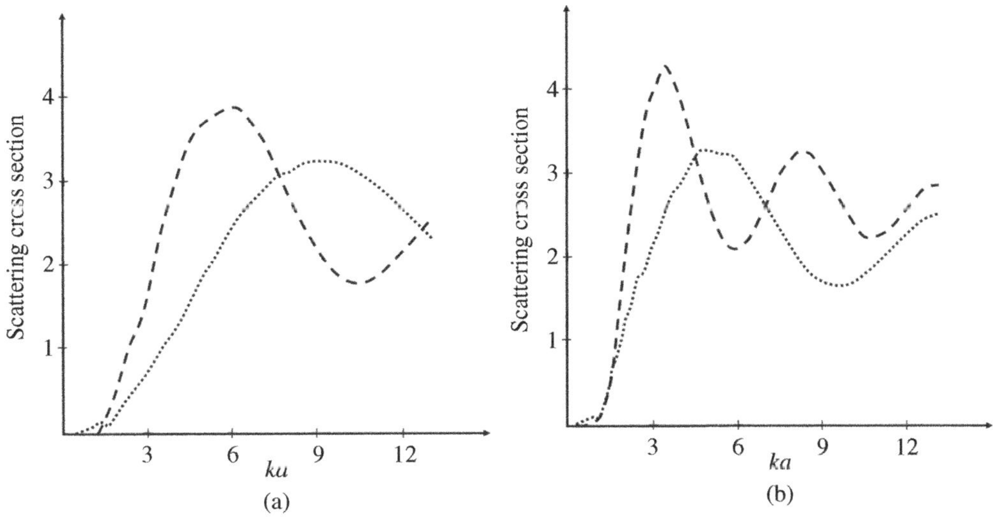

Figure 8.10 Scattering cross sections for prolate (dots) and oblate (dashes) spheroids with aspect ratios of 2. (a) is for *L*-wave incidence; (b) is for *SV*-wave incidence. Plane waves are incident at 30° with respect to the polar axes of the spheroids. The inclusion has velocities and density about 25% less than the background material. Source: Adapted from [11].

DC-10 time with United. Fitch had learned of a 1985 crash of Japan Air Lines Flight 123, which was caused by a catastrophic loss of hydraulic control, and had wondered if it was possible to control an aircraft using throttles only so he practiced that with similar conditions on a simulator. The short answer is he went to the cockpit to try it for real, and *almost* got the plane landed safely. You've probably seen the video of the fiery crash taken through a fence.

So there's lots of literature on elastic wave scattering from penny-shaped cracks, which are a limiting case of an oblate-spheroidal cavity. In principle, such a problem can be addressed with exactly the method discussed above. In practice, the elastic wave scattering method is frustrated by the same issues that made electromagnetic wave scattering from a perfectly conducting spheroid a whole lot more complex than the corresponding acoustic case. The issue seems to be orthogonality. For the acoustic case, there are no transverse waves and it just happens to work out that orthogonality allows the modal coefficients to be solved for without the pesky infinite summations over l and m. Things don't fall out so easily for the electromagnetic and elastic wave cases, because there are transverse waves involved. You can still write the systems of boundary conditions equations as a matrix, but you can't just invert it term-by-term. Sorry.

It appears that a complete solution for the scattering of elastic waves by a spheroidal inclusion has only recently appeared in the literature. In [11], the method that we've discussing is applied to the scattering of elastic waves by prolate and oblate spheroidal inclusions. As per usual, the problem is solved in the frequency domain where separation of variables leads to a solution involving spheroidal wave functions of the angular and radial kind. The problem is that for spheroids, the boundary equations remain coupled with respect to one of the separation indices. Expanding the angular spheroidal wave functions in terms of associated Legendre functions and using their orthogonality properties does lead to a set of linear equations that can be solved to simultaneously obtain solutions for all coupled modes of both scattered and interior fields. Figure 8.10 illustrates the normalized scattering cross sections of [11] for L and SV wave incident at 30° with respect to the polar axes of prolate and oblate spheroids.

References

1 Duhem, P. (1892). Emile Mathieu, his life and works. *Bulletin of the American Mathematical Society* 1: 156168. https://doi.org/10.1090/S0002-9904-1892-00067-5.

2 Burke, J.E. and Twersky, V. (1964). On scattering of waves by an elliptic cylinder and by a semielliptic protuberance on a ground plane. *Journal of the Optical Society of America* 54: 732–744.

3 Bowman, J.J., Senior, T.B.A., and Uslenghi, P.L.E. (1987). *Electromagnetic and Acoustic Scattering by Simple Shapes (Revised edition)*. New York: Hemisphere Publishing.

4 Yeh, C. and Kim, C.S. (1963). Erratum: The diffraction of waves by a penetrable ribbon [J. Math. Phys. 4, 65 (1963)]. *Journal of Mathematical Physics* 29: 721–722. https://doi.org/10.1063/1.528013.

5 McLachlan, N.W. (1947). *Theory and Applications of Mathieu Functions*. Oxford University Press.

6 Brimacombe, C., Corless, R.M., and Zamir, M. (2020). Computation and applications of Mathieu functions: a historical perspective. *SIAM Review* 63: 653–720.

7 Pao, Y.H. and Mow, C.C. (1973). *Diffraction of Elastic Waves and Dynamic Stress Concentrations*. New York: Krane Russak.

8 Spence, R.D. and Granger, S. (1951). The scattering of sound from a prolate spheroid. *The Journal of the Acoustical Society of America* 23 (6): 701–706. https://doi.org/10.1121/1.1906827.

9 Sinha, B.P. and MacPhie, R.H. (1977). Electromagnetic scattering by prolate spheroids for plane waves with arbitrary polarization and angle of incidence. *Radio Science* 12 (2): 171–184. https://doi.org/10.1029/RS012i002p00171.

10 Kerker, M. (2016). *The Scattering of Light and Other Electromagnetic Radiation*. Elsevier.

11 Johnson, L.R. (2018). Scattering of elastic waves by a spheroidal inclusion. *Geophysical Journal International* 212 (3): 18291858. https://doi.org/10.1093/gji/ggx482.

9

Scattering from Parallelepipeds

So we went down a bit of a rabbit hole in Chapter 8. We started out quite hopeful that our old stand-by separation of variables approach was going to work for elliptic cylinders and spheroids. We were somewhat tentative because the necessary special functions were unfamiliar, but we didn't let that totally scare us off. We were even amazed that people could do the by-hand numerical work necessary to get plots in the days before computers became useful tools for research. We may even have coded up some results and made our own plots that matched the classic results that were included. But then orthogonality gave us some tough love, which Brittany S. Pierce from Glee says "feels a lot like mean." Back in the mid-1900s when people figured out that for many shapes the Mie-scattering approach was a bit of a dead end, they didn't just give up and go do something else with their lives. The Cold War had heated up and understanding in excruciating detail the interaction of acoustic, electromagnetic, and elastic waves with realistic scatterers was a matter of national security. There was also the matter that billions of dollars of research money was going to be available for academics and their PhD students to write their esoteric equations.

9.1 Integral Equations

A hot minute ago, there were no computers big enough to grid up 3D space and do FDTD simulations of anything realistic. The best you could hope to do was some sorts of 2D approximations, which might even work pretty well if the problem you were trying to model was axisymmetric or something. 3D? Oh. Hell. No. You probably can't imagine a world where teenagers aren't all walking around with semidisposable supercomputers in their pockets. But then you're not about to turn 60 and missed the following txt message from your wife yesterday: "Any chance you would want to go axe throwing at 5:00pm? A group of people I work with (including Julie) are going to practice so we don't embarrass ourselves when we do this with the Board in October." I happen to carry a ruggedized flip phone with the camera deleted, and sometimes don't notice it buzzing in my pocket when I get a txt. My wife thought I was ignoring her. I thought she was just hangry when she got home. She's probably going to insist on getting me an iPhone for my birthday and set it up, so it will buzz me loudly whenever it's her messaging even if I have the ringer turned off or whatever.

Anyway, if you can only ever hope to discretize things in two dimensions, the approach that you'll likely take to scattering is integral equations. You'll call this family of methods "low-frequency" techniques because you need a certain number of grid points per wavelength and you also have to reproduce the geometry of the scatterer, so that is also going to be a constraint on your

A Data Engineering Approach to Wave Scattering Analysis: with Applications in Radar, Sonar, Medical Diagnostics, Structural Flaw Detection and Intelligent Robotics, First Edition. Mark K. Hinders.

computational requirements, but the good news is you'll be able to consider scattering from a cube. If the cube isn't too big, of course.

I'm going to include a fairly standard derivation of integral equations for scattering, but it's more for completeness than anything. When I lecture on this for my graduate students, I lead with the comment that these methods are somewhat passé, but there's such a huge literature out there that they should know about them.[1]

The three most common forms of Green's theorem are

$$\begin{aligned} \int_V \nabla \cdot \vec{F} dv &= \int_S \vec{F} \cdot \hat{n} dS \\ \int_V \nabla \phi dv &= \int_S \phi \hat{n} dS \\ \int_V \left(\nabla \times \vec{F}\right) dv &= \int_S \left(\hat{n} \times \vec{F}\right) dS \end{aligned} \tag{9.1}$$

where $\vec{F}$ is a vector field in some volume V and ϕ is a vector field in V. Here $\hat{n}$ is the normal vector of the surface S which encloses the volume V. Green's theorem will be the magic answer to simulating a 3D scattering problem when the biggest computer you have access to can only handle things that are discretized in 2D. In words, Green's theorem relates a volume integral (3D) to a corresponding surface integral (2D). The basic idea is that we consider as our volume V the region inside some spherical surface, excluding the source volume and the scatterer, and then use Green's theorem to convert to three 2D integrals over the interior surface of the bounding sphere as well as the exterior surface of the source and the scatterer. We'll then let the bounding sphere tend to infinity and happily note that the scattered field tends to zero in that limit. We, of course, know what the source term is, so that surface integral isn't really a problem, and mostly we're just left with a surface integral over the scatterer whose shape is known. The surface of the scatterer is discretized and the surface integral is replaced with a discrete summation. If we need to, we can perform some mathematical gymnastics with Green's theorem in order to pivot back and forth between various volume and surface integrals in order to get things into a convenient form for numerical implementation. We usually need to do that, but I'm going to proceed with the equations before talking about that just a bit.

Consider two scalar fields ϕ and ψ and construct the vector fields $\phi\nabla\psi$ and $\psi\nabla\phi$. Green's theorem gives

$$\int_V \left[\phi\nabla^2\psi + \nabla\phi \cdot \nabla\psi\right] dV = \int_S \phi\hat{n} \cdot \nabla\psi dS \tag{9.2}$$

$$\int_V \left[\psi\nabla^2\phi + \nabla\phi \cdot \nabla\psi\right] dV = \int_S \psi\hat{n} \cdot \nabla\phi dS \tag{9.3}$$

Subtract these to get

$$\int_V \left[\psi\nabla^2\psi - \psi\nabla^2\phi\right] dV = \int_S \hat{n} \cdot \left[\phi\nabla\psi - \psi\nabla\phi\right] dS \tag{9.4}$$

---

1 Roger F. Harrington started out majoring in electrical engineering in 1943 at Syracuse University, but his studies were interrupted by World War II, and he served as an instructor under the Electronics Training Program at the U.S. Naval Radio Materiel School while working as an electronics technician. He completed his studies after the war, receiving his PhD in 1952 at Ohio State. Now Professor Emeritus at Syracuse University, he is best known for his contributions to computational electromagnetics with his development of method of moments. Harrington's 1968 book, *Field Computation by Moment Methods*, is regarded as a pivotal textbook on the subject. You'll want to get a copy of his 1961 book, "Time-Harmonic Electromagnetic Fields" which was reissued by IEEE in 2001 because it's a classic.

Now assume that $\phi(\vec{r})$ satisfies the scalar wave equation in V

$$\left(\nabla^2 + k^2\right)\phi(\vec{r}) = -a(\vec{r}) \tag{9.5}$$

where $a(\vec{r})$ is the source term. Choose $\psi(\vec{r})$ as the Green's function, which satisfies

$$\left(\nabla^2 + k^2\right)G(\vec{r},\vec{r}') = \delta(\vec{r}-\vec{r}') \tag{9.6}$$

and write, for the LHS terms

$$\int_V \left[\phi(\vec{r})\nabla^2 G(\vec{r},\vec{r}') - G(\vec{r},\vec{r}')\nabla^2\phi(\vec{r})\right] dv = \int_V \left[-\phi(\vec{r})\delta(\vec{r}-\vec{r}') + a(\vec{r})G(\vec{r},\vec{r}')\right] dv \tag{9.7}$$

where the $k^2\phi G - k^2 G\phi$ term cancels. We can then write, since

$$\left(\nabla^2 + k^2\right)\phi(\vec{r}) = -a(\vec{r}) \tag{9.8}$$

and

$$\int_V \phi(\vec{r})\delta(\vec{r}-\vec{r}')dV = \begin{cases} \phi(\vec{r}')\,\text{for}\,\vec{r}'\text{ in}\,S \\ 0\,\text{for}\,\vec{r}'\text{ outside}\,S \end{cases} \tag{9.9}$$

the relation

$$\int_V a(\vec{r})G(\vec{r},\vec{r}')dv + \int_S \left[G(\vec{r},\vec{r})\nabla\phi(\vec{r}) - \phi(\vec{r})\nabla G(\vec{r},\vec{r}')\right]\cdot\hat{n}dS = \begin{cases} \phi(\vec{r}')\,\text{for}\,\vec{r}'\ in\,S \\ 0\,\text{for}\,\vec{r}'\text{ outside}\,S \end{cases} \tag{9.10}$$

To use this relation, we are given the source function $a(\vec{r})$ inside V and the values of $\phi(\vec{r})$ and $\hat{n}\cdot\nabla(\vec{r})$ on S. Keep in mind that the surface S that we're talking about here is going to be just the surface of the scatterer because both the finite source term and the scattered field term are going to tend to zero when the surface enclosing the volume V tends to infinity. Also note that in this scalar case, we're left with the field and the normal derivative of the field on that surface. In acoustics, the field is zero at the surface of a hard scatterer and the normal derivative of the field is zero at the surface for a soft scatterer. That's, of course, why things were arranged this way.

For vector fields, we again start with the divergence theorem

$$\int_V \nabla\cdot\vec{F}dv = \int_S \vec{F}\cdot\hat{n}ds \tag{9.11}$$

First, however, consider two vector fields $\vec{F}_1$ and $\vec{F}_2$ and the constructions

$$\nabla\cdot\left[(\nabla\cdot\vec{F}_1)\vec{F}_2\right] = (\nabla\cdot\vec{F}_1)(\nabla\cdot\vec{F}_2) + \vec{F}_2\cdot\nabla(\nabla\cdot\vec{F}_1)$$

$$\nabla\cdot\left[\vec{F}_2\times(\nabla\times\vec{F}_1)\right] = (\nabla\times\vec{F}_1)\cdot(\nabla\times\vec{F}_2) - \vec{F}_2\cdot(\nabla\times(\nabla\times\vec{F}_1))$$

Interchange $\vec{F}_1$ and $\vec{F}_2$ and subtract to get the pair of equations

$$\nabla\cdot\left\{\left[(\nabla\cdot\vec{F}_1)\vec{F}_2\right] - \left[(\nabla\cdot\vec{F}_2)\vec{F}_1)\right]\right\} = \vec{F}_2\cdot\nabla(\nabla\cdot\vec{F}_1) - \vec{F}_1\cdot\nabla(\nabla\cdot\vec{F}_2)$$

$$\nabla\cdot\left\{\left[\vec{F}_2\times(\nabla\times\vec{F}_1)\right] - \left[\vec{F}_1\times(\nabla\times\vec{F}_2)\right]\right\} = \vec{F}_1\cdot\left(\nabla\times(\nabla\times\vec{F}_2)\right)\vec{F}_2\cdot\left(\nabla\times(\nabla\times\vec{F}_1)\right)$$

The divergence theorem then gives

$$\int_V \left[\vec{F}_1\cdot\nabla(\nabla\cdot\vec{F}_2) - \vec{F}_2\cdot\nabla(\nabla\cdot\vec{F}_1)\right] dv = \int_S \left[(\nabla\cdot\vec{F}_2)\vec{F}_1 - (\nabla\cdot\vec{F}_1)\vec{F}_2\right]\cdot\hat{n}ds \tag{9.12}$$

and

$$\begin{aligned} &\int_V \left[-\vec{F}_1\cdot\left(\nabla\times(\nabla\times\vec{F}_2)\right) + \vec{F}_2\cdot\left(\nabla\times(\nabla\times\vec{F}_1)\right)\right] dv \\ &\quad = \int_S \left[\vec{F}_1\times(\nabla\times\vec{F}_2) - \vec{F}_2\times(\nabla\times\vec{F}_1)\right]\cdot\hat{n}ds \end{aligned} \tag{9.13}$$

Now use my favorite vector identity

$$\nabla \times \nabla \times () = \nabla(\nabla \cdot ()) - \nabla^2()$$

and add the two equations to get

$$\int_V \left[\vec{F}_1 \cdot \nabla^2 \vec{F}_2 - \vec{F}_2 \cdot \nabla^2 \vec{F}_1\right] dv = \int_S \left[(\nabla \cdot \vec{F}_2)\vec{F}_1 - (\nabla \cdot \vec{F}_1)\vec{F}_2 + \vec{F}_1 \times (\nabla \times \vec{F}_2) - \vec{F}_2 \times (\nabla \times \vec{F}_1)\right] \cdot \hat{n} ds \tag{9.14}$$

Almost there. The electric and magnetic (vector) fields satisfy

$$(\nabla^2 + k^2)\,\vec{E} = \vec{A}_E \qquad (\nabla^2 + k^2)\,\vec{H} = \vec{A}_H \tag{9.15}$$

and the corresponding Green's function equation is

$$(\nabla^2 + k^2)\,\vec{G}(\vec{r}, \vec{r}') = \delta(\vec{r} - \vec{r}')\hat{x} \tag{9.16}$$

where $\vec{G}(\vec{r}, \vec{r}') \equiv \hat{x} \cdot \underline{\underline{G}}(\vec{r}, \vec{r}') = \vec{b}G(\vec{r}, \vec{r}')$. In this notation, $\underline{\underline{G}}$ is the usual dyadic Green's function for electromagnetic fields, G is the scalar Green's function, and $\vec{G}$ is the dot product of the dyadic Green's function with a constant vector (unit) in the direction of the source vector. Here $\vec{b}$ is a constant vector.

I can't help but point out at this stage that the notational complexities arise out of a stubborn refusal to use index notation, which is so rarely needed in electromagnetics that most radar scattering experts never bothered learning it. Elastic wave scattering n00bs all have to use index notation to write their basic equations, so they get rather peeved when highly educated colleagues who are comfortable with all these Green's-function gymnastics refuse to exploit tensor notation. This particular issue caused a problem for me back in the day. My colleague was an expert in electromagnetic scattering, having just done a PhD on that subject at a top program. I was doing radar scattering by accident and had just completed a PhD on elastic waves scattering. My colleague was a major; I was a first lieutenant. I was comfortable with index notation; he wasn't. We were working to understand a new class of meta-materials and needed to do Green's function analysis. The analysis was complex enough that we agreed to each do it separately and then compare results. I did it Saturday morning while I was having a cup of coffee and on Monday morning was ready to compare results. My colleague didn't have a chance to work on it over the weekend. Kid's soccer practice or some such thing. Throughout that week and the one that followed, he was hard at work in his office with stacks of books containing the necessary vector identities and whatnot, and I figured out pretty quickly that I should stop stopping by and asking how it was going. After two weeks, we compared our results and they matched. I tried not to make a big deal of it, but we never really worked together after we presented those results at a conference.

Exercise 9.1 Go back through the analysis with index notation.

Now, write, using countless vector identities

$$\vec{b} \cdot \int_V \left(G\nabla^2\vec{F}_2 - \vec{F}_2\nabla^2 G\right) = \int_S \left[\vec{b}(\nabla \cdot \vec{F}_2) - (\vec{b} \cdot \nabla G)\vec{F}_2 + \vec{b} \times (\nabla \times \vec{F}_2)G - \vec{F}_2 \times (\nabla G \times \vec{b})\right] \cdot \hat{n} ds$$

and with even more vector identities

$$= \vec{b} \cdot \int_S \left[(\nabla \times \cdot \vec{F}_2)G\hat{n} - (\hat{n} \cdot \vec{F}_2)\nabla G - \hat{n} \times (\nabla \times \vec{F}_2)G - (\hat{n} \times \vec{F}_2) \times \nabla G\right] ds \tag{9.17}$$

Since $\vec{b}$ is an arbitrary vector, we write

$$\int_V \left[G\nabla^2\vec{F}_2 - \vec{F}_2\nabla^2 G\right] dv = \int_S \left[(\nabla\cdot\vec{F}_2)G\hat{n} - (\hat{n}\cdot\vec{F}_2)\nabla G - \hat{n}\times(\nabla\times\vec{F}_2)G - (\hat{n}\times\vec{F}_2)\times\nabla G\right] ds$$

With $(\nabla^2 + k^2)\vec{F} = 0$ and $(\nabla^2 + k^2)\vec{b}G(\vec{r},\vec{r}') = -\vec{A}(\vec{r}')$, we have

$$\int_V \vec{A}Gdv + \int_S \left[(\nabla\cdot\vec{F}G\hat{n} - (\hat{n}\cdot\vec{F})\nabla G - \hat{n}\times(\nabla\times\vec{F})G - (\hat{n}\nabla\vec{F})\times\nabla G\right] ds = \begin{cases}\vec{F}(\vec{r}')\,\vec{r}' \text{ inside } S\\ 0 \quad \vec{r}' \text{ outside } S\end{cases} \tag{9.18}$$

Usually $\vec{F}$ will be $\vec{E}$ or $\vec{H}$, so we need to know the source in V and the field $\vec{E}$ or $\vec{H}$ on the surface S in order to find $\vec{E}$ or $\vec{H}$ at $\vec{r}'$.

So that was all preamble. Let's go back to the notationally simpler acoustics problem and see a bit more about how to use these equations. Then I'll show you some results for electromagnetic scattering from a cube and maybe spill a little more tea from back in the day.

Consider an acoustics scattering problem where there are two finite regions defined by surfaces S_i and S_0 inside of which the fluid parameters may vary in an arbitrary way. Outside S_i and S_0, the fluid parameters are constant. Assume that the source is inside S_i while a (passive) scatterer with differing material parameters is in S_0.

In the integral equations that we derived a few pages ago, the surface S_R doesn't have to be simply connected, so we take that to be S_i, S_0, and a sphere S_R of radius R, which encloses S_i and S_0. The volume V is exterior to S_i and S_0 but inside S_R, as sketched in Figure 9.1. In the various integral equations, the unit normal vector was always outward-pointing from V and we have to consider $\hat{n} = -\hat{n}_i$ and S_i and $\hat{n} = -\hat{n}_0$ on S_0. Since we have assumed that the only sources are in S_i the volume V is otherwise source-free and homogeneous. Thus our integral equation is

$$\int_{S=S_i+S_0+S_R} \left[G(\vec{r},\vec{r}')\nabla\phi(\vec{r}) - \phi(\vec{r})\nabla G(\vec{r},\vec{r}')\right]\cdot\hat{n}ds = \phi(\vec{r}') \tag{9.19}$$

which allows us to calculate $\phi(\vec{r}')$ in V if we know ϕ at the surfaces S_i, S_0, and S_R. With this, we can write the "incident" pressure field as:

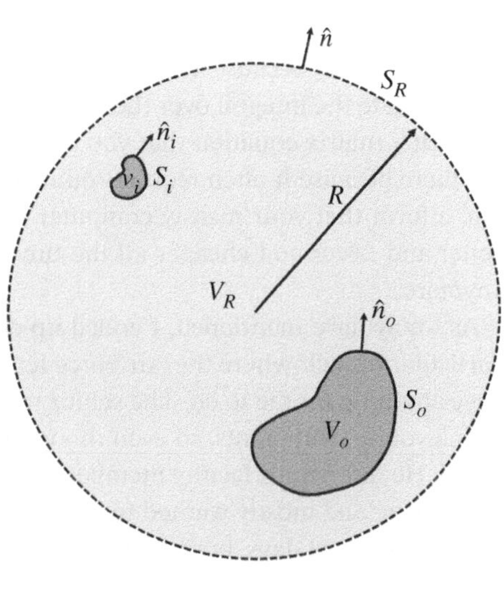

Figure 9.1 Source term is in the volume V_i while passive scatterer is in volume V_0. The surface S_R contains the spherical volume V_R.

$$p^{inc}(\vec{r}') = \int_{S_i} \left[G(\vec{r},\vec{r}')\nabla p(\vec{r}) - p(\vec{r})\nabla G(\vec{r},\vec{r}') \right] \cdot \hat{n} ds \tag{9.20}$$

or, changing signs

$$p^{inc}(\vec{r}') = \int_{S_i} \left[p(\vec{r})\nabla G(\vec{r},\vec{r}') - G(\vec{r},\vec{r}')\nabla p(\vec{r}) \right] \cdot \hat{n}_i ds \tag{9.21}$$

It turns out that the fields tend to zero on the surface S_R when $R \to \infty$ since we require that the radiation condition be satisfied. All we have left is

$$p^{S_0}(\vec{r}') = \int_{S_0} \left[p(\vec{r})\nabla G(\vec{r},\vec{r}') - G(\vec{r},\vec{r}')\nabla p(\vec{r}) \right] \cdot \hat{n}_o ds \tag{9.22}$$

and since $p = p^i + p^{S_0}$, we write

$$p(\vec{r}') = \int_{S_i} \left[p(\vec{r})\nabla G(\vec{r},\vec{r}') - G(\vec{r},\vec{r}')\nabla p(\vec{r}) \right] \cdot \hat{n}_i ds + \int_{S_0} \left[p(\vec{r})\nabla G(\vec{r},\vec{r}') - G(\vec{r},\vec{r}')\nabla p(\vec{r}) \right] \cdot \hat{n}_0 ds$$

These equations don't look too useful since the pressure we are after appears in every term. The story gets a little better if we assume that p^i is a known incident field, which we could pedantically name p^{inc}. Don't judge. We've always assumed a form for the incident field. Quite often it was a plane wave of unit amplitude. We didn't actually work out any of those scattering cases, but it's often quite simple to have a point source instead of a plane wave for the incident wave, and then any more complex incident field can be synthesized from a bunch of point sources. In any event, we assume that p^{inc} is known. We therefore write

$$p(\vec{r}') = p_{inc} + \int_{S_0} \left[p(\vec{r})\nabla G(\vec{r},\vec{r}') - G(\vec{r},\vec{r}')\nabla p(\vec{r}) \right] \cdot \hat{n}_o ds \tag{9.23}$$

and the second term we'll call the scattered field. Thus, the total pressure is the incident field plus the scattered field, and the scattered field is that part due to the presence of the region enclosed by S_0, which we call the scatterer.

OK fine, you might be saying. How do I use this? The first thing to do is to note that in acoustics, for hard scatterers, the normal derivative of the pressure is zero at the surface and for soft scatterers, the pressure is zero at the surface. I hope you see that in each these two cases one of the terms in the integral over the surface of the scatterer is going to be zero. The next thing to remember is that the whole point of this exercise was to come up with a formulation that could be handled computationally because a hot minute ago we could only ever contemplate 2D discretizations. So, approximate the integral over the surface of the scatterer as a summation and you might just have yourself a matrix equation that you can numerically invert or whatever. There's a huge literature out there because it often requires quite a bit more mathematical gymnastics to get the equations into a form that your measly computer can handle. But Moore's Law means that computers get better and faster and cheaper all the time, so it's not even clear whether all this math is needed anymore.

As I may have mentioned, I ended up doing radar scattering analysis entirely by accident. The particular branch where the Air Force let me assign myself wasn't expecting me and didn't really have anything for me to do. The senior researcher in the branch wasn't all that interested in mentoring young lieutenants, so even though my desk was next to his office, we didn't work together much. He had been a faculty member in Virginia for a short time, but apparently wasn't interested in teaching and mostly wanted to be left alone to do his integral equation research. He was pleasant enough most days, but I figured out pretty early on that he was perhaps the most antimilitary

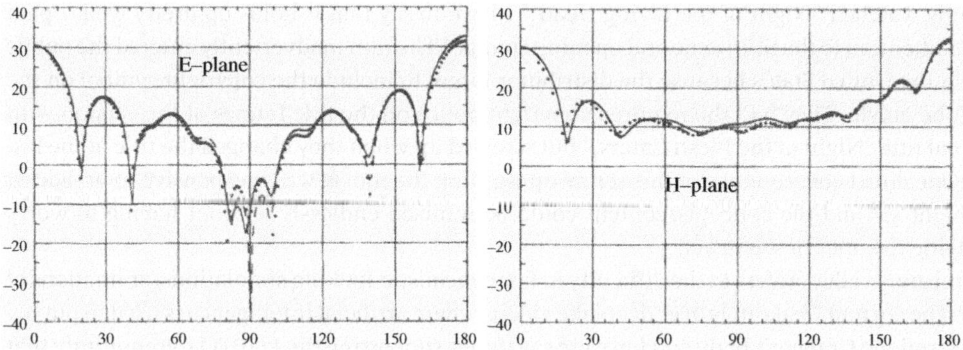

Figure 9.2 Scattering cross section for a perfectly conducting cube with a perimeter 12λ meaning that it's three wavelengths per side. The solid line is the integral equation solution, the dashed line is the high frequency solution, and the dots are measured backscatter cross section.

person I had ever met and so the cognitive dissonance of drawing his paycheck from the Air Force would build up until he needed to yell at someone. When that someone was the branch chief I could hear the whole thing because of a common air vent. One time it was the directorate chief who got yelled at in front of the entire directorate. That was a good lesson in civil service protections for government employees for the young officers who would be punished severely by court martial or whatever for insubordination directed toward a more senior officer.

In Figure 9.2 are shown some results for EM scattering from a perfectly conducting cube from [1]. The 1728×1728 complex matrix of integral equations took up to 117 iterations and 84 min of CPU time on a Vax 8690 computer. The computer time became considerable for cubes much larger than several wavelengths on a side. Interestingly, the three coauthors of this paper had quite distinct roles in the work. The senior author[2] wrote the equations, but didn't do any programming. Indeed, he didn't even have a computer in 1988. The second author was a support services contractor who did the computer programming and made the plots. The first author did the measurements in an anechoic chamber radar scattering facility on a lovely seaside hill in Ipswich.

9.2 High Frequency Scattering and Diffraction Coefficients

You might wonder what became of radar scattering experts after the Cold War ended. One answer is video games. The very same equations that were used to make better and better predictions of the radar cross sections of airplanes, missiles and such, can be implemented to render better and more realistic imaginary worlds. But first zombies.

---

2 Dr. Arthur Yaghjian is currently an independent consultant in electromagnetics. His research in electromagnetics has led to the determination of electromagnetic fields in materials and "metamaterials"; the development of exact, numerical, and high-frequency methods for predicting and measuring the near and far fields of antennas and scatterers; the characterization and design of electrically small antennas and supergain arrays; and the reformulation of the classical equations of motion of charged particles. Dr. Yaghjian received his BS, MS, and PhD degrees in electrical engineering from Brown University in 1964, 1966, and 1969. After teaching mathematics and physics at Hampton University, VA, in 1971, he joined the research staff of the Electromagnetics Division of the National Institute of Standards and Technology (NIST), Boulder, CO. He transferred in 1983 to the Electromagnetics Directorate of the Air Force Research Laboratory (AFRL), Hanscom AFB, MA, where he was employed as a Research Scientist until 1996.

I recently watched "Night of the Living Dead." I have to say that it holds up pretty well. I paid particular attention to the title scene and opening credits.[3] The film inadvertently entered the public domain in the United States because the distributor forgot to include the copyright symbol on the prints of the movie. They had the required copyright notice on the title frames of the version with the original title "Night of the Flesh Eaters," but screwed up when they changed the title at the last minute. One direct consequence of this screw up was that the movie was inexpensive to broadcast on late-night TV and the concepts/content could be remixed endlessly without having to worry about paying royalties or whatever.

The "uncanny valley" refers to the difficulty in faking realistic-looking simulations/animations of humans. The current freakout is over deepfake videos where artificial intelligence is used to put my face on Hermione Granger's body and have me say things supporting the LGBTQ+ community that J.K. Rowling never wrote. But for a long time, the technical issue has been trying to render humans that don't look just a bit (or quite a lot) off somehow. One reason for the popularity of zombies in video games is they're supposed to look just a bit off because they're undead. A second reason for the popularity of zombies in video games is that moms think they're disgusting and so they'll leave you alone. A third reason for the popularity of these video games is that it's OK to shoot zombies because they're already dead. But the main reason for the popularity of zombie video games is that dark, dirty, dystopian vistas are computationally efficient to render. I'll explain.

When the PC game "Rollercoaster Tycoon" introduced water features, every kid everywhere crashed their computers. Water reflects light. All the light from all the sources of light and other reflections of light and so on. All those many and various rays have to be traced to render the scene. That's not much of an issue if there isn't much of anything that reflects light, but one too many awesome water slides and your game crashes. Happy, shiny princess scenes will tax your computational resources. Dark, dingy postapocalyptic scenes don't reflect much at all. Even zombie blood isn't shiny. Hence, zombie video games could be made much more realistic, which is ironic. Now back to the Cold War.

The integral equation approach is classified as a low-frequency method, because the surface of the scatterer was discretized, so the surface integral equation was approximated by a matrix system to be inverted. Computational limitations restricted the "number of wavelengths across" the scatterer and if the airplane or missile or whatever has to be only a few (or several) wavelengths in size that is going to be a pretty low-frequency radar. The radar threat band of interest during the height of the Cold War was 2–18 GHz. Do the math on what that means for the wavelengths and you might even include that low frequency approaches are hopeless.

Let's instead consider high-frequency EM scattering from perfectly conducting objects located in isotropic, homogeneous media. That sounds rather restrictive, but remember the motivation is to predict the radar cross section of airplanes and missiles and such. At radar frequencies, it's an excellent approximation to assume that aluminum is a perfect conductor. Maxwell's equations give

$$\nabla^2\vec{E} + k^2\vec{E} = 0 \qquad \nabla\cdot\vec{E} = 0 \tag{9.24}$$

where $k = \omega\sqrt{\mu\epsilon}$ and $e^{i\omega t}$ time dependence is assumed. For large ω, the asymptotic high-frequency solution to these equations is

3 Wikipedia sez: "Night of the Living Dead is a 1968 American independent horror film that introduced the flesh-eating ghouls that would become synonymous with the term zombie. The story follows seven people trapped in a farmhouse in rural Pennsylvania, under assault by reanimated corpses. The movie was directed, photographed, and edited by George A. Romero, written by Romero and John Russo, and produced by Russell Streiner and Karl Hardman. It stars Duane Jones and Judith O'Dea."

$$\vec{E}(\vec{R},\omega) \sim \exp\left[-ik\psi(\vec{R})\right] \sum_{n=0}^{\infty} \frac{\vec{E}_n(\vec{R})}{(i\omega)^n} \tag{9.25}$$

Since ω is large, we'll retain only $n = 0$ to write

$$\vec{E}(\vec{R},\omega) \sim \exp\left[-ik\psi(\vec{R})\right] \vec{E}_0(\vec{R}) \tag{9.26}$$

Note that $\nabla^2 = \nabla \cdot \nabla$ and $\hat{s} = \nabla\psi$, which gives the eikonal equation

$$|\nabla\psi|^2 = 1 \tag{9.27}$$

together with the first-order transport and conditional equations

$$\frac{\partial \vec{E}_0}{\partial s} + \frac{1}{2}\left(\nabla^2\psi\right)\vec{E}_0 = 0 \qquad\qquad \hat{s}\cdot\vec{E}_0 = 0 \tag{9.28}$$

The unit vector $\hat{s}$ is in the direction of the ray path and s is the distance along the ray path from a reference point O chosen for convenience. Integrating the transport equation gives

$$E(s) \sim E(0)\left(\frac{\rho_1\rho_2}{(\rho_1+s)(\rho_2+s)}\right)^{1/2} e^{-iks} \tag{9.29}$$

where the field at the reference point $s = 0$ is

$$E(0) = E_0(0)e^{-ik\psi(0)} \tag{9.30}$$

Note also that ρ_1 and ρ_2 are the principal radii of curvature of the wavefront at $s = 0$. In addition, please be aware that at $s = -\rho_1$ or $s = -\rho_2$, $E(s)$ is infinite and thus our simple geometric optics (GO) approximation breaks down. Because $\hat{s}\cdot\vec{E}_0 = 0$, the electric field in GO is perpendicular to the ray path. From

$$\nabla \times \vec{E} = -i\omega\mu\vec{H}$$

we can write the magnetic field as:

$$\vec{H} \sim \sqrt{\frac{\epsilon}{\mu}}\hat{s}\times\vec{E} \tag{9.31}$$

Now consider scattering from the smooth, curved, perfectly reflecting surface shown in Figure 9.3. We choose Q_R as the reference point, and $\hat{e}_\perp$ is the unit vector perpendicular to the plane of incidence, while $\hat{e}^i_\parallel$ and $\hat{e}^r_\parallel$ are the unit vectors parallel to the plane of incidence. The reflected field is given by

$$\vec{E}^r(0) = \vec{E}^i(Q_R)\cdot\underline{\underline{R}} = \vec{E}^i(Q_r)\cdot\left[\hat{e}^i_\parallel\hat{e}^r_\parallel - \hat{e}^i_\perp\hat{e}^r_\perp\right] \tag{9.32}$$

where the dyadic reflection coefficient is

$$\underline{\underline{R}} = \begin{pmatrix} 1 & 0 \\ 0 & -1 \end{pmatrix}$$

We then write

$$\vec{E}^r(s) = \vec{E}^i(Q_R)\cdot\underline{\underline{R}}\left(\frac{\rho^r_1\rho^r_2}{(\rho^r_1+s)(\rho^r_2+2)}\right)^{1/2} e^{-iks} \tag{9.33}$$

where ρ^r_1, ρ^r_2 are the principal radii of curvature of the reflected wavefront at Q_R. We can write

$$\frac{1}{\rho^r_1} = \frac{1}{2}\left(\frac{1}{\rho^i_1} + \frac{1}{\rho^i_2}\right) + \frac{1}{f_1}$$

$$\frac{1}{\rho_2^r} = \frac{1}{2}\left(\frac{1}{\rho_1^i} + \frac{1}{\rho_2^i}\right) + \frac{1}{f_2}$$

where

$$\frac{1}{f_{1,2}} = \frac{1}{\cos\theta_i}\left(\frac{\sin^2\theta_2}{R_1} + \frac{\sin^2\theta_1}{F_2}\right) \pm \left[\frac{1}{\cos^2\theta_1}\left(\frac{\sin^2\theta_2}{R_1} + \frac{\sin^2\theta_1}{R_2}\right) - \frac{4}{R_1 R_2}\right]^{1/2}$$

In this θ_1 and θ_2 are the angles between $\hat{s}^i$ and the principal directions associated with the principal radii of curvature R_1, R_2.

That was all pretty abstract, so consider that for plane wave incidence $\sqrt{\rho_1^r \rho_2^r} = \sqrt{R_1 R_2}/2$ and for an incident spherical wave $\rho_1^i = \rho_2^i$. Note that for a flat plate, a cylinder or a cone R_1 or R_2 is infinite and this last simplification is invalid. I'm not sure that clarification really helped.

Geometrical optics (GO) gives the high-frequency scattered field via the law of reflection. To make it better we could, in principle, include the $n = 1, 2, \ldots$ terms in the original high-frequency solution to Maxwell's equations. That turns out not to be a good approach to take. When we say "high-frequency" what we really mean is there are lots of wavelengths across any scattering feature. That can't ever be true for corners. This is a chapter about scattering from parallelepipeds after all.

I feel like at this point that I should remind everybody that the goal of the work and the reason it was being funded so lavishly back in the day was to be able to predict the radar cross section of things like Figure 9.4. You might be amused to know that most of us doing radar scattering research for the Air Force were unaware of the details of the F117 because classified information is always tightly compartmentalized and you have to have both the appropriate level of security clearance and a need to know. You don't know what you don't know because before you can know somebody who already knows has to decide that you have a need to know and decide to read you into the program. The stealth fighter program had the code name Have Blue. What little classified R&D we did in our branch was under the code name Have Nots, a joke about how little funding we had. It

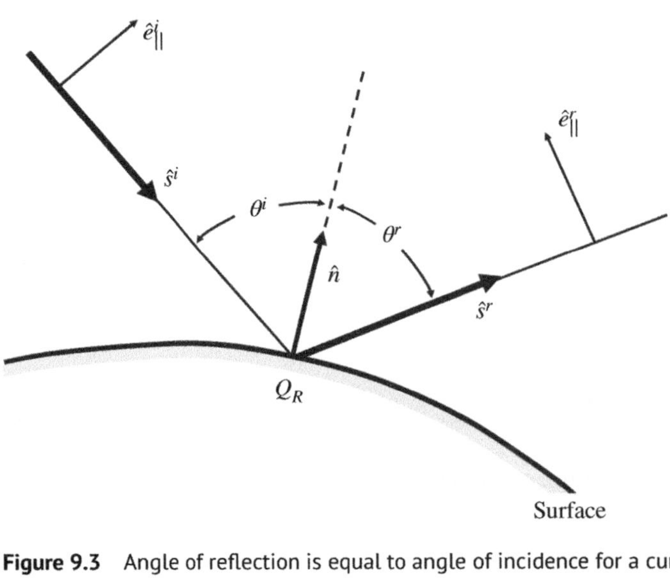

Figure 9.3 Angle of reflection is equal to angle of incidence for a curved surface.

(a) (b)

Figure 9.4 Front view of the F117 stealth fighter (a) from which you can see that there are lots of corners and edges where simple GO approaches are going to break down. The facets will be fine, but corners and edges will cause diffraction. Based on what it purported to be "inside" information, Testors released this conceptual F-19 in 1986 (b). Garnering massive media attention, the design became the shape of the mysterious Stealth Fighter in the public's mind until the actual Stealth – the F-117 Nighthawk – was unveiled in 1990. As it turned out, Testors' sleek and low-profile Stealth looked not a thing like a highly angular, faceted F-117 it was meant to portray. Source: U.S Air Force/Public Domain.

may also have been a bit of a joke, but Testors put out a stealth fighter model,[4] which I may have purchased at the mall, also shown in Figure 9.4

There are two primary defects of Geometrical Optics. First, the field is zero in the shadow region. This isn't true for the shadows due to sunlight that you're used to, which actually is high frequency. There's no way such a simplification holds for radar, where the wavelengths are perhaps a centimeter or even a bit larger. Second, GO assumes that the field is discontinuous at shadow boundaries. You may have never looked closely at a shadow boundary. Do that sometime. Don't be self-conscious. Get down on your hands and knees and look closely. If you're self-conscious about it, casually toss a penny down so you can pick it up and pretend that it's lucky. Losing an earring is also a good excuse for looking closely at the ground, but when passersby try to help you find your earring, their giant melon heads will block the sun, so go with the penny ruse if you need to. What you'll see is that the shadow boundary is a little fuzzy. That's diffraction, folks.

Joe Keller fixed the two primary defects of GO with his geometric theory of diffraction (GTD). The scientific papers we're talking about here are circa 1960, so we know that the work was radar-motivated and so the wavelengths are relatively large and metals are perfectly conducting. There are three *Postulates of GTD*:

1. Diffracted fields propagate along rays determined by a generalization of Fermat's principle to include points on the boundary surface in the ray trajectory.
2. Diffraction is a local phenomenon at high frequencies, depending only on the nature of the boundary surface and the incident field in the immediate neighborhood of the point of diffraction.
3. The diffracted field propagates along its ray so that (a) power is conserved in a tube and (b) the phase delay along the path is the product of the wavenumber and the distance.

Since GTD includes diffracted fields, we can write

$$\vec{E} = \vec{E}^{GO} + \vec{E}^{D}$$

It is important to note that the diffracted fields can't be calculated with GO just by considering the $n = 1, 2, \ldots$ terms since corners aren't ever large WRT wavelength.

4 https://fantastic-plastic.com/lockheed-f-19-stealth-fighter-concept-by-testors.html.

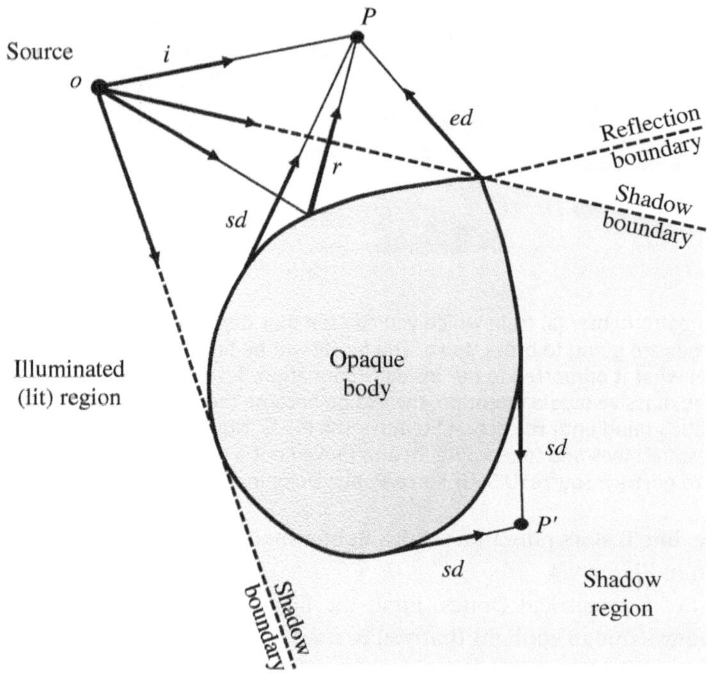

Figure 9.5 Opaque body with source at point O and observation at point *P*, where we have to account for the direct ray *i* as well as the reflected ray *r*. There will also be an edge diffracted ray *ed* and a surface diffracted ray *sd*. Also note that the shadow region is not dark and the shadow boundaries are not sharp. At point *P*′, there will be surface diffracted rays, as shown.

Consider the rays reflected and diffracted from an opaque body as sketched in Figure 9.5. The source is at O and two field points, P in the lit region and P' in the shadow region, are indicated. Diffracted rays are initiated at points on the boundary surface where the GO field is discontinuous, that is, at shadow or reflection boundaries. Since diffracted rays follow paths that minimize the optical distance, they will be straight lines through space or geodesics along smooth surfaces. The total field is the sum of all rays through a point. The good news is that the diffracted field looks familiar

$$\vec{E}^{d}(s) \sim \vec{E}(0)\left(\frac{\rho_1\rho_2}{(\rho_1+s)(\rho_2+s)}\right)^{1/2} \exp\left[-i(ks - l\pi/2)\right] \tag{9.34}$$

where the reference point is the point of diffraction. The bad news is that we have to derive "dyadic diffraction coefficients" for different cases, and there are lots of different cases. Of course, that bad news turns out to be good news if you have in mind making a career out of this. Each of those dyadic diffraction coefficients is a dissertation project for one of your PhD students and because the Cold War means money is no object when the outcome of your work is better stealth aircraft for your side and simultaneously better ways to detect the increasingly stealthy aircraft of the other side. If the timing just happens to work out for you, you and your colleagues could spend your entire professional lives cranking out more and more refined estimates of diffraction coefficients for all

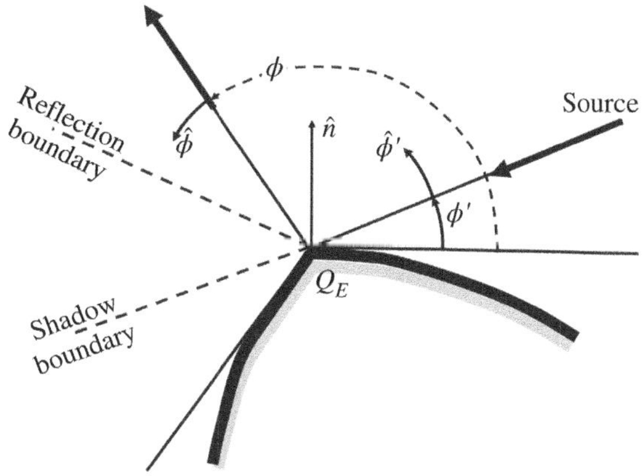

Figure 9.6 Diffraction at a curved edge. The source is shown coming in from the right at an angle ϕ', with the ray directed at the sharp edge at Q_E. Note the dashed line extending that ray defines the shadow boundary. The surface normal at the edge is indicated by $\hat{n}$, which defines the reflection boundary indicated by a dashed line. The reflected ray at angle ϕ is indicated.

the geometries of interest.[5] If the timing just happened to not work for yours truly, I could spend a few years learning quite a bit about all of this just before the Cold War ended. COVID-19 wasn't the first lifequake.

This is all still rather abstract, so consider the specific case of diffraction from a curved edge, with the incident ray striking obliquely, making an angle β_0' with the edge. Diffracted rays are on the cone defined by β_0' but I find the 3D aspects confusing enough so consider the 2D geometry in Figure 9.6.

The diffracted field is written

$$\vec{E}^d(s) = \vec{E}^i(Q_E) \cdot \underline{\underline{D}}(\phi, \phi'; \beta_0') \left(\frac{\rho}{s(\rho + s)} \right)^{1/2} e^{-iks} \tag{9.35}$$

where

$$\underline{\underline{D}}(\phi, \phi'; \beta_0') = -\hat{\beta}_0' \hat{\beta}_0 D_s(\phi, \phi'; \beta_0') - \hat{\phi}' \hat{\phi} D_n(\phi, \phi'; \beta_0')$$

The unit vectors $\hat{\phi}'$ and $\hat{\phi}$ are perpendicular to the edge-fixed planes of incidence and diffraction, respectively. The unit vectors $\hat{\beta}'$ and $\hat{\beta}_0$ are parallel to these planes:

$$\hat{\beta}_0' = \hat{s}' \times \hat{\phi}' \qquad \hat{\beta}_0 = \hat{s} \times \hat{\phi}$$

Thus, the coordinates of the diffracted ray (s, β_0, ϕ) are spherical coordinates and so are the coordinates of the incident ray (s', β_0', ϕ') except that the radial unit vector points toward the origin Q_E.

The quantities D_s and D_n are the scalar diffraction coefficients

$$D_{s,n}(\phi, \phi'; \beta_0') = \frac{e^{-i\pi/4} \sin(\phi/n)}{n(2\pi k)^{1/2} \sin \beta_0'} \times \left[\frac{1}{\cos(\pi/n) - \cos[(\phi - \phi')/n]} \mp \frac{1}{\cos(\pi/n) - \cos[(\phi + \phi')/n]} \right]$$

if the field point is not close to a shadow or reflection boundary and $\phi' \neq 0$ or $n\pi$. The general expression is more complicated. Other diffracting geometries are found separately. I typed those last two sentences with a bit of a smirk on my face. I assume that you found this rather

5 Robert Kouyoumjian (1923–2011) was a professor of electrical engineering at Ohio State for over 40 years after serving in World War II as a captain in the Air Force where he received training in meteorology. His main research was conducted at the ElectroScience Laboratory; ESL included students who obtained over 300 PhD degrees and 500 MSc degrees. Prof. Kouyoumjian's work included the development of accurate solutions for several canonical wire and plate structures and studying the polarization properties of antennas and the study of thermal properties of electromagnetic waves. His research was critical for future antenna and radar system design and analysis. In 1995, he was inducted into the National Academy of Engineering for his work on Uniform Geometrical Theory of Diffraction.

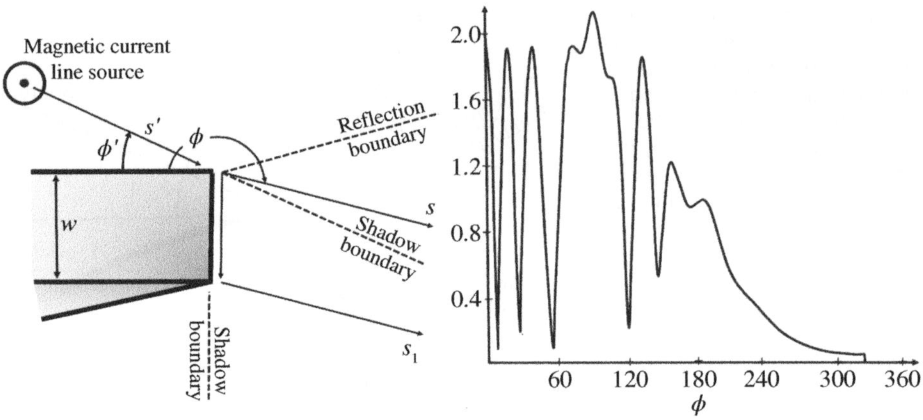

Figure 9.7 Geometry for magnetic line current illuminating a truncated wedge of thickness w and opening angle of 30°. Total field of a magnetic line current is sketched.

confusing. Most of the aforementioned mumbo-jumbo is just keeping track of the geometry because the incident ray could be coming from pretty much any direction relative to the point on the curved edge that we want to know the diffraction coefficient(s) for, and it's inherently 3D even though I declined to attempt a 2D rendition of that 3D geometry because I'm self-aware enough to know that even if a diagram I drew for that made sense to me it would be rather unlikely to resonate with you. Instead, I'll show you a specific case and then a plot or two adapted from [2].[6]

We'll consider a 2D, 30° truncated wedge (Figure 9.7), which is $w = 1.5\lambda$ thick and is illuminated by a magnetic current line source a distance of $s' = 3\lambda$ from the vertex at an angle $\phi' = 30°$. Sorry that it's not a better diagram. You're seeing the wedge from the side, which is sort of bluish, and also a bit up from the bottom, which is the shaded triangular portion. Note that there are two shadow boundaries because we're accounting for the "doubly diffracted" fields. Don't brainlock on my inadequate diagram or the (still) rather confusing equations.

The incident magnetic field is

$$H_i = e^{-ik\rho}/(k\rho)^{1/2} \tag{9.36}$$

and the singly diffracted field is calculated from

$$H^{d1} = \frac{e^{-iks'}}{(ks')^{1/2}} D_n(\phi, \phi', kL, n) \frac{e^{-iks}}{s^{1/2}} \tag{9.37}$$

and the doubly diffracted field is

$$H^{d2} = \frac{e^{-iks'}}{(ks')^{1/2}} D_n(\phi, \phi', kL, n) \frac{e^{-ikw}}{w^{1/2}} \frac{1}{2} D_n(\phi_1, \phi_1', kL_1, n_1) \frac{e^{-iks_1}}{s_1^{1/2}} \tag{9.38}$$

6 Vasundara Venkatraman Varadan studied physics at the University of Kerala and earned her bachelor's degree in 1967. Her sisters studied physics and business. She remained there for her graduate studies and completed her master's degree in 1969. She moved to the University of Illinois for her doctorate, which she defended in 1974. In 1974, Varadan joined Cornell University, where she worked on wave propagation and materials science. She moved to Ohio State University in 1977, where she was an associate professor in wave physics and composite materials. She served on the faculty at Pennsylvania State University for over 20 years, working as codirector of the Center for Electronic Engineering. She was promoted to professor in 1986. In 2002, Varadan was made National Science Foundation Division Director of Electrical & Communications Systems. She moved to the University of Arkansas in 2005, where she served as the Billingsley Chair and distinguished professor, now emerita.

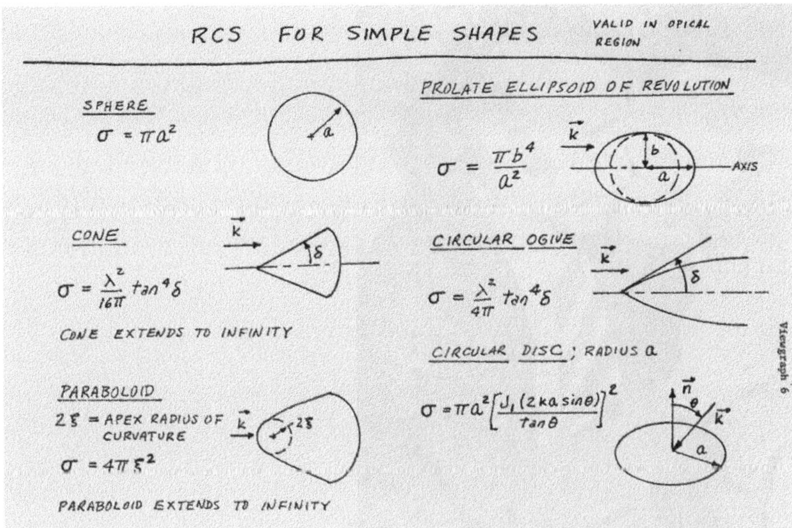

Figure 9.8 The radar cross sections of simple shapes in the high-frequency approximation are well known. This is a slide from a famous set of lectures by Prof. Allen E. Fuhs of the Naval Postgraduate School. Source: Adapted from [3].

The total field then can be computed, as sketched in Figure 9.7.

In Figure 9.8, I've included Viewgraph 6 of [3], which I have a paperback copy of, but you should be able to find online. Most of the lectures you've been to during your lifetime were probably PowerPoint. Some professors still write on the whiteboard during class and think it's some sort of multi-media extravaganza if they use different colors of dry-erase markers. Some professors know which classrooms on campus still have excellent slate chalkboards and will try to get their department administrators to assign them those rooms. Such dinosaurs have a cache of colored chalk for when they want to get really fancy, and may comment that their colleagues who have professional-looking PowerPoints are probably just using the slide deck that came with the instructor's version of the textbook. I may be a little salty about all of this, because I remember the days when hand-drawn viewgraphs were perfectly acceptable at a scientific conference or a graduate-level lecture on something highly mathematical like electromagnetic scattering. Back in those days, in order to get fancy viewgraphs, you had to send your hand-drawn draft "slide deck" to the Photo Shop on base and they would make your slides for you. Professors couldn't afford that, unless they had won a Nobel Prize (Figure 9.9). Prof. Fuhs includes the following bullet points to go along with Viewgraph 6, RCS for Simple Shapes:

- The direction of the incident wave is specified by $\vec{k}$, which is usually parallel to an axis for the simple cases considered here.
- The equations are valid only in optical region where $ka \gg 1$.
- The cone and paraboloid extend to infinity. σ is due to scattering at the tip for a cone and blunt nose for a paraboloid.
- Compare the RCS for a sphere and a paraboloid. What do you notice?
- The prolate (cigar-shaped) ellipsoid of revolution has an RCS less than a sphere of radius b. Rewrite formula for σ as $\sigma = (\pi b^2)(b/a)^2$ = (RCS OF SPHERE OF RADIUS b)$(b/a)^2$ As ratio b/a decreases, the radius of curvature at the nose decreases; σ decreases. Interpret the result in terms of $\sigma = \rho_1\rho_2$.

Figure 9.9 My advisor, Asim Yildiz (right) talking with his advisor, Julian Schwinger. The envelope in Schwinger's lap is likely his vugraphs.

- The circular ogive is tangent to a cylinder. The cylinder must tend to infinity. Note RCS is less for a cone than for an ogive with the same angle, δ. RCS is due to scattering by the tip.

In Figure 9.10 which I've adapted from Viewgraph 15 of [3], it's shown when polarization matters to the radar cross sections of wires, rods, cylinders, and discs. The phase space shown here has size parameter ka for the horizontal axis and length (thickness) compared to wavelength L/λ for the vertical axis. The long wire has large L/λ and small ka and the radar cross section depends strongly on polarization:

$$\sigma_{\parallel} = \frac{\pi L^2}{(\phi/2)^2 + \left[\ln(\lambda/1.78\pi a)\right]^2}$$

$$\sigma_{\perp} = \frac{9}{4}\pi L^2 (ka)^4$$

The short rod has small L/λ and ka, so polarization isn't important in the shaded gray area near the origin. The radar cross section is given by $\sigma \sim L^2(ka)^4$. The finite cylinder of radius a and height L, where both L/λ and ka are large, is in the other regime where polarization isn't important. The radar cross section is $\sigma = kaL^2$. For the disc of radius a and thickness L which has large ka but small L/λ, polarization is important when illuminated edge on:

$$\sigma_{\parallel}/\pi a^2 < -40\,\text{db} \qquad \sigma_{\perp} \simeq \frac{a}{k}$$

I don't know if it was obvious to you that the cone in Figure 9.8 was infinite. In the equations leading up to Figure 9.7, there was a double-dyadic diffraction coefficient or some such thing, which you probably skipped over because the geometry and math and such are rather confusing. That second diffraction coefficient referred to the diffracted rays from the top edge of the wedge which travel down the side to the bottom edge and diffract from that even though that bottom edge isn't illuminated by the source. That's why there was a second shadow boundary. I know. It's confusing, and again my apologies for my inadequate 2D diagram.

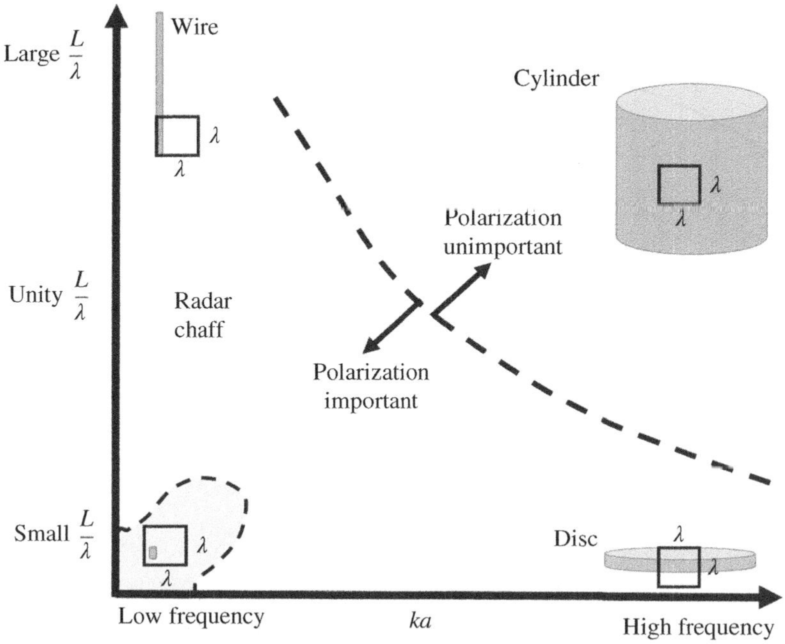

Figure 9.10 The RCS of wires, rods, cylinders, and discs depend on ka and the ratio of length (thickness) to wavelength L/λ. I've drawn this from Viewgraph 15 of [3]. Squares showing the wavelength λ are indicated by each of the shapes. The L/λ-ka plane has been divided into three regions. In the upper right where $L/\lambda \gg 1$ and $ka \gg 1$, the polarization of the wave is not important. In the corner near the origin (shaded), where $L = a$ and $ka \ll 1$, polarization is not important. In between these two regions, polarization is important and one needs both $\sigma_{\parallel}$ and $\sigma_{\perp}$ to be complete. Source: [3]/AIAA American Institute of Aeronautics and Astronautics/Public Domain.

Let me paint you a word picture for a finite cone, rather than trying to PowerPoint one for you. There will be diffraction from the tip of the cone, which makes good sense, I suppose. But if the cone is finite, there will be rays which are diffracted from the tip and travel along the cone surface and then when they get to the back edge of the cone, they will diffract from that. So, you need one diffraction coefficient for the tip, and a second one for the tip-to-base diffraction. As your aircraft get stealthier and stealthier, you might just need to incorporate more and more of these things in order to match up your models with measurements. You and your PhD students could keep coming up with new diffraction coefficients forever!

One or both of two things might happen to ruin that gravy train. One is that the Cold War might end. The other is that we might start making our stealth aircraft primarily out of advanced composite materials rather than aluminum. Both happened. Composite materials, like graphite-epoxy, are not perfect conductors. Indeed, Boeing uses quite a lot of graphite-epoxy in the primary structure of their passenger jets these days, and so they have to embed wire mesh in them to protect against lightning strikes. The B2 Stealth Bomber is a smoothly-shaped flying wing, which also uses quite a lot of advanced composite materials. The F117 Stealth Fighter was primarily aluminum, and you can assume that its faceted shape was decided upon based on the limited ability to predict radar cross section with Keller's

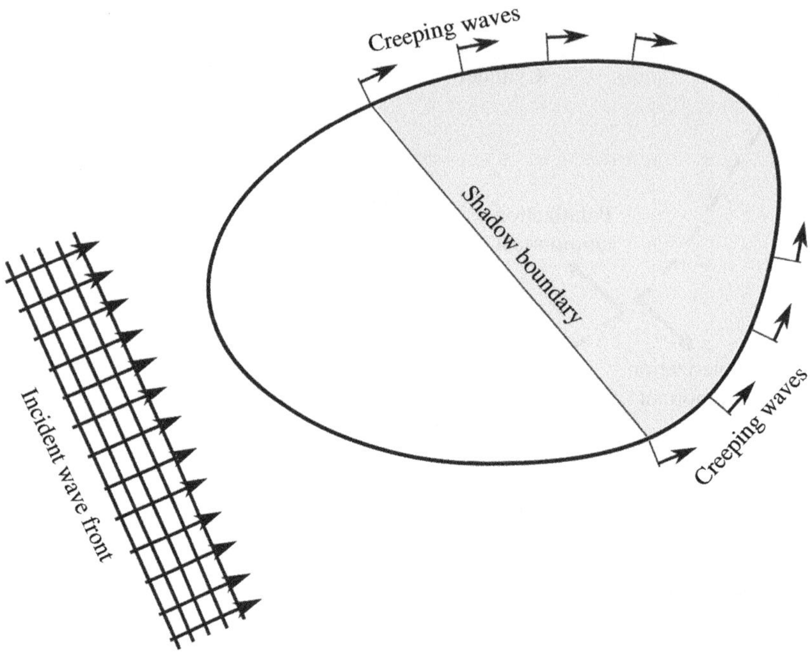

Figure 9.11 Creeping waves are generated at the shadow boundary for any rays, which are tangential to the surface there. Creeping waves are associated with currents on the body in the shadow region, and emerge at the opposite shadow boundary.

physical theory of diffraction[7] [4] and subsequent improvements, like the uniform theory of diffraction [5].

Figure 9.11 shows a wavefront incident on a smooth body, which is adapted from Viewgraph 24 in the *No See Um Book*. Although creeping waves usually yield smaller RCS than specular reflections, one of the first design changes made in the design of stealthier aircraft was to try to minimize the effects of specular reflections. They can't be done away with, of course, but they can be controlled. Indeed, the design philosophy of the faceted F117 was to have the specular reflections all go in just a few directions, none of which were the backscattering direction. It's not your imagination, the various facets do line up in a few directions. In principle only the corner and edge diffractions contribute much to the RCS, and those can be modeled with a small number of painfully derived diffraction coefficients. Creeping waves are important for smooth blunt bodies such as spheres, cylinders, ellipsoids, and orange VW Beetles with pumpkin stems (Figure 1.17).

7 Joseph Keller was considered by many the *Dean of Applied Mathematics*. He was best known for his Geometrical Theory of Diffraction, a method for describing the propagation, scattering and diffraction of waves, especially as they bend around the edges and corners of an obstacle. The theory, developed while he was on the faculty at New York University, built on work he had done during and after World War II using sonar to determine the presence and location of submarines and underwater mines. The theory can be applied whether the waves are acoustic, electromagnetic, elastic, or fluid, and has become an indispensable tool for engineers and scientists working on applications such as radar, stealth technology, and antenna design. His intellectual curiosity and humor were recognized in two Ig Nobel Prizes for "research that makes you laugh and then makes you think." The first of these, in 1999, honored his work explaining why teapots dribble and how to avoid it. The second, in 2012, recognized his discussion of the physical forces that make a jogger's ponytail swing horizontally even though the jogger is oscillating vertically. He died at age 93 in 2016.

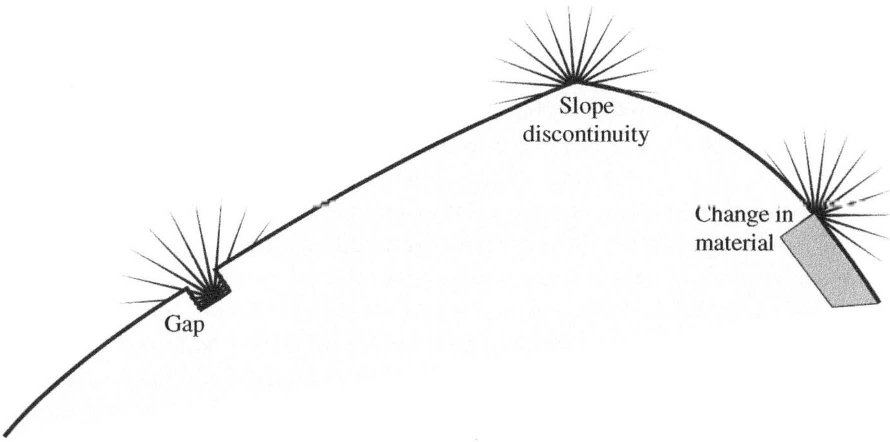

Figure 9.12 Creeping waves will scatter from any discontinuity of surface-topography or material.

For long thin bodies such as wires, prolate ellipsoids, and ogives near head-on incidence, traveling waves will be excited that scatter from the back end. Bodies with dielectrics favor the excitation of traveling waves. In Figure 1.21, I sketched traveling waves diffracting from the aft of a Volvo 240 station wagon. In Figure 1.20, I indicated a variety of different scattering centers on a generic airplane. In Figure 9.12, I've drawn a generic surface and note that creeping waves will diffract from any gap, surface-slope discontinuity, or change in materials. We now know that part of the final preparations for flight in the F117 was taping over and/or filling in as many gaps in the surface one the pilots were in and the plane ready to take off. The reason we know that is the materials were toxic when burned and after landing they were removed and burned in an open pit where workers breathed the secretly hazardous smoke from burning the secret materials of aircraft that the Air Force didn't acknowledge publicly they had at a base in the desert they denied even existed. Lots of secret things still go on at Groom Lake, but it shows up on Google Earth, so there's no denying it exists.[8] When the Cold War ended and all those young PhDs who had become experts in calculating dyadic diffraction coefficients had to find something else to do, at least some of them went into video games. Accounting for diffraction and then secondary diffraction and so forth and so on is what you need to do to make video games look realistic. The mathematical machinery and expertise that kept careful track of all the rays which originate at a light source anywhere in the scene and then reflect, refract, diffract, etc. before contributing to the total field at any other point anywhere in the scene, turned out to be just what was needed to render increasingly realistic scenes. Video games have gotten realistic enough by now that oldsters sometimes think they're watching a movie rather than a game their grandson is playing. Worlds don't have to be dark and dystopian anymore. They can be bright and shiny and happy, if your GPUs can handle all those extra rays. At some point along the way, games could even include princesses with blonde hair. That was a real milestone

8 I'm skeptical that there are crashed flying saucers there because if there were I think we'd all have flying cars by now. In addition, Kenneth Arnold didn't say that the UFOs he saw in 1947 were saucer shaped, he said they moved like saucers skipping over water. Some reporter got that a bit wrong and that's why people think UFOs are flying saucers. Oh, and the Roswell Crash was a Project Mogul balloon. The scattered balloon wreckage seemed odd in 1947 because Mylar was an unfamiliar material, and they had used some decorative tape to reinforce the balsa wood structure. The UFO cover story that was issued and then retracted was designed to keep the Ruskies from realizing that we had microphones and whatnot floating high up in the atmosphere listening for nuclear bomb tests. The Roswell mystery has persisted primarily in order to facilitate tourism. Why else, if not for Nessie, would anyone go on holiday at Loch Ness? Same for Roswell, NM.

because to make blonde hair look bright and shiny, you have to accurately account for some of the light reflecting from the hair (strand) and some going through it. It's an electromagnetic wave scattering from a dielectric cylinder problem.

One final application of this sort of analysis. The same sorts of people who think there are aliens and flying saucers at Area 51 also claim that the Apollo moon landing was faked. We didn't have the technology to fake that in 1969.[9] "Look at the shadows in the footage. If the light source were a nearby spotlight, the shadows would originate from a central point. But because the source is so far away, the shadows are parallel in most places rather than diverging from a single point. That said, the sun isn't the only source of illumination – light is reflected from the ground too. That can cause some shadows to not appear parallel. It also means we can see objects that are in the shadow." Analysis of the various light rays, including reflection and diffraction is key to debunking this sort of conspiracy. In addition, I tease my friends at NASA that if we are getting our advanced aerospace technology by reverse-engineering crashed flying saucers, it follows logically that NASA is a front organization designed to hide that. They never think this is a funny joke. It's a little funny.

Here's something that's not even a little bit funny. Those of us who write scattering equations and whatnot for a living[10] don't expect to be at the center of this kind of real-life drama.[11] Generally speaking, when we're asked by our significant others what we did at work that day, honey, the answer isn't interesting to narrate as my wife can attest.

Exercise 9.2 Is there a well-developed version of GTD but for underwater sound, that is, sonar? Do the wavelengths for sonar allow us to make high-frequency assumptions like in radar? Are there creeping waves and traveling waves on submarines that re-radiate into the shadow zone?

In the aforementioned discussion about creeping waves and traveling waves and corner diffraction and such, I presume that you remembered that there was a whole chapter on guided waves. In

9 https://www.pbs.org/newshour/science/apollo-landing-footage-would-have-been-impossible-to-fake-a-film-expert-explains-why.

10 Professor Ajit Mal received his PhD in Applied Mathematics/Mechanics from Calcutta University in 1964, did postdoctoral research at UCLA and UC Berkeley during 1964–1966, and joined the faculty at UCLA as an Assistant Professor of Engineering in 1967. He became a full professor in 1974. His research interest is in the general area of mechanics of solids with specialization in wave propagation. He has made major research contributions in scattering and diffraction of elastic waves from inclusions, cracks, and corners; strong earthquake ground motion; micromechanical theories of wave propagation in fiber-reinforced composites; quantitative NDE of composites, thin films and bonded joints; and characterization of materials degradation due to corrosion and fatigue in structural components.

11 "Mainak Sarkar did not impress in class: UCLA Indian-origin professor" Times of India (5 June 2016) LOS ANGELES: An Indian-American professor has said Mainak Sarkar, who was behind the UCLA murder-suicide, left little impression as a student in his class and never used to greet him when they passed each other despite both hailing from West Bengal. Professor Ajit Mal was in his University of California, Los Angeles, office on Wednesday getting ready to teach his engineering class when IIT-Kharagpur alumni Mainak Sarkar shot and killed 39-year-old professor William Klug, who he had accused of stealing his computer code and giving it to someone else.

Mal praised another UCLA professor Christopher Lynch for his quick action that kept the 38-year-old UCLA gunman from escaping and potentially shooting more people. Both Mal and Lynch were quoted by the Los Angeles Times as saying that Sarkar's allegation that Klug had stolen his computer code was groundless.

Lynch said all UCLA employees and graduate students sign over any intellectual property developed there to the university and, if it is subsequently licensed, enter royalty agreements to share in the profits. Both men said that Sarkar had enrolled in their classes several years earlier but left little impression. Mal said Sarkar was quiet and reserved and would not even greet him when the two men passed each other, which the professor found somewhat odd since both hail from West Bengal and speak the same language. He also said it was likely that Klug never knew of Sarkar's animosity toward him. If he had, Mal said, Klug would probably have consulted him for his Indian cultural insights and years of experience.

particular, we considered Rayleigh waves in some detail and discussed their usefulness in acoustic microscopy at very high (for acoustics) frequencies. You may even have gotten a little excited about a straightforward-but-tedious method to model Rayleigh wave diffraction at cracks and other near-surface discontinuities. Recall that in the simplest possible situation, where a Rayleigh wave, which is incident on the edge of a quarter space, it will reflect and transmit Rayleigh waves, but will also scatter bulk L and SV waves. If that quarterspace was immersed in a liquid, such as water used for coupling in high-frequency ultrasound, the surface waves would be leaking sound waves into the fluid as they propagate. Whether there's any GTD-type diffraction at the corner is a question that's probably best answered by doing an FDTD simulation. I wouldn't expect any such diffraction to be a significant effect, but then the whole point of GTD in radar applications was to quantify increasingly small contributions to the radar cross section as the dominant ones were systematically eliminated.

A general groundwork for 3D GTD applied to ultrasonic detection of idealized cracks in solids was developed in the late 1970s. It's a lot more complicated than the corresponding radar problem because the diffracted waves will be both longitudinal and shear bulk waves in addition to Rayleigh waves propagating along each face of the crack. The diffracted bulk waves will have characteristic cones with half angles determined by Snell's law of edge diffraction. It's a bit unwieldy, frankly. A heroic effort, though [6] Prof. Achenbach.[12]

Instead, consider two adjoining quarterspaces as shown in Figure 9.13, where Rayleigh waves will be reflected and transmitted, as will both L and T waves in each of the quarter spaces. There might also be guided waves that travel down the interface between the two quarterspaces. Of course, if there was coupling liquid there would be longitudinal waves leaking up into that fluid at the critical angle and presumably some diffracted acoustic waves which may or may not be useful for nondestructive evaluation.

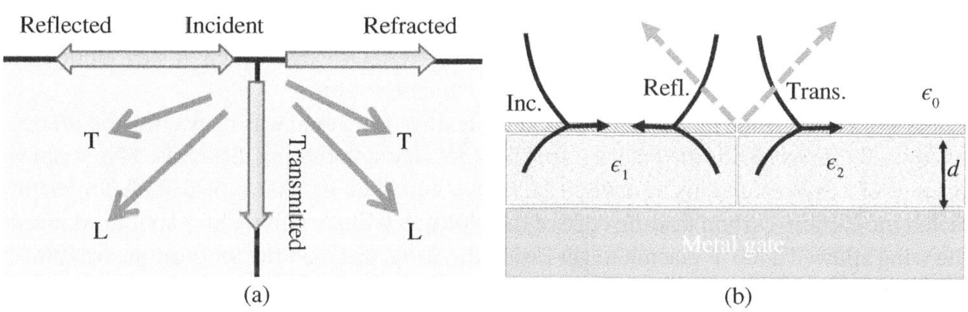

Figure 9.13 A Rayleigh wave incident on the boundary of two adjoining quarterspaces will reflect and transmit Rayleigh waves, but there will also be scattered transmitted L and T waves (a). Very similar behavior is exhibited by plasmonic polaritons (b) where a thin conducting layer, perhaps of graphene, is atop dielectric layers. The guided wave in the conducting layer is evanescent in both the dielectrics ϵ_1, ϵ_2, and free space ϵ_0 with reflected and transmitted fields shown at the boundary between the two dielectric slabs as both guided waves and scattered bulk waves that are likely mid-IR/THz for graphene plasmonics.

---

12 Jan Drewes Achenbach was born in the northern region of the Netherlands. He studied aeronautics at Delft University of Technology and then earned a PhD at Stanford in 1962. After a year as a preceptor at Columbia he was appointed assistant professor at Northwestern. Prof. Achenbach and his students developed methods for flaw detection and characterization by using contact transducers and laser-based ultrasonics, and also methods for thin-layer characterization by acoustic microscopy. His work was both analytical and experimental in nature, with extensive cooperation with other universities and industry via the Center for Quality Engineering and Failure Prevention, which he founded. The Walter P. Murphy Professor and Distinguished McCormick School Professor, emeritus, he died in 2020.

But now look at Figure 9.13b where a rather similar problem is sketched that will yank us right back to the modern day. There is a class of important optics problems, which might just allow us to apply our building knowledge of surface waves and diffraction and such. The magic word of the day is *plasmonics*. But first[13] let's go back to some simple reflection and refraction problems in electromagnetics.

In 1609, a fleet of ships was on its way to the fledgling Jamestowne Colony (where things weren't going especially well) when they got caught in a hurricane [7]. NASA hadn't launched any weather satellites quite yet, so the tempest caught them by surprise. The lead ship, *Sea Venture*, was very close to sinking but miraculously got wedged in a small gap in the dangerous reefs that surround the island of Bermuda. Everybody was able to get ashore, and they were able to salvage their soggy supplies and all usable ship fittings. Nobody lived on the island because the reefs made it too dangerous to try to go there, but Spaniards had left some hogs there back in the day, so there'd be plenty of food for any subsequent castaways. The island was also full of delicious birds that were so docile you could walk right up to them and club them to get a snack; those birds screeched like the devil, which is how the island became known as Devil's Island. This is the literal Bermuda Triangle origin story, BTW. It's also the inspiration for Shakespeare's last play, *The Tempest*.

The *Sea Venture* castaways were stranded in Bermuda for about ten months. Some wanted to stay there. Some wanted to go back to England. The Virginia Company executives insisted that everybody pitch in to build two small ships and finish the trip to Jamestowne, where they showed up in 1610 just after The Starving Time. It didn't smell great. Sir Thomas Gates took over as Governor until Lord De La Warr and his resupply fleet arrived a few months later.

I've traced the genealogy and Dr. William P. Winfree of NASA Langley Research Center in Hampton, VA seems to be a direct descendant of Sir Thomas Gates. When Hurricane Katrina was bearing down on New Orleans in 2005, Bill Winfree saw it coming thanks to weather satellites. He was at the Michoud Assembly Facility in the eastern section of New Orleans, which is surrounded by water on all sides. He caught the last flight out before the tempest hit.

The critical work being done by Dr. Winfree and his team at Machoud was inspecting the orange foam that covered the Space Shuttle external fuel tank for flaws, including disbonds. That foam is the consistency of a Styrofoam cup, and when *Columbia* launched in 2003, a piece of foam fell off and impacted the carbon–carbon leading edge of the Shuttle's wing. Upon reentry, that un-noticed flaw in the wing allowed 2400°F plasma to get inside the wing and melt the aluminum structure.

None of the standard NDE technologies were suitable for inspecting the foam. An emerging technology that exploited the heretofore unavailable TeraHertz part of the electromagnetic spectrum, was being brought to bear to ensure that the Space Shuttle could be returned to flight safely. T-ray scanners can now be small enough to fit on a desktop, and they have enabled widespread terahertz use in medical, corporate, manufacturing, and security settings across the world, with more developments to come,[14] but in those days the only industrial use of T-ray scanning was for

13 A google search for "What is Plasmonics in simple terms?" gives: Plasmonics (or nanoplasmonics) is a young topic of research, which is part of nanophotonics and nano-optics. Plasmonics concerns the investigation of electron oscillations in metallic nanostructures and nanoparticles (NPs). Surface plasmons have optical properties, which are very interesting. Similarly, "Is graphene a plasmonic material?" gives: With relatively low loss, high confinement, flexible feature, and good tunability, graphene can be a promising plasmonic material alternative to the noble metals.

14 https://venturebeat.com/datadecisionmakers/t-ray-technology-is-moving-from-sci-fi-to-mainstream. These days you can buy a T-Ray imaging system from Luna Innovations, Inc., which can measure thickness down to 50 μm at a rate of 1000 measurements per second without nuclear or ionizing radiation.

cigarette paper inspection at Philip Morris in Richmond. The THz range lies right at the intersection of what was historically possible with electronics and optics – slightly too high in frequency for conventional electronics to reach, and too low in frequency for optical approaches to be effective. Maxwell's equations work from DC to light, so modeling T-ray inspections is pretty straightforward. The wavelength is high enough to resolve the sorts of thin discontinuities of interest in ensuring homogenous, well-bonded foam. The primary technical challenge is that the external fuel tanks are enormous so both scanning and data analysis can take quite some time.

9.3 Reflection/Transmission by a Slab

Reflection and transmission by a planar slab are problems of considerable practical interest in optics, radar, wireless communications, etc. For such a simple geometry, there's a surprising amount of complexity, especially once we let the material properties be complex and frequency dependent. Consider three arbitrary homogeneous media characterized by the wave numbers k_1, k_2, k_3, where $k = \omega\sqrt{\mu\epsilon}$ as usual. Since electromagnetic waves are transverse and the directions of propagation are normal to the interfaces, we don't have to concern ourselves with a vector formulation of the problem. The boundary conditions of continuity of tangential electric and magnetic fields at $z = 0$ and $z = d$ need simply the field components that we will have by assuming

$$\begin{aligned}
E^{inc} &= E_0 e^{ik_1 z - i\omega t} & H^{inc} &= \frac{k_1}{\omega\mu_1} E^{inc} \\
E^{ref} &= E_1 e^{-ik_1 z - i\omega t} & H^{ref} &= \frac{-k_1}{\omega\mu_1} E^{ref} \\
E^{slab} &= \left(E^{(+)} e^{ik_2 z} + E^{(-)} e^{-ik_2 z}\right) e^{-i\omega t} & H^{slab} &= \frac{k_2}{\omega\mu_2}\left(E^{(+)} e^{ik_2 z} + E^{(-)} e^{-ik_2 z}\right) \\
E^{tran} &= E_3 e^{ik_2 z - i\omega t} & H^{tran} &= \frac{k_3}{\omega\mu_3} E^{tran}
\end{aligned} \tag{9.39}$$

It's convenient both in terms of simplifying the expressions and emphasizing the important physical quantities to introduce the impedance $Z = \sqrt{\mu/\epsilon}$ and since $k = \omega\sqrt{\mu\epsilon}$ and so we have $\frac{k}{\omega\mu} = \frac{1}{Z}$ and here we go. The boundary conditions are continuity of the E- and H-fields at $z = 0$

$$\begin{aligned}
E_0 + E_1 &= E_2^{(+)} + E_2^{(-)} \\
E_0 - E_1 &= \frac{Z_1}{Z_2}\left(E_2^{(+)} - E_2^{(-)}\right)
\end{aligned} \tag{9.40}$$

and also at $z = d$

$$\begin{aligned}
E_2^{(+)} e^{ik_2 d} + E_2^{(-)} e^{-ik_2 d} &= E_3 e^{ik_3 d} \\
E_2^{(+)} e^{ik_2 d} - E_2^{(-)} e^{-ik_2 d} &= \frac{Z_2}{Z_3}\left(E_3 e^{ik_3 d}\right)
\end{aligned} \tag{9.41}$$

which can be solved for the reflection and transmission coefficients

$$\begin{aligned}
R = \frac{E_1}{E_0} &= \frac{\left(1 - Z_1/Z_2\right)\left(1 + Z_2/Z_3\right) + \left(1 + Z_1/Z_2\right)\left(1 - Z_2/Z_3\right) e^{2ik_2 d}}{\left(1 + Z_1/Z_2\right)\left(1 + Z_2/Z_3\right) + \left(1 - Z_1/Z_2\right)\left(1 - Z_2/Z_3\right) e^{2ik_2 d}} \\
T = \frac{E_3}{E_0} &= \frac{4e^{-ik_3 d}}{\left(1 + Z_1/Z_2\right)\left(1 + Z_2/Z_3\right) + \left(1 - Z_1/Z_2\right)\left(1 - Z_2/Z_3\right) e^{2ik_2 d}}
\end{aligned}$$

Exercise 9.3 Check my algebra. Twice. I could have made a mistake when I did it, or I could have typed something wrong. I've been following Stratton[15] (1941) and he could have had a typo, since things were heating up at the Rad Lab right about then. When you're confident in the expressions for the reflection and transmission coefficients, choose some materials and make some plots like Figure 9.14. But first, look at the structure of the expressions because making plots takes time and

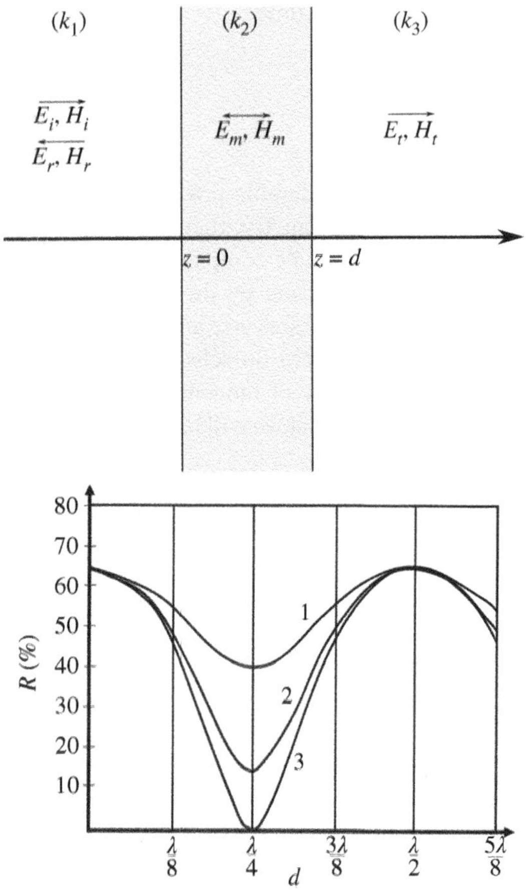

Figure 9.14 Plane wave reflection and transmission by a slab of thickness d. Percentage reflection for: (1) $\epsilon_2/\epsilon_1 = 2$, (2) $\epsilon_2/\epsilon_1 = 4$, (1) $\epsilon_2/\epsilon_1 = 9$ plotted as a function of the slab thickness, d. Note the minimum at one quarter-wavelength thickness in the slab and maximum at half-wavelength thickness in the slab. I've mimicked Figure 97 in Stratton here, but you should be able to make many such plots quite easily.

15 Born in Seattle, WA on 18 May 1901, Julius A. Stratton attended the University of Washington for one year until he went to M.I.T. to obtain his bachelor's degree in 1923 and master's degree in 1925. He then did graduate study in Grenoble and Toulouse, France, and the Technische Hochschule of Zurich, Switzerland, awarded him the degree of Doctor of Science in 1927. Stratton joined the staff of M.I.T. in 1928 and served in the electrical engineering and physics departments for 20 years. In 1945, he was appointed Director of the Research Laboratory of Electronics, was named Provost of M.I.T. in 1949, and in 1951 he became vice-president as well. In 1956, he was named to the newly created post of Chancellor, acted as deputy to the president, and served as general executive officer. Dr. Stratton was president of MIT from 1959–1966 and chairman of the board of the Ford Foundation from 1966 to 1971. During World War II he served as Expert Consultant in the Office of the Secretary of War, and was awarded its Medal for Merit in 1946. The Radiation Laboratory demonstrated impressively the value of interdisciplinary research and, as the end of the war neared, Dr. Stratton and others sought a way in which its momentum and program methods could be sustained for peacetime research. The Research Laboratory of Electronics provided a pattern for interdisciplinarity, and its example was followed at other institutions. Dr. Stratton was a Life Trustee and Member of the Corporation of the Boston Museum of Science, and championed the importance of science at all levels of education and the need for humanistic studies in undergraduate scientific and engineering curricula. He died at the age of 93.

effort. It sure did in 1941. It still did in 1991. Note that the impedances always show up in ratios, which are dimensionless. Note that the thickness shows up everywhere as $k_2 d$ in the reflection coefficient, which is also dimensionless and gets the name electrical thickness. Hence, plot R vs. $k_2 d$ for a bunch of different values of relative impedance ratios.

A common simplification to this problem is to consider $k_1 = k_3$ and then of course $Z_1 = Z_3$, which simply means that the media on either side of the slab is the same. Maybe the slab is a window and by adjusting the properties of the glass you tune the amount of light that gets transmitted or what pretty colors the transmitted light makes. Careful, though, if some impurity in the glass accidentally makes it turn pale purple and you sold that batch of glass to make windows for fancy-schmancy houses on Beacon Hill in Boston, then everybody is going to want some of that suddenly fashionable purple window glass of just the right tint and you could easily go bankrupt trying to reproduce it. Take the Duck Boat or Beantown Trolley tour in Boston and they'll point out a few panes of that glass across from the Common. It turns out that the colors of stained glass are due to the finite conductivity of metallic particles added to molten glass which ends up giving a frequency dependence to k_2 and Z_2. Gustav Mie found that he couldn't explain these physical phenomena without finite conductivity of the spherical scatterers in his model.

Exercise 9.4 These days plasmonics folks allude to stained glass because surface plasmon polaritons can be generated in spherical particles. Does that mean that G. Mie was doing plasmonics more than 100 years ago? Can we finally explain the pale purple Beacon Hill windowpanes?

Another application of this is understanding why you get no cell phone reception in some buildings. Now we're talking RF frequencies instead of optical wavelengths, but remember that we've got things parameterized according to dimensionless electrical thickness, $k_2 d$. You should be able to look up what frequencies your phone is using and the properties of typical building materials and estimate the thickness of walls and such. Make some plots. Then investigate a bit what's different for 5G wireless compared to 4G. As cell phones move up to mm-wave frequencies, there will be more issues associated with larger electrical thicknesses of walls. If you do get a 5G signal, you'll be able to stream better, though.

A different special case is when medium 3 is a perfect reflector, which is pretty much the situation for metals in the microwave regime. Radar absorbing materials are sometimes applied to metallic structures in order to make them stealthy. Medium 1 can be plain old air. Medium 2 will be some fancy RAM with complex permeability and permittivity, $\mu_2 = \mu_2' + i\mu_2''$ and $\epsilon_2 = \epsilon_2' + i\epsilon_2''$. Introducing conductivity means that $k_3^2 = \omega^2\epsilon_3\mu_3 + i\omega\sigma_3\mu_3$ and if we let $\sigma_3 \to \infty$, the transmission coefficient should vanish. Do that limit carefully and see what you get, paying attention also to the Z_2/Z_3 terms. Compare the behavior of your simpler expressions with what you get by using large, but finite, conductivity in the full expressions. If things are behaving strangely, go back through the above analysis, but with the second set of boundary conditions reduced to enforcing tangential electric field zero at $z = d$.

$$E_2^{(+)} e^{ik_2 d} + E_2^{(-)} e^{-ik_2 d} = 0 \tag{9.42}$$

So, with this important knowledge, what sort of material could you apply to a moving target to minimize the reflection from a stationary radar operating in X- and Ka-bands? Is that something that a civilian could buy? Would it survive the car wash?

9.4 Reflection at Conducting Halfspace

Following the notation of Stratton (1941), we consider the reflection and refraction of a plane electromagnetic wave at the boundary between two halfspaces. Note in Figure 9.15 that the boundary is $x = 0$ and the incident, reflected, and transmitted waves are in the xz-plane. We first consider the TE case where the x and z components of the electric fields are zero. We thus write

$$\begin{aligned} E_y^{inc} &= E_0 e^{ik_2(-x\cos\theta_0 + z\sin\theta_0)} \\ E_y^{tran} &= E_1 e^{ik_1(-x\cos\theta_1 + z\sin\theta_1)} \\ E_y^{refl} &= E_2 e^{ik_2(x\cos\theta_2 + z\sin\theta_2)} \end{aligned} \tag{9.43}$$

Since $H_x = \frac{1}{i\omega\mu}\frac{\partial E_y}{\partial z}$, $H_y = 0$ and $H_z = \frac{-1}{i\omega\mu}\frac{\partial E_y}{\partial x}$ we also have

$$\begin{aligned} H_z^{inc} &= \frac{k_2\cos\theta_0}{\omega\mu_2} E_0 e^{ik_2(-x\cos\theta_0 + z\sin\theta_0)} \\ H_z^{tran} &= \frac{k_1\cos\theta_1}{\omega\mu_1} E_1 e^{ik_1(-x\cos\theta_1 + z\sin\theta_1)} \\ H_z^{refl} &= \frac{-k_2\cos\theta_2}{\omega\mu_2} E_2 e^{ik_2(x\cos\theta_2 + z\sin\theta_2)} \end{aligned} \tag{9.44}$$

Boundary conditions are continuity of tangential electric and magnetic fields at $x = 0$ for all values of z

$$H_z^{inc} + H_z^{refl} = H_z^{tran} \qquad E_y^{inc} + E_y^{refl} = E_y^{tran} \tag{9.45}$$

which give

$$\frac{k_2\cos\theta_0}{\mu_2} E_0 e^{ik_2 z\sin\theta_0} - \frac{k_2\cos\theta_2}{\mu_2} E_2 e^{ik_2 z\sin\theta_2} = \frac{k_1\cos\theta_1}{\mu_1} E_1 e^{ik_1 z\sin\theta_1} \tag{9.46}$$

and

$$E_0 e^{ik_2 z\sin\theta_0} + E_1 e^{ik_2 z\sin\theta_2} = E_1 e^{ik_1 z\sin\theta_1} \tag{9.47}$$

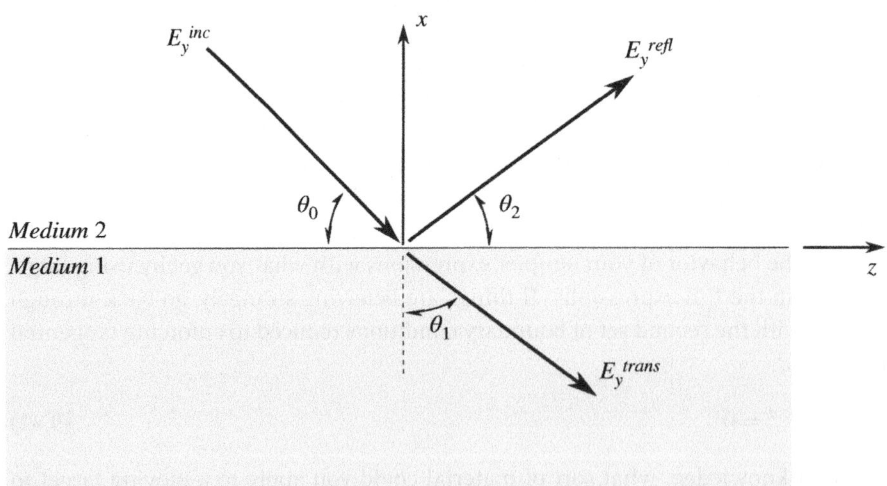

Figure 9.15 Incident TE plane wave reflection and refraction at a halfspace.

Because these must hold for all values of z, we conclude that

$$k_2 \sin\theta_0 = k_2 \sin\theta_2 = k_1 \sin\theta_1 \tag{9.48}$$

so the boundary condition equations become

$$\cos\theta_0 E_0 - \cos\theta_2 E_1 = \frac{k_1}{k_2}\frac{\mu_2}{\mu_1}\cos\theta_1 E_1 \tag{9.49}$$

and

$$E_0 + E_2 = E_1 \tag{9.50}$$

These two equations can be solved to give the transmission and reflection coefficients

$$T = \frac{E_1}{E_0} = \frac{2\frac{\mu_1}{\mu_2}\cos\theta_0}{\frac{\mu_1}{\mu_2}\cos\theta_0 + \sqrt{\left(\frac{k_1}{k_2}\right)^2 - \sin^2\theta_0}} \tag{9.51}$$

$$R = \frac{E_2}{E_0} = \frac{\frac{\mu_1}{\mu_2}\cos\theta_0 - \sqrt{\left(\frac{k_1}{k_2}\right)^2 - \sin^2\theta_0}}{\frac{\mu_1}{\mu_2}\cos\theta_0 + \sqrt{\left(\frac{k_1}{k_2}\right)^2 - \sin^2\theta_0}} \tag{9.52}$$

Exercise 9.5 Pick some materials and plot R and T as a function of the angle of incidence. Note that you'll need μ and ϵ for both materials, but the way that we've written things, you won't have to worry about units if you use material parameters relative to free space. Note that nonmagnetic materials have a relative permeability of 1.

Things get slightly more interesting if medium 2 is a dielectric but medium 1 is a conductor, in which case

$$k_1^2 = \omega^2\epsilon_1\mu_1 + i\omega\sigma_1\mu_1 \qquad k_1 = \alpha_1 + i\beta_1$$

$$k_2^2 = \omega^2\epsilon_2\mu_2 \qquad k_2 = \alpha_2 = \omega\sqrt{\epsilon_2\mu_2}$$

so that k_1 is complex and depends on frequency in perhaps a rather complicated way. Considering the transmission coefficient for the simple case of nonmagnetic materials where $\mu_1 = \mu_2 = 1$, we still have the factor of k_1/k_2 under the square root in the denominator. In addition, it's important to note that we have

$$E_y^{tran} = E_1 e^{i(\alpha_1 + i\beta_1)(-x\cos\theta_1 + z\sin\theta_1)} \tag{9.53}$$

where α_1 and β_1 are the real and imaginary parts of k_1 that you get to work out as an exercise. When you do that algebra, it will be necessary to invoke a constraint on β to avoid nonphysical results. In particular, the sign of β has to result in a transmitted wave that is finite for large negative values of x and large positive values of z. Think about it. We have an incident plane wave as well as a reflected plane wave in medium 2. If medium 1 is a perfect conductor, then all of the energy is reflected and the reflected wave is the same amplitude and angle as the incident wave. If medium 1 is a dielectric or an imperfect conductor, then some fraction of the energy ends up in the transmitted wave. For the imperfect conductor, energy will be lost in that medium due to dissipation, so as z gets large, the transmitted wave has to tend to zero, or at least it would if we weren't talking about an infinite time-harmonic plane wave. It's also apparent that as negative-x gets large, the

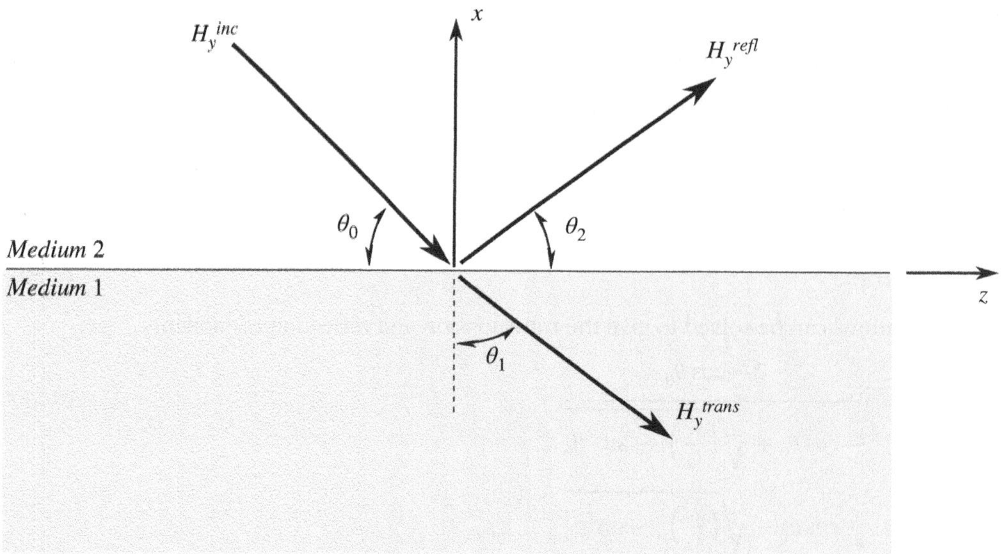

Figure 9.16 Incident TM plane wave reflection and refraction at a halfspace.

transmitted wave has to go to zero. According to my quick bit of algebra, this means we need to have β be positive, but do check that yourself and see what that means in terms of actual material properties of actual materials. In terms of wave propagation, it means that the transmitted wave might be what we call a surface wave. Check the relative magnitudes of α and β for some conductors and some frequencies to get a feel for what's happening. In addition, note that the reflection coefficient depends on the ratio of k_1 to k_2 in a quite complicated manner, so I wonder if there are some combinations of incident angle, frequency, and/or material parameter(s), where funny things happen to the reflected wave?

Now consider again the reflection and refraction of a plane electromagnetic wave at the boundary between two halfspaces. Note in Figure 9.16 that the boundary is $x = 0$ and the incident, reflected, and transmitted waves are in the xy-plane as before, but for the TM case, the x and z components of the magnetic fields are zero. We thus write

$$
\begin{aligned}
H_y^{inc} &= H_0 e^{ik_2(-x\cos\theta_0 + z\sin\theta_0)} \\
H_y^{tran} &= H_1 e^{ik_1(-x\cos\theta_1 + z\sin\theta_1)} \\
H_y^{refl} &= H_2 e^{ik_2(x\cos\theta_2 + z\sin\theta_2)}
\end{aligned}
\tag{9.54}
$$

Since $E_z = \frac{-1}{i\omega\epsilon(\omega)}\frac{\partial H_y}{\partial z}$, $E_y = 0$ and $E_z = \frac{1}{i\omega\epsilon(\omega)}\frac{\partial H_y}{\partial x}$, we also have

$$
\begin{aligned}
E_z^{inc} &= \frac{\omega\mu_2\cos\theta_0}{k_2} H_0 e^{ik_2(-x\cos\theta_0 + z\sin\theta_0)} \\
E_z^{tran} &= \frac{-\omega\mu_1\cos\theta_1}{k_1} H_1 e^{ik_1(-x\cos\theta_1 + z\sin\theta_1)} \\
E_z^{refl} &= \frac{\omega\mu_2\cos\theta_2}{k_2} H_2 e^{ik_2(x\cos\theta_2 + z\sin\theta_2)}
\end{aligned}
\tag{9.55}
$$

Note that we've explicitly written $\epsilon(\omega)$ to remind ourselves that conductivity makes permittivity complex and frequency dependent. We can still utilize $k^2 = \omega^2\epsilon\mu$ as long as we keep this in mind.

Boundary conditions are again continuity of tangential electric and magnetic fields at $x = 0$ for all values of z

$$H_z^{inc} + H_z^{refl} = H_z^{tran} \qquad E_y^{inc} + E_y^{refl} = E_y^{tran} \tag{9.56}$$

which give

$$\frac{-\omega\mu_2 \cos\theta_0}{k_2} H_0 e^{ik_2 z \sin\theta_0} + \frac{\omega\mu_2 \cos\theta_2}{k_2} H_2 e^{ik_2 z \sin\theta_2} = \frac{-\omega\mu_1 \cos\theta_1}{k_1} H_1 e^{ik_1 z \sin\theta_1} \tag{9.57}$$

and

$$H_0 e^{ik_2 z \sin\theta_0} + H_1 e^{ik_2 z \sin\theta_2} = H_1 e^{ik_1 z \sin\theta_1} \tag{9.58}$$

Because these must hold for all values of z, we conclude that

$$k_2 \sin\theta_0 = k_2 \sin\theta_2 = k_1 \sin\theta_1 \tag{9.59}$$

so the boundary condition equations become

$$\cos\theta_0 H_0 - cos\theta_2 H_1 = \frac{\mu_1}{\mu_2}\frac{k_2}{k_1}\cos\theta_1 H_1 \tag{9.60}$$

and

$$H_0 + H_2 = H_1 \tag{9.61}$$

These two equations can be solved to give the transmission and reflection coefficients

$$T = \frac{H_1}{H_0} = \frac{2\cos\theta_0}{\cos\theta_0 + \frac{\mu_1}{\mu_2}\sqrt{\left(\frac{k_2}{k_1}\right)^2 - \sin^2\theta_0}} \tag{9.62}$$

$$R = \frac{H_2}{H_0} = \frac{\cos\theta_0 - \frac{\mu_1}{\mu_2}\sqrt{\left(\frac{k_2}{k_1}\right)^2 - \sin^2\theta_0}}{\cos\theta_0 + \frac{\mu_1}{\mu_2}\sqrt{\left(\frac{k_2}{k_1}\right)^2 - \sin^2\theta_0}} \tag{9.63}$$

Pick some materials and plot R and T as a function of the angle of incidence. Note again that you'll need μ and ϵ for both materials, but the way that we've written things, you won't have to worry about units if you use material parameters relative to free space. Note that nonmagnetic materials have a relative permeability of 1, and for conductors $k = \omega\sqrt{\mu\epsilon(\omega)}$, where $\epsilon(\omega) = \epsilon - \sigma/i\omega$.

9.5 Surface Plasmon Polaritons

Maxwell's equations are valid over an amazing space of frequencies, from DC to light, as they say.[16] Nano-optics is using light to investigate nanomaterials or nanoscale features in things. Plasmonics is a name for this interaction of light with materials. A quick bit of Googling will confirm that this is a rather hot topic these days. It also seems to be a new and exciting application of electromagnetic

16 Jin Au Kong was a 74th-generation lineal descendant of Confucius. At MIT, where he started as an assistant professor in 1969, and later Zhejiang University, Prof. Kong supervised about 50 PhD theses and 90 Master theses. Kong was also the founding chair of the Progress In Electromagnetics Research Symposium (PIERS), and the author of the excellent book "Electromagnetic Wave Theory" published in several editions until his death in 2008. PIERS is now the PhotonIcs and Electromagnetics Research Symposium, which provides a forum for reporting recent advances in electromagnetics, photonics, and applications.

scattering, which means that people like us should probably pay attention to it. If someone ever asks you whether you know something about any subject, you should resist the temptation to reply that that's not what you do. Instead, you should respond that you know a bit about that or ask them to tell you a bit more. Often what you want to do is stall for time so you can look into the subject a bit. An excellent response is, "Let's talk next week sometime." Between now and next week, you should be able to come up to speed. You don't need a comprehensive literature review; you just need a toe hold to get started. Here are some papers that I found that seem to indicate to me that we can contribute to plasmonics via our knowledge of scattering [8–23]. I'm not pretending that this is anything approaching a proper review of the literature, of course, because this field is moving rather fast. Reading several of these papers and looking at the papers they reference gives a sense of where the literature is, and what keywords to search for. Use Google Scholar. Sort by date. Focus on review articles. In a few days, you should be able to come up to speed and understand how the kind of scattering analysis you know how to do relates to this literature.

Exercise 9.6 Consider how to model surface plasmon polaritons scattering from a hole and whether an eigenfunction solution approach can be used. You'll have to click through to their supplementary material in [19] to see the cylindrical-hole, guided-wave scattering analysis.

With that as preamble, let's consider a semi-infinite metal slab characterized by a complex, frequency-dependent permittivity $\epsilon_r = \epsilon(\omega)$ for $z < 0$ with some superstrate material of dielectric constant ϵ_{sup} for $z > 0$. We're going to write solutions for surface waves traveling along the interface $z = 0$, but first let's take a step back and consider Maxwell's equations for a system containing no naturally magnetic components

$$\begin{aligned} \epsilon_0 \frac{\partial \vec{E}}{\partial t} &= \nabla \times \vec{H} - \frac{\partial \vec{P}}{\partial t} \\ \mu_0 \frac{\partial \vec{H}}{\partial t} &= -\nabla \times \vec{E} \end{aligned} \tag{9.64}$$

where for many ordinary dielectric media $\vec{P} = \epsilon_0 \chi_r \vec{E}$ and if we write $\epsilon_r = 1 + \chi_r$ we have, in frequency domain

$$\begin{aligned} -i\omega\epsilon_r(\omega)\epsilon_0\vec{E}(\omega) &= \nabla \times \vec{H}(\omega) \\ -i\omega\mu_0\vec{H}(\omega) &= -\nabla \times \vec{E}(\omega) \end{aligned} \tag{9.65}$$

which can be combined to give

$$\nabla \times \nabla \times \vec{E}(\omega) - \epsilon_r(\omega)\left(\frac{\omega}{c}\right)^2 \vec{E}(\omega) = 0$$

where we have used $c = \omega\sqrt{\mu_0\epsilon_0}$ and of course the relative permeability of nonmagnetic materials is just 1. In addition, since $\nabla \cdot \vec{E} = 0$ we can write this as:

$$\nabla^2\vec{E} + k(\omega)^2\vec{E} = 0 \tag{9.66}$$

For plasmonic materials, the complex, frequency-dependent permeability $\epsilon(\omega)$ is what gives rise to the physical behavior we're interested in. Looking back through the derivation we just did, this could mean that χ_r is frequency dependent and complex, or maybe plain old conductivity is where some or all of that comes from.

You can verify that a solution of Maxwell's equations is

$$\vec{E}(\omega) = A\left(\hat{x} - \frac{k_x}{k_z}\hat{z}\right) exp[i(k_z z + k_x x)] \tag{9.67}$$

with

$$k_x(\omega) = \frac{\omega}{c}\sqrt{\frac{\epsilon_{sup}\epsilon(\omega)}{\epsilon_{sup} + \epsilon(\omega)}}$$

and

$$k_z(\omega) = \frac{\omega}{c}\sqrt{\frac{\epsilon_{sup}}{\epsilon_{sup} + \epsilon(\omega)}} \qquad z > 0$$

$$k_z(\omega) = \frac{\omega}{c}\sqrt{\frac{\epsilon(\omega)}{\epsilon_{sup} + \epsilon(\omega)}} \qquad z < 0$$

Surface plasmon polaritons arise from the peculiar optical properties of metals in certain spectral regimes, where $\epsilon(\omega)$ has a negative real part and a positive imaginary part, which makes $k_x(\omega)$ and $k_z(\omega)$ complex, but we care about the particular situation where k_x is real and positive while k_z is imaginary. This gives a wave mode that is propagating in the x-direction but whose amplitude decays exponentially away from the interface $z = 0$, which is what we call a surface wave. What's most interesting about these surface plasmon polariton waves is that when they hit a discontinuity in either topography or material properties, they will scatter. Some of that energy will be into reflected/refracted/diffracted surface wave modes, but some of it will also be in propagating wave modes that diffract up into the dielectric superstrate.

Now consider a metal halfspace characterized by a real dielectric function $\epsilon(\omega)$ in the region $z < 0$ where $z > 0$ is vacuum, except that there is a 2D surface defect described by $\zeta(x)$ that is assumed to be differentiable and nonzero only over a finite portion of the x-axis near the origin. $\zeta(x)$ might be described as an indentation or an excrescence.

A surface plasmon polariton of frequency ω propagates in the $+x$-direction and scatters from the nonzero part of $\zeta(x)$, so we can write the total magnetic field above $\zeta(x)$ as the incident plus the scattered fields, where the former is written such that it is a propagating wave in the x-direction but evanescent in the z-direction, while the latter is written as a superposition of plane waves. The field below $\zeta(x)$ we can similarly write as an incident plus transmitted fields. What we're really going to say, mathematically, is that we know the magnetic fields in the dielectric and the metal are predominantly propagating surface waves whose amplitudes fall off exponentially away from the surface, but there are also fields diffracted by the local surface irregularity. We'll call those diffracted fields the scattered and transmitted fields and we'll solve for them using the boundary conditions, Figure 9.17.

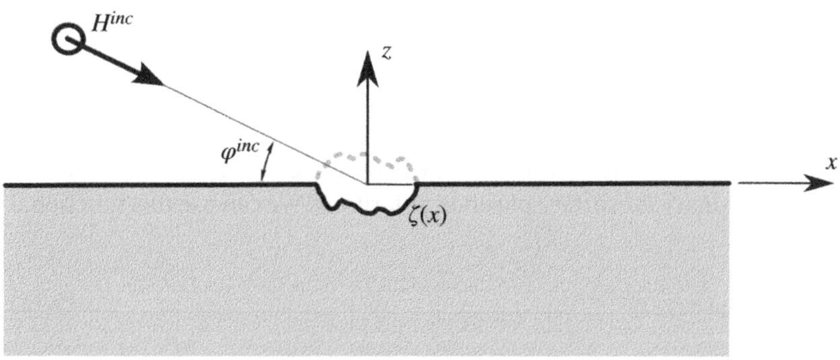

Figure 9.17 Incident TE plane wave reflection and diffraction by a defect.

Rather than just writing it all down and proceeding, it's probably helpful to take a step back and do the little derivation that is usually omitted.

Consider the Green's function for the scalar Helmholtz equation

$$\left(\nabla^2 + K^2\right) G = \delta(R) \qquad\qquad R = r - r' \tag{9.68}$$

If we Fourier transform with the kernel $\nabla^2 \to -p^2$ we have $\left(K^2 - p^2\right)\tilde{G} = 1$ so that $\tilde{G} = \frac{1}{K^2 - p^2}$ and we then inverse Fourier transform to get

$$G = \frac{1}{(2\pi)^3} \iiint d^3p \frac{e^{i\vec{p}\cdot\vec{R}}}{K^2 - p^2} \tag{9.69}$$

The Weyl formula gives the Green's function representation in the form of a superposition of plane waves propagating upward at $z > 0$ and downward at $z < 0$, which is what we need for this scattering problem. Since the essence of the Green's function method of solution is that we can convolve the source function with the Green's function solution to get the physical field of interest, we write

$$H = \int_{-\infty}^{\infty}\int_{-\infty}^{\infty}\int_{-\infty}^{\infty} d^3p H_0 \frac{e^{i\vec{p}\cdot\vec{r}}}{K^2 - p^2} \tag{9.70}$$

and then evaluate the p_z integral by the method of residues.

$$H = \int_{-\infty}^{\infty}\int_{-\infty}^{\infty} dp_x dp_y \int_{-\infty}^{\infty} dp_z H_0 \frac{e^{i\vec{p}\cdot\vec{r}}}{K^2 - p^2} \tag{9.71}$$

where $\vec{p}\cdot\vec{r} = p_x x + p_y y + p_z z$. Hence, we have an integral of the form

$$I = \int_{-\infty}^{\infty} dp_z H_0 \frac{e^{ip_z z}}{p_z^2 - \gamma^2}$$

where $\gamma^2 = K^2 - p_x^2 - p_y^2$ so in the complex p_x-plane, there are simple poles at $p_z = \pm\gamma$ and the residue is $e^{i\gamma z}/2\gamma$. So we have

$$I \propto H_0 \frac{e^{i\gamma z}}{\gamma} \tag{9.72}$$

and we'll chose the sign in $\gamma = \pm\sqrt{K^2 - p_x^2 - p_y^2}$ to match the physical situation we're modeling. Putting the pieces back together, we can write

$$H_{2D} \propto \int_{-\infty}^{\infty}\int_{-\infty}^{\infty} dp_x dp_y \left(H_0/\gamma\right) e^{ip_x x} e^{ip_y y} e^{-\gamma z} \tag{9.73}$$

In the case of a two-dimensional flaw and perpendicular incidence, the coordinate system can be arranged so there is no variation with respect to y and this reduces to

$$H_{1D} \propto \int_{-\infty}^{\infty} dp_x \left(H_0/\gamma\right) e^{ip_x x} e^{-\gamma z} \tag{9.74}$$

with $\gamma = \pm\sqrt{p_x^2 - K^2}$.

We can conclude that the magnetic field is $H_y \propto e^{ik_x x} e^{-\beta z}$, where we've written $k_z = i\beta$ since we want k_x to be real and k_z to be imaginary for surface plasmon polaritons. We can use this functional form together with H_{1D} to write

$$H^{(sup)} = e^{ik_x x} e^{-\beta_0 z} + \int_{-\infty}^{\infty} dp_x \,(R/\gamma)\, e^{ip_x x} e^{-\gamma z} \tag{9.75}$$

$$H^{(met)} = e^{ik_x x} e^{-\beta_1 z} + \int_{-\infty}^{\infty} dp_x \,(T/\gamma)\, e^{ip_x x} e^{-\gamma z} \tag{9.76}$$

where we could just as easily have folded a factor of $1/\gamma$ into the definitions of R and T because those are the unknowns that we want to solve for using the boundary conditions of continuous tangential electric and magnetic fields at the interface $z = \zeta(x)$. Continuity of the tangential electric field is written

$$\frac{\partial}{\partial n} H^{(sup)} = \frac{1}{\epsilon(\omega)} \frac{\partial}{\partial n} H^{(met)} \tag{9.77}$$

where

$$\frac{\partial}{\partial n} = \frac{1}{\sqrt{1 + (\zeta'(x))^2}} \left(-\zeta'(x) \frac{\partial}{\partial x} + \frac{\partial}{\partial z} \right)$$

and $\zeta'(x)$ is the slope of the surface function, that is, its first derivative.

$$\begin{aligned} &\int_{-\infty}^{\infty} dp_x R(p_x|k(\omega)) e^{ip_x x - \beta_1(p_x,\omega)\zeta(x)} - \int_{-\infty}^{\infty} dp_x T(p_x|k(\omega)) e^{ip_x x + \beta_2(p_x,\omega)\zeta(x)} \\ &= -e^{ik(\omega)x - \beta_1(\omega)\zeta(x)} + e^{ik(\omega)x + \beta_2(\omega)\zeta(x)} \end{aligned} \tag{9.78}$$

$$\begin{aligned} &\int_{-\infty}^{\infty} dp_x R(p_x|k(\omega)) \left[-ip_x\zeta'(x) - \beta_1(p_x) \right] e^{ip_x x - \beta_1(p_x,\omega)\zeta(x)} \\ &- \frac{1}{\epsilon(\omega)} \int_{-\infty}^{\infty} dp_x T(p_x|k(\omega)) \left[-ip_x\zeta'(x) + \beta_2(p_x) \right] e^{ip_x x + \beta_2(p_x,\omega)\zeta(x)} \\ &= - \left[-ip_x\zeta'(x) - \beta_1(p_x) \right] e^{ik(\omega)x - \beta_1(\omega)\zeta(x)} \\ &+ \frac{1}{\epsilon(\omega)} \left[-ip_x\zeta'(x) + \beta_2(p_x) \right] e^{ik(\omega)x + \beta_2(\omega)\zeta(x)} \end{aligned} \tag{9.79}$$

In principle, we now have two equations that can be solved for the two unknown functions, once the surface profile is specified. In practice, we're only ever going to need R because that's what would be measured optically as light diffracted when the surface plasmon polariton scattered from the localized surface irregularity. So now what? Well, one approach could be to try some functional forms of $\zeta(x)$ to see if there are any that simplify the equations somehow. Go ahead and try it. Good luck with that. The more fruitful approach seems to be manipulating the pair of already complex boundary condition integral equations to eliminate T and get a single somewhat more complex integral equation containing only R that can then be evaluated numerically. That might mean that people who know how to solve integral equations to calculate the RCS of a small metal cube can dust off their old FORTRAN codes and update their LinkedIn profiles to include the keyword *plasmonics.*

References

1 Cote, M.G., Woodworth, M.B., and Yaghjian, A.D. (1988). Scattering from the perfectly conducting cube. *IEEE Transactions on Antennas and Propagation* 36 (9): 1321–1329. https://doi.org/10.1109/8.8612.

2 Varadan, V.K. and Varadan, V.V. (eds.) (1986). *Low and High Frequency Asymptotics: Acoustic, Electromagnetic and Elastic Wave Scattering*, vol. 2. North Holland.

3 Fuhs, A.E. (1984). *Radar Cross Section Lectures: The No-See-Um Book*. New York: AIAA. ISBN 0-915928-88-4.

4 Keller, J.B. (1962). Geometrical theory of diffraction. *Journal of the Optical Society of America* 52: 116–130.

5 Kouyoumjian, R.G. and Pathak, P.H. (1974). A uniform geometrical theory of diffraction for an edge in a perfectly conducting surface. *Proceedings of the IEEE* 62 (11): 1448–1461. https://doi.org/10.1109/PROC.1974.9651.

6 Achenbach, J.D., Gautesen, A.K., and McMaken, H. (1982). *Ray Methods for Waves in Elastic Solids: With Applications to Scattering by Cracks*. Pitman Advanced Publishing Program. https://books.google.com/books/about/Ray_Methods_for_Waves_in_Elastic_Solids.html.

7 Strachey, W. and Jourdain, S. (2013). *A Voyage to Virginia in 1609: Two Narratives: Strachey's "True Reportory" and Jourdain's Discovery of the Bermudas* (ed. L.B. Wright). University of Virginia Press, newly republished, June 17, 2013.

8 Wei, B., Mao, X., Liu, W. et al. (2023). Recent progress of surface plasmon–enhanced light trapping in GaAs thin-film solar cells. *Plasmonics* 18 (6): 2009–2029. https://doi.org/10.1007/s11468-023-01902-0.

9 Huo, Z., Li, Y., Chen, B. et al. (2022). Recent advances in surface plasmon resonance imaging and biological applications. *Talanta* 255: 124213. https://doi.org/10.1016/j.talanta.2022.124213.

10 Zhang, H., Ijaz, M., and Blaikie, R.J. (2023). Recent review of surface plasmons and plasmonic hot electron effects in metallic nanostructures. *Frontiers of Physics* 18 (6): 63602. https://doi.org/10.1007/s11467-023-1328-9.

11 Ogawa, S., Fukushima, S., and Shimatani, M. (2020). Graphene plasmonics in sensor applications: a review. *Sensors* 20 (12): 3563. https://doi.org/10.3390/s20123563.

12 Stockman, M.I., Kneipp, K., Bozhevolnyi, S.I. et al. (2018). Roadmap on plasmonics. *Journal of Optics* 20 (4): 043001. https://doi.org/10.1088/2040-8986/aaa114.

13 Chaves, A.J., Amorim, B., Bludov, Y.V. et al. (2018). Scattering of graphene plasmons at abrupt interfaces: an analytic and numeric study. *Physical Review B* 97: 035434. https://doi.org/10.1103/PhysRevB.97.035434.

14 Yu, R., Liz-Marzán, L.M., and de Abajo, F.J.G. (2017). Universal analytical modeling of plasmonic nanoparticles. *Chemical Society Reviews* 46 (22): 6710–6724. https://doi.org/10.1039/C6CS00919K.

15 Frezza, F. and Tedeschi, N. (2015). Electromagnetic inhomogeneous waves at planar boundaries: tutorial. *Journal of the Optical Society of America A* 32: 1485–1501. https://doi.org/10.1364/JOSAA.32.001485.

16 Polanco, J., Fitzgerald, R.M., and Maradudin, A.A. (2013). Scattering of surface plasmon polaritons by one-dimensional surface defects. *Physical Review BCondensed Matter and Materials Physics* 87: 155417. https://doi.org/10.1103/PhysRevB.87.155417.

17 Gray, S.K. (2013). Theory and modeling of plasmonic structures. *The Journal of Physical Chemistry C* 117 (5): 1983–1994. https://doi.org/10.1021/jp309664c.

18 Avouris, P. and Freitag, M. (2013). Graphene photonics, plasmonics, and optoelectronics. *IEEE Journal of Selected Topics in Quantum Electronics* 20 (1): 72–83. https://doi.org/10.1109/JSTQE.2013.2272315.

19 Rotenberg, N., Spasenovi?, M., Krijger, T.L. et al. (2012). Plasmon scattering from single subwavelength holes. *Physical Review Letters* 108: 127402. https://doi.org/10.1103/PhysRevLett.108.127402.

20 Maier, S.A. and Atwater, H.A. (2005). Plasmonics: localization and guiding of electromagnetic energy in metal/dielectric structures. *Journal of Applied Physics* 98 (1): 011101. https://doi.org/10.1063/1.1951057.

21 Zayats, A.V., Smolyaninov, I.I., and Maradudin, A.A. (2005). Nano-optics of surface plasmon polaritons. *Physics Reports* 408 (3–4): 131–314. https://doi.org/10.1016/j.physrep.2004.11.001.

22 Evlyukhin, A.B. (2005). Cross sections for surface plasmon polariton scattering from a nanoparticle in the dipole approximation. *Technical Physics Letters* 31: 817–820. https://doi.org/10.1134/1.2121825.

23 Pincemin, F., Maradudin, A.A., Boardman, A.D., and Greffet, J.-J. (1994). Scattering of a surface plasmon polariton by a surface defect. *Physical Review B* 50: 15261. https://doi.org/10.1103/PhysRevB.50.15261.

10

Inverse Scattering

My group is pushing hard into the Internet of Things (IoT) right now. The sorts of low-power sensor hubs that are in wearables have an amazing amount of capability, both in terms of sensors and local processing of sensor data which enables what's starting to be called distributed machine learning. For the last 30 years, my students and I have been modeling the scattering of radar, sonar, and ultrasound waves from objects, tissues, materials, and structures. The reflection, transmission, refraction and diffraction of light, and the conduction of heat, are also included in this body of work. As computers have become more and more capable, three-dimensional simulations of these interactions have become a key aspect of sorting out very complex behaviors. Typically, our goal has been to solve inverse problems where we know what the excitation source is, and some response of the system is measured. Success is being able to automatically and in (near) real time deduce the state of the object(s), tissue(s), material(s) and/or structure(s). We, of course, want quantitative outputs with a resolution appropriate for each particular use case. I assume that you've noticed that the mathematical modeling techniques and numerical simulation methods are the same across a wide range of physical situations and fields of application. Why the sky is blue and how to make a bomber stealthy both utilize Maxwell's equations. Seismic waves and ultrasonic NDE employ identical equations once feature sizes are normalized by wavelength. Sonar of whales and submarines is identical to that of echolocating bats and obstacle avoiding robots. We have centrally supported HPC capabilities here in the Integrated Science Center[1] with PhD-level sysadmins down the hall from my office who are charged with helping students get their codes up and running in parallel. We don't charge internal users for computing time or expert help. I should note that these days my students all do theory *and* experiments *and* high-performance computing.

The question is whether a dramatic proliferation of IoT and edge computing will finally enable inverse scattering as imagined conceptually in Figure 10.1. The basic idea is that if we know the source and then measure the scattered field on some surrounding surface, we'll be able to use formal mathematical methods to determine the characteristics of the scatterer. Tomography seemed to be one path toward inverse scattering, although to make it work, we had to have many sources and for each innumerable receivers. In most cases, though, what people actually mean by inverse scattering is *a source* and *a few receivers*. Since I'm an engineer and not a mathematician, I feel that it's my duty to point out that if you go looking for a unicorn, you're likely to have to settle for a rhinoceros and to save face, you'll try to convince everybody that the unicorn is just a little chubby, but with diet and exercise and several more years of research... Woody might even conclude that what you call flying is merely falling with style.

1 https://www.youtube.com/watch?v=aX-glqG8HWk.

A Data Engineering Approach to Wave Scattering Analysis: with Applications in Radar, Sonar, Medical Diagnostics, Structural Flaw Detection and Intelligent Robotics, First Edition. Mark K. Hinders.

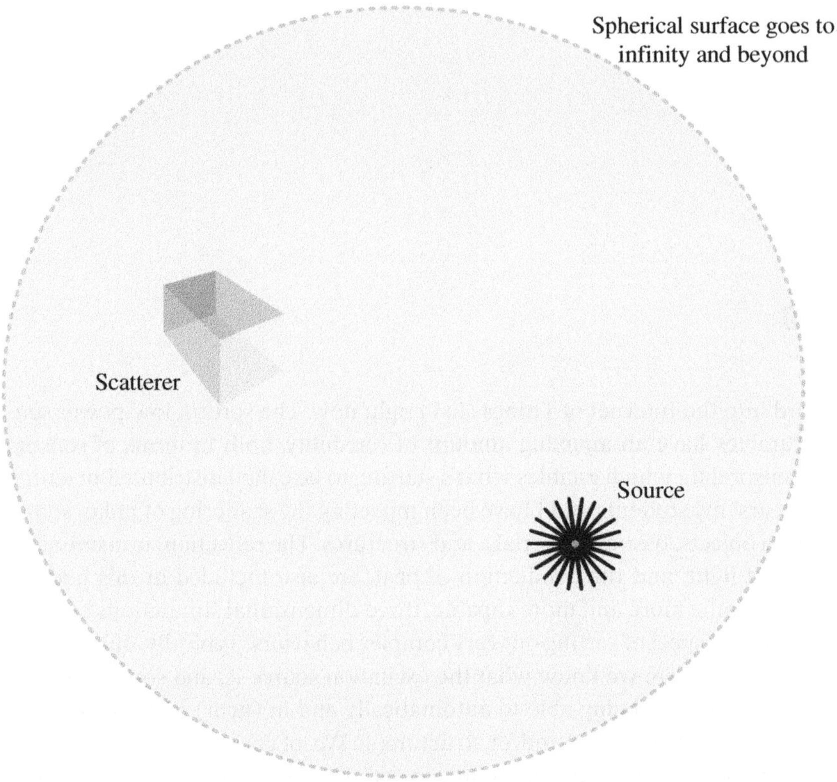

Figure 10.1 For inverse scattering, we know the source and measure the scattered field on some surrounding surface. The goal is to determine the location, orientation, size, shape, materials, etc. of the scatterer(s). IoT seems to mean that we can have enough sensors within that spherical surface that we can imagine solving inverse problems. The data engineering challenge is always going to be deciding what features to extract from the sensor data streams.

In general, inverse scattering problems are too hard, but bats do avoid obstacles and find food. Radar detects inbound threats and remotely estimates precipitation. Cars park themselves and (usually) stop before impact, even if the human driver is not paying close attention. Medical imaging is now so good that we find ourselves in the midst of an overdiagnosis dilemma. Computed tomography is familiar to most people from X-ray CT scanners used in medicine and baggage screening. In a different configuration, CT is a workhorse method for seismic wave exploration for natural resources. We adapted these methods for ultrasound-guided-wave characterization of flaws in large engineered structures like aircraft and pipelines. By extracting features from signals that have gone through the region of interest in many directions, the reconstruction algorithms output a map of the quantities of interest, for example, tissue density or pipe wall thickness. The key is understanding which features to extract from the signals for use in tomographic reconstruction, but that understanding comes from an analysis of the signal energy interacting with the tissue, structure or material variations that matter, which is scattering.

For many years now, we've been developing the underlying physics to enable robots to navigate the world around them. We tend to focus on those sensors and imagers where the physics is interesting, and then develop the machine learning algorithms to allow a robot to autonomously interpret its sensors and imagers. For autonomous vehicles, we now say IoV instead of IoT, and there the focus is sensors/interpretation, but also communication among vehicles, which is how an autonomous vehicle sees around corners and such.

Machine learning is the modern lingo for what we've always been trying to do, and lately, I've been having my graduate students take formal CS coursework in machine learning during their first semester. Sometimes we call it pattern classification, but the key issue is determining what's what from measured data streams. Machine learning is usually divided into supervised and unsupervised learning. Supervised learning requires training data with known class labels, and unless one has a sufficient amount of relevant training data these methods will return erroneous classifications. Unsupervised learning can reveal structures and interrelations in data with no class labels required for the data. It can be used as a precursor for supervised learning, but can also uncover the hidden thematic structure in collections of documents, phone conversations, emails, chats, photos, videos, etc. This is important because more than 90% of all data in the digital universe is unstructured. Most people have an idea that every customer-service phone call is now monitored, but that doesn't mean that a supervisor is listening in; it means that computers are hoovering up everything and analyzing the conversations. Latent topics are those unknown unknowns in unstructured data, and contain the most challenging insights for humans to uncover. Topic modeling can be used as a part of the human–machine teaming capability that leverages both the machine's strengths to reveal structures and interrelationships, and the human's strengths to identify patterns and critique solutions using prior experiences. Listening to calls and/or reading documents will always be the limiting factors if we depend on humans with human attention spans, but topic modeling allows the machines to plow through seemingly impossible amounts of data to uncover the unknown unknowns that could lead to actionable insights.[2] That ancillary information can be exploited to better interpret scattering patterns from radar, sonar, etc. Human operators[3] have always done this, but now machine learning holds the promise of doing it better, faster, and cheaper.

I've watched as the hardware to acquire, digest and share sensor data plummeted from US $\$10^4$ to US $\$10^1$ and that logarithmic trend-line seems to be extending. Music and video streaming on-demand are the norm, even in stadiums where people have paid real money to experience live entertainment. Battery fires are what make the news, but my new smoke detector lasts a decade without needing new batteries. Even radar has gotten small enough to put on a COTS drone, and not just a little coffee can radar but a real phased array radar using meta-materials to sweep the beam in two directions with an antenna-lens combination about the size and weight of an iPad and at a cost already an order of magnitude lower than before. For this drone radar to be able to sense-and-avoid properly, though, the algorithms are going to need to improve because small drones operate in a cluttered environment. This explains why Echodyne's output looks a lot like B-mode ultrasound.[4] While it's true that power lines and barbed-wire fences look pretty similar to a radar, the fence will always have bushes, trees, cows, etc. that will backscatter at least as much. This clutter problem is also a key issue for autonomous ground robots because

2 Dr. William L. Fehlman II is a Data Scientist at USAA, after retiring as a Lieutenant Colonel having served 24 years in the US Army. Bill holds an MS degree in Applied Mathematics from RPI and a PhD in Applied Science from W&M. He is the inventor of US Patent 10,444,945 involving a Natural Language Processing innovation. He built and directed the first ever Data Science practice at USAA that supports Contact Center Operations and was Principal Investigator of artificial intelligence and autonomous unmanned systems projects at NASA.

3 It used to be that data wasn't actually recorded in many situations. For example, NDT inspectors would assess the signals in real time and either accept or reject the part based on their expert judgment of the absence or presence of flaws. "Stamping off" was the *fraudulent* practice of pretending a part had been inspected when it had not, and then marking it as having passed and then doctoring the paperwork. See, for example: Norwood, D. (2022). *Diary of An Aerospace Whitsleblower*. https://www.amazon.com/gp/product/B0BMT22RXW.

4 https://www.echodyne.com.

there's no clear distinction between targets and clutter and some would say pedestrians are neither.

In current commercial work on the development of 5G wireless, one of the most widely considered options is the mmWave because the high frequency (3–300 GHz) spectrum is plentiful and lightly licensed. However, there remain unanswered technical questions about this largely unexplored spectrum. The fundamental properties of the mmWave channel differ from current cellular models in that RF interaction with the environment is strongly frequency dependent in this band. As we've seen, scattering and diffraction are controlled primarily by object size compared to wavelength, and changing the frequency just a bit can alter the character of the scattering dramatically. Material properties (complex permittivity and permeability) are frequency dependent, and spread spectrum signals will need to consider this via scattering/diffraction models and numerical simulations of RF interactions.

Data engineering is necessary for this sort of work, both because the signals are quite complex and because covering large areas or volumes with a minimum number of sensors means propagation distances are going to be large and the "fingerprint" of the scattering will usually be quite subtle. Because scattering is frequency dependent, the most promising signal processing approaches include joint time-frequency and time-scale methods.

A 2D color image time-frequency representation (TFR) typically has time delay on the horizontal axis and frequency on the vertical axis. The simplest way to form a spectrogram is via a boxcar FFT, where an FFT is performed inside of a sliding window to give the spectrum at a sequence of time delays. Boxcar FFT is almost never the optimal TFR, however, since it suffers rather badly from an uncertainty effect. Making the time window a lot shorter to better localize the frequency content in time usually means there often aren't enough sample points to accurately form the FFT. Lengthening the window to get a more accurate spectrum doesn't solve the problem, since then the time localization is imprecise. Alternative TFRs have been developed to overcome many of the deficiencies of the traditional spectrogram, and since our probing signals are typically finite pulses, pings, etc. it is natural to explore TFRs that use basis functions with compact support.

Wavelets are very useful for analyzing time series data because the wavelet transform allows us to keep track of both time and frequency, or scale, features. Whereas Fourier transforms break down a signal into a series of sines and cosines in order to identify the frequency content of the entire signal, wavelet transforms keep track of local frequency features in the time domain. In 2002, we developed a tool that rendered the time series data in 2D time-scale binary images. Since then we've applied this technique to multimode extraction of Lamb wave signals for tomographic reconstruction, time domain reflectometry (TDR) signals wiring flaw detection, acoustic microscopy, an ultrasonographic periodontal probing device, 5G wireless, and so on. Our preferred method, which we call dynamic wavelet fingerprints, is discussed briefly later. Here's the TL; DR: It's important to intelligently downselect the features you're going to be using for machine learning.[5]

You'd be within your rights to argue that this isn't really a chapter about inverse scattering. That's true in the way that term is typically used, of course, but the traditional methods of inverse scattering haven't really worked out very well. I suppose you could argue that tomography is a way to do inverse scattering and I'd have to agree with you that that has been rather successful in various applications with various sorts of interrogation energy. We certainly had quite a lot of fun developing Lamb wave tomography, and the particular data engineering techniques we developed in order to make that work started us down the path to what we now know to call machine learning.

---

5 See Hinders, M.K. (2020). *Intelligent Feature Selection for Machine Learning Using the Dynamic Wavelet Fingerprint*, 346 pp. Springer Nature. https://www.springer.com/us/book/9783030493943.

That even turns out to be why I stand by calling this a chapter about inverse scattering. Machine learning holds the promise of using measurements and models of forward scattering processes to actually do inverse scattering out in the real world. With IoT sensors, though, we come back to what is to be done with the signals that are recorded and then what features are going to be extracted from signals to feed into a machine learning system. As you'll see later, I almost called this a chapter about wavelets.

Wavelets[6] are often ideally suited to analyzing nonstationary signals, especially since there are a wide variety of mother wavelets that can be evaluated to find those that most parsimoniously represent a given class of signals. The wavelet transform coefficients can be rendered in an image similar to a spectrogram, except that the vertical axis will now be "wavelet scale" instead of frequency. The horizontal axis will still be time delay because the "wavelet shift" corresponds to that directly. Nevertheless, these somewhat abstract time-scale images can be quite helpful for identifying subtle signal features that may not be resolvable via other TFR methods.

The *continuous wavelet transform* (CWT) of a square-integrable, continuous function $s(t)$ can be written

$$C(a,b) = \int_{-\infty}^{+\infty} \psi^*_{a,b}(t)s(t)dt \tag{10.1}$$

where $\psi(t)$ is the mother wavelet, * denotes the complex conjugate, and $\psi_{a,b}(t)$ is given by

$$\psi_{a,b}(t) = |a|^{-p}\psi\left(\frac{t-b}{a}\right) \tag{10.2}$$

Here, the constants $a, b \in \mathbb{R}$, where a is a scaling parameter with $p \geq 0$, and b is a translation parameter related to the time localization of ψ. The choice of p is dependent only upon which source is being referred to, much like the different conventions for the Fourier transform, so we choose the most common value of $p = 1/2$. The mother wavelet can be any square-integrable function of finite energy, and is often chosen based on its similarity to the inherent structure of the signal(s) being analyzed. The scale parameter a relates to different frequency components of the signal. Small values of a give a compressed mother wavelet, which will emphasize the signal's high-frequency components. Large values of a result in stretched mother wavelets, returning approximations of the signal related to low-frequency aspects.

10.1 Wavelet Fingerprinting

Once a raw signal has been filtered, we then pass it through the DWFP algorithm. Originally developed by Jidong Hou,[7] the DWFP applies a wavelet transform on the original time domain data,

6 Ingrid Daubechies is the Meryl Streep of mathematics, according to the NY Times:https://www.nytimes.com/2021/09/14/magazine/ingrid-daubechies.html You should get yourself a copy of her classic book, "Ten lectures on wavelets." Daubechies, Ingrid. Society for industrial and applied mathematics, 1992. Currently, the James B. Duke Distinguished Professor of Mathematics and Electrical and Computer Engineering at Duke, her work "is of tremendous importance in image compression, medical imaging, remote sensing, and digital photography" and therefore she was awarded the 2023 Wolf Prize in Mathematics for her work in wavelet theory and applied harmonic analysis. Along with the Abel Prize and the Fields Medal, the Wolf Prize is considered one of the most prestigious awards in mathematics.

7 Jidong Hou is Sr. Scientist, Research and Advanced Development at Natus Medical where he develops deep learning-based signal and image processing algorithms for applications in neurology and ophthalmology as well as hearing and balance. He was previously a DSP Algorithm Engineer at SleepIQ Labs and a Research Scientist at Philips. He holds a PhD in Applied Science from W&M and was a Postdoctoral Fellow at U. Maryland School of Medicine. While completing his PhD, he worked at Sonix, Inc. in Springfield, VA.

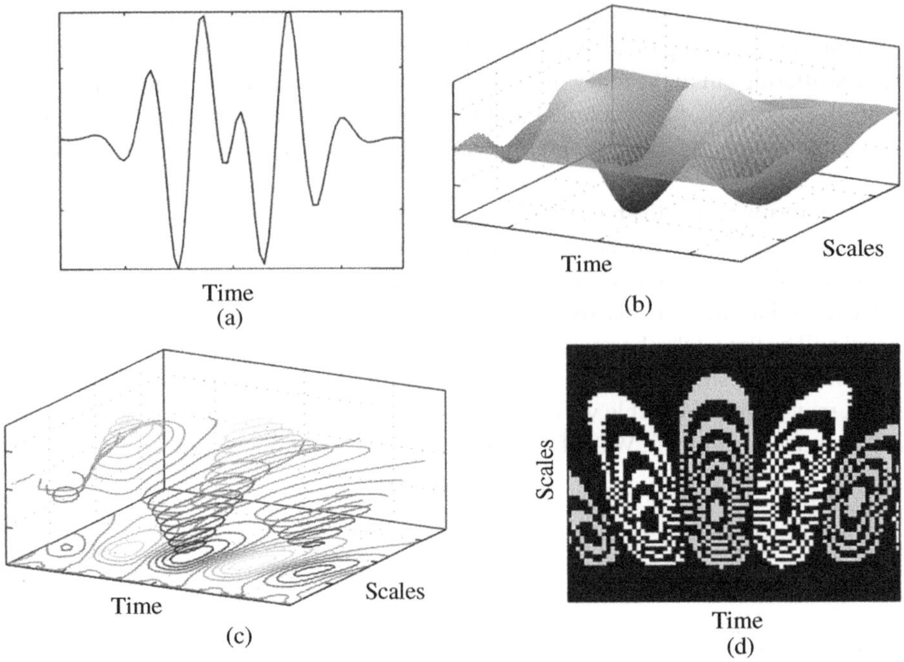

Figure 10.2 A visual summary of the DWFP algorithm. A time-domain signal (a) from which a set of wavelet coefficients is generated via the continuous wavelet transform (b). The coefficients are then "thickly" sliced (c) and projected onto the time-scale plane, resulting in two-dimensional images, shown (d) with white peaks and gray valleys.

resulting in an image containing "ridge" features that resemble fingerprints. The wavelet transform coefficients can be rendered in an image similar to a spectrogram, except that the vertical axis will be scale instead of frequency. These time-scale image representations can be quite helpful for identifying subtle signal features that may not be resolvable via other time-frequency methods. As I often say, it's worked for everything we've tried it on.

In forming the fingerprints, the values of slice thickness and number of slices can be varied to alter the appearance, as can changing which mother wavelet is used. In practice, the mother wavelets used are often chosen based on preliminary analysis of a new set of signals as well as experience with similar signals. In Figure 10.2, we've deliberately shown this process at a "resolution" where the pixilated nature of the wavelet fingerprint is obvious. This is important because each of the pixels is either black or white: *it is a binary image*. We sometimes do peaks and valleys in white and gray, respectively, so technically that's not a binary image, but still....

The problem is thus transformed from a one-dimensional signal identification problem to a 2D image recognition scenario. The binary matrix is easily stored and transferred, and is amenable to edge computing implementations. There is also robustness to the method since different mother wavelets emphasize different features in the signals.

The last piece of the DWFP technique is recognition of those binary image features that correspond to some waveform features of interest. We've found that using simple-to-implement metrics, like ridge counting in the 2D images, is often a helpful way to identify some features of interest. Once a feature has been identified in the time-scale space, we know its arrival in the time domain as well and we can then draw conclusions about the location based on our knowledge of wave velocity.

The inherent advantage of traditional TFRs is to transform a time-trace signal into a two-dimensional image, which then allows for application of powerful *image* processing methods to be brought to bear. False-color spectrogram images also happen to be visually appealing, but this turns out to be somewhat of a disadvantage when the goal is to automatically identify the features in the image that carry the information about the scatterer(s). High-resolution color imagery is computationally expensive to both store and process, and segmentation is always problematic. This latter issue is particularly difficult for scattering problems because it's going to be something about the shape of the signal(s) in the TFR image that we're searching for automatically via image processing algorithms. A binary image requires much less computer storage than does a gray-scale or color image, and segmentation isn't an issue at all because it's a trivial matter to decide between black and white. Binary fingerprint images can be formed from any TFR, of course, although wavelets seem to work quite well for the wide variety of applications that we have investigated.

As candidate TFRs are considered and binary fingerprint images are formed from them, the primary *data engineering* task is to downselect those that seem to best highlight the signal features of interest while minimizing the clutter from signal features that aren't of interest. We typically perform this in an interactive manner, since we've implemented the wavelet fingerprint method via a GUI. This allows us to easily read in signals, perform denoising and windowing as needed, and then form wavelet fingerprints from any of the dozens of mother wavelets, selecting a variety of other parameters/options particular to the method. This works well because the human visual system is naturally adapted to identifying features in this sort of binary imagery (think reading messy handwriting) and we're interested initially in naming features and describing qualitatively how the fingerprint features change as the flaws or other physical features in the measurement setup are varied. For example, a triangular feature might be the signature of a particular kind of flaw, with the number of ridges indicating the severity of the flaw and the position in time corresponding to the flaw's physical location. A feature of interest might instead be a circle or an oval, with the "circularity" or eccentricity and orientation as quantitative measures of severity. The point is that once such features are identified it is a relatively simple matter to write algorithms to track them in fingerprint images. Indeed, there is quite a large literature on fingerprint classification algorithms dating back to the telegraph days when fingerprint images were manually converted to strings of letters and numbers so they could be transmitted over large distances to attempt to identify criminal suspects.

10.2 Wavelet Fingerprints Applied

Here are a few applications of this method that we've worked on. We've only recently begun using the terminology "data engineering," but that's what we've been using the wavelet fingerprints for all along. Recall that I decided to call this a chapter on inverse scattering, with the logic that extracting sufficient features from wavelet fingerprints can allow one to use machine learning to actually solve inverse scattering problems. Modeling and simulation of the forward scattering is key to the data engineering aspects, of course.

10.2.1 Roof Fall Detection

One of the best field trips I've ever been on was a tour of a coal mine in Big Stone Gap, VA. It was high coal, so we could walk around comfortably and it was surprisingly not claustrophobic. I was there with one of my engineers at the invitation of the state geologist, who I had an office next to for some years when he was at the College. The tour was led by the mine superintendent, and we were with two inspectors from the state regulatory agency (MME). We got to see an active mine face up close and ask the miners questions. We even got to see the roof bolter get fussed at by MME for going inby. The point of the visit was to understand the miners' workflow in order to refine a concept we were proposing to develop for preventing roof falls.

In a typical room-and-pillar coal mine, over the course of a few days, the geologic stresses redistribute as material is removed. During that time there is an increased danger that small chunks of the roof will fall, which can cause the issuance of a "fatal gram" to everybody on the email list. As the mining proceeds, the roof bolter drills holes up into the roof of the mine and glues meter-long roof bolts in place to prevent this. Our concept was to have the roof bolter drill an extra hole at each intersection in the room-and-pillar grid and glue up in there a battery-powered accelerometer module with wireless connectivity. Roof-fall precursor microseismic vibrations in the rock would be picked up by the nearest few accelerometers and that information would be relayed by the adaptive mesh wireless network to a central processing unit back over by the microwave, coffee pot, and porta-potty. A machine-learning system could then give warning to the miners to move back from the active face before any collapse.

The problem we faced immediately was systematic denial that there might-could be any problem. Nobody was willing to admit that they needed technologies to improve mine safety because that would be an admission that their operations weren't already safe enough. We had to refine our argument by leading with the following. The continuous miner will be generating strong vibrations that propagate out ahead into the mine face and if there are any abandoned works ahead of the miner, those will backscatter those vibrations and our system will pick that up. Backscattering from a void is different than backscattering from a water-filled void, so we can easily give warning before breaking through into flooded abandoned works and thereby prevent fatal grams due to drowning. Although coal mines are all carefully preplanned and mapped, everybody seems to take out a little extra coal before shutting down and these days everybody is mining up in between abandoned works.

The other refinement to our argument was that if miners happen to get trapped because of roof falls, the automatically adaptable wireless network of accelerometer modules might be able to be used by the trapped miners to communicate with the outside. All it would take is for the miners to have a compatible radio system and to be able to connect to any surviving nearby node in the network. Barring that, they could bang on the roof in Morse code and those very subtle vibrations would be picked up by nearby accelerometer nodes and relay messages to the surface.

Once we had refined our arguments and gotten operators on board who were willing to let us install a proof-of-concept system, we began looking for funding sources. Our geology department was in a building named after an alum who happened to be a Virginia coal baron, but that didn't pan out. Federal mine safety funding comes via PHS, and we were all geared up to propose this concept when research money rained from the sky as a part of the Obama Stimulus.

Figure 10.3 shows typical wavelet fingerprints from limestone mine data that proved the basic concept. The features selected for classification included features specific to the shape of the wavelet fingerprints as well as features related to spectral source parameters. The k-means clustering technique resolved significant events and motivated the calibration of the Roof Fall Index. Events with a Roof Fall Index above 30 predict roof falls by as much as several hours or even a few days, which can be used to mitigate mining hazards. We were all (shovel) ready to do this, but there didn't seem to be a single dollar of coal-mine safety funding in the stimulus because the Obama administration had decided to kill off coal and hence coal mine safety wasn't worth investing your tax dollars in.

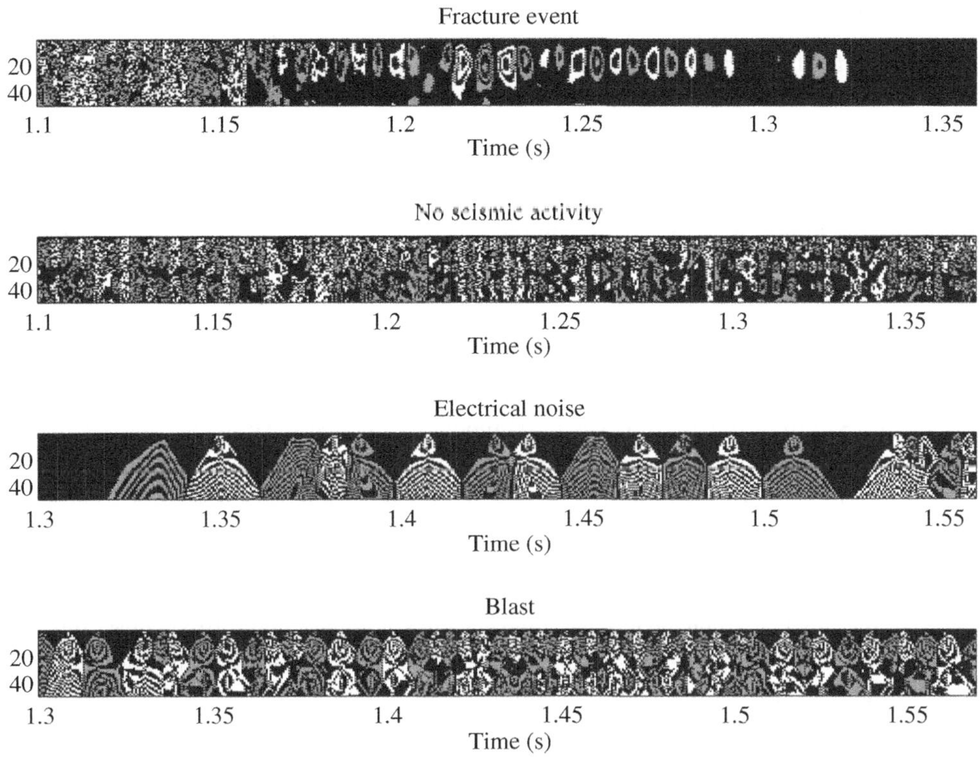

Figure 10.3 The top fingerprint, labeled as a fracture event, shows the kind of signals that indicate shear stresses leading to roof falls. The second shows no visible event, whether seismic or noise. The third fingerprint demonstrates the geophone response to electrical noise, and the bottom one shows a recorded blast event in the mine. The event shown in the first plot demonstrates the kind of fingerprints that the classification algorithm needs to identify and separate from the other three sample fingerprints, which are still triggered by the system as significant but do not reflect seismic activity. Previous studies have shown that rock fracture events produce a signal that is sharply defined, with large amplitude and short duration. Spatial patterns in fingerprints can distinguish these events and highlight the ones most of interest.

Given that our preliminary results seemed to show that we were able to identify precursor signals of roof fall events, we were rather disappointed that we didn't then get training data to be able to implement a real-time system that would save lives.[8] It was a time when the sky was falling (economically) and so I naturally wondered whether these methods could help find precurser signals in financial data that indicated impending falls. We looked into it a bit, of course, and found that the sorts of math academic econometrics folks (and presumably Wall Street quants) were doing were rather rudimentary. We never actually pursued that, but we did begin to consider other sorts of time series data streams that these methods might be applied to. For example, augmenting the analysis of medical imaging with information extracted from text-based medical files using natural language processing, something we've been wanting to be able to do since our prostate cancer detection project in the late 1900s.

8 Dr. Crystal (Bertoncini) Acosta is currently Senior Director at Capital One and Head of US Card Fraud Modeling, including over 20 internal models aimed at defending customers against fraud across customer management stages. She was previously a Research Physicist at the Naval Research Laboratory, where she was Principal Investigator and team lead for several research projects within the Tactical Electronic Warfare Division, but the Navy won't let you plug in your electric car at work and private industry pays quite a lot more. She holds a BA in Physics and Mathematics from Vassar as well as MS and PhD in Physics from W&M working in the NDE Lab of Prof. Hinders.

10.2.2 RF Scattering from a Food Truck

Next, consider a hip food truck that blocks the signal for your fancy new 5G smartphone that you're about to post about on social media. I don't do social media and I don't carry a smartphone, but if I did, I'd be more likely to post about food safety issues rather than how photogenic my fish taco was. A food truck only carries so much potable water, so I'm skeptical that the workers wash their utensils enough. In restaurants, I always go wash my hands after I'm done touching their germy laminated menus, and I peek into the kitchen on the way back to see if it looks clean. My wife is a generous tipper.

We used pairs of software-defined radios (USRPs) to collect RF scattering data around the front and sides of a campus food truck. One node was placed at the front of the vehicle 1.15 m from the right front corner, while the other node was located on the right side of the truck 1.5 m from the right front corner. Signals were collected using center frequencies of 2.5 and 4.0 GHz, with both USRPs positioned about 0.5 m from the ground. The CAD model we created to simulate wave propagation in this environment consisted of a simplified geometry using the dimensions and panel angles defining the actual vehicle, with hollow areas in the main body of the truck and front seats. The corresponding simulations used between 25 and 30 GB of memory to store the $4 \times 7 \times 3$ m computational space and ran for nearly 60 hours each on a parallel cluster. Systematic simulations of complex 3D scattering can often be used to develop data engineering approaches that intelligently downselect features which are then fed into machine learning modules. Figure 10.4 shows a typical case for that food truck, where the signal(s) of interest are buried in the noise in the time domain signal. We have found that wavelet fingerprints are quite useful for extracting signals of interest from very noisy signals, and since the number of ridges in the fingerprints correspond to the amplitude of the signal, we can use ridge count as a function of time to pull the signal out of the noise. Various mother wavelets and then various geometric features in those fingerprints can be used to form candidate feature vectors for machine learning. With a sufficient training data set, augmented by simulations and insight from scattering analyses, the optimal combination of features can often be determined.

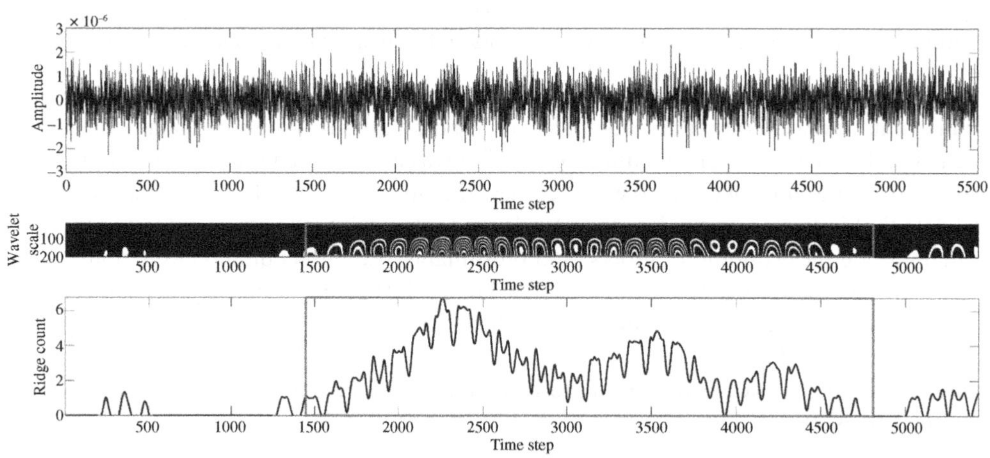

Figure 10.4 (Top) The time domain plot of a pulse in the received signal for the 10 GHz frequency simulation with significant noise. A typical fingerprint generated for this segment of signal is shown directly below the time domain plot, with the box outlining the area containing the pulse. Bottom is a plot of the ridge count feature for fingerprint, showing that the areas containing a pulse are composed of more complex fingerprint objects with a higher number of ridges.

10.2.3 Time Domain Reflectometry

Although our work on guided waves has primarily been for structural health monitoring using ultrasonics to ensure the integrity of aircraft structures and piping systems, we've also applied our knowledge of elastodynamics to determine the need for tamping of rail beds of the robotic container-stacking cranes at the Port of Virginia and transmitting sensor/image data sets through the walls of sealed containers in order to detect problems with shipments *en route*. We've even exploited the same knowledge of reflection/refraction at the boundary between water and ice floes to design a communication scheme to alert distant stakeholders to the precise location of oil spilled under arctic ice. I feel a little bad that I didn't talk about electromagnetic or optical-guided waves, though, so I'll include a bit of that here.

"What did you do in the war, Grandpa?" is a question you may have once asked. Neither of mine served in either World War. I wrote radar equations on the chalkboard during the first Gulf War; my father was called up briefly during the Cuban Missle Crisis and may have gone briefly AWOL for the birth of my older brother. Neither of those really count, though. I do have half a dozen ancestors who fought on the winning side of the Civil War, and my wife and I each have a dozen ancestors who fought in the Revolutionary War. She has one who was on the losing side, although that wasn't really his fault. His family were French Hugonout refugees in Holland and when he came of age he was conscripted into the Army and sold to England. That's how he ended up a drum major at Cornwallis' surrender at Yorktown. He was paroled as a POW by General Washington and then apprenticed as a glove maker in Fredericksburg before moving to Ohio in the early 19th century. A century before the American Revolution I have ancestors in New England who fought on both sides of King Philip's War, the deadliest war per capita in America.

I'm the grand-student of Julian Schwinger, so naturally, I was curious about what he did during WWII. It turns out that the answer is waveguide scattering. He worked at the Radiation Lab at MIT. Schwinger is recognized as one of the greatest physicists of the 20th century, responsible for much of modern quantum field theory, including a variational approach, and the equations of motion for quantum fields. Having developed the habit of working until late at night, he went further and made the day/night switch more complete, working at night and sleeping during the day, a habit he would carry throughout his career. During WWII, he would typically be arriving for work as everybody else was leaving for the night, and vice versa. I always tell my own graduate students to work whatever schedule makes them most productive, and since I typically arrive at the office quite early in the morning, students and I cross paths like this at times. Iterating a report or manuscript draft can be very efficient; if they click send before, they go to bed I will have done my edits before they wake up, and vice versa. We do need to overlap some IRL, though.

Waveguide scattering turned out to be critical to the war effort because waveguides are how the radar signal gets from the magnetron to the antenna, and that process needs to be optimized. You can safely assume there is a huge literature[9] on the subject that you can draw upon if you've got a problem to solve involving Maxwell's equations and waveguides. Don't forget that Maxwell's

---

9 Two classic texts that cover this are Collin, R.E. (1990). *Field Theory of Guided Waves*. Wiley and Felsen, L.B. and Marcuvitz, N. (1994). *Radiation and Scattering of Waves*. Wiley. Earlier versions of these are often available for download or second hand. I should warn you that this math is really mathy, and most RF engineers simply use finite-element modeling these days because all modern FEM systems do multiphysics.

Leopold B. Felsen was born in Munich, Germany in 1924. He received the BEE, MEE, and DEE degrees from the Polytechnic Institute of Brooklyn, NY, in 1948, 1950, and 1952, respectively, having emigrated to the United States in 1939 and served in the US Army during WWII. He remained with the Polytechnic until 1994, when he was granted the status of University Professor Emeritus. I met him briefly when he was Professor of Aerospace and Mechanical Engineering and Professor of Electrical and Computer Engineering at Boston University in a sort of semi-retirement, which meant that he wanted to mentor young faculty rather than graduate students. Dr. Felsen equations apply from DC to light, and fiber optics are therefore electromagnetic waveguides. But I'd like to talk about coaxial cables and time-domain reflectometry.

The goal is to determine the state of damaged wires in cases where the damage (flaw) causes something other than a simple open or short circuit. Of particular interest are subtle flaws, such as broken shielding, crushed sections, slightly cut or frayed conductor, dielectric damage, and so on. These fault classes may be undetectable by conventional TDR, but wavelet fingerprints offer a better way to extract the structure of information while having the added benefit of preserving information about time so that if the cable is deemed faulty, the location of the fault along the cable can be easily determined.

We collected systematic TDR data on coaxial cables of type RG58, using an Agilent 86100A TDR oscilloscope in tests that damaged the shielding with increasing levels of severity. The impedance discontinuity caused by a broken shield is inductive in nature with a characteristic positive pulse in the TDR signal. Such flaw effects were compared to flaws caused by crushing the cable, again with varying levels of severity. In contrast to the broken shield, a crush causes an impedance discontinuity that is capacitive, with a characteristic downward pulse in the TDR. Data sets were collected for flaws located at various locations over cables of three different lengths: 5-ft, 25-ft, and 50-ft. Figure 10.5 shows some typical results, which were quite promising.

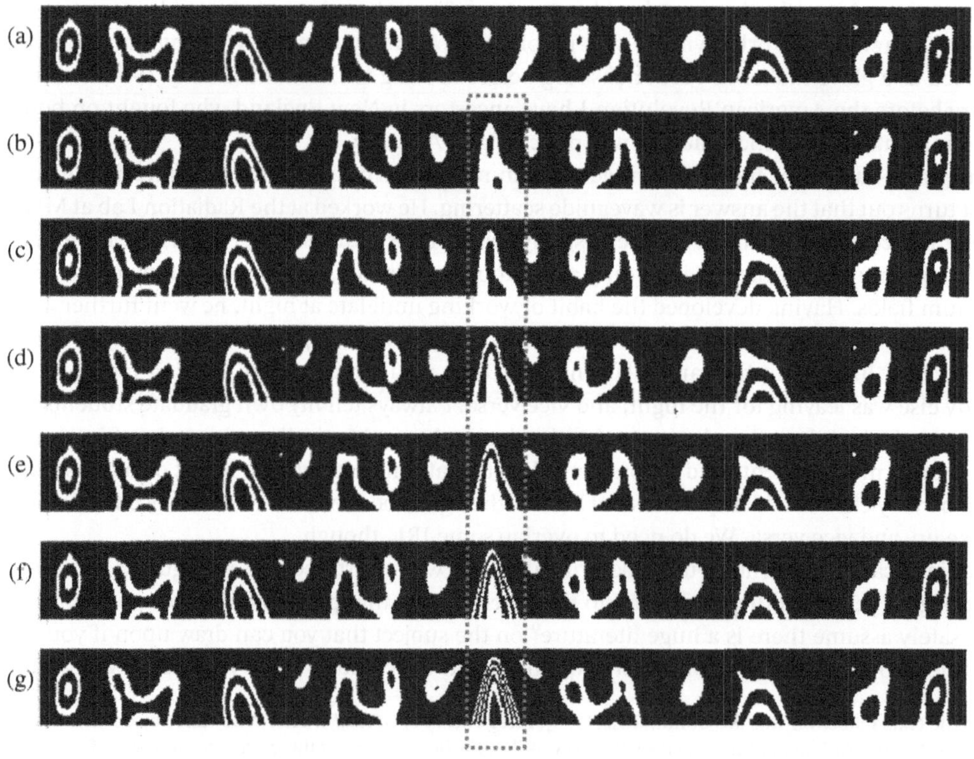

Figure 10.5 Wavelet fingerprints from TDR waveforms. The top fingerprint (a) is for an unflawed coaxial cable, with (b)–(g) for increasing sizes of chafing flaws in the outer conductor. The dotted rectangle indicates the location of that flaw, and the number of ridges in the triangular feature corresponds to its severity.

---

was the author or coauthor of over 350 papers and of several books, including the classic *Radiation and Scattering of Waves*. His research interests encompassed wave propagation and diffraction in complex environments and in various disciplines, high-frequency asymptotic, and short-pulse techniques, and phase-space methods with an emphasis on wave-oriented data processing and imaging. His Poet's Corner appeared in the IEEE AP-S Magazine for more than 20 years. He died in 2005.

10.2.4 Counterfeit Routers

The unhelpful dumbasses in IT just convinced my administration that it would be a good idea to yank all the office phones and have everybody use MS Teams instead. They forgot to order headsets. When people dial my office number it now immediately kicks over to vmail and any message left there is transcribed and forwarded to my email spam folder where I may find it in a few days. I don't like to talk on the phone anyway, and I only use MS Teams if one of my sponsors wants to chat that way. It does make it kind of hard for prospective employers to call me up and ask for a reference for a student. It also makes it hard for people to call me up and give me money, which happens surprisingly often.

For example, back in the day, I was sitting in my office and the phone rang. "Hi, my name's Frank and I'd like to come talk to you about counterfeit Cisco routers." I responded, "I don't really know anything about that." Frank said, "I read one of your papers and I think you can help me." I flipped to the next page on my desk calendar and said, "Would you like to come by sometime next week?" "I want to come tomorrow," said Frank.

So a strange guy, who I'm going to continue to call Frank, showed up in my office the next morning and told me about counterfeit Cisco routers. The basic idea is that in the global supply chain, there's no way to know where the various chips in a router (which the unhelpful dumbasses in IT are about to insert into your secure network) came from. Malicious actors can (and apparently do) swap out components with nominally identical ones which have some additional functionality you certainly don't want. Bad guys will apparently intercept your shipment, carefully open the package and perform the chip switcheroo, seal things back up and doctor the shipping documents, and then they are all up in your private business. As Frank was telling me this, I was rather skeptical, but then, when the Wikileaks dump happened, there was a formerly super-secret PowerPoint slide of exactly this happening. Yikes!

We agreed to start with a less terrifying, nonsecret application. Chips are sometimes recycled (which seems good) except that now that new solder doesn't have led in it (also good), the new solder requires higher reflow temperatures. When chips are removed from old devices and reused in new devices, if all the moisture isn't carefully baked out of them, they can develop "popcorn" delamination flaws which affect their reliability. Since high-frequency ultrasound is routinely used to inspect microchip packaging, finding delaminations in chips seemed quite doable. The only wrinkle was that Frank wanted a device that could be used on the loading dock before routers were installed or even while routers were already installed in a rack. The chips had to be inspected for delaminations without removing them from the board and immersing in a couplant bath.

Figure 10.6 shows the device that we designed, built, and demonstrated in a project that only lasted several months. It worked quite well, and the wavelet fingerprint technique easily identified features in the backscattered A-lines that indicated delaminations. The prototype was a deliverable of the contract, so after we demonstrated it in the conference room of a generic office park north of the DC beltway, they took it back to a part of the building where we weren't allowed and we came home. I was kind of glad to be done and out of there, and I live where The Farm is just across I64 from Williamsburg.

10.2.5 Bladder Distension Monitor

Figure 10.7 shows ultrasound A-lines and corresponding wavelet fingerprints for a prototype bladder distension monitor. John Companion had first built a prototype while at NASA, with preliminary data provided by his co-workers as they used the Branch restroom. A key finding of that

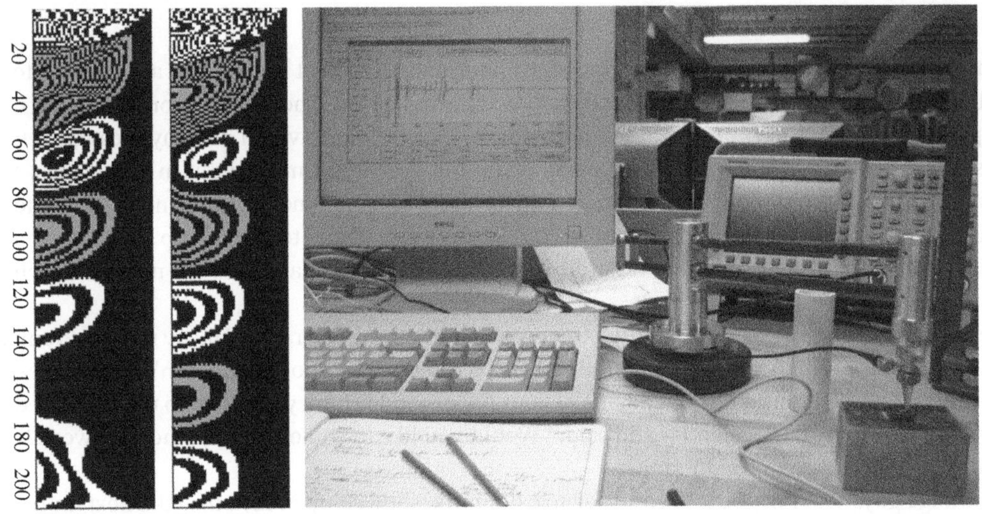

Figure 10.6 The 100 MHz ultrasonic pulse-echo instrument detects delaminations in chips using a tapered quartz delay line and a custom stabilizer unit, which allows a user to spot-check across the area of a chip while installed in a router. Wavelet fingerprints with and without delaminations are shown at the left. Note the presence/absence of the gray feature at a depth of 160 samples.

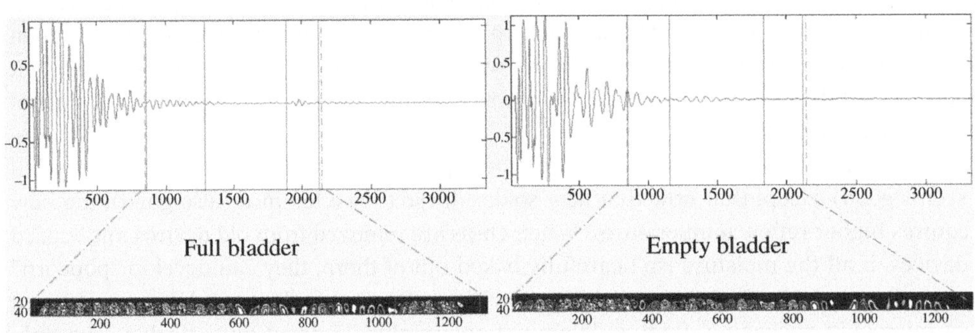

Figure 10.7 Typical ultrasound backscattering results for full (left) and empty (right) bladders where wavelet fingerprints are used to isolate differences in the front-wall and back-wall signals from a human bladder. Transducer is placed low on the abdomen and the backscatter is recorded before and after voiding.

preliminary research was a strong correlation between maximum bladder capacity and career progression, with the obvious conclusion that professional advancement requires sitting through long, coffee-fueled meetings. The device was patented, but the licensee sat on the invention so as not to displace their existing (highly profitable) product(s) from the marketplace. When that patent lapsed, John revisited the project leveraging new ultrasonics technology and wavelet fingerprints for improved data interpretation. The intended commercialization path targeted nursing homes, so the bladder distension monitor would use WiFi to send a signal to the nursing station that a patient will be needing assistance. Assisting patients to the restroom in time to not have to change the bedding affects the economics and quality of care rather significantly.[10]

10 Slate magazine had a reporter test out various adult diapers (http://www.slate.com/id/2199722) with the conclusion that the best adult diapers are from Europe. Apparently, European manufacturers don't have to cater to institutional purchasers' demands (i.e. Medicare/Medicaid), so they're more likely to sell on quality rather than cost.

10.2.6 RF Occlusion by Building

Machine learning is the modern terminology for what we've been trying to do all along, that is, making sense of the complex signals in radar and sonar and medical ultrasound and structural health monitoring. In some of the example applications earlier, we were able to identify fingerprint features directly that distinguished different classes, for example, flaw vs. no flaw. In some cases, we identify some number of candidate features and use the formal methods of machine learning to downselect the feature vectors to use, using our knowledge of the situation to guide the learning of the machine. Deep learning is fashionable right now and those sorts of black-box approaches are effective if there is a sufficient volume and quality of training data. However, when we have appropriate physical and mathematical models of the scattering of the radar, sonar, lidar, ultrasound, etc. from the materials, tissues, and/or structures of interest, it seems odd not to harness that hard-won knowledge. Sometimes we're able to draw on a deep literature developed over decades in order to address a problem of current interest.

The higher frequencies used by 5G wireless systems mean that RF interaction with the built environment will be more of an issue. Shorter propagation at these higher frequencies also leaves open opportunities for cognitive radio approaches that take advantage of locally unused bandwidth. Using pairs of USRPs, augmented with FDTD simulations, we have studied the subtleties of signals near the corners of buildings. The key question is to be able to distinguish signal transients due to occlusion by a building corner from signal transients due to starting/stopping a signal transmission. The problem requires isolating the beginning/ending portion of a signal, and then identifying features, which are characteristic of the transient occlusion event(s). Scattering!

We used three-dimensional EM FDTD simulations to model signal propagation in different environments for frequencies in the range from 2 to 5 GHz, which were compared using the dynamic wavelet fingerprint technique to measurements made with a set of USRPs placed in similar surroundings. These models, which characterized the interactions of RF signals with the landscape and the built environment, took into account absorption, diffraction, reflection, refraction, multipath, and fading effects in a range of scenarios and were simulated with high-fidelity using a high performance computing cluster.

In a systematic set of outdoor data collection, we positioned the USRPs on adjacent sides of a building, placed 50 cm both from the ground and away from the wall. Datasets were collected every 25 cm starting at 1 m from the corner and continuing until the USRPs were each 2 m from the corner, with transmissions at a center frequency first of 4 GHz and then of 5 GHz. The simulations modeling the data-collection environments also used input signals modulated to center frequencies of 4 or 5 GHz and ran for a total of 10,000 time steps, which corresponds to about 200 ns. The computational space consisted of a $450 \times 450 \times 200$ cm section of the exterior of a building with 20 cm thick walls set within a $550 \times 550 \times 200$ cm grid, with the placements of the transmit and receive nodes mirroring those in the physical signal propagation scenario as closely as possible. Each simulation ran for 20–25 hours on our HPC cluster, using nearly 18 GB of memory to store the simulation space.

After modeling situations in which the propagation of a signal is impacted by a scatterer moving into an environment, we also modeled scenarios where one of the nodes changes position in relation to a feature in the environment. In the first simulation, the Tx node was set 2.0 m above the ground and 1.5 m in from the corner of the side of a building. The receive node, at a height of 1.0 m and initially 1.0 m away from the Tx node on the same side of the building, gradually moves away from the source, turns around the corner of the building, and proceeds to move along the adjacent wall until it stops 3.5 m down from the corner. Every 0.25 m, the Rx node records a new

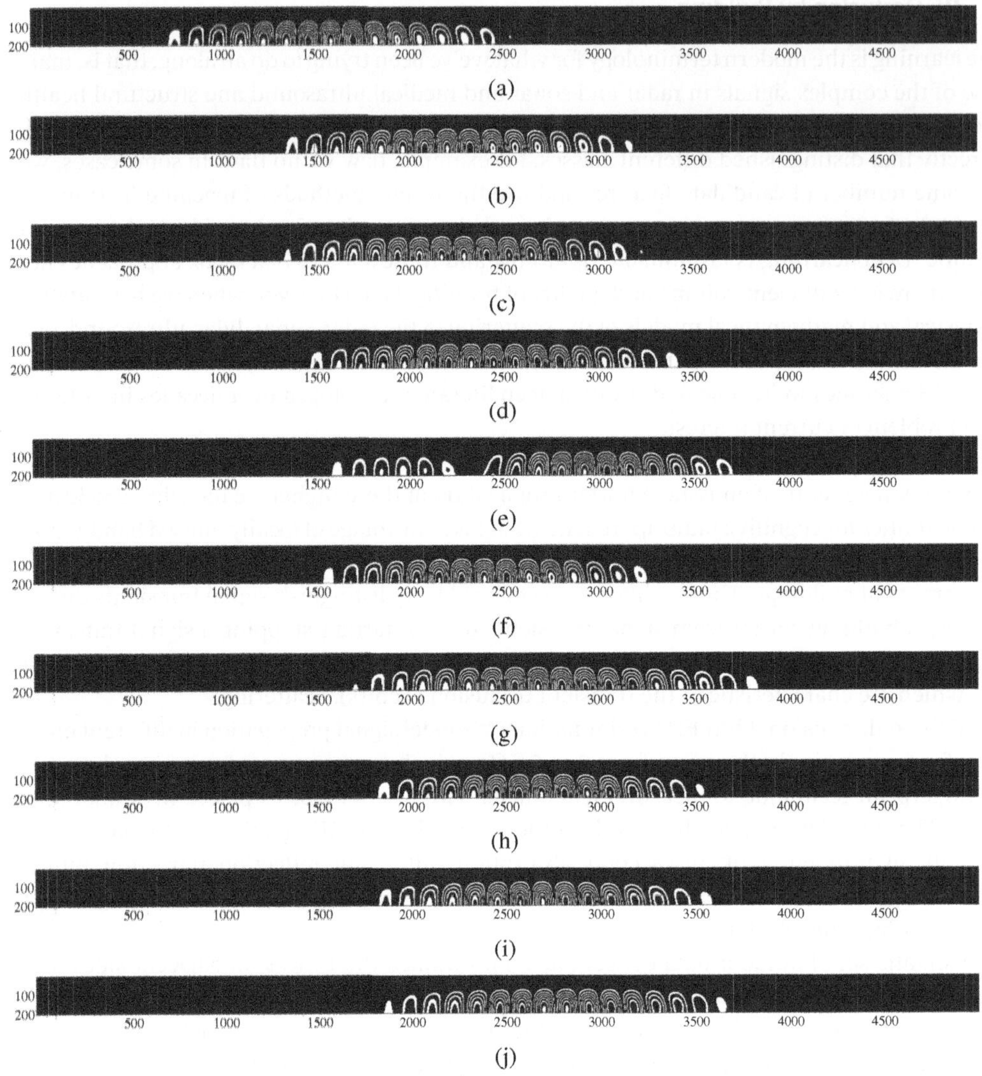

Figure 10.8 Fingerprints of the pulse, demodulated from a 5 GHz center frequency, recorded at the Tx node and at the Rx node for several positions around the corner of a building. The vertical axis shows the wavelet scale, while the horizontal axis shows the time. (a) Fingerprints of the pulse transmitted at the source node, (b) fingerprints of the signal recorded at Rx_0, (c) fingerprints of the signal recorded at Rx_1, (d) fingerprints of the signal recorded at Rx_2, (e) fingerprints of the signal recorded at Rx_3, (f) fingerprints of the signal recorded at Rx_4, (g) fingerprints of the signal recorded at Rx_5, (h) fingerprints of the signal recorded at Rx_6, (i) fingerprints of the signal recorded at Rx_7, and (j) fingerprints of the signal recorded at Rx_8.

pulse. Figure 10.8 shows fingerprints generated from the recorded signal at various positions for simulations using pulses modulated to 5 GHz center frequency.

Since the dynamic wavelet fingerprint technique is indifferent to the overall amplitude of an input signal, this method of analysis provided a natural conduit for comparison between the results from our FDTD simulations and the signals recorded with the set of USRPs. We began our signal analysis by testing several wavelet families including Daubechies, Coiflets, and Symlets for the signal filtering and fingerprint generation stages.

Once the signals were aligned in the fingerprint domain, we focused on performing an analysis of the sets of fingerprints that extended beyond a cursory visual comparison of binary images. Using image processing techniques, features relating to the shape, complexity, and area covered by the fingerprints were calculated from each significant object in the sets of fingerprints and plotted as a function of time. Prior to feature extraction, all fingerprints in the binary object were labeled as individual connected objects. This was performed by categorizing as a single unit any connected components whose centers of mass differed by less than a set threshold, since in many cases, the connected object function in MATLAB identifies each ridge of a fingerprint as a distinct object.

After extracting features from the set of fingerprints generated from each pulse recorded in a particular scattering environment, we compared plots of the feature values related to the USRP data to those related to the FDTD results. For the data recorded with the USRPs, since each of the received signals contained approximately 200–300 transmission events, the plotted feature values reflect the average over all the sets of fingerprints at each time step. The plotted average feature values for the FDTD results are obtained from the sets of fingerprints created from the several pulses transmitted within each simulation scenario. The vast majority of the features were in very good agreement across the physical signals and the simulated signals for the various propagation environments, with some of the most closely correlated features between the data sets being those related abstractly to the breadth of the frequency content of the signals.

Analysis of the signals using the wavelet fingerprint method allowed us to obtain information about the received signal in greater detail than we would have been able to gather from a simple FFT, and provided a conduit for direct comparison between signals of significantly different amplitudes in the time domain. After verifying in several scenarios that the fingerprints generated from measured signals and the simulated signals contained the same significant features, we simulated scenarios to predict signal reception in scattering environments which would be likely to induce hidden node problems. These simulations used some of the lower mmWave frequencies under consideration for 5G and examined the frequency dependency of environments and the potential limitations of utilizing mmWave bands.

10.3 Conclusions

For a wide variety of applications, we have found time-domain signal processing to be relatively inaccurate compared to 2D techniques. In particular, we favor joint time-frequency and time-scale transforms, which convert the 1D time trace echoes into 2D representations. The best-known example of this family of techniques is the "spectrogram," which computes an FFT inside a sliding window to get a 2D plot that shows both the frequency content of the reflection and how that changes over time. The goal is to find the optimal time-frequency kernel for the particular signals of interest (it's almost never the spectrogram), so that artifacts are minimized while features of interest are enhanced. Similarly, a wavelet transform can be constructed from a variety of mother wavelets to give a scale vs. time 2D representation. One can then do a 2D version of template matching or cross correlation to compare one or many of these 2D parameter images with those stored in a library.

Two decades ago we implemented a dynamic wavelet fnger print (DWFP) algorithm to interpret echoes from an ultrasonographic periodontal probe, extract multimode waveform properties from a guided wave tomography system, and identify deeply buried delaminations in microchip packaging via acoustic microscopy. We found that time-domain processing was not accurate enough for these purposes, but by performing a CWT on the A-lines and then a simple contouring operation on the resulting 2D data maps we obtain 2D black and white images that look remarkably like fingerprints. Comparing binary (black and white) images to templates allows us to draw on optical character recognition technology, of course, but our real goal is to make use of the sorts of extensive fingerprint classification systems that were developed in the early 20th century.[11] Although those systems eventually became unwieldy due to the inability to cross check larger and larger databases by hand, they did result in a variety of classification schemes that may be straightforward to utilize now with machine learning. The schemes turned fingerprint images into strings of numbers and letters, which could be written down on a card and carried, or sent via letter or telegraph. Fingerprint (and facial) identification technology is currently advancing rapidly[12] with some of the most successful algorithms operating by identifying the relative coordinates of characteristic features rather than by comparing one image to another in the sort of privacy-invasive database tyrants all love.

Here's a quote by Bruce Thompson[13] in Volume I of the Review of Progress in Quantitative Evaluation, p. 27: "A second use of the forward scattering models is through the empirically based, adaptive learning techniques.... In this case, the scattering models are used to produce 'theoretical data' which serves as a training set to develop empirical relationships between the experimental observables and the flaw parameters. In this case, a detailed reconstruction of the flaw shape is not obtained, but rather specific parameters are estimated. The successful application of this approach requires that the training set of flaws contain representatives of all flaw types and sizes to be encountered in practice." That was written in 1981.

Twenty years later, in 2001 or so, here's what my sense of the state-of-the art was and where we were heading. You can judge for yourself how precient I may or may not have been. These paragraphs are from a proposal that wasn't funded, but sometimes if you're thinking a bit too far ahead, your proposal won't be well received. At least that's what I tell myself when reviewers hate something I've proposed.

Until recently, ultrasound transducers were almost entirely PZT and similar crystals, with appropriate backing layers, matching layers, and discrete leads attached to opposite (metallized) faces of the crystal. Arrays of up to 128 elements were made by dicing the crystals and/or patterning the metallization layer. Although 1.5D arrays are now common on high-end medical ultrasound systems and a few true 2D arrays are in the clinical testing stage, this is still the same technology that has been around (and successful) for decades. New chip-based technologies of several types that offer very large arrays of ultrasonic elements are now being developed. It is reasonable to expect them to track pretty much the IR sensor-array development path since there is no consumer electronics killer app that would stimulate the ever-accelerating growth paradigm. These sensor packages are sometimes set up to provide real-time video imagery (as in medical imaging) or they can be fully incorporated into computer-controlled data acquisition and processing (as in NDT), but either way, time-evolving digital imagery often makes its way into the computer. Two things distinguish the

11 Fingerprints found widespread use because the sorts of body-measurement systems being employed in the days before mugshots were both inaccurate and expensive (Cole, S.A. (2009). *Suspect Identities: A History of Fingerprinting and Criminal Identification*. Harvard University Press). Plus they were invasive enough that it was considered bad manners to use them on prostitutes; fingerprints preserved the modesty of prostitutes.

12 https://www.techdirt.com/2023/01/04/tsas-opt-in-facial-recognition-program-doesnt-seem-all-that-optional-in-real-life.

13 R. Bruce Thompson was a world leader in many aspects of nondestructive evaluation, including his pioneering work on the development of model-assisted techniques for determining the probability of detection of flaws in materials that could lead to failure. Thompson earned his PhD in applied physics from Stanford University in 1971. He then joined the staff at Rockwell International Science Center, becoming a group leader before leaving for Iowa State. He served in many leadership and advisory roles both within and outside Iowa State, including director of the Center for NDE from 1997 until his death in 2011.

ultrasound situation from either visible or IR imagery. The first is that ultrasound can provide 3D data plus time (=4D). In pulse-echo, the arrival delays of the echoes give depth information in a quantitative way. IR and visible typically show us 2D only. Even when stereoscopic systems are used to give binocular 3D images, what we see are 2D surfaces only. Ultrasound "sees through" opaque media and so is naturally 3D in a quite new way for those used to video taken either in the IR or visible. The second distinguishing point is that it's usually not obvious how ultrasound data should best be rendered. Image segmentation is a real problem in ultrasound. Today it is usually a labor-intensive manual process done after the data is acquired. Medical 3D ultrasound systems provide "electronic scalpels" that allow the user to trim away unwanted features to be able to render the needed views of the anatomy of interest. The simplest rendering trick is to rectify the echo trains and then assign grayscale values directly. That approach discards quite a lot of useful information, and there are a number of ways to extract parameters from the waveforms before rendering "parameter images." For example, one can gate (window) short time segments and then pick out the peak amplitude in each. Changing the lengths of the gates used changes pretty significantly the appearance of the resulting C-scans. Instead of peak detecting in each window one could FFT the windowed signal to make a different type of image. Sliding the window continuously while recording the FFT gives a spectrogram representation that illustrates the frequency content variation in lateral space and depth. A spectrogram at each point in 3D space is a 4D data set, so often one will integrate each windowed spectrum to render an "integrated backscatter" parameter image. Of course, which representation is appropriate depends on the application of interest, but the point is that it's not at all obvious *a priori* what to do with the huge amount of data chip-based ultrasonic array output. No matter what "parameter image" one renders, showing it to a human is an unnecessary limit on its usefulness.

Fortunately, by sending the data through AI modules instead of rendering it for display, one is often able to "make the diagnosis" automatically in the sense that the necessary smarts are "in the box" itself. We teach the computer to search for the characteristic diagnostic features in the data that indicate high suspicion for pathology in tissues or flaws in structures. Of course, in practice, the computer is "slaved" to a physician or engineer "master" and merely reports suspicious areas for closer inspection by the human expert, but it's a small subsequent step to fully autonomous process control using high-volume, dense 3D sensor data from visible, IR and ultrasound sensor arrays. The limiting case is a subsystem that is able to continuously interrogate its surrounding and "decide" for itself what's going on around it. It turns out that the same things necessary to build a useful household robot to clean up the playroom (or dangerous waste site) and do the yardwork are critical to process control automation and fail safe NDT and real-time medical screening. The computer hardware and sensor arrays are improving rapidly. What is needed is attention paid to physics-based modeling and interpretation algorithms to effectively deal with the data streams and make the assessments on the fly. The key enabling step to automatic interpretation seems to be tomographic reconstruction because it outputs cross-sectional maps that are readily correlated to the physical properties of interest.

The goal here is to automate the interpretation of NDE inspection and remote sensing data for a variety of applications. We feel that it is necessary to use multiple sensing modalities in order to be able to make the AI interpretation algorithms robust and useful for most real-world applications, and we have found that tomographic reconstruction techniques often provide the necessary correlation between the recorded data and the physical properties of interest.

I think my predictions held up pretty well, but nobody really foresaw what was going to happen with machine learning in the early 2020s. These days, TensorFlow makes it easy for beginners and experts alike to create machine learning models. A common situation is when a couple of n00bs have a modest data set which they use to train a TensorFlow model which works quite well. They sometimes rush to publish their results without testing their model on a related, but distinct, set of data where the results are typically a coin flip. That's because most people underestimate by an order of magnitude (or three) the amount of training data that's going to be needed. If you have insufficient training data, your machine learning system isn't learning, it's just rememberizing your data. For most applications, you will be at least somewhat data starved, which is why scattering models and numerical simulations of scattering behavior can be so important. They can help you to understand which parameters are important for the scattering behavior and then enable you to explore the range of scattering behavior. In some cases, models and simulations can be used to augment your training data set. In almost all cases, synthetic test data sets can be formed from scattering models and simulations that allow you to test your machine learning system because you have a good understanding of what the output should be.

Since TensorFlow and the like are designed to deal with image data, wavelet fingerprints can feed into them pretty easily. Because scattering behavior is so strongly frequency dependent, a common approach for time-domain signals is to form spectrograms and feel those images into the ML system. Go ahead and try that as a first cut. If it works your problem is probably too easy. My question in such cases is, "Why would you expect a boxcar FFT to be the optimal time-frequency representation?" It almost never is, but there are innumerable TFRs so you might as well try some of them too. Then consider various wavelet transforms, which can be used to form time-scale images, which are in some ways equivalent to spectrograms. If you've got the computational horsepower and data storage necessary to do it, try all the time-frequency and time-scale representations you can think of and then let the machine learning system downselect the best one(s) or combos of them for your problem.

Part of what we like about the wavelet fingerprint is that they are presegmented, binary images. That both minimizes computer storage requirements compared to full-color images like spectrograms and it makes it straightforward to identify shapes and numerous spatial properties in them. In addition, the human eye is naturally suited to identifying shapes in this kind of black-and-white imagery. The human brain sees patterns even when they're not there, thanks to our distant ancestors who thought they saw a saber-tooth tiger crouching behind a bush when there wasn't one. Ancient humans who didn't notice the tiger behind the bush didn't pass on their genes, of course. When we're starting a new application with a new class of signals, we typically try various mother wavelets to see which ones look promising. That helps to narrow things down a bit before we start feeding things into pattern classification systems.

We've recently started forming color images from wavelet fingerprints. In a structural health monitoring application, we're using three-axis accelerometers to record vibrations in rail-mounted robotic cranes to detect flaws in rails and wheels, dips in the roadbed, etc. We first form traditional binary (black & white) wavelet fingerprints from the output of each of the three accelerometer axes. We then form RGB images by coloring the three fingerprints (as R, B, and G) and those we can feed directly into things like TensorFlow. First, however, we collected an obscene amount of accelerometer data (7 GB per day) over the course of about a year in known-good and known-bad and after-repair situations. We're pretty confident that we're not starved for data, but with small data collection units installed on the cranes, we can collect as much data as we could ever want.

In 1912 when the unsinkable Titanic hit an iceberg at night in a fog, Sir Hiram Maxim lamented whether science had reached the end of its tether. Nope. In the last 112+ years, an amazing amount of science and technology has ensued to actually do what Maxim sketched in Figure 1.1. I've made some attempts throughout to highlight the lives of many notables I've known personally, or by reputation, who spent their professional lifetimes developing the scattering methods we've discussed. I ended up doing scattering by accident. I mean that literally because if the space shuttle Challenger hadn't exploded in 1986, I wouldn't have ended up doing radar scattering in the Air Force just before the Cold War ended or almost doing seismology, but then switching to ultrasonic nondestructive evaluation because the Geophysics Lab was a tenant organization at Hanscom, AFB. The Challenger explosion also led to a closer connection between NASA Langley and William & Mary, which was the impetus for the faculty position that I've held for 62 semesters and counting. Life is funny like that.

Lifequakes happen all the time, although Loomis seeing the Great Depression coming is pretty unusual. We all just lived through the COVID-19 pandemic, which nobody saw coming and was particularly devastating for the elderly. If you were paying close-enough attention, you may have noticed that many of the notables I've highlighted died during those years. You can honor them and their contributions to human knowledge by building on their work to solve new problems in the coming decades. None of us know quite what those new problems are going to be, but we can take heart that the math and physics we've been discussing here will carry over directly. Knowing what's already been solved and how to adapt those known solutions to new problems is always going to be valuable.

Index

A Data Engineering Approach to Wave Scattering Analysis: with Applications in Radar, Sonar, Medical Diagnostics, Structural Flaw Detection and Intelligent Robotics, First Edition. Mark K. Hinders.

Printed and bound by CPI Group (UK) Ltd, Croydon, CR0 4YY
17/09/2026
14964335-0004